Condensed MATTER THEORIES

VOLUME 7

A Continuation Order Plan is available for this series. A continuation order will bring
delivery of each new volume immediately upon publication. Volumes are billed only
upon actual shipment. For further information please contact the publisher.

Condensed
MATTER
THEORIES

VOLUME 7

Edited by

Araceli Noemi Proto

and

Jorge Luis Aliaga

University of Buenos Aires
Vicente López, Argentina

Springer Science+Business Media, LLC

Proceedings of the Fifteenth International Workshop on Condensed Matter Theories,
held July 1-6, 1991, in Mar del Plata, Argentina

Library of Congress Catalog Card Number 87-656591 (ISSN 0893-861X)

ISBN 978-1-4613-6478-8   ISBN 978-1-4615-3352-8 (eBook)
DOI 10.1007/978-1-4615-3352-8

© 1992 Springer Science+ Business Media New York
Originally published by Plenum Press, New York in 1992

FOREWORD

The XV International Workshop on Condensed Matter Theories was held at the beautiful seaside resort of Mar del Plata, Argentina, during the first week of July 1991.

The first meeting of this workshop took place at the Instituto de Física Teorica, Sao Paulo, Brazil, in 1977, as the first Panamerican Workshop on Condensed Matter Theories. Its purpose was to bring together scientists from the Western countries, to work on many different topics related to the manifold aspects of condensed matter theories. The Workshop was so successful in facilitating exchanges of ideas and techniques pertaining to different areas of scientific endeavour that it quickly transformed itself into a broadly based, interdisciplinary forum for the informal discussion of the interrelation and mutual connections that naturally arise between the diverse disciplines encompassed under the common name condensed matter theories. From the green-house effect to neural networks, all theoretical efforts are interwined in a very complex fashion.

The next five workshops were held at Trieste, Italy (1978); Buenos Aires, Argentina (1979); Caracas, Venezuela (1980); Mexico City, Mexico (1981); and St. Louis, Missouri (1982). At the last meeting, in view of the truly international dimension reached by these gatherings, it was decided to substitute the word "International" for "Panamerican", and hold the next meeting, appropriately enough, in Europe. Altenberg, Germany was the place chosen for the 1983 edition, that was followed by Granada, Spain (1984); San Francisco, California (1985); Oulu, Finland (1987); Taxco, Mexico (1988); Campos do Jordao, Brazil (1989) and Elba, Italy (1900). The 1991 Mar del Plata Workshop has gathered together several scientists from diverse areas of Condensed Matter, such as Classical and Quantum Fluids, Atoms and Molecules, High Tc-Superconductivity and Particle and Nuclear Phisics. Both formal methods and applications were discussed in a friendly and relaxed atmosphere.

As was the case in previous editions, the participants, specialists of different areas, were able to establish a common language that showed the great similarities that exist in apparently unrelated topics of theoretical physics. A sense of wonder is irrescapable in these interdisciplinary meetings, when one is able to feel, even fleetingly, that there might be some truth in the great philosophical tradition that has been telling us, for several centuries, that "The One is the Whole"

This seventh volume includes all the invited talks heard at Mar del Plata, and it is to be hoped that the reader may also participate, through these pages, in the sense of wonder that contemporary physics continues to create, decade after decade, as it probes deeper and deeper into the marvellous and ever more complex texture of Nature.

Angel Plastino

ACKNOWLEDGEMENTS

The Editors have the pleasure to thank Prof. Angel Plastino for writing the Foreword to these proceedings. We would like to thank the members of the International Advisory Committee, of the Editorial Board and of the Program Committee for their assistance. We want to acknowledge the financial support of Consejo Nacional de Investigaciones Científicas y Técnicas (CONICET), International Centre for Theoretical Physics (ICTP), Workshop Trust Foundation, Centro Latino Americano de Física (CLAF), National Science Foundation, NASA, and the U.S. Army Research Office. We are also indebted to the Universidad Nacional de Mar del Plata (UNMdP), Universidad Nacional del Centro de la Provincia de Buenos Aires (UNCPBA) and Universidad Nacional de La Plata (UNLP) for sponsoring the workshop.

We wish to thank Dr. Ricardo Rapacioli, President of the Instituto de Investigaciones Científicas y Técnicas de las Fuerzas Armadas (CITEFA) for his encouragement. We express our special thanks to the other members of the Argentine Organizing Committee, Prof. S. Hernández and Prof. B. Alascio, and to the members of the Local Organizing Committee, Lic. C.M. Arizmendi and Ing. J.R. Sanchez. We are also grateful to Lic. Gustavo Grespo and Lic. Luis Irastorza whose assistance was essential in many aspects of the organization of the workshop.

Araceli N. Proto
Jorge Aliaga

CONTENTS

I. FORMAL METHODS

Requantization in the Time-Dependent Hartree-Fock Method
M.C. Cambiaggio and F.P. Taddei 1

A Reaction Theory and the Relation Between Widths and Energy Shifts
F. Bary Malik 11

The Control of Chaos
Bernardo A. Huberman.. 23

Berry Phase and its Unitarily Invariant Generalization
Donald H. Kobe 39

Natanzon Potentials and Spectrum Generating Algebras
Patricio Cordero and Sebastián Salomó 49

Ground States of Finite Systems and Information Theory
L. Arrachea, N. Canosa, A. Plastino, M. Portesi and R. Rossignoli 63

Information Theory and Quantum Wave Functions
N. Canosa, R. Rossignoli and A. Plastino 69

Decomposition of Signals and Information Theory
A. Plastino, L. Rebollo Neira and F. Zyserman 79

Finite Temperature Correlated Mean Field Treatments and Information Theory
R. Rossignoli and A. Plastino 87

Dynamics of Quantum Inhibited Processes Through Repeated Measurements
G. Crespo, Hilda A. Cerdeira and A.N. Proto 97

II. CLASSICAL AND QUANTUM FLUIDS

Features of HNC/0 Solutions for Symmetric Films of Liquid ^{4}He at $T = 0$ K
L. Szybisz and R.O. Vallejos 107

Compressible Rayleigh-Benard Convection in a Hard Disks System
Dino Risso and Patricio Cordero 119

Positron Annihilation as a Probe of Localized States in Fluids
Bruce N. Miller 131

III. ELECTRONIC SYSTEMS, ATOMS AMD MOLECULES

Lower Bound Aspects of Fermion Density Functionals
J.K. Percus 143

Solution of the Ornstein-Zernike Equation for a Mixture of Sticky
Hard Spheres and Yukawa Closure
J.N. Herrera, L. Blum and Fernando Vericat 153

Dynamical Properties of Strongly Coupled Coulomb Systems
G. Kalman 163

Z-Dependent Perturbation Theory and Complex Rotation Method in
Autoionizing States of Atoms
Lonnie W. Manning and Frank C. Sanders 179

Selfconsistent Semiclassical Mean Field
M. Casas, H. Krivine and A. Puente 189

A Simple Approach to Surface States in Covalent Crystals
D. Mirabella, R. Deza and C.M. Aldao 201

Scattering of Atomic Beams in Semiquantum Gases
Eugene P. Bashkin and Sergey B. Stepanyantz 207

Attempts to Calculate the Structure and Dynamics of Macromolecules
Clas Blomberg 215

The N-Representability Problem and the Local-Scaling Version
of Density Functional Theory
Eugene S. Kryachko and Eduardo V. Ludeña 229

Atomic and Ionic Information Entropies
M.C. Donnamaria and A.N. Proto 243

IV. HIGH-TC SUPERCONDUCTIVITY

Superconductivity in $Ba_{1-x} K_x Bi O_3$ Cubic Oxides
Wei Jin, M.H. Degani, Rajiv K. Kalia, Priya Vashishta
and C.-K. Loong 253

Non Conventional Superconductivity in $BaPb_{1-x}Bi_xO_3$ and $Ba_{1-x}K_xBiO_3$
A.A. Aligia and M. Baliña 269

The Essential Singularity in BCS Superconductivity Theory and
the Gap-to-T_c Ratio
V.C. Aguilera-Navarro and M. de Llano 287

Search of Superconductivity in Metal Clusters
M. Barranco, E.S. Hernández, R.J. Lombard and Ll. Serra 303

The Role of Spin Occupancy in the Electron Gas Problem
A. Calles, A. Cabrera, F. Ramos-Gómez, M.L. Marquina,
E. Yépez and J.J. Castro 313

Local Pair Model for the Analysis of Tunneling and Photoemission
Experiments in High Tc Superconductors: Superconducting Phase
C.I. Ventura, B.R. Alascio and R. Allub 325

Grand-Canonical Description of the Hubbard Hamiltonian
    J. Aliaga, A.N. Proto and V. Zunino 335

V. STOCHASTIC PROCESSES

Reactant Segregation: The Effect of Strong Space Disorder in
    Diffusion-Limited Bimolecular Reactions
    H.S. Wio, M.A. Rodriguez, L. Pesquera and C.B. Briozzo 345

Interaction Effects in the Toda Soliton Statistics
    C. Lucheroni and F. Marchesoni 351

Effects of Nonlinear Surface Diffusion on the Scaling of Rough Surfaces
    C.M. Arizmendi and J.R. Sanchez 361

VI. NUCLEAR AND PARTICLE PHYSICS

A Direct Approach to the Tamm-Dancoff Approximation
    Roberto C. Bochicchio and Horacio Grinberg ...._............. 367

Finite Size Effects in the Deconfinement Transition
    H.G. Miller, N.J. Davidson, R.M. Quick and A. Plastino 373

The Hyperspherical Harmonic Method Applied to ^{12}C and ^{16}O in the
    α-Particle Model
    A. Kievsky, M. Viviani and S. Rosati ...._............. 387

Atomic Small Clusters and Their Correspondence to Nuclear Physics
    G.S. Anagnostatos ...._............. 399

VII. OVERWIEW TALK

Some Considerations on the Greenhouse Effect and Related Problems
    V.M. Canuto.. 413

Contributors ...._........ 417

Index 423

REQUANTIZATION IN THE TIME-DEPENDENT HARTREE-FOCK METHOD

M.C.Cambiaggio and F.P.Taddei

Departamento de Física, Comisión Nacional de Energía Atómica
Avda. del Libertador 8250
1429 Buenos Aires, Argentina

INTRODUCTION

The problem of extracting quantum information from a time-dependent variational solution has attracted considerable attention for many years. In particular, the Time-Dependent Hartree-Fock approach (TDHF) is a time-dependent variational method, widely used in nuclear physics, which provides classical equations whose solutions need to be requantized for being able to extract quantum results. This problem has been attacked from many points of view and it s fairly straightforwardly solved in systems with one degree of freedom in which the non-trivial classical TDHF equations are one-dimensional[1-5]. In these cases approximate energy levels are obtained through a Bohr-Sommerfeld-like quantization rule applied to the action calculated integrating over a periodic solution of the TDHF equations of motion. Another requantization procedure based on the Fourier analysis of the time-dependent mean value of a relevant operator has been applied for obtaining the corresponding matrix elements[6,7].

In more realistic cases in which the TDHF equations are of dimension two or greater the systems turn out to be non-integrable and the quantization procedure to be used is not well defined. Some different quantization prescriptions for energy levels have been proposed[8,9] based on periodic trajectories and Poincaré sections. However, the procedures are not straightforward and become increasingly difficult with increasing excitation energy or number of degrees of freedom. Even in the simplest non-integrable system, the two-dimensional one, only the few lower-lying states are obtained. Moreover, these prescriptions are limited to the regular region of phase space. As it is well-known, in systems of two or higher dimensions chaotic behaviours appear which require a different quantization method but we will not be concerned with this problem here.

In this work we propose an alternative quantization procedure based on the Fourier analysis of the classical action as a function of time. For illustrating it we work in a simple model consisting of L shells of equal degeneracy and fermions interacting through a pairing force.

THE MODEL

The model used consists of L non-degenerate shells of the same degeneracy 2Ω and single particle energies ϵ_i and particles interacting via a pairing force. The hamiltonian is

$$H = \sum_{i=1}^{L} 2\epsilon_i K_i^0 - \frac{G}{2}(K^+ K^- + K^- K^+) \tag{1}$$

where G is the interaction strength and

$$K_i^0 = \frac{1}{2}\sum_{m_i}(b_{j_i m_i}^\dagger b_{j_i m_i} - \frac{1}{2})$$

$$K^+ = \sum_{i=1}^{L} K_i^+ = \sum_{i=1}^{L} \frac{1}{2}\sum_{m_i}(-1)^{j_i - m_i} b_{j_i m_i}^\dagger b_{j_i - m_i}^\dagger \tag{2}$$

The exact results are obtained diagonalizing the hamiltonian (1) in the basis determined by the occupation numbers in each shell[6].

TIME-DEPENDENT MEAN FIELD APPROXIMATION

For implementing the time-dependent variational approach we take into account that the operators K_i^0, K_i^+ and $K_i^- = (K_i^+)^\dagger$ defined in (2) are SU(2) generators. The vacuum state $|0>$ is characterized by

$$K_i^- |0> = 0$$

$$K_i^0 |0> = -\frac{\Omega}{2}|0> \tag{3}$$

and the coherent state in this representation is

$$|Z> = |Z_1 ... Z_L> = e^{\sum_{i=1}^{L} \bar{Z}_i K_i^+} |0> \tag{4}$$

The equations of motion are obtained through the time-dependent variational principle appropriate for non-normalized states[10] with an action defined as

$$S = \int dt \left[\frac{1}{2} i \frac{<\psi|\dot{\psi}> - <\dot{\psi}|\psi>}{<\psi|\psi>} - \frac{<\psi|H|\psi>}{<\psi|\psi>} \right] \tag{5}$$

A detailed derivation is given in ref.6. Here we will only say that although
all calculations could be performed in terms of the variables $Z_i, \bar{Z}_i$, it is more convenient to introduce new variables which are canonical[6,11]. This s done with the
transformation

$$\omega_i = \sqrt{\frac{\Omega}{1 + Z_i \bar{Z}_i}} Z_i \tag{6}$$

In terms of ω_i the variational equations look like the ordinary Hamilton
equations

$$i\dot{\bar{\omega}}_i = \frac{\partial \mathcal{H}}{\partial \omega_i} \qquad and \quad c.c. \tag{7}$$

where $\mathcal{H}$ is the mean value of the hamiltonian divided by the norm.

In the two-level case (L=2), the dynamical problem has two degrees of free-
dom, the complex variables ω_1 and ω_2. Therefore the existence of two constants of
motion, the energy and the number of particles, makes the system integrable. The
most adequate variables for the integration are

$$m = \frac{\omega_1 \bar{\omega}_1 + \omega_2 \bar{\omega}_2}{\Omega} - 1; \qquad \varphi = \frac{1}{2}(\varphi_1 + \varphi_2)$$
$$n = \frac{\omega_2 \bar{\omega}_2 - \omega_1 \bar{\omega}_1}{\Omega}; \qquad \alpha = \frac{1}{2}(\varphi_2 - \varphi_1) \tag{8}$$

where $\varphi_i = arg(\omega_i)$. Noting that $\omega_i \bar{\omega}_i$ is the mean number of pairs n level i one gets
that m is conserved and that the range of n is $-1 \leq n \leq 1$.

For $\epsilon_1 = -\frac{\epsilon}{2}, \epsilon_2 = \frac{\epsilon}{2}$ the energy function becomes[6]

$$\mathcal{E} = \frac{\mathcal{H}}{\epsilon \Omega} = n + \frac{1}{4}\chi n^2 - \frac{1}{4}\chi \left[(1 - m^2) + cos(2\alpha)\sqrt{[(m+1)^2 - n^2][(m-1)^2 - n^2]} \right] \tag{9}$$

3

where $\chi = 2G\Omega/\epsilon$. The action takes the standard form

$$S = \Omega \int dt \, [\dot{\varphi}m + \dot{a}n - \mathcal{E}] \tag{10}$$

In the three-level case (L=3) the system is non-integrable. The most adequate variables for the integration are

$$
\begin{aligned}
m_1 &= \frac{\omega_1 \bar{\omega}_1 + \omega_2 \bar{\omega}_2 + \omega_3 \bar{\omega}_3}{\Omega}; & \theta_1 &= \frac{1}{3}(\varphi_1 + \varphi_2 + \varphi_3) \\
m_2 &= \frac{\omega_1 \bar{\omega}_1 - \omega_2 \bar{\omega}_2}{\Omega}; & \theta_2 &= \frac{1}{2}(\varphi_1 - \varphi_2) \\
m_3 &= \frac{\omega_1 \bar{\omega}_1 + \omega_2 \bar{\omega}_2 - 2\omega_3 \bar{\omega}_3}{\Omega}; & \theta_3 &= \frac{1}{6}(\varphi_1 + \varphi_2 - 2\varphi_3)
\end{aligned}
\tag{11}
$$

As before, m_1 is conserved. For $m_1 = 1$ and $\epsilon_2 = 0$ the energy function becomes

$$
\begin{aligned}
\frac{\mathcal{H}}{\Omega} =& -\frac{1}{3}(\epsilon_1 + \epsilon_3) - \frac{2}{3}G\Omega + \epsilon_1 m_2 + \frac{1}{3}(\epsilon_1 - 2\epsilon_3)m_3 + G\Omega\left(\frac{m_2^2}{2} + \frac{m_3^2}{6}\right) \\
& - 2G\Omega \cos(2\theta_2)\sqrt{h_{12}\left(\frac{1}{3} - \frac{m_3}{3} + h_{12}\right)} \\
& - 2G\Omega \cos(\theta_2 + 3\theta_3)\sqrt{h_{13}\left(\frac{1}{3} - \frac{m_2}{2} + \frac{m_3}{6} + h_{13}\right)} \\
& - 2G\Omega \cos(-\theta_2 + 3\theta_3)\sqrt{h_{23}\left(\frac{1}{3} + \frac{m_2}{2} + \frac{m_3}{6} + h_{23}\right)}
\end{aligned}
\tag{12}
$$

with

$$
\begin{aligned}
h_{12} &= \frac{1}{9} - \frac{m_2^2}{4} + \frac{m_3^2}{36} + \frac{m_3}{9} \\
h_{13} &= \frac{1}{9} - \frac{m_3^2}{18} + \frac{m_2}{6} - \frac{m_3}{18} - \frac{m_3 m_2}{6} \\
h_{23} &= \frac{1}{9} - \frac{m_2}{6} - \frac{m_3}{18} + \frac{m_2 m_3}{6} - \frac{m_3^2}{18}
\end{aligned}
\tag{13}
$$

Again the action takes the standard form

$$S = \Omega \int dt \left(\sum_i \dot{\theta}_i m_i - \frac{\mathcal{H}}{\Omega}\right) \tag{14}$$

Inserting the derivatives of the mean value of the hamiltonian given by eq.(9) or (12) in the equations of motion (7) and solving the latter one gets the TDHF solutions, i.e. the variables (8) or (11) as functions of time.

RESULTS AND DISCUSSION

As a first step we work in the two-level model. In this case, as we have already said, the system is integrable and all the trajectories are periodic[6]. Consequently, it is easy to apply the Bohr-Sommerfeld- like quantization rule for the energy levels. As m and $\mathcal{E}$ are constant one works with the reduced action

$$S_\alpha = \Omega \int \dot{\alpha} dn \tag{15}$$

which is calculated as the area enclosed by the different trajectories and imposes the condition $S_\alpha = 2\pi s$ with s an integer. The energies for which this condition is fulfilled are selected as the energy levels.

For implementing the new requantization procedure we propose we calculate the reduced action but as a function of time and Fourier transform it. As the trajectories are periodic the Fourier transforms present a peak whose position in frequency corresponds to an excitation energy. Then a self- consistency prescription is applied. The energy at which the calculation is performed corresponds to a certain excitation energy with respect to the preceding level, if this excitation energy is equal to the one obtained from the Fourier transform the energy is accepted as an energy level. For example, for obtaining the first excited state one begins with an energy $\mathcal{E}$ a little higher than the Hartree-Fock ground state energy, solves the TDHF equations, calculates the action as a function of time, Fourier transforms it and compares the resulting excitation energy provided by the Fourier transform with the one entered in the calculation which is the difference between $\mathcal{E}$ and the ground state energy. If these two excitation energies are not equal one increases $\mathcal{E}$ and repeats the procedure until self-consistency is achieved. Once the first excited state has been fixed one gets the second excited state proceeding in the same way calculating in each step the excitation energy that enters in the calculation as the difference between $\mathcal{E}$ and the first excited state energy. In this way one is able to construct the entire spectrum.

In fig. 1 we show the results obtained for 40 pairs of particles in a two-level model with $\Omega = 40$ and $\chi = 1.2$. Only the fifteen lower energy levels are shown for being able to clearly see the differences but the entire spectrum is obtained in all the three calculations. Both requantization procedures provide comparable results.

Once our requantization procedure has been checked in the simple two-level model we turn to a more realistic system as the non-integrable one provided by the three-level model. In this case the trajectories are not all periodic and, consequently, the Fourier transforms present several peaks. We show a typical example in fig. 2. For deciding which peaks are the relevant ones to be used for applying the self-consistency prescription we perform an RPA calculation and identify the peaks corresponding to the RPA frequencies in the Fourier transform obtained for an energy very near the Hartree-Fock ground state energy. Provided the energy increase in each step of the self-consistent procedure is small one is able, in principle, to follow these peaks throughout the spectrum. Taking into account the conservation of the number of particles the three degrees of freedom of the dynamical problem are reduced to two and, consequently, one gets two RPA frequencies and has to achieve self-consistency independently for each of the two excitation energies provided by the Fourier transform.

Another important point to be considered is that now there is not only one trajectory for each energy. Different initial conditions provide different trajectories with the same energy and the variation on the positions of the peaks in the corresponding Fourier transforms determines the error in the approximate energy levels. It is important to remark that this variation is very small for low excitation energy and even if it increases with the energy it is not so big as to invalidate the

Figure 1. Lower fifteen energy levels for 40 pairs of particles in a two-level model with $\Omega = 40$ and $\chi = 1.2$.

self-consistent requantization procedure, at least in the three-level model studied here.

The fifteen lower energy levels obtained for 10 pairs of particles in a three-level model with $\varepsilon_1 = 1.73$, $\varepsilon_2 = 0$, $\varepsilon_3 = -1$, $\Omega = 10$ and $G = 0.01$ are shown in fig. 3. The agreement between the exact values and the requantized TDHF results is very good. The errors of the latter are of the order of 0.1% at low excitation energy and of 3% at the higher levels shown. For this small value of the interaction strength it is possible to construct the entire spectrum.

The results obtained for $G=0.01$ are promising although they are not conclusive because for small interaction strength the system behaves almost as two uncoupled anharmonic oscillators. Therefore, the next step will be to perform the calculations for a greater coupling constant corresponding to a strongly coupled regime where the other requantization procedures[8,9] can only provide few low energy levels.

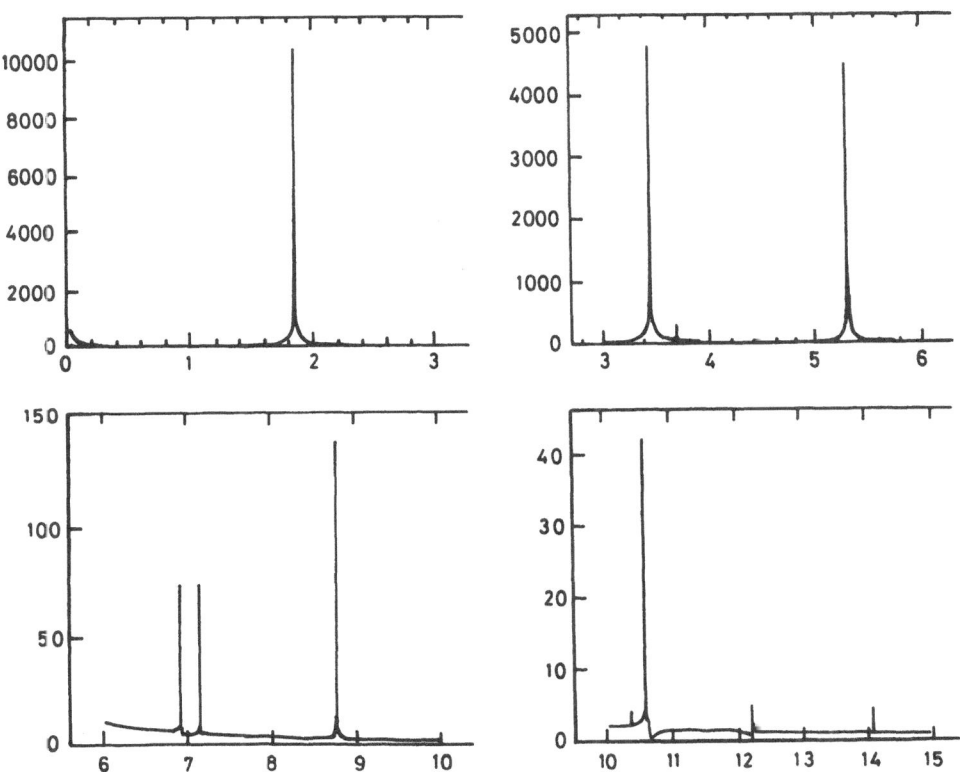

Figure 2. Fourier transform of one of the trajectories at $\mathcal{H} = -23$ for 10 pairs of particles in a three-level model with $\varepsilon_1 = 1.73$, $\varepsilon_2 = 0$, $\varepsilon_3 = -1$, $\Omega = 10$ and $G = 0.01$.

7

EXACT   RTDHF

Figure 3. Fifteen lower energy levels for 10 pairs of particles in a three-level model with $\epsilon_1 = 1.73, \epsilon_2 = 0, \epsilon_3 = -1, \Omega = 10$ and $G = 0.01$

ACKNOWLEDGEMENTS

This work has been supported in part by Carrera del Investigador Científico y Técnico (MCC) of the CONICET and by Fundación Antorchas, Argentina.

REFERENCES

1.- K.K.Kan, J.J.Griffin,P.C.Lichtner and M.Dworzecka, Nucl. Phys. *A332*:109 (1979).

2.- S.Levit, J.W.Negele and Z.Paltiel, Phys. Rev. *C21*:1603 (1980).

3.- R.Shankar, Phys. Rev. Lett. *45*:1088 (1980).

4.- D.H.Feng, R.Gilmore and S.R.Deans, Phys. Rev. *C23*:1254 (1981).

5.- A.E.L.Dieperink and O.Scholten, Nucl. Phys. *A346* :125 (1980).

6.- M.C.Cambiaggio, G.G.Dussel and M.Saraceno, Nucl.Phys. *A415*:70 (1984).

7.- M.C.Cambiaggio, G.G.Dussel and J.A.Ramirez, to be published in Phys. Rev. C.

8.- R.D.Williams and S.E.Koonin, Nucl. Phys. $\underline{A391}$:72 (1982).

9.- E.Caurier, M.Ploszajczak and S.Drozdz, Phys. Lett. $\underline{160B}$:357 (1985).

10.- P.Kramer and M.Saraceno, Geometry of the Time Dependent variational principle in quantum mechanics, Lecture notes in physics, vol. 140 (Springer, Berlin, 1981).

11.- K.K.Kan, Phys. Rev. $\underline{C24}$:109 (1981).

A REACTION THEORY AND THE RELATION BETWEEN WIDTHS AND ENERGY SHIFTS

F. Bary Malik

Physics Department, Southern Illinois University
Carbondale, Illinois, 62901, U.S.A.

ABSTRACT

We present here a complete reaction theory that incorporates the Pauli principle explicitly and is devoid of projection operators or matching radii. We provide explicit expression for widths, reduced widths and energy shift and proper relations between them. For a series of nearby overlapping resonances we develop explicit expressions showing that they cannot be, in general, considered as a series of non-interacting isolated resonances. The analysis indicates that the location of resonance energies are affected in the complex-rotation method and examine the situation when this could be neglected. Finally we examine the importance of including the Pauli principle for the calculation of auto-ionization widths of He-like atomic system.

I. INTRODUCTION

Currently there has been a lot of interest to use atomic physics in the "complex rotation method" to calculate the decay widths[1] in auto-ionization processes. The basis for this interest is the original investigation of the scattering by a spherical symmetric potential by Combes [2,3] and his coworker using Schrödinger equation. The crux of the method lies in the proposition that for a wide class of spherical symmetric potentials the location of the resonances in the complex energy plane remains unaffected if one rotates the radial coordinate r through an angle θ, using the transformation $r = ye^{i\theta}$ (ref. 4). One may then obtain the decay widths from the imaginary part of the energy.

The actual situation, however, involves usually the scattering of a particle by a composite system e.g., the scattering of an electron by He-like ion is at least a three-body problem. Such cases must be analyzed within the framework of a proper reaction theory. In this paper we develop a reaction theory by generalizing the early work of Mustafa and

Malik[5] which has its root in the investigation of the dielectronic recombination by Trefftz[6]. This analysis indicates that there is a distinct relation between decay-widths in a particular channel and their corresponding resonance energies which are shifted from their unperturbed positions. This energy shift is already well known for the Breit-Wigner resonances[7]. In fact, Haider and Malik[8] used the relation between the energy shift and widths to analyze the fine structures seen in the fusion cross section of the (^{12}C + $^{28,29,30}Si$) system.

The reaction theory developed in ref. 5 is limited by its lack of explicit incorporation of the anti-symmetrization and continuum-continuum coupling. In section II, we propose to remove these limitations and present a completely general theory devoid of any explicit use of projection operators which are basic to the theory of Feshbach[9] and of characteristic matching radii which underlie R-matrix type of theories[7,10,11]. In section III we analyze the relation between this theory and the theory of Feshbach. We further discuss the importance of including the exchange term explicitly by discussing the computations of the widths of autoionizing states for He-like systems by Seminario and Sanders[12].

II. THE THEORY

A. The Equations

In this paper, we consider the scattering of a fermion by a composite system of fermions e.g., the scattering of an electron by atoms or a nucleon by nuclei. All the coodinates of the scattered fermion are designated by "i" and those of the particles of the target by $1,2,...i-1, i+1, ... n+1$. Thus, the total Hamiltonian $H(1,2,. i-1,i, i-1...n-1)$ can be partitioned as follows:

$$H(1,2,...i-1,i,i+1...n+1)$$
$$= H_o(1,2,...i-1,i+1,...n) + t(i) + \sum_{i\neq j=1}^{n} v(i,j) \tag{1}$$

In the above, H_o is the Hamiltonian of the target, $v(i,j)$ is the interaction of the incident particle with each of the fermions of the composite system, and $t(i)$ is the one-body operator describing the intrinsic state of the incident particle, which is usually, but not necessarily, the kinetic energy operator. In case H does not contain time explicitly the total wavefuction $\Psi(1,2...i-1,i,i+1...n+1)$ satisfies the equation.

$$H\Psi(1,2,...i-1,i,i+1...n+1) = E\Psi(1,2...i-1,i,i+1....n+1) \tag{2}$$

with E, the total energy of the system. Following the general procedure for writing explicitly anti-symmetric products[13,14], we may expand Ψ into an infinite series of products of a single particle wavefunction, $\phi_\alpha(i)$, and $\Phi_\beta(1,2....i-1, i+1...n+1)$ which is the eigenfunction of H_o with eigenvalues ε_α:

$$\Psi(1,2....i-1,i,i+1....n+1)$$
$$=\sum_{\alpha\beta} \sum_i (-1)^i \phi_\alpha(i) \, \Phi_\beta(1,2,...i-1,,i+1,...n+1) \tag{3}$$

with

$$H_o\Phi_\beta = \varepsilon_\beta\Phi_\beta \tag{4}$$

(3) involves ordered permutation and Φ_β is anti-symmetric with respect to interchange of two successive coordinates. Without any loss of generality, one may impose the following orthonormality conditions:

$$(\Phi_\alpha,\Phi_{\alpha'}) = \delta_{\alpha\alpha'} \, ; \; (\Phi_E,\Phi_{E'}) = \delta(E-E') \, ; (\phi_\alpha,\phi_{\alpha'}) = \delta_{\alpha\alpha'} \, ; (\phi_E,\phi_{E'}) = \delta(E-E') \tag{5}$$

In the above, the scaler products involve integration over all coordinates.

We obtain the following set of equations for $\phi_\alpha(i)$ by constructing the scaler product $(\Phi_\alpha, (H - E)\Psi) = 0$ where the integration is over the coordinates $1,2, 3...i-1, i+1,...n+1$

$$[T(i) - (E-\varepsilon_\alpha) + V_{\alpha\alpha}(i)-M_{\alpha\alpha}(i)]\phi_\alpha(i)=-\sum_{\alpha'\neq\alpha} [V_{\alpha\alpha'}-M_{\alpha\alpha'}()]\phi_{\alpha'}(i) \tag{6}$$

In (6), the matrices $V_{\alpha\alpha'}$ and $M_{\alpha\alpha'}$ are defined as

$$V_{\alpha\alpha'} = (\phi_\alpha, \sum_{i\neq j,j=1}^n v(i,j) \, \Phi_{\alpha'}) \tag{7a}$$

$$M_{\alpha\alpha'}(i) \, \phi_\alpha(i) = (\phi_\alpha, \sum_{i\neq j,j=1}^n v(ij) \, \phi_{\alpha'}(j)\Phi_{\alpha'}) \tag{7b}$$

In the above, we have adopted the notation that "α" refers to a set $\phi_\alpha\,\Phi_\beta$ in order to use a compact form. The integration in (7a) and (7b) is over the coordinates $1,2...i-1,i+1,....n+1$ and $\Phi_{\alpha'}$ in (7b) is a function of $1,2,...i-1,i,i+1,j-1,j+1....n+1$. (6) is not a single equation, but a set of infinite number of coupled integro-differential equations. Basically, the states ϕ_α with $\alpha = 1,2....n$ can be divided into two classes of states: a set of states for which $(E-\epsilon_\alpha) > 0$ that characterizes the continuum and a set of states $(E-\varepsilon_\alpha) \leq 0$ which is characteristic of bound states.

For further discussion we may consider only the spherical symmetric potentials and expand ϕ_α in spherical harmonics $Y_{\ell,m}$

$$\phi_\alpha(i) = (X_\alpha(i)/r_i)\, Y_{\ell,m}(\theta_i,\phi_i) \tag{8}$$

In the above $X_\alpha(i)$ is r times the radial wavefunction for the state α. The set of equations (6) can now be reduced to

$$[T(i) + k_\alpha^2 - (U_{\alpha\alpha}(i) - K_{\alpha\alpha}(i)]X_\alpha(i) = \sum_{\alpha'\neq\alpha} (U_{\alpha\alpha'}(i) - K_{\alpha\alpha'}(i))X_{\alpha'}(i) \tag{9a}$$

Here,

$$T_\alpha(i) = d^2/dr_i^2 + \ell_\alpha(\ell_\alpha+1)/r_i^2 + u_\alpha(i) \tag{9b}$$

where $u_\alpha(i)$ may be zero or an external potential in $T_\alpha(i)$. $U_{\alpha\alpha'}$ and $K_{\alpha\alpha'}$ are radial parts of the potentials multiplied by $(2\mu|\hbar^2)$ (μ: reduced mass in the channel α).

$$k_\alpha^2 = (2\mu|\hbar^2)\,[E-\varepsilon_\alpha] \tag{10}$$

For further discussion it is convenient to consider explicitly continuum states noted by indices c and c' and bound states noted by sub-scripts b and b'. Thus, we have

$$[T_c(i) + k_c^2 - W_{cc}(i)]\,X_c(i) = \sum_{b\neq c} W_{cb}(i)X_b(i)$$
$$+ \sum_{c'\neq c} W_{cc'}(i)\,X_{c'}(i) \tag{11a}$$

and

$$[T_b(i) + k_b^2 - W_{bb}(i)]\,X_b(i)$$
$$= \sum_{c\neq b} W_{bc}(i)X_c(i) + \sum_{bb'} W_{bb'}(i)\,X_{b'}(i) \tag{11b}$$

In (11a) and (11b) we have used the following abbreviation

$$W_{\alpha\alpha'} = U_{\alpha\alpha'} - K_{\alpha\alpha'} \tag{12}$$

One may write down the formal solution[15] of (11a)

$$X_c = X_c^{(0)} + \int dr'\, G_c(r,r')\, [\sum_{c'\neq c} W_{cc'}\, X_{c'} + \sum_{b\neq c} W_{cb}\, X_b] \tag{13}$$

where $G_c(r,r')$ is the Green-function

$$G_c(r,r') = -(1/k_c)\,[X_c^{(0)}(r_<)\,X_c^{(1)}(r_>) + i\, X_c^{(0)}(r_<)\, X_c^{(0)}(r_>)] \tag{14}$$

In (14), $X_c^{(0)}$ and $X_c^{(1)}$ are, respectively, the regular and singular solutions of the homogeneous part of (11a) near the origin and $r_>$ and $r_<$ refer to the larger and smaller of r and r', respectively. In case, the external potential $u_\alpha(i)$ in (9b) contains a Coulomb potential $Z_1 Z_2 e^2 r / r_i$ and $W_{cc'} X_c$ and $W_{cb} X_b$ fall off faster than $1/r^2$, $X_c^{(0)}$ and $X_c^{(1)}$ have the following symptotic forms:

$$X_c^{(0)} \sim \sin(k_c r - \pi \ell_c / 2 - \eta_c \ln(2k_c r) + \sigma_c + \delta_c) \tag{15a}$$

$$X_c^{(1)} \sim \cos(k_c r - \pi \ell_c / 2 - \eta_c \ln(2k_c r) + \sigma_c + \delta_c) \tag{15b}$$

In (15a) and (15b), the Sommerfeld parameter is

$$\eta_c = \mu Z_1 Z_2 e^2 / (\hbar^2 k_c) \tag{16}$$

and σ_c and δ_c are, respectively Coulomb and non-Coulomb phase shifts for the channel "c".

B. CONDITION FOR RESONANCES

Without any loss of generality we can expand the solution of (11b), X_b in a complete orthonormal set of functions X_{nb} and X_{qb} which are eigensolutions of he homogeneous part of (11b) with eigenvalues k_{nb}^2:

$$X_b = \sum_n a_{nb} X_{nb} + \int dq \, a_{qb} X_{qb} \tag{17}$$

with the condition

$$(X_{nb}, X_{n'b'}) = \delta_{nn'} \delta_{bb'} \quad (X_{qb}, X_{q'b}) = \delta(q - q') \tag{18}$$

Substituting (17) in (11b) and taking a scaler product with respect to X_{nb}, we obtain the equations for the amplitudes

$$(k_b^2 - k_{nb}^2) a_{nb} = (X_{nb}, \sum_{b' \neq b} W_{bb'} X_{b'}) + (X_{nb}, \sum_{c \neq b} W_{bc} X_c) \tag{19a}$$

$$- (q^2 + k_{qb}^2) a_{qb} = (X_{qb}, \sum_{b' \neq b} W_{bb'} X_{b'}) + (X_{qb}, \sum_{c \neq b} W_{bc} X_c) \tag{19b}$$

Unless the matrix elements of the potential $W_{bb'}$ and W_{bc} are highly singular, the amplitude a_{nb} could become singular only when k_b^2 approaches k_{nb}^2 but this cannot occur for a_{qb}. The singular points of a_{qb} in the analytic plane of k_b^2 are thus responsible for the resonant behavior of X_c given by (13).

However, (13) also involves $X_{c'}$ which may also be expanded in an orthonormal set

$$X_{c'} = \sum_n b_{nc'} X_{nc'} + \int dq\, b_{qc'} X_{qc'} \qquad (20)$$

In (20), the bound and continuum states are defined by $k_{c'}^2 \leq 0$ and $k_{c'}^2 > 0$, respectively.

This leads to a set of equations similar to (19a) and (19b) for $b_{nc'}$ and $b_{qc'}$. Once again only $b_{nc'}$ could lead to resonances in the channel "c". However, this does not alter the structure of the subsequent discussion but adds additional terms.

C. EXPRESSION FOR WIDTHS, ENERGY-SHIFT AND S-MATRIX

Resonances are characterized by the fact that near a given energy one or a number of amplitudes a_{nb} in (17) or b_{nc} in (20) dominates. In that case, X_c given by (13) around that energy will be determined basically by these dominating terms. At this stage we do not consider $b_{nc'}$ explicitly, but note that resonance like behavior of $b_{nc'}$ can explicitly be incorporated in a fashion similar to our subsequent treatment of resonance-like behavior of a_{nb}. Since at the resonance energies, we need only to incorporate dominating amplitudes, we may replace (17) by a finite sum,

$$X_b(r) = \sum_{b'=1}^{N} a_{nb'} X_{b'} \qquad (21)$$

From (13), (14) and (19a) we obtain,

$$\sum_{b'} \{ (k_{b'}^2 - k_{nb}^2)\, \delta_{bb'} - \int dr\, X_{nb}(r)\, W_{bb'}\, X_{nb'}(r) - \int\int dr dr' \sum_{c \neq b} W_{bc}\, [Re G_c(r,r') - i\, Im\, G_c(r,r) \\ W_{cb'}(r')\, X_{nb'}(r')] \}\, a_{nb'} = \sum_c \int dr\, X_{nb}(r)\, W_{bc}\, X_c^0(r) \qquad (22)$$

The resonant behavior of $a_{nb'}$ is explicit in (22) and more apparent if by noting that (22) may be written as

$$\sum_{b'} [(E_{b'} - E_{nb'})\delta_{bb'} + \Delta E_{nbb'} + (i/2)\Gamma_{bb'}]\, a_{nb'} \\ = \sum_c \gamma_{bc} \qquad (23)$$

where

$$\Gamma_{bb'} = \sum_c (4/\hbar v_c)\, \gamma_{bc}\, \gamma_{b'c} \qquad (24)$$

with the reduced width defined by

$$\gamma_{bc} = (\hbar^2/2\mu) \int dr\, X_{nb}(r)\, W_{bc}\, X_c^{(0)}(r) \tag{25}$$

and

$$\Delta E_{nbb'} = -(\hbar^2/2\mu) \int dr\, X_{nb}(r)\, W_{bb'}\, X_{nb'}(r)$$
$$+ \sum_c (\hbar/\mu v_c) \int dr dr'\, X_{nb}(r)\, W_{bc}\, X_c^{(0)}(r_<)\, X_c^{(1)}(r_>)\, W_{cb'}(r')\, X_{nb'}(r) \tag{26}$$

The resonant structure of the amplitude involves an energy-shift. We note again that the inclusion of resonant amplitudes $b_{nc'}$ in expansion (19a) and (19b) would not change the structure of (23) but affect (24) and (25). In case of well separated isolated resonances, only one amplitude dominates at a given energy and the expression for this resonating amplitude is

$$a_{nb} = \frac{\gamma_{b\lambda}}{E_b - (E_{nb} - \Delta E_{nbb}) + i\Gamma_{bb}/2} \tag{27}$$

This leads to the expression of Breit-Wigner[10] for cross section.

The general structure of (23), indicates that the decay widths $\Gamma_{bb'}$ and the energy shift $\Delta E_{nbb'}$ associated with a particular resonance at E_{nb} are strongly influenced by nearby resonances and cannot easily be extracted from data, since they are coupled. One may, however, find a similarity transformation T such that

$$T^{-1}(E_n - i\Gamma)T = \gamma \tag{28}$$

$$(E_n - i\Gamma) = T\gamma \tag{29}$$

i.e., $\Sigma_{b'} (E_{nbb'} - i\,\Gamma_{bb'})\, T_{b'd} = T_{b'd}\,\gamma_d$

In the above $E_{nbb'} = (E_{b'} - E_{nb'})\,\delta_{bb'} + \Delta E_{nbb'}$. Since E and Γ are symmetric matrices, T is an orthogonal matrix i.e., $\Sigma_a\, T_{ac} T_{ac'} = \delta_{cc'}$ and $\Sigma_c\, T_{ac}\, T_{bc} = \delta_{ab}$ i.e., T^{-1} = its transpose.

Because of the Kronecker delta in (23), the transformation (28) reduces to

$$(\Delta E_n - i\Gamma)T = T\gamma \tag{30}$$

i.e. $\Sigma_{b'} (\Delta E_{nbb'} - i\,\Gamma_{bb'})\, T_{b'd} = t_{bd}\,\gamma_d$ where the matrix ΔE_n has elements $\Delta E_{nbb'}$. The final expression for the matrix A_n which has a_{nb} as its elements is

$$A_n = T(E_n - i\ \Gamma)^{-1}\ T^{-1}\ \gamma \tag{31}$$

$$a_{nb} = \sum_{db'} \frac{T_{bd}\ T_{b'd}\ \gamma_{b'}}{(E_b - E_{nb}) + \Delta\epsilon_{nd} + i\Gamma_d} \tag{32}$$

where $\Delta\epsilon_{nd}$ and Γ'_d are, respectively, elements of $T^{-1}\ (\Delta E_n)\ T$ and $T^{-1}\ \Gamma T$.

The expression for S-matrix elements can be derived following ref. 8,

$$
\begin{aligned}
S_{cc'} = \delta_{cc'} &- (2i/k_{c'}) \\
&[\textstyle\int dr\ X_c^{(0)}(r)\ [W_{cc}^{NC}(r)\ X_c(r) \\
&+ \sum_{c' \ne c}\ W_{cc'}(r)\ X_{c'}(r) + \sum_{b \ne c}\ W_{bc}\ X_b(r)]
\end{aligned}
\tag{33}
$$

where W^{NC} is the non-Coulomb part of W_{cc} and $U_c^{(0)}(r)$ is the regular solution of the homogeneous part of the open channel ,c, having only the Coulomb interaction.

III. DISCUSSION

A. Relation to the Feshbach Theory

Feshbach's reaction theory[9] introduces projection operators such that

$$P\Psi = \Psi(k)\ ;\ Q\Psi = 0 \tag{34}$$

Thus, in the usual approach of the Feshbach theory, one projects the wavefunction onto a single continuum state. This corresponds to the omission of the second term on the right hand side of (11a) i.e., setting $W_{cc'} = 0$. In essence, therefore, the direct coupling between continuum states is avoided. As a result, there is only one continuum state "c" and the summations over c in (23), (24), (25) and over c' in (33) are missing. One has essentially the results of ref. 5, except that, one does not have in ref. 5 the exchange term $K_{aa'}$ in the potential. In most of the analyses involving the Feshbach approach, one has not considered the case of overlapped resonances and as such has not derived the key equation (23) which is the key difference between a series of overlapped and isolated resonances. The most of the analyses approximates a series of overlapped resonances as superposition of non-interacting isolated resonances.

B. Relation Between Widths and Energy Shifts

(23) clearly exhibits that widths and energy shifts are indeed intimately connected. They originate from the real and imaginary part of the same Green function that determines the solution of the continuum states and therefore depends on the nature of the potential. A transformation $r = ye^{i\theta}$ done in refs. (1-3) affects the expression (24) for the width as well as (25) for the energy shift which determine the location of the

resonance. Strictly speaking, the assumption that such a transformation does not affect the location of resonance energies[1,4] is not valid. This assumption is only a suitable one if one neglects the continuum-bound state coupling W_{bc} and $W_{cb'}$ in (26). The term containing $W_{bb'}$ in the expression of $\Delta E_{nbb'}$ then can be adsorped by diagonalizing (11b) (with $W_{bc} = 0$) using a suitable basis set limited to bound states only.

C. Some General Applications

(a) Fusion Cross Section in heavy-ion heavy-ion scattering:

Experimental information on widths and resonance energies can only be obtained by a suitable diagonalization of (23) in case of overlapping or closely spaced resonances. Such a procedure has been adopted by Haider and Malik[8] in analyzing the structure seen in the fusion cross sections of the (^{12}C + ^{28,29,30}Si) system.

(b) Auto-ionization from He-like system:

Sanders and his collaborators[12] (and see this volume) have calculated decay widths of auto-ionized states for He-like ions using the complex-rotation method assuming that the transformation $r=ye^{i\theta}$ does not affect resonance energies. However, they have incorporated correlation energy between two electrons using Z-dependent perturbation theory in calculating energies of bound states i.e., in a sense, they have partially diagonalize (11b) with $W_{bb'}$ and neglected W_{bc}. Thus, the $W_{bb'}$ term in (26) is already included in calculation of eigenvalues. The success of their result indicates that either $W_{bc} \approx 0$ or a state independent for this kind of system.

Table 1. Calculated decay widths in Hartree atomic unit for the $2s2p\,^1P$ (col.2) and $2s2p^3P$ (col. 3) for atomic number Z = 2 to 10 compiled from tables VII and IX of ref. 12.

Z	Singlet	Triplet
2	0.00133	0.00052
3	0.00214	0.00049
4	0.00267	0.00045
5	0.00304	0.00044
6	0.00330	0.00042
7	0.00350	0.00041
8	0.00366	0.00040
9	0.00378	0.00039
10	0.00388	0.00039

One can also explain the general trend of their result showing that the decay widths for triplet states are always considerably less than those for singlet states as noted in table 1 taken from ref. 12 and also presented at Sanders and Manning's article in this volume. For two-electron systems, the spin independent part of the singlet and the triplet state are given by a sum of the states of the type

$$(1/\sqrt{2})[\phi_1(1)\,\phi_2(2) \pm \phi_1(2)\,\phi_2(1)] \tag{35}$$

(+ for singlet and - for triplet). Since W in (24) and (25) is a sum of the contribution of both the direct and the exchange terms, it is considerably smaller for the triplet states compared to the singlet ones due to large concellation originating from the minus sign in (35). This general conclusion is borne out in table 1 and emphasizes the importance of including the Pauli principle in the calculation of widths.

IV. ACKNOWLEDGEMENT

It is my pleasure to thank Mr. L. Manning for many stimulating discussions, and to acknowledge a travel grant from the U.S. Army Research Office which made my participation to the workshop possible.

REFERENCES

1. See the review article, Y. K. Ho, Phys. Rep. 99, 1 (1983).

2. J. Aguilera and J. M. Combes, Comm. Math. Phys. 22, 269 (1971).

3. E. Balslev and J. M. Combes, Comm. Math. Phys 22, 280 (1971).

4. B. Simon, Ann. Math. 97, 247 (1973).

5. M. G. Mustafa and F. B. Malik, Ann. Phys. (N.Y.) 83, 340 (1974).

6. E. Trefftz, Z. Astrophys. 65, 299 (1967).

7. G. Breit and E. P. Wigner, Phys. Rev. 49, 519, 642 (1936).

8. Q. Haider and F. B. Malik, Phys. Rev. C 28, 2328 (1983).

9. H. Feshbach, Ann. Phys. (N.Y.) 5, 357 (1958); ibid. 19, 287 (1982).

10. E. P. Wigner and L. Eisenbud, Proc. Nat'l. Acad. Sci, (U.S.A.) 27, 281 (1941); Phys. Rev. 72, 29 (1947).

11. P. L. Kapur and R. Peierls, Proc. Roy. Soc. (Lond.) A166, 277 (1938).

12. J. M. Seminario and F. C. Sanders, Phys. Rev. A 42, 2562 (1990).

13. J. Y. Shapiro and F. B. Malik, Condensed Matter Theories, 3, 11 (1988).

14. F. B. Malik, R. H. Richardson and J. Y. Shapiro, J. Phys. G: Nucl. Phys. $\underline{14}$, 535 (1988).

15. N. F. Mott and H.S. W. Massey, *Theory of Atomic Collisions* (Oxford University Press, third edition, 1965) Chap. XIII.

16. Q. Haider and F. B. Malik, Phys. Rev. C $\underline{26}$, 989 (1982).

THE CONTROL OF CHAOS

Bernardo A. Huberman

Dynamics of Computation Group
Xerox Palo Alto Research Center
Palo Alto, CA 94304

Abstract

This paper discusses the problem of controlling nonlinear systems with chaotic dynamics. In the low dimensional case, one can achieve efficient regulation by introducing error-driven dynamics for the control parameter. For systems with many degrees of freedom, as encountered in distributed computation, there is a simple and robust procedure for freezing out chaotic behavior even when imperfect and delayed information cannot be avoided. It is based on a reward mechanism whereby the relative number of computational agents following effective strategies is increased at the expense of the others. This procedure is able to control chaos through a series of dynamical bifurcations into a stable fixed point. Stability boundaries are computed and the minimal amount of diversity required in the system is established.

1. Introduction

Over the past ten years, a veil has been lifted over much of previously unexplored territory. Nonlinear dynamical problems that only a decade ago seemed beyond grasp have yielded elegant and general answers, along with new notions such as bifurcation sequences, strange attractors, fractal basin boundaries, scaling laws, and universality classes. Among the most notable developments, I would single out the understanding that we have gained about the universal behavior of many dissipative systems that are pervasive in condensed matter physics, chemistry and engineering. Examples of such structures are Josephson junctions in the presence of electromagnetic radiation, oscillatory and chaotic chemical reactions, hydrodynamic instabilites, and many nonlinear electronic circuits. As a result, the appearance of randomness in systems such as the Josephson junction is no longer considered the consequence of an interplay among many degrees of freedom, but the unavoidable chaos brought about by the intrinsic nonlinearity of deterministic systems.

Condensed Matter Theories, Vol. 7, Edited by A.N. Proto
and J.L. Aliaga, Plenum Press, New York, 1992

Impressive as it is, this addition to our knowledge is just the beginning of the nonlinear enterprise. Many open issues remain to be explored, that range from the behavior of field equations to the application of this new knowledge to many nonlinear systems. And when I speak of applications I have in mind both the use of dynamical systems theory for understanding previously unexplained behavior, and the use of nonlinearity to make a system perform according to set goals. It is in this latter realm of control where, if understanding could be translated into design, we stand to gain much. Phase locked loops, stressed fluids, adaptive systems and computer networks, are a few instances of the many systems where desired, stable, performance is not always easy to enforce.

The existence of chaos in nonlinear systems poses serious problems for their control. The consequent unpredictability of their behavior implies that the observation of an error signal at a given time cannot be translated into an obvious procedure to reduce it to zero later on. Nevertheless, it is well established that a number of natural and artificial systems manage to operate in regular and smooth fashion in spite of their intrinsic nonlinearities, a fact which raises the interesting question about the mechanisms underlying their stability. In many cases, equilibria are achieved through adaptive controls in which feedback signals are used to produce stable outputs within a range of parameter values. As to applications, it is also the case that in many processes stable outputs are required in spite of the presence of nonlinear effects in either the plant itself or its controls. While an extensive literature exists on the subject of linear controls, little is known about their nonlinear counterparts[1,2,3].

Another important instance of controlling chaos is provided by large distributed computational systems for which centralized resource allocation is not feasible. As we recently showed, when computational agents in these systems compete for resources in the presence of delays in information and imperfect knowledge about the state of the system, their dynamics can become extremely complex, giving rise to nonlinear oscillations and chaos [4,5]with a consequent degradation in overall performance. Part of the complexity in behavior is due to the coevolving nature of the system, i.e. the value of an agent's decision depends on the choices made by the all the others. Since complicated dynamics can lead to significantly lowered performance, it is important to devise methods which allow for the control of such systems while maintaining their responsiveness and the robustness of local decisions. The task is complicated by the fact that existing procedures for controlling chaotic systems with very few degrees of freedom[6,7] are inapplicable to distributed computation, which is characterized by a very large number of interacting processes over long distances.

Now, it is not obvious at all that chaos is something that should always be avoided. In the case of computational ecosystems, for example, the system's very stability may prevent it from adjusting to time-dependent constraints. A recent study that we have performed with N. Glance and T. Hogg of the behavior of such systems in a changing environment, showed that if the ecosystem behaves chaotically in the absence of any environmental modulations, then it is possible for it to perform better (albeit only transiently) than its stable counterpart when there is periodic change in the utility functions imposed on the system. Nevertheless this paper will deal with ways of controlling chaos in situations where stability, and not necessarily adaptability, is the desired goal.

2. Controlling Few Degrees of Freedom

Consider a general adaptive system with feedback. Its behavior is specified by the values of an output variable X, and an error signal which is equal to the difference between the actual output of the system and a desired state value, X_s. This error signal can be used to change the control parameters of the system so as to reduce the error to zero. This is the case in a wide variety of systems, ranging from self-regulating biological units to artificial adaptive structures such as recursive filters, controls[8] and neural networks[9].

The control problem one faces is simple to state. If the system is nonlinear and when can undergo transitions into undesirable chaotic regimes, how does one keep it operating within desired constraints in the presence of strong perturbations?

We have recently proposed a simple adaptive control mechanism which, besides providing efficient regulation, allows for a straightforward quantitative analysis of its dynamics[6]. Basically, it relies on the fact that if one has a general N-dimensional dynamical system of the form

$$\frac{dX}{dt} = F(X; \mu; t) \tag{1}$$

where $X \equiv (X_1, X_2, ...X_n)$ are the dynamical variables and $\mu \equiv (\mu_1, \mu_2, ...\mu_m)$ are parameters that determine the behavior of the system, one can achieve effective control through additional dynamics given by

$$\frac{d\mu}{dt} = \epsilon(X - X_s) \tag{2}$$

where ϵ indicates the stiffness of the control mechanism.

This procedure is not only effective in controlling the fixed points of a variety of systems, but also generates novel dynamical behavior. In particular, we showed that sudden perturbations in the system's parameters can make it degenerate into chaotic bursts with no precursors. When such bursts occur, the system first reverberates wildly and then recovers in times that are inversely proportional to the stiffness of the control mechanism. We also exhibited a general control principle which provides a quantitative relation between the maximum amplitude of a perturbation from which a system can recover, and the speed at which it does so.

These results, which were first obtained in the context of one-dimensional dynamical systems, have recently been extended to higher dimensional systems by Sinha et al. [7].

Through a set of relevant examples, they showed that not only can one use this mechanism to provide efficient control of fixed points in higher dimensions, but also (with suitable modifications) to regulate limit cycles as well. One can therefore anticipate that the actual implementation of this simple control in real world systems will take place in the not so distant future.

3. Chaos in Computational Systems

A useful view of distributed computer systems considers them as a collection of processes or agents that choose among various resources with which to accomplish their tasks[4]. Since decisions aren't centrally controlled, the agents independently and asynchronously select among the available choices based on their perceived payoff. These payoffs are actual computational measures of performance, such as the time required to complete a task, accuracy of the solution, amount of memory required, etc. In general, the payoff G_r for using resource r depends on the number of agents already using it. In a purely competitive environment, the payoff for using a particular resource tends to decrease as more agents make use of it. Alternatively, the agents using a resource could assist one another in their computations, as might be the case if the overall task could be decomposed into a number of subtasks. If these subtasks communicate extensively to share partial results, the agents will be better off using the same computer rather than running more rapidly on separate machines and then being limited by slow communications. As another example, agents using a particular database could leave index links that are useful to others. In such cooperative situations, the payoff of a resource would then increase as more agents use it, until it became sufficiently crowded.

Imperfect information about the state of the system causes each agent's perceived payoff to differ from the actual value, with the difference increasing when there is more uncertainty in the information available to the agents. This type of uncertainty concisely captures the effect of many sources of errors such as some program bugs, heuristics incorrectly evaluating choices, errors in communicating the load on various machines and mistakes in interpreting sensory data. Specifically, the perceived payoffs are taken to be normally distributed, with standard deviation σ, around their correct values. In addition, information delays cause each agent's knowledge of the state of the system to be somewhat out of date. Although for simplicity we will consider the case in which all agents have the same effective delay, uncertainty, and preferences for resource use, we should mention that the same range of behaviors is also found in more general situation[5].

As a specific illustration of this approach, we consider the case of two resources, so the system can be described by the fraction f of agents which are using resource 1 at any given time. Its dynamics is then governed by[4]

$$\frac{df}{dt} = \alpha(\rho - f) \tag{3}$$

where α is the rate at which agents reevaluate their resource choice and ρ is the probability that an agent will prefer resource 1 over 2 when it makes a choice. Generally, ρ is

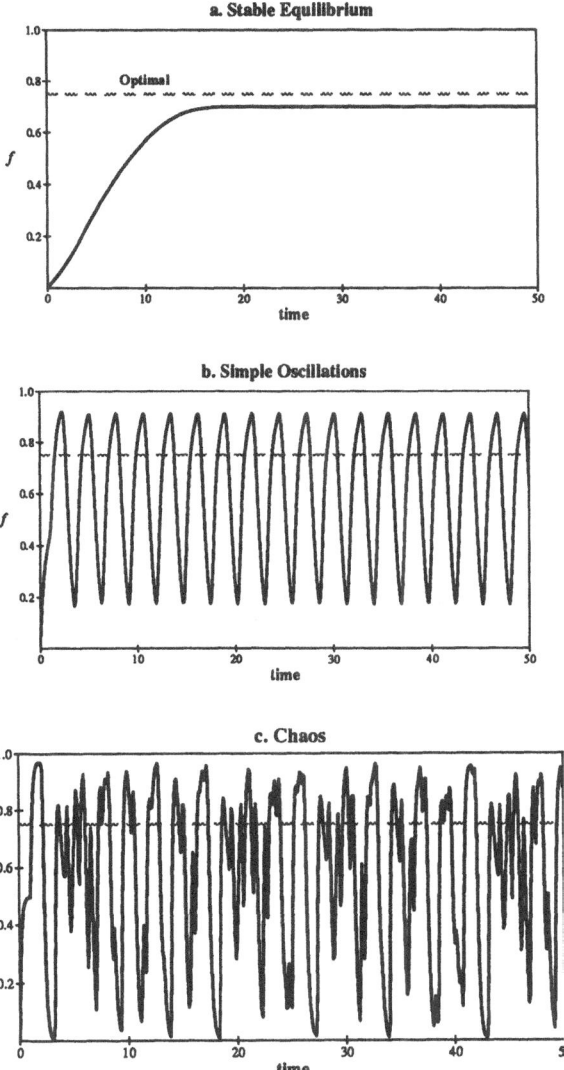

Fig. 1. Typical behaviors for the fraction f of agents using resource 1 as a function of time for successively longer delays: a) relaxation toward stable equilibrium, b) simple persistent oscillations, and c) chaotic oscillations. The payoffs are $G_1 = 4 + 7f - 5.333f^2$ for resource 1 and $G_2 = 4 + 3f$ for resource 2. The time scale is in units of the delay time τ, $\sigma = 1/4$ and the dashed line shows the optimal allocation for these payoffs.

a function of f through the density dependent payoffs. In terms of the payoffs and uncertainty, we have

$$\rho = \frac{1}{2}\left(1 + \operatorname{erf}\left(\frac{G_1(f) - G_2(f)}{2\sigma}\right)\right) \tag{4}$$

where σ quantifies the uncertainty. Notice that this definition captures the simple requirement that an agent is more likely to prefer a resource when its payoff is relatively large. Finally, delays in information are modeled by supposing that the payoffs that enter into ρ at time t are the values they had at a delayed time $t - \tau$.

For a typical system of many agents with a mixture of cooperative and competitive payoffs, the kinds of dynamical behaviors exhibited by the model are shown in Fig. 1. When the delays and uncertainty are fairly small, the system converges to an equilibrium point close to the optimal obtainable by an omniscient, central controller. As the information available to the agents becomes more corrupted, the equilibrium point moves further from the optimal value. With increasing delays, the equilibrium eventually becomes unstable, leading to the oscillatory and chaotic behavior shown in the figure. In these cases, the number of agents using particular resources continues to vary so that the system spends relatively little time near the optimal value, with the consequent drop in its overall performance.

4. Rewarding Performance

We now describe an effective procedure for controlling chaos in distributed systems[10]. It is based on a mechanism that rewards agents according to their actual performance. As we shall see, such an algorithm leads to the emergence of a diverse community of agents out of an essentially homogenous one. This diversity in turn eliminates chaotic behavior through a series of dynamical bifurcations which render chaos a transient phenomenon.

The actual performance of computational processes can be rewarded in a number of ways. A particularly appealing one is to mimic the mechanism found in biological evolution, where fitness determines the number of survivors of a given species in a changing environment. This mechanism is used in computation under the name of *genetic algorithms*[11]. Another example is provided by computational systems modelled on ideal economic markets[12,13], which reward good performance in terms of profits. In this case, agents pay for the use of resources, and they in turn are paid for completing their tasks. Those making the best choices collect the most currency and are able to outbid others for the use of resources. Consequently they come to dominate the system.

While there is a range of possible reward mechanisms, their net effect is to increase the proportion of agents that are performing successfully, thereby decreasing the number of those who do not do as well. It is with this insight in mind that we now develop a general theory of effective reward mechanisms without resorting to the details of their implementations. Since this change in agent mix will in turn change the choices made by every agent and their payoffs, those that were initially most successful need not be so in the future. This leads to an evolving diversity whose eventual stability is by no means obvious.

Before proceeding with the theory we point out that the resource payoffs that we will consider are instantaneous ones (i.e. shorter than the delays in the system), e.g. work actually done by a machine, currency actually received, etc. Other reward mechanisms, such as those based on averaged past performance, could lead to very different behavior from the one exhibited in this paper.

In order to investigate the effects of rewarding actual performance we generalize the previous model of computational ecosystems by allowing agents to be of different types, a fact which gives them different performance characteristics. Recall that the agents need to estimate the current state of the system based on imperfect and delayed information in order to make good choices. This can be done in a number of ways, ranging from extremely simple extrapolations from previous data to complex forecasting techniques. The different types of agents then correspond to the various ways in which they can make these extrapolations.

Within this context, a computational ecosystem can be described by specifying the fraction of agents, f_{rs} of a given type s using a given resource r at a particular time. We will also define the total fraction of agents using a resource of a particular type as

$$f_r^{res} = \sum_s f_{rs}$$
$$f_s^{type} = \sum_r f_{rs}$$

(5)

respectively.

As mentioned previously, the net effect of rewarding performance is to increase the fraction of highly performing agents. If γ is the rate at which performance is rewarded, then Eq. 3 is enhanced with an extra term which corresponds to this reward mechanism. This gives

$$\frac{df_{rs}}{dt} = \alpha\left(f_s^{type}\rho_{rs} - f_{rs}\right) + \gamma\left(f_r^{res}\eta_s - f_{rs}\right)$$

(6)

where the first term is analogous to that of the previous theory, and the second term incorporates the effect of rewards on the population. In this equation ρ_{r_s} is the probability that an agent of type s will prefer resource r when it makes a choice, and η_s is the probability that new agents will be of type s, which we take to be proportional to the actual payoff associated with agents of type s. As before, α denotes the rate at which agents make resource choices and the detailed interpretation of γ depends on the particular reward mechanism involved. For example, if they are replaced on the basis of their fitness it is the rate at which this happens. In a market system, on the other hand, γ corresponds to the rate at which agents are paid. Notice that in this case, the fraction of each type is proportional to the wealth of agents of that type.

Since the total fraction of agents of all types must be one, a simple form of the normalization condition can be obtained if one considers the relative payoff, which is given by[1]

$$\eta_s = \frac{\sum_r f_{rs}G_r}{\sum_r f_r^{res}G_r}$$

(8)

---

[1] This form assumes positive payoffs, e.g. they could be growth rates. If the payoffs can be negative (e.g. they are currency changes in an economic system), one can use instead the difference between the actual payoffs and their minimum value m. Since the η_s must sum to 1, this will give

$$\eta_s = \frac{\sum_r f_{rs}G_r - m}{\sum_r f_r^{res}G_r - Sm}$$

(7)

which reduces to the previous case when $m = 0$.

Note that the numerator is the actual payoff received by agents of type s given their current resource usage and the denominator is the total payoff for all agents in the system, both normalized to the total number of agents in the system.

Summing Eq. 6 over all resources and types gives

$$\frac{df_r^{res}}{dt} = \alpha\left(\sum_s f_s^{type}\rho_{rs} - f_r^{res}\right)$$

$$\frac{df_s^{type}}{dt} = \gamma\left(\eta_s - f_s^{type}\right)$$

(9)

which describe the dynamics of overall resource use and the distribution of agent types, respectively. Note that this implies that those agent types which receive greater than average payoff (i.e. types for which $\eta_s > f_s^{type}$) will increase in the system at the expense of the low performing types.

Note that the actual payoffs can only reward existing types of agents. Thus in order to introduce new variations into the population an additional mechanism is needed (e.g. corresponding to mutation in genetic algorithms or learning).

5. Results

In order to illustrate the effectiveness of rewarding actual payoffs in controlling chaos, we examine the dynamics generated by Eq. 6 for the case in which agents choose among two resources with cooperative payoffs, a case which we have shown to generate chaotic behavior in the absence of rewards[5]. As in the particular example of Fig. 1c, we use $\tau = 10$; $G_1 = 4 + 7f_1 - 5.333f_1^2$; $G_2 = 7 - 3f_2$; $\sigma = 1/4$, and an initial condition in which all agents start by using resource 2.

One kind of diversity among agents is motivated by the simple case in which the system oscillates with a fixed period. In this case, those agents that are able to discover the period of the oscillation can then use this knowledge to reliably estimate the current system state in spite of delays in information. Notice that this estimate does not neccesarily guarantee that they will keep performing well in the future, for their choice can change the basic frequency of oscillation of the system.

In what follows, we take the diversity of agent types to correspond to the different past horizons, or extra delays, that they use to extrapolate to the current state of the system. These differences in estimation could be due to having a variety of procedures for analyzing the system's behavior. Specifically, we indentify different agent types with the different assumed periods which range over a given interval. Thus, we take agents of type s to use an effective delay of $\tau + s$ while evaluating their choices.

The resulting behavior is shown in Fig. 2 which should be contrasted with Fig. 1c. We used an interval of extra delays ranging from 0 to 40. As shown, the introduction of actual payoffs induces a chaotic transient which, after a series of dynamical bifurcations, settles into a fixed point that signals stable behavior. Furthermore, this fixed point is exactly that obtained in the case of no delays. That this equilibrium is stable against perturbations can seen by the fact that if the system were perturbed again (as shown in Fig 3), it rapidly returns to its previous value. In additional experiments, with a smaller range of delays, we found that the system continued to oscillate without achieving the fixed point.

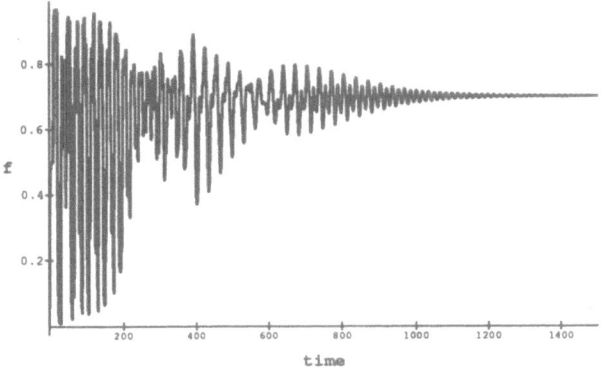

Fig. 2. Fraction of agents using resource 1 as a function of time with adjustment based on actual payoff. These parameters correspond to Fig. 1c so without the adjustment the system would remain chaotic.

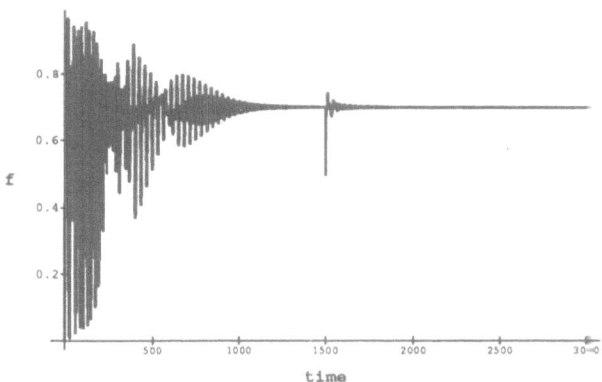

Fig. 3. Behavior of the system shown in Fig. 2 with a perturbation introduced at time 1500.

This transient chaos and its eventual stability can be understood from the distribution of agents with extra delays as a function of time. As can be seen in Fig. 4 actual payoffs lead to a highly heterogeneous system, characterized by a diverse population of agents of different types. It also shows that the fraction of agents with certain extra delays increases greatly. These delays correspond to the major periodicities in the system.

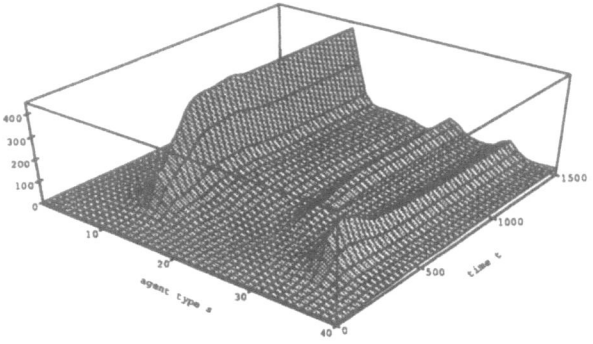

Fig. 4. Ratio $f_s^{type}(t)/f_s^{type}(0)$ of the fraction of agents of each type, normalized to their initial values, as a function of time. Note there are several peaks, which correspond to agents with extra delays of 12, 26 and 34 time units. Since $\tau = 10$, these match periods of length 22, 36 and 44 respectively.

6. Stability and Minimal Diversity

As we showed in the previous section, rewarding the performance of large collections of agents engaging in resource choices leads to a highly diverse mix of agents that stabilize the system. This suggests that the real cause of stability in a distributed system is that provided by sufficient diversity, and that the reward mechanims is an efficient way of automatically finding a good mix.

This raises the interesting question of the minimal amount of diversity needed in order to have a stable system.

The stability of a system is determined by the behavior of a perturbation around equilibrium, which can be found from the linearized version of Eq. 6.

In our case, the diversity is related to the range of different delays that agents can have. For a continuous distribution of extra delays, the characteristic equation is obtained by assuming a solution of the type $e^{\lambda t}$ in the linearized equation, giving

$$\lambda + \alpha - \alpha \rho' \int ds \, f(s) \, e^{-\lambda(s+\tau)} = 0 \tag{10}$$

Stability requires that all the values of λ have negative real parts, so that perturbations will relax back to equilibrium. As an example, suppose agent types are uniformly distributed in $(0, S)$. Then $f(s) = 1/S$, and the characteristic equation becomes

$$\lambda + \alpha - \alpha \rho' \frac{1 - e^{-\lambda S}}{\lambda S} e^{-\lambda \tau} = 0 \tag{11}$$

Defining a normalized measure of the diversity of the system for this case by $\eta \equiv S/\tau$, introducing the new variable $z \equiv \lambda \tau (1 + \eta)$, and multiplying Eq. 11 by $\tau(1 + \eta) z e^z$ introduces an extra root at $z = 0$ and gives

$$(z^2 + az)e^z - b + be^{rz} = 0 \qquad (12)$$

where

$$a = \alpha\tau(1 + \eta) > 0$$

$$b = -\rho'\frac{\alpha\tau(1 + \eta)^2}{\eta} > 0 \qquad (13)$$

$$r = \frac{\eta}{1 + \eta} \in (0, 1)$$

The stability of the system with uniform distribution of agents with extra delays thus reduces to finding the condition under which all roots of Eq. 12, other than $z = 0$, have negative real parts. This equation is a particular instance of an *exponential polynomial*, whose terms consist of powers multiplied by exponentials. Unlike regular polynomials, these objects generally have an infinite number of roots, and are important in the study of the stability properties of differential-delay equations[14]. Established methods can then be used to determine when they have roots with positive real parts. This in turn defines the stability boundary of the equation. The result for the particular case in which $\rho' = -3.41044$, corresponding to the parameters used in Sec. 5, is shown in the left half of Fig. 5.

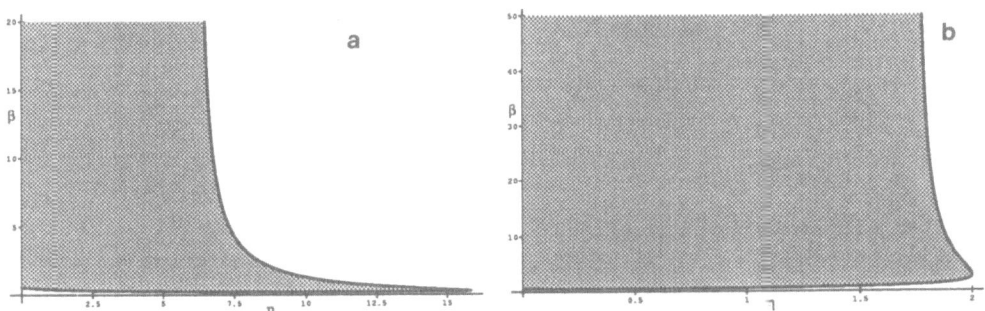

Fig. 5. Stability as a function of $\beta = \alpha\tau$ and $\eta = S/\tau$ for two possible distributions of agent types: a) $f(s) = \frac{1}{S}$ in $(0, S)$, and b) $f(s) = \frac{1}{S}e^{-s/S}$. The system is unstable in the shaded regions and stable to the right and below the curves.

Similarly, if we choose an exponential distribution of delays, i.e. $f(s) = \frac{1}{S}e^{-s/S}$ with positive S, the characteristic equation acquires the form

$$(z^2 + pz + q)e^z + r = 0 \qquad (14)$$

where

$$p = \alpha\tau + 1/\eta > 0$$

$$q = \alpha\tau/\eta > 0 \qquad (15)$$

$$r = -\alpha\tau\rho'/\eta > 0$$

and $z \equiv \lambda\tau$. An analysis similar to that for the uniform distribution case leads to the stability diagram shown in the right hand side of the figure.

Although the actual distributions of agent types can differ from these two cases, the similarity between the stability diagrams suggests that regardless of the magnitude of β one can always find an appropriate mix that will make the system stable. This property follows from the vertical asymptote of the stability boundary. It also illustrates the need for a minimum diversity in the system in order to make it stable when the delays aren't too small.

Having established the right mix that produces stability one may wonder whether a static assignment of agent types at an initial time would not constitute a simpler and more direct procedure to stabilize the system without resorting to a dynamic reward mechanism. While this is indeed the case in a non-fluctuating environment, such a static mechanism cannot cope with changes in both the nature of the system (e.g. machines crashing) and the arrival of new tasks or fluctuating loads. It is precisely to avoid this vulnerability by keeping the system adaptive that a dynamic procedure is needed.

Having seen how sufficient diversity stabilizes a distributed system, we now turn to the mechanisms that can generate such heterogeneity, as well as the time that it takes for the system to stabilize.

In particular, the details of the reward procedures determine whether the system can even find a stable mix of agents. In the cases describe above, reward was proportional to actual performance, as measured by the payoffs associated with the resources used. One might also wonder whether stability would be achieved more rapidly by giving greater (than their fair share) increases to the top performers.

We have examined two such cases: a) rewards proportional to the square of their actual performance, and b) giving all the rewards to top performers (e.g., those performing at the 90th percentile or better in the population). In the former case we observed stability with a shorter transient, whereas in the latter case the mix of changes continued to change through time, thus preventing stable behavior. This can be understood in terms of our earlier observation that whereas a small percentage agents can identify oscillation periods and thereby reduce their amplitude, a large number of them can no longer perform well.

Note that the time to reach equilibrium is determined by two parameters of the system. The first is the time that it takes to find a stable mix of agent types, which is governed by γ, and the second the rate at which perturbations relax, given the stable mix. The latter is determined by the largest real part of any of the roots, λ, of the characteristic equation.

7. Discussion

In this paper we exhibited several control mechanisms that can actually be used to stabilize nonlinear systems with oscillatory and chaotic behavior. For systems with few degrees of freedom, one can achieve effective control by introducing simple error—driven dynamics for the control parameter. In the case of distributed computational systems, rewards based on the actual performance of agents in a distributed computational system can stabilize an otherwise chaotic or oscillatory system. This leads in turn to greatly improved system performance.

In all these cases, stability is achieved by making chaos a transient phenomena. In the case of distributed systems, the addition of the reward mechanism has the effect of dynamically changing the control parameters of the resource allocation dynamics in such a way that a global fixed point of the system is achieved. This brings the issue of the length of the chaotic transient as compared to the time needed for most agents to complete their tasks. Even when the transients are long, the results of this study show that the range gradually decreases, thereby improving performance even before the fixed point is achieved.

A particularly relevant question for distributed systems is the extent to which these results generalize beyond the mechanism that we studied. Since we only considered the specific situation of a collection of agents with different delays in their appraisal of the system evolution, it is of interest to inquire whether using rewards to increase diversity works more generally than in the case of extra delays.

That this is the case in a more general sense is suggested by the following. Suppose there is a variety of techniques for evaluating the state of the system, and that each of these techniques is used by a particular type of agent. In the previous case, these techniques differed in the amount of extra delays used in the evaluation of the system. Consider now the case where these techniques involve the use of the same delay but make different systematic errors or biases in the evaluations. For instance, a positive bias would correspond to a heuristic that tends to overvalue a resource. Together with T. Hogg we have investigated this situation and discovered that once again the introduction of rewards rapidly stabilizes the system through a set of dynamical bifurcations that lead to a fixed point[10]. Moreover, this fixed point is optimal in the sense that it corresponds to the resource usage obtained when there are no delays or uncertainty in the system.

Since we only considered agents choosing between only two resources, it is important to understand what happens when there are many resources the agents can choose from. One may argue that since diversity is the key to stability, a plurality of resources provides enough channels to develop the necessary heterogeneity, which is what we observed in situations with three resources. Another note of caution has to do with the effect of fluctuations on a finite population of agent types. While we have shown that sufficient diversity can, on average, stabilize the system, in practice a fluctuation could wipe out those agent types that would otherwise be successful in stabilizing the system. Thus, we need either a large number of each kind of agent or a mechanism, such as mutation, to create new kinds of agents.

One problem that one may encounter when trying to apply these results to general computational and market systems has to do with individual performance measures. In any heterogeneous system, agents can have very different kinds of preferences (megaflops, currency, prestige, etc) which are difficult to reduce to a single scalar. In our theory, we avoided this problem by assuming that all payoffs are comparable so that we could reward the agents based on a simple comparison of their actual payoffs. This amounts to a single measure of fitness for the agents performance. Although the applicability of this method to more general situations remains an open problem, we note that economic systems provide a universal currency system that reduces everything to a single quantity, to which our reward mechanism can then be applied.

A final issue concerns the time scales over which rewards are assigned to agents. In our treatment, we assumed the rewards were always based on the performance at the time they were given. Since in many cases this procedure is delayed, there is the question of the extent to which rewards based on past performance are also able to stabilize chaotic distributed systems.

Nevertheless, I feel encouraged by the fact that one can freeze out chaotic behavior in distributed computation, while opening the way for further applications of this approach to complex nonlinear systems.

8. Acknowledgement

This work has benefitted from discussions with Tad Hogg and Eric Lumer.

9. References

[1]. I. M. Y. Mareels and R. R. Bitmead, "Bifurcation effects in robust adaptive control", IEEE Trans. Circuits Syst., vol. 35, pp. 835–841 (1988), and "On the dynamics of an equation arising in adaptive control", in Proc. 25th. IEEE Conf. on Decision and Control, Athens, Greece, pp. 1161–1166, (1986).

[2]. F. M. A. Salam and S. Bai, "Complicated dynamics of a prototype continuous-time adaptive control system", IEEE Trans. Circuits Syst., vol. 35, pp. 842–849 (1988).

[3]. A. Hubler and E. Luscher, "Resonant stimulation and control of nonlinear oscillators", TUM preprint LR 3895, (1988). This work presents a procedure for controlling chaos that relies on statistical forecasting techniques. Since by its very nature this method implies a delay between the detection of an error signal and the system's prediction, it does not allow for inmediate damping of errors after their appearance.

[4]. B. A. Huberman and T. Hogg, "The behavior of computational ecosystems", in The Ecology of Computation, B. A. Huberman, editor, pp. 77–115, North Holland (1988).

[5]. J. O. Kephart, T. Hogg and B. A. Huberman, "Dynamics of computational ecosystems", Physical Review A, vol. 40, pp. 404–421, (1989).

[6]. B. A. Huberman and E. Lumer, "Dynamics of Adaptive Systems", IEEE Trans. Circuits Syst., vol 37, pp. 547–550 (1990).

[7]. S. Sinha, R. Ramaswamy, and J. Subba Rao, Physica D vol. 43, 118–128 (1990).

[8]. B. Widrow and S. D. Stearns, Adaptive Signal Processing, Prentice-Hall (1985).

[9]. L. B. Almeida "Backpropagation in perceptrons with fedback" NATO ASI Seris, vol. F41, in Neural Computers, R. Eckmillier and Ch. Marsburg, editors, Springer-Verlag (1988).

[10]. T. Hogg and B. A. Huberman, "Controlling Chaos in Distributed Systems", Technical Report P-90–00133, Xerox PARC, Palo Alto, CA. (1990).

[11] D. E. Goldberg, *Genetic Algorithms in Search, Optimization and Machine Learning*, Addison-Wesley, (1989).

[12]. M. S. Miller and K. E. Drexler, "Markets and Computation: Agoric open systems", in *The Ecology of Computation*, B. A. Huberman, editor, pp. 133–176 North-Holland (1988).

[13]. C. A. Waldspurger, T. Hogg, B. A. Huberman, J. O. Kephart and S. Stornetta, "Spawn: A distributed computational economy" Technical Report P89–00025, Xerox PARC, Palo Alto, CA. (1989).

[14]. R. Bellman and K. L. Cooke, *Differential-Difference Equations*. Academic Press, (1963).

BERRY PHASE AND ITS UNITARILY

INVARIANT GENERALIZATION

Donald H. Kobe

Department of Physics
University of North Texas
Denton, Texas   76203-5368
U.S.A.

ABSTRACT

The geometrical phase in quantum mechanics discovered by Berry is
not invariant under unitary transformations. It is genera_ized to be
unitarily invariant and applicable to nonadiabatic and nonzyclic systems.
Examples of a spin-1/2 particle in a precessing magnetic f_eld and a time-
dependent generalized harmonic oscillator are considered.

1.   INTRODUCTION

In 1984 a phase in quantum mechanics was discovered by Berry,[1] who
showed that in the adiabatic and cyclic case it has a geometrical significance. This geometrical phase, now called the "Berry phase," can be
observed in interference type experiments.[2] The following year it was
shown by Hannay[3] that a corresponding classical angle existed with a
geometrical significance, to which the Berry phase reduced in the classical limit.[4] The Berry phase is a consequence of the Schrödinger equation, so it does not constitute any correction to quantum mechanics.
Similarly, the Hannay angle follows directly from Hamilton's equations in
terms of action and angle variables, so it does not constitute a
correction to classical mechanics.

The phase introduced by Berry is not invariant under gauge transformations or, more generally, unitary transformations.[5] An observable
quantity, like the relative Berry phase, must be gauge invariant, since
its value cannot depend on the gauge chosen.

A generalized form of the Berry phase which is gauge invariant
emerges naturally in quantum mechanics when the distinction is made
between the Hamiltonian and the energy operator.[6] The Hamiltonian
describes the time development of the system, while the energy is a
function whose time derivative gives the power transferred between the
system and its environment. It is not necessary to consider only
adiabatic and cyclic processes, except to give the phase a geometrical
interpretation. The generalized Berry phase arises because the operator
which induces transitions between states has a diagonal matrix element.
The diagonal matrix element gives a phase which is the generalized Berry
phase. This phase was originally discovered by Yang,[7] so it is appropriate to call it the "Yang phase."[6]

In order to illustrate the concept of the generalized Berry phase,
two examples are considered. The first one is a spin-1/2 particle in a
precessing magnetic field.[8] The generalized Berry phase, which is the
flux of a "curvature" in parameter space, is equal to one-half the solid
angle subtended by the surface whose boundary is the curve swept out by

the magnetic field vector in one period. This result is the standard one obtained by Berry.[1]

The second example is a generalized time-dependent harmonic oscillator, which has a cross term in the Hamiltonian in addition to the usual quadratic terms. A nonzero Berry phase was obtained by a number of workers.[3,4,9] On the other hand, an ambiguity in the Berry phase for this problem was reported by Jackiw[10] and Gerbert.[11] The generalized Berry phase resolves the ambiguity and gives zero for the generalized Berry phase,[6] except for the case of an oscillator with damping.[12]

The outline of this paper is the following. In Sec. 2 some background to the Berry phase is given. In Sec. 3 the generalized Berry phase is described, and its geometrical interpretation in the adiabatic and cyclic case is given. An example of a spin-1/2 particle in a precessing magnetic field is given in Sec. 4. A generalized harmonic oscillator is given as an example in Sec. 5. The conclusion is given in Sec. 6.

2. BACKGROUND

The Berry phase[1] could have been discovered any time after 1925 when the Schrödinger equation was discovered. In the well-known quantum mechanics text by Schiff,[13] the geometrical phase was given and then eliminated through a misuse of a gauge transformation on the wave function.

2.1. Adiabatic Approximation

The Schrödinger equation with the time-dependent Hamiltonian $H(t)$ and wave function $\psi(t)$ is

$$H(t)\psi(t) = i\partial\psi(t)/\partial t, \qquad (2.1)$$

where natural units such that $\hbar = 1$ and the speed of light is unity are used. In the adiabatic approximation, where the Hamiltonian is slowly varying in time, Schiff[13] uses eigenstates $u_n(t)$ of the Hamiltonian $H(t)$

$$H(t)u_n(t) = E_n(t)u_n(t), \qquad (2.2)$$

with eigenvalues $E_n(t)$. The wave function $\psi(t)$ can be expanded in terms of these eigenstates at time t as

$$\psi(t) = \sum_n b_n(t)\, u_n(t), \qquad (2.3)$$

where $b_n(t) = \langle u_n(t)|\psi(t)\rangle$ are time-dependent amplitudes. When Eq. (2.3) is substituted into the Schrödinger equation in Eq. (2.1), an equation is obtained for the amplitudes $b_n(t)$,

$$i\dot{b}_n - E_n b_n + \langle u_n|(i\partial/\partial t)u_n\rangle b_n = -{\sum_k}' \langle u_n|(i\partial/\partial t)u_k\rangle b_k \quad , \qquad (2.4)$$

where the sum is restricted so that it does not include the term $k = n$. We can transform Eq. (2.4) to the interaction picture with

$$b_n(t) = B_n(t)\, \exp\{-i\int_0^t dt' E_n(t') + i\gamma_n(t)\}, \qquad (2.5)$$

where the phase $\gamma_n(t)$ is

$$\gamma_n(t) = \int_0^t dt'\, \langle u_n|(i\partial/\partial t')u_n\rangle. \qquad (2.6)$$

When Eq. (2.5) is substituted into Eq. (2.4), the equation for the amplitude $B_n(t)$ in the interaction picture is

$$i\dot{B}_n = -{\sum_k}' \langle u_n|(i\partial/\partial t)u_k\rangle\, B_k$$

$$\times \exp\{-i\int_0^t dt'(E_k - E_n) + i(\gamma_k - \gamma_n)\} \quad . \qquad (2.7)$$

In the adiabatic approximation, the transitions to other states described by the right-hand side of Eq. (2.7) are neglected. For a cyclic process with period T the Berry phase is $\gamma_n(T)$.

2.2. Gauge Transformations

Schiff[13] tries to remove the phase $\gamma_n(t)$ by making a gauge transformation on the eigenstates,

$$u'_n = \exp\{i\alpha(t)\}\, u_n .\qquad(2.8)$$

The new phase $\gamma'_n(t)$ is

$$\gamma'_n(t) = \int_0^t dt' \, \langle u'_n | (i\partial/\partial t') u'_n \rangle$$

$$= \gamma_n(t) - \alpha(t).\qquad(2.9)$$

By choosing $\alpha(t) = \gamma_n(t)$, Schiff[13] can set $\gamma'_n(t) = 0$ for *one* given n. But he goes on to say "we assume that all the phases have been chosen in this way ...."[13] It is clear that *all* the $\gamma_n(t)$ cannot be eliminated in this way. The $\gamma_n(t)$ depends on the state n, but the gauge function $\alpha(t)$ does not depend on the state. How many thousands of students (and their professors) have read this remark since 1949 when the first edition was published and not realized the error!

Berry[1] realized that the phases $\gamma_n(t)$ could not be removed by a gauge transformation. He pointed out that the phase had a geometrical significance in the adiabatic and cyclic case, and that it could be observed in some circumstances.

Under the time-dependent gauge transformation in Eq. (2.8) the phase in Eq. (2.9) is changed, but not in such a way that the difference of two phases at time t is changed. A more general gauge transformation,

$$u'_n(r,t) = \exp\{i\Lambda(r,t)\}\, u_n(r,t),\qquad(2.10)$$

can be made, where the gauge function $\Lambda = \Lambda(r,t)$ depends both on space r and time t coordinates. Then the new phase is

$$\gamma'_n(t) = \int_0^t dt' \, \langle u'_n | (i\partial/\partial t') u'_n \rangle$$

$$= \gamma_n(t) - \int_0^t dt' \, \langle u_n | (\partial\Lambda/\partial t') u_n \rangle .\qquad(2.11)$$

Since the gauge function $\Lambda(r,t)$ is arbitrary, the phase γ_n can be changed in an arbitrary way by making a gauge transformation. Clearly, if the Berry phase is to be an observable it cannot depend on the gauge.

The generalization of the phase $\gamma_n(t)$ so that it is invariant under gauge transformations or, more generally, under unitary transformations is discussed in Sec. 3.

3. UNITARILY-INVARIANT GENERALIZED BERRY PHASE

The usual discussion of the Berry phase in Sec. 2 was shown to be gauge dependent, which is unacceptable for an observable quantity. In this section the Berry phase is generalized so that it is gauge invariant and applicable to systems that are neither adiabatic nor cyclic.

3.1. Hamiltonian and Energy Operator

The basic idea in this section is to use eigenstates of a gauge-invariant energy operator, rather than the gauge-dependent Hamiltonian as is done in Sec. 2. To be specific, consider the Hamiltonian H for a particle of mass m and charge q in an external electromagnetic field characterized by the vector potential A and the (nonconservative) scalar potential A_0,

$$H = (2m)^{-1}(p - qA)^2 + V(r) + qA_0(r,t).\qquad(3.1)$$

The conservative potential energy $V(\mathbf{r})$ is determined by the physical system. The kinetic (or mechanical) momentum is $m\mathbf{v} = \mathbf{p} - q\mathbf{A}$, where $\mathbf{p} = -i\nabla$ is the canonical momentum operator. For this time-dependent problem, the Hamiltonian is not in general the energy.[14]

The energy is the sum of the kinetic and the (conservative) potential energy

$$E = (2m)^{-1}(\mathbf{p} - q\mathbf{A})^2 + V(\mathbf{r}). \tag{3.2}$$

To verify that Eq. (3.2) is indeed the energy, the total time derivative of its expectation value can be taken, which gives

$$d\langle\psi|E\psi\rangle/dt = \langle\psi|P\psi\rangle . \tag{3.3}$$

The Hermitian quantum mechanical power operator P is

$$P = \tfrac{1}{2}q(\mathbf{E}\cdot\mathbf{v} + \mathbf{v}\cdot\mathbf{E}), \tag{3.4}$$

where the velocity operator $\mathbf{v}$ is

$$\mathbf{v} = m^{-1}(\mathbf{p} - q\mathbf{A}) . \tag{3.5}$$

The electric field $\mathbf{E}$ and the magnetic field $\mathbf{B}$ are

$$\mathbf{E} = -\nabla A_0 - \partial\mathbf{A}/\partial t, \qquad \mathbf{B} = \nabla \times \mathbf{A} . \tag{3.6}$$

The magnetic field does not contribute to the power in Eq. (3.4). Equation (3.3) is a form of Ehrenfest's theorem, which states that the time rate of change of the average energy is equal to the expectation value of the quantum mechanical power operator. From Eqs. (3.1) and (3.2) the energy operator E is

$$E = H - qA_0 . \tag{3.7}$$

Therefore, in a gauge in which $A_0 = 0$ the Hamiltonian is the energy, when the canonical momentum is written as a function of the velocity.

3.2. Generalized Berry Phase

The eigenvalue problem for the energy operator is

$$E\psi_n = \epsilon_n\psi_n , \tag{3.8}$$

where ψ_n is the energy eigenstate and ϵ_n is the energy eigenvalue. Since the energy operator depends on the time, both the energy eigenstate and eigenvalue in general depend on time as a parameter.

The wave function ψ in Eq. (2.1) can be expanded in terms of the complete set of energy eigenstates

$$\psi(t) = \sum_n c_n(t)\, \psi_n(t), \tag{3.9}$$

where $c_n(t) = \langle\psi_n(t)|\psi(t)\rangle$ is the probability amplitude for finding the system in the energy eigenstate $\psi_n(t)$ at time t. When Eq. (3.9) is substituted into the Schrödinger equation in Eq. (2.1), the resulting equation for the probability amplitudes is

$$i\dot{c}_n - \epsilon_n c_n + \dot{\gamma}_n c_n = - \sum_k{}' \langle\psi_n|(i\partial/\partial t - qA_0)\psi_k\rangle c_k , \tag{3.10}$$

where the time derivative of γ_n is the diagonal matrix element of the operator which induces transitions,

$$\dot{\gamma}_n(t) = \langle\psi_n|(i\partial/\partial t - qA_0)\psi_n\rangle . \tag{3.11}$$

The time dependence in the problem has three effects in Eq. (3.10). First, it gives in general a time-dependent energy eigenvalue $\epsilon_n(t)$, due to the effect of the external field on the energy eigenstates. Second, it causes transitions to other energy eigenstates, because the right-hand side of Eq. (3.10) is not zero. Third, it gives a time-dependent phase $\gamma_n(t)$ due to the diagonal matrix element in Eq. (3.11) of the operator $i\partial/\partial t - qA_0$ which produces transitions.

If we make a transformation to the interaction picture, the amplitude is

$$c_n(t) = C_n(t) \exp\{-i \int_0^t dt' \epsilon_n(t') + i\gamma_n(t)\} . \tag{3.12}$$

If Eq. (3.12) is substituted into Eq. (3.10), it becomes

$$i\dot{C}_n = - \sum_k{}' \langle\psi_n|(i\partial/\partial t - qA_0)\psi_k\rangle C_k$$

$$\times \exp\{-i \int_0^t dt' (\epsilon_k - \epsilon_n) + i(\gamma_k - \gamma_n)\} . \tag{3.13}$$

For the nondegenerate case, where $\epsilon_n \neq \epsilon_k$ for $n \neq k$, the off-diagonal matrix elements in Eq. (3.13) are

$$\langle\psi_n|(i\partial/\partial t - qA_0)\psi_k\rangle = -i(\epsilon_n - \epsilon_k)^{-1} \langle\psi_n|P\psi_k\rangle , \tag{3.14}$$

where P is the power operator in Eq. (3.4). If the matrix element of the power operator between two different states is zero, then no direct transition between these two states can take place.

The generalized Berry phase, or Yang phase, in Eq. (3.12) is

$$\gamma_n(t) = \int_0^t dt' \langle\psi_n|(i\partial/\partial t' - qA_0)\psi_n\rangle , \tag{3.15}$$

and is present in the nonadiabatic and noncyclic case. This generalized Berry phase has been shown to be invariant under a general unitary transformation.[6]

3.3. Geometrical Phase for Adiabatic and Cyclic System

For an adiabatic and cyclic system the generalized Berry phase reduces to the usual geometrical phase of Berry.[1] In the adiabatic case, transitions are neglected. The right-hand side of Eq. (3.13) vanishes, and hence $C_n(t) = C_n(0)$ is constant in time. If we also choose a gauge in which $A_0 = 0$, then Eq. (3.15) reduces to

$$\gamma_n(t) = \int_0^t dt' \langle\psi_n|(i\partial/\partial t')\psi_n\rangle , \tag{3.16}$$

which is the expression Berry[1] obtained. We further assume that the time dependence is due to some external parameters $X(t) = \{X_1(t), X_2(t), X_3(t), ..., X_m(t)\}$, and that they are periodic with a period T, $X(T) = X(0)$. The energy eigenstates have time dependence through $X(t)$, so $\psi_n(x,t) = \psi_n(x,X(t))$. For one period T Eq. (3.16) for the Berry phase be written as

$$\gamma_n(T) = \int_0^T dt' \langle\psi_n|i\partial\psi_n/\partial X\rangle \cdot \dot{X}(t') = \oint_C dX \cdot A_n , \tag{3.17}$$

where the integral is over a complete cycle C in parameter space. The "connection" A_n in parameter space X is defined as

$$A_n = \langle\psi_n|i\partial\psi_n/\partial X\rangle , \tag{3.18}$$

which is an m-dimensional vector depending on the energy eigenstate n. By Stokes's theorem, the integral in Eq. (3.17) can be transformed to

$$\gamma_n(T) = \int_s dS \cdot B_n , \tag{3.19}$$

where dS is an element of area in parameter space and the integral is over the surface S which has the boundary $\partial S = C$, the curve swept out by $\mathbf{X}(t)$ in one cycle in parameter space. For m = 3 the "curvature" B_n in parameter space is the curl of the connection A_n,

$$B_n = (\partial/\partial \mathbf{X}) \times A_n . \tag{3.20}$$

The geometrical interpretation of Eq. (3.19) is that the Berry phase $\gamma_n(T)$ is the flux of the curvature B_n through a surface S in parameter space that is bounded by the closed path C swept out by the parameters in one cycle. This general geometrical interpretation is rather abstract, so it is worthwhile to illustrate it with examples.

4. SPIN-1/2 PARTICLE IN A MAGNETIC FIELD

In this section the Berry phase for a spin-1/2 particle in a precessing magnetic field is calculated.[8] The energy and the Hamiltonian for a spin-1/2 particle in a magnetic field $\mathbf{B}(t)$ is

$$E = H = -\mu\sigma\cdot\mathbf{B}(t) , \tag{4.1}$$

where μ is the Bohr magneton and $\sigma = (\sigma_1, \sigma_2, \sigma_3)$ are the Pauli matrices. The magnetic field $\mathbf{B}$ is

$$\mathbf{B} = (B_1, \ B_2, \ B_3)$$

$$= (B \sin\theta \cos\phi, \ B \sin\theta \sin\phi, \ B \cos\theta), \tag{4.2}$$

where $B = |\mathbf{B}(t)|$ is the magnitude of the magnetic field, and θ and ϕ are the polar and azimuthal angles, respectively.

The energy eigenvalue problem in Eq. (3.8) gives the two energy eigenvalues $\epsilon_1 = -\mu B$ and $\epsilon_2 = \mu B$, with energy eigenstates

$$\psi_1 = \cos(\theta/2) \exp(-i\phi)\alpha_+ + \sin(\theta/2)\alpha_- \tag{4.3}$$

and

$$\psi_2 = -\sin(\theta/2) \exp(-i\phi)\alpha_+ + \cos(\theta/2)\alpha_- , \tag{4.4}$$

respectively. The states $\alpha_\pm$ are eigenstates of σ_3 with eigenvalues ±1.

In the parameter space ($\mathbf{X} = \mathbf{B}$) the connection A_2 in Eq. (3.18) is

$$A_2 = i \langle\psi_2|\partial\psi_2/\partial\mathbf{B}\rangle . \tag{4.5}$$

When Eq. (4.4) is used in Eq. (4.5) the connection is

$$A_2 = \hat{\phi}(B \sin \theta)^{-1} \sin^2(\theta/2), \tag{4.6}$$

where $\hat{\phi}$ is a unit vector in the azimuthal direction. The curvature B_2 in Eq. (3.20) is

$$B_2 = (\partial/\partial\mathbf{B}) \times A_2 \ = \ \tfrac{1}{2}\hat{B}/B^2 , \tag{4.7}$$

where $\hat{B}$ is a unit vector in the $\mathbf{B}$ direction. In the parameter space the curvature has the same form as the magnetic (or electric) field in real space $\mathbf{r}$ of a magnetic (or electric) charge 1/2 located at the origin, i.e., $\tfrac{1}{2}\hat{r}/r^2$, where $\hat{r}$ is a unit vector in the $\mathbf{r}$ direction.

The flux of the curvature B_2 in Eq. (4.7) in parameter space is the Berry phase in Eq. (3.19),

$$\gamma_2(T) = \int_S (\tfrac{1}{2}\hat{B}/B^2)\cdot\hat{B}B^2 d\Omega = \tfrac{1}{2}\Omega , \tag{4.8}$$

where $d\Omega$ is an element of solid angle and Ω is the solid angle subtended by the surface S, whose boundary C is swept out by $\mathbf{B}$ in one period T.

If the polar angle θ is a constant, then the solid angle subtended in Eq. (4.8) is

$$\Omega = 2\pi(1 - \cos\theta) \ . \tag{4.9}$$

From Eq. (4.8) we see that the Berry phase, which is the flux of a curvature in parameter space, is proportional to the solid angle enclosed by $\mathbf{B}(t)$ in one period, which is a completely geometrical concept.[1]

5. GENERALIZED TIME-DEPENDENT HARMONIC OSCILLATOR

Another example of the Berry phase can be given, which is a generalized time-dependent harmonic oscillator.[6]

5.1. Hamiltonian and Equation of Motion

The classical Hamiltonian for a time-dependent generalized harmonic oscillator is

$$H = \tfrac{1}{2}\{a(t)p^2 + 2b(t)px + c(t)x^2\} \ , \tag{5.1}$$

where x is a generalized coordinate and p is the canonical momentum conjugate to it. The functions a, b, and c are all time dependent. The oscillator is called "generalized" because of the presence of the cross term px in the Hamiltonian. From Hamilton's equations, the velocity is

$$\dot{x} = \partial H/\partial p = ap + bx \ , \tag{5.2}$$

and the time rate of change of the canonical momentum is

$$\dot{p} = -\partial H/\partial x = -bp - cx \ . \tag{5.3}$$

When Eq. (5.2) is solved for p and substituted into Eq. (5 3), the Newtonian equation of motion is

$$d(m\dot{x})/dt = -kx \ , \tag{5.4}$$

where the time-dependent mass $m = a^{-1}$ and the time-dependent spring "constant" is

$$k(t) = c - mb^2 - d(mb)/dt \ , \tag{5.5}$$

which must be positive. The system described by the Hamiltonian in Eq. (5.1) is a harmonic oscillator with a time-dependent mass and spring "constant."

5.2. Energy

The energy E can be obtained for this time-dependent problem by multiplying Eq. (5.4) by $\dot{x}$ and rearranging. Then the energy E can be written as the sum of the kinetic energy plus the potential energy,

$$E = \tfrac{1}{2}m\dot{x}^2 + \tfrac{1}{2}kx^2 \ , \tag{5.6}$$

where the spring "constant" k in Eq. (5.6) is the same as in the equation of motion in Eq. (5.4). The time rate of change of Eq. (5.6) is

$$dE/dt = \tfrac{1}{2}\dot{m}\dot{x}^2 + \tfrac{1}{2}\dot{k}x^2 + (-\dot{m}\dot{x})\dot{x} = P. \tag{5.7}$$

The power P consists of three terms. The first term on the right-hand side in Eq. (5.7) is the power required to change the mass at constant velocity. The second term is the power required to change the spring "constant" at constant displacement. The third term is the power due to the "fictitious force" $-\dot{m}\dot{x}$, which comes from the left-hand side of Eq. (5.4), $m\ddot{x} + \dot{m}\dot{x}$. Each term in Eq. (5.7) has a physical interpretation as a power term. That the fictitious force also does work can be seen when $k = 0$. From Eq. (5.4) the kinetic (or mechanical) momentum $m\dot{x}$ is conserved. The energy in this case is purely kinetic, $(m\dot{x})^2/2m$, and so its time derivative is $-\tfrac{1}{2}\dot{m}\dot{x}^2$, which is the sum of the first and third terms on the right-hand side of Eq. (5.7). Thus both the first and third terms in Eq. (5.7) are required in the general case.

The energy in Eq. (5.6) can be written in terms of the canonical momentum in Eq. (5.2) as

$$E = (2m)^{-1}(p + mbx)^2 + \tfrac{1}{2}kx^2 . \tag{5.8}$$

In this form, the system is suitable for canonical quantization.

5.3. Canonical Quantization

A system can be canonically quantized by replacing the canonical momentum p by the operator $-i\partial/\partial x$, so that the canonical commutation relations are satisfied. Then the energy eigenvalue problem is Eq. (3.8), with the energy given by Eq. (5.8). The term mbx can be removed by a gauge transformation with the gauge function $\Lambda = \tfrac{1}{2}mbx^2$. The transformed energy operator is

$$E' = \exp(i\Lambda)E\exp(-i\Lambda) = p^2/2m + \tfrac{1}{2}kx^2 , \tag{5.9}$$

and the transformed eigenstate is

$$\psi_n' = \exp(i\Lambda)\ \psi_n . \tag{5.10}$$

The transformed energy eigenvalue problem in Eq. (3.8) is

$$E'\psi_n' = \epsilon_n\psi_n' , \tag{5.11}$$

with the same energy eigenvalue. By inspection of Eq. (5.9), the energy eigenvalue ϵ_n is

$$\epsilon_n = \omega(n + \tfrac{1}{2}) , \tag{5.12}$$

where n = 0,1,2,3, ... and the time-dependent angular frequency is

$$\omega(t) = (k/m)^{1/2} . \tag{5.13}$$

The energy eigenstates in Eq. (5.11) are real because the energy operator is Eq. (5.9).

5.4. Generalized Berry Phase

The generalized Berry phase in Eq. (3.15) can be calculated because we have solved the energy eigenvalue problem. The scalar potential term in Eq. (3.7) is

$$qA_0 = H - E = \tfrac{1}{2}[d(mb)/dt]\ x^2 = \partial\Lambda/\partial t , \tag{5.14}$$

for $\Lambda = \tfrac{1}{2}mbx^2$. The generalized Berry phase in Eq. (3.15) is

$$\gamma_n(t) = \int_0^t dt'\ \langle\psi_n|(i\partial/\partial t' - qA_0)\psi_n\rangle$$

$$= \int_0^t dt'\ \{\langle\psi_n'|(\partial\Lambda/\partial t')\psi_n'\rangle + \langle\psi_n'|(i\partial/\partial t')\psi_n'\rangle - \langle\psi_n'|(\partial\Lambda/\partial t')\psi_n'\rangle\}$$

$$= 0 . \tag{5.15}$$

Because ψ_n' is real, $\langle\psi_n'|\dot\psi_n'\rangle = \langle\psi_n'|(\partial/\partial t)\psi_n'\rangle$ is real. On the other hand, because $\langle\psi_n'|\psi_n'\rangle = 1$, $\langle\psi_n'|\dot\psi_n'\rangle$ is pure imaginary. Therefore $\langle\psi_n'|\dot\psi_n'\rangle = 0$ in Eq. (5.15), and the generalized Berry phase is zero. This result is reasonable, since we know that the Berry phase for a simple harmonic oscillator with a time-dependent frequency is zero. The Hamiltonian in Eq. (5.1) is gauge equivalent to the Hamiltonian for a simple harmonic oscillator with angular frequency given by Eq. (5.13).[10,11] The Berry phase should be gauge invariant, since the relative Berry phase is in principle observable. Thus, the Berry phase for the generalized harmonic oscillator is also zero. This statement is in contrast to results obtained for this problem by Berry[1] and Hannay,[3] who assume that the Hamiltonian is the energy.[12]

6. CONCLUSION

Berry[1] pointed out that a phase in quantum mechanics, which had previously been neglected, has a geometrical interpretation in the adiabatic and cyclic case and also has observable consequences. On the other hand, the phase introduced by Berry is not invariant under gauge or, more generally, unitary transformations.

It is possible to generalize the Berry phase in a way previously done by Yang.[7] When the distinction is made between the Hamiltonian and the energy operator, a gauge-invariant phase emerges as a result of the diagonal matrix element of the operator which induces transitions between states. In the adiabatic and cyclic case the generalized Berry phase reduces to the flux of a "curvature" in parameter space through a surface whose boundary is swept out by the parameters in one period.

Two examples of the Berry phase are given. A particle of spin-1/2 in a precessing magnetic field is considered.[8] The Berry phase for the excited state is 1/2 times the solid angle subtended by the surface whose boundary is swept out by the time-dependent magnetic field in one cycle.

The other example considered is the generalized harmonic oscillator.[6] It is shown that when the distinction between the energy and the Hamiltonian is made, the generalized Berry phase is zero. This result is in contrast to the results of Berry[4] and Hannay.[3] The generalized harmonic oscillator is gauge equivalent to a simple harmonic oscillator with time-dependent frequency. Since the Berry phase of the simple harmonic oscillator is zero, gauge invariance implies that the Berry phase for the generalized harmonic oscillator is also zero.

ACKNOWLEDGEMENTS

I would like to thank Professor A. H. Zimerman, Professor V. C. Aguilera-Navarro, Mr. ·S. Kurcbart, and Mr. W. Chyla for helpful discussions concerning this work. I am grateful to Professor R. Aldrovandi, Professor G. Francisco, and the other faculty and students at the Instituto de Física Teórica in São Paulo, Brazil for their hospitality during my visit there in the summer (northern hemisphere) of 1991. I would like to acknowledge and express my thanks to the U.S. Army Research Office for partial support of my travel.

REFERENCES

1. M. V. Berry, *Proc. R. Soc.* A 392:45 (1984).
2. J. W. Zwanziger, M. Koenig, and A. Pines, *Annu. Rev. Phys. Chem.* 41:601 (1990).
3. J. Hannay, *J. Phys.* A 18:221 (1985).
4. M. V. Berry, *J. Phys.* A 18:15 (1985).
5. G. Giavarini, E. Gozzi, D. Rohrlich, and W. D. Thacker, *Phys. Lett.* A 137:235 (1989); *J. Phys.* A 22:3513 (1989).
6. D. H. Kobe, *J. Phys.* A 23:4249 (1990).
7. K.-H Yang, *Ann. Phys. (N.Y.)* 101:62 (1976).
8. D. H. Kobe, V. C. Aguilera-Navarro, S. M. Kurcbart, and A. H. Zimerman, *Phys. Rev.* A 42:3744 (1990).
9. X. Gao, J. B. Xu, and T. Z. Qian, *Ann. Phys. (N.Y.)* 204:235 (1990).
10. R. Jackiw, *Int. J. Mod. Phys.* A 3:285 (1988).
11. P. de S. Gerbert, *Ann. Phys. (N.Y.)* 189:155 (1989).
12. D. H. Kobe, *J. Phys.* A 24:2763 (1991).
13. L. I. Schiff, "Quantum Mechanics," McGraw-Hill, New York (1968), 3rd ed., pp. 289-291.
14. D. H. Kobe and K.-H. Yang, *Eur. J. Phys.* 8:236 (1987 .

NATANZON POTENTIALS AND
SPECTRUM GENERATING ALGEBRAS

Patricio Cordero*

Departamento de Física
Facultad de Ciencias Físicas y Matemáticas
Universidad de Chile, Casilla 487, Santiago 3, Chile
e-mail: pcordero@uchcecvm.bitnet

Sebastián Salamó

Departamento de Física
Universidad Simón Bolívar
Apartado Postal 89000, Caracas, Venezuela

ABSTRACT

It is shown that the bound state problem posed by the Schrödinger equation can be solved by means of spectrum generating algebra techniques for all potentials for which it can be reduced either to a confluent hypergeometric form or to a general hypergeometric form. The connection with supersymmetric quantum mechanics is analysed in the case of the Morse potential.

1. INTRODUCTION

Since the beginning of quantum mechanics a standing question has been which are the most general potentials for which the Schrödinger equation can be solved analytically. Such question is too wide to have a definite answer. If we ask however which are the most general potentials for which the Schrödinger equation can be reduced to a hypergeometric (or a confluent hypergeometric) form then the most general answer — as far as we know — corresponds to the potentials found by Natanzon [1]. The Natanzon potentials can be divided in two classes. The *confluent Natanzon potentials*

---

*Partially supported by grant *FONDECYT* 90-1240

which are those for which the Schrödinger equation is reduced to a confluent hyperge-ometric form and the *hypergeometric Natanzon potentials* which are those for which the Schrödinger equation is reduced to a hypergeometric form.

Some time ago a spectrum generating algebra (SGA) [2] [3] technique that gives the analytic solution to the Schrödinger equation for several potentials and even the whole class of confluent Natanzon potential case [4] [5] was developed. A related technique de-veloped some time later [6] has allowed us to solve the hypergeometric Natanzon poten-tials [7] case. Both techniques are based on the use of realizations of the infinitesimal generators J_μ ($\mu = 0, 1, 2$) of the group SO(2,1) in terms of differential operators.

There is another algebraic language, the so called supersymmetric quantum me-chanics [8] or SUSYQM which has given a different formalism to deal with known potentials [9] [10], in particular the Natanzon potentials [11]. This formalism has also induced finding new potentials for which an analytic solutions can be found [11] [12]. If the analytic bound state solution to the Schrödinger equation is known a standard procedure gives a whole sequence of potentials for which the analytic solution is au-tomatically known. These are the supersymmetric sequences. In some cases however this SUSY procedure does not produce really new potentials but only the same func-tion with different values for their defining parameters. Such potentials are called *shape invariant* [13] [14]. SUSYQM allows to solve the bound state problem in a purely algebraic way only in the case of shape invariant potentials.

Using the *potential group* approach the Morse and Pöschl-Teller potential prob-lem (bound and scattering states) were shown to be connected with unitary repre-sentations of $SU(2)$ (bound state sector) and $SU(1,1) \approx SO(2,1)$ (scattering sector) [15]. See also [16]. More recently it was shown that making variable and operator transformation on shape invariant potentials it was possible to solve more general Hamiltonians and in this way they [17] obtained the solution for the general hyperge-ometric Natanzon potential starting from the Pöschl-Teller potential and also from the 3D harmonic oscillator they solved confluent Natanzon potentials. On the Morse po-tential see also ref.18. Quite recently a generalization of the potential group approach has been proposed [19] to deal both with the confluent and general hypergeometric related Schrödinger equations.

In all the above methods the Hamiltonian is written directly in terms of the operators of the algebraic structure involved. Typically quadratic expressions on the elements of the algebra are used. In our case, on the contrary we make an identification of the form $\mathcal{G}(r)(H - E)\Psi(r) = [aJ_0 + bJ_1]\Psi(r)$. The right hand side is linear on the generators J_μ of $SO(2,1)$.

The generators of $SO(2,1)$ satisfy the commutation relations,

$$[J_0, J_1] = iJ_2, \quad [J_2, J_0] = iJ_1, \quad [J_1, J_2] = -iJ_0, \quad (1.1)$$

which are similar to angular momentum commutation relations except that here the last relation has a minus sign. This important difference implies that $SO(2,1)$ is a noncompact group and hence all the nontrivial unitary representations of it are infinite dimensional. The operator J_0 is the only compact generator, meaning that it has a discrete spectrum.

Defining the Casimir operator Q,

$$Q = J_0^2 - J_1^2 - J_2^2 \tag{1.2}$$

the irreducible representations are partially characterized by an eigenvalue q of Q. There are different unitary irreducible representations but we need to know only those for which J_0 has a spectrum bounded below, condition related to Hamiltonian operators with the same property as we will see. For details see, for example, the book by Wybourne [20]. The spectrum of J_0 for these representations is,

$$j_0 = \nu + \frac{1}{2} + \sqrt{\frac{1}{4} + q} = \nu + \frac{\gamma}{2} \quad \text{with} \quad \nu = 0, 1, 2 \ldots \tag{1.3}$$

where γ is defined by

$$q = \frac{\gamma}{2}\left(\frac{\gamma}{2} - 1\right) \quad \text{or} \quad \gamma = 1 + \sqrt{1 + 4q}. \tag{1.4}$$

Since γ gives the same value for q as $\gamma' = 1 - \gamma$ its definition is completed requiring an extra condition that we choose to be

$$\gamma \geq 1. \tag{1.5}$$

(In angular momentum quantum mechanichs the analogous choice is $\ell \geq 0$ but one could equally well prefer $\ell \leq -1$ since $\ell(\ell + 1)$ is invariant to $\ell \rightarrow -\ell - 1$.)

In section 2 we describe the spectrum generating algebra techniques, in particular the method for solving the Schrödinger equation with confluent Natanzon potentials. This technique is applied in section 3 to solve the bound state problem for the all the confluent cases. Section 4 has the description of the spectrum generating algebra for the hypergeometric Natanzon potentials and it gives the method to solve the bound state problem with these potentials. In section 5 there is a brief and partial summary of results in supersymmetric quantum mechanics with an emphasis on the idea of supersymmetric sequences of Hamiltonians. Section 5 also includes a table of simple confluent Natanzon potentials. Section 6 is dedicated to show a peculiar intertwining between SGA and SUSYQM in the case of the Morse supersymmetric sequence.

2. SPECTRUM GENERATING ALGEBRA TECHNIQUES

We will be interested in two SGA methods that differ in that in the first one the Schrödinger operator is linearly related to the generators of the $SO(2,1)$ algebra and in the second one the Schrödinger operator is related to the Casimir operator of the algebra:

$$\mathcal{G}(r)(H-E)\Psi(r) = [2(1+\beta)J_0 + 2(1-\beta)J_1 - \delta]\Psi(r) = 0 \qquad (2.1a)$$

$$\mathcal{G}(r)(H-E)\Psi(r) = (Q-q)\Psi(r) = 0 \qquad (2.1b)$$

where $\mathcal{G}(r)$ is a factor which is always trivially determined.

These two forms are useful to solve the confluent and hypergeometric cases respectively. In this section we give in detail the way to deal with (2.1a) namely with the confluent Natanzon potentials. The SGA method to solve the bound state problem with hypergeometric Natanzon potentials is described in §4.

Once the generators J_μ are realized as differential operators and the potential function $V(r)$ is given there is a first identification of terms from (2.1) that begins to determine the parameters of the algebraic problem. We give a simple example a few paragraphs below.

To find the energy spectrum it is necessary to tilt the basis of the algebra, changing to operators $\tilde{J}_\mu = \exp[i\theta J_2]J_\mu \exp[-i\theta J_2]$. With $\theta = \frac{\beta-1}{\beta+1}$ equation (2.1a) can be rewritten as

$$\tilde{J}_0\Psi_{\nu,\gamma} = \frac{\delta}{4\sqrt{\beta}}\Psi_{\nu,\gamma}. \qquad (2.2)$$

$\Psi_{\nu,\gamma}$ is the Schrödinger wave function. From the above expressions and (1.3) it follows that

$$\gamma = -2\nu + \frac{\delta}{2\sqrt{\beta}} \qquad \text{and} \qquad \frac{\delta}{2\sqrt{\beta}} \geq 1 \qquad (2.3)$$

which is our first basic equation for solving the eigenvalue problem.

The following realization for the generators J_μ in terms of an arbitrary function $h(r)$ will be used [2],

$$J_0 = -\frac{h}{h'^2}\frac{d^2}{dr^2} - \frac{h'''h}{2h'^3} + \frac{3}{4}\frac{h''^2 h}{h'^4} + \frac{q}{h} + \frac{h}{4},$$

$$J_1 = -\frac{h}{h'^2}\frac{d^2}{dr^2} - \frac{h'''h}{2h'^3} + \frac{3}{4}\frac{h''^2 h}{h'^4} + \frac{q}{h} - \frac{h}{4}, \qquad (2.4)$$

$$iJ_2 = \frac{h}{h'}\frac{d}{dr} + \frac{1}{2}\frac{h''h}{h'^2}.$$

Equation (2.1a) — using (2.4) for the generators — yields [2] the following relationship between the basic function $h(r)$ and the potential $V(r)$,

$$E - V(r) = \frac{1}{2}\frac{h'''}{h'} - \frac{3}{4}(\frac{h''}{h'})^2 - q(\frac{h'}{h})^2 - \frac{\beta}{4}h'^2 + \frac{\delta}{4}\frac{h'^2}{h} \qquad (2.5)$$

which is our second and last basic equation for solving the eigenvalue problem.

It may be noticed that at least one of the constants γ, β or δ depend on the energy eigenvalue and therefore on the discrete index ν. In general then, these quantities will carry a subindex ν. Similarly, the Casimir eigenvalue q sometimes needs an subindex ν. See (3.5) below.

Example: Take $h(r) = 2r$ and $V(r) = -\frac{e^2}{r} + \frac{\ell(\ell+1)}{r^2}$. From (2.5) it directly follows that $\delta = 2e^2$, $\beta = -E$ and $q = \ell(\ell+1)$. The last of these relations implies that $\gamma = 2\ell + 2$ which we use in (2.3) to obtain that

$$E_\nu = -\frac{e^4}{4(\ell + \nu + 1)^2}.$$

(2.6)

This is all the effort it takes to obtain an energy spectrum.

Notice that obtaining the energy spectrum is a purely algebraic task. It does not require the knowledge of the representation space or the wave functions.

To know the eigenfunctions we need to know that the carrier space of the representation (2.4) is

$$\Psi_{\nu,\gamma}(h(r)) = \frac{h^{\gamma/2}(r)}{\sqrt{h'(r)}} \exp[-\frac{h(r)}{2}] \, {}_1F_1(-\nu, \gamma, h(r))$$

(2.7)

where ${}_1F_1(-\nu, \gamma, h(r))$ are the standard confluent hypergeometric functions. This is easily derived, for example, from the self-contained §5 of ref.3.

The Schrödinger wave function is an eigenfunction of the tilted operator $\tilde{J}_0$ as it was seen in (2.2) and to know the carrier space of the tilted representation one has to tilt the operators (2.4). It turns out that the answer is extremely simple. The tilted generators have the same form as (2.4) but with $f(r)$ playing the role of $h(r)$,

$$f(r) = \sqrt{\beta} h(r).$$

(2.8)

Namely, the tilt simply rescales by a factor $\sqrt{\beta}$ the basic function h acting as a dilatation. The wave functions therefore are the functions (2.7) but with argument $f(r)$.

For future reference we state that from the commutation relations it follows that the tilted generators have the effect:

$$\tilde{J}_+\Psi_{\nu,\gamma}(f(r)) = (\gamma + \nu)\Psi_{\nu+1,\gamma}(f(r)),$$
$$\tilde{J}_-\Psi_{\nu,\gamma}(f(r)) = \nu\Psi_{\nu-1,\gamma}(f(r))$$

(2.9)

where $\tilde{J}_\pm = \tilde{J}_1 \pm \tilde{J}_2$.

In the following section we define the confluent Natanzon potentials and describe the way to solve the associated bound state problem using the SGA method already defined.

3. ALGEBRAIC SOLUTION FOR THE CONFLUENT NATANZON POTENTIALS

The confluent Natanzon potentials

$$V(r) = \frac{g_2 h^2 + g_1 h + \eta}{R} + \frac{\sigma_1 h - \sigma_2 h^2}{R^2} - \frac{5}{4}\frac{\Delta h^2}{R^3} \tag{3.1a}$$

are defined in terms of six parameters g_1, g_2, σ_1, σ_2, c_0, η and a function $h(r)$ satisfying

$$\frac{dh}{dr} = \pm\frac{2h}{\sqrt{R}} \tag{3.1b}$$

where

$$R(r) = \sigma_2 h^2 + \sigma_1 h + c_0 \tag{3.1c}$$

and

$$\Delta = (\sigma_1^2 - 4\sigma_2 c_0). \tag{3.1d}$$

The $\pm$ in (3.1b) will be discussed later on.

The algebraic method to solve the Schrödinger problem for the confluent Natanzon potentials is now presented step by step [4]. First notice that R satisfies,

$$\frac{dR}{dr} = \pm(2\sigma_2 h + \sigma_1)\frac{2h}{\sqrt{R}} \tag{3.2}$$

and use this relation and (3.1b) to discover that (2.5) becomes,

$$E_\nu - V(r) = \frac{\delta_\nu h - \beta_\nu h^2 - 4q - 1}{R} - \frac{4\sigma_2 h^2 + \sigma_1 h}{R^2} + \frac{5(\sigma_2 h^2 + \frac{1}{2}\sigma_1 h)^2}{R^3}. \tag{3.3}$$

If h is eliminated in favor of R solving the quadratic equation (3.1c) (which root to choose is immaterial) and the Natanzon potential is added, an expression which should be E_ν itself is obtained.

But really three type of terms come out: a) r-independent terms, b) terms proportional to $R^{-1}\sqrt{4R\sigma_2 - \Delta}$ and c) terms proportional to R^{-1}; hence three conditions emerge

$$E_\nu = \frac{g_2 - \beta_\nu}{\sigma_2},$$

$$0 = \frac{\delta_\nu + g_1}{2\sigma_2} + \frac{\sigma_1(\beta_\nu - g_2)}{2\sigma_2^2}, \tag{3.4}$$

$$0 = -\frac{2c_0(\beta_\nu - g_2) - \sigma_1(\delta_\nu + g_1)}{2\sigma_2} + \frac{(g_2 - \beta_\nu)\sigma_1^2 \eta}{2\sigma_2^2} - 1 - 4q_\nu.$$

The last two expressions represent a linear system for β_ν and δ_ν, which can be replaced back in the first equation. They yield

$$
\begin{aligned}
q_\nu &= \frac{1}{4}(\eta - 1 - c_0 E_\nu), \\
\delta_\nu &= -g_1 + \sigma_1 E_\nu, \\
\beta_\nu &= g_2 - \sigma_2 E_\nu.
\end{aligned}
\tag{3.5}
$$

Eliminating q_ν in favor of γ_ν,

$$
\gamma_\nu = 1 + \sqrt{\eta - c_0 E_\nu}
$$

and demanding that (2.3), namely that the expression $\delta_\nu = 2\sqrt{\beta_\nu}(\gamma_L + 2\nu)$ is identified with the second expression in (3.5) it follows that

$$
\frac{g_1 - \sigma_1 E_\nu}{2\sqrt{g_2 - \sigma_2 E_\nu}} - \sqrt{\eta - c_0 E_\nu} = 1 + 2\nu.
\tag{3.6}
$$

This is the expression that determines the energy spectra for the family of confluent Natanzon potentials.

As already explained, the wave functions are the functions $\Psi(f)$ with argument f namely,

$$
\Psi_{\nu,\gamma_\nu} \propto R^{1/4} h^{(\gamma_\nu - 1)/2} \exp[-\sqrt{\beta_\nu} h/2] \; {}_1F_1(-\nu, \gamma_\nu; \sqrt{\beta_\nu} h).
\tag{3.7}
$$

The confluent Natanzon bound state problem is totally solved.

The present formalism is invariant to the simultaneous change

$$
\begin{aligned}
h &\to -h \\
g_1 &\to -g_1 \\
\sigma_1 &\to -\sigma_1
\end{aligned}
\tag{3.8}
$$

which implies $\delta_\nu \to -\delta_\nu$ and changing $\sqrt{\beta_\nu} \to -\sqrt{\beta_\nu}$. In particular $f(r)$ in (2.8) and γ_ν in (2.3) remain invariant. Given this freedom there is no loss of generality choosing h positive. The sign in (3.1b) (and therefore in (3.2)) has to be chosen according to the true sign of dh/dr. For example, for the harmonic and the Coulomb cases the + sign is necessary while for the Morse potential the minus sign is compulsory in (3.1b) since in this case $h = \exp[-\alpha r]$ is a negative slope function.

4. THE HYPERGEOMETRIC NATANZON POTENTIALS ALGEBRAIC SOLUTION

The hypergeometric Natanzon potentials too are defined in terms of six parameters: a, f, h_0, h_1, c_0, c_1 and a function $z(r)$

$$V(r) = \frac{fz(z-1) + h_0(1-z)h_1 z + 1}{R} + \frac{a + [a + (c_1 - c_0)(2z-1)]}{R^2} z(1-z)$$
$$- \frac{5\Delta z^2(1-z)^2}{4R^3},$$

where

$$R = az^2 + (c_1 - c_0 - a)z + c_0, \tag{4.1b}$$
$$\Delta = (a - c_0 - c_1)^2 - 4c_0 c_1, \tag{4.1c}$$

and

$$\frac{dz}{dr} = \frac{2z(1-z)}{\sqrt{R}}. \tag{4.2}$$

This time the realization of the $SO(2,1)$ generators in terms of an arbitrary function $z(r)$ and the corresponding Casimir operator is

$$J_0 = -i\frac{\partial}{\partial y} \tag{4.3a}$$

$$J_\pm = e^{\pm iy}[\pm i\frac{z^{3/2}}{z'}\frac{\partial}{\partial r} \mp i\frac{\sqrt{z}}{z'}\frac{\partial}{\partial r} + \frac{1}{2\sqrt{z}}\frac{\partial}{\partial y} + \frac{\sqrt{z}}{2}\frac{\partial}{\partial y} \mp \frac{z''z}{8z'} \mp \frac{z''}{8z'z}$$
$$\pm \frac{z''}{4z'} \mp \frac{z'}{4z} \pm \frac{3}{16}\frac{z'}{z^2} \pm \frac{z'}{16} - \frac{ip}{2\sqrt{z}} + \frac{ip}{2}\sqrt{z} \mp \frac{i}{4\sqrt{z}} \mp \frac{i}{4}\sqrt{z}] \tag{4.3b}$$

$$Q = \frac{z(1-z)^2}{z'^2}\frac{\partial^2}{\partial r^2} + \frac{(1-z)^2}{4z}\frac{\partial^2}{\partial y^2} + \frac{p}{2iz}(1-z^2)\frac{\partial}{\partial y} + \frac{zz'''}{2z'^3}(1-z)^2$$
$$- \frac{3zz''^2}{4z'^4}(1-z)^2 - \frac{p^2}{4}\frac{(1-z)^2}{z} + \frac{z}{4} + \frac{1}{4z} - \frac{1}{2} \tag{4.4}$$

p being an arbitrary constant. This time the Casimir operator is a differential operator suitable to be related to the Hamiltonian through (2.1b).

Equation (2.1b) becomes

$$\frac{z'^2}{z(1-z)^2}(Q-q)e^{imy}\Psi(r) = (E-H)e^{imy}\Psi(r) \tag{4.5}$$

where q represents as usual the eigenvalues of Q while j_0 is again given by (1.3)

$$j_0 = \nu + \frac{1}{2} + \sqrt{\frac{1}{4} + q}$$

leading to

$$E - V(r) = \frac{z'''}{2z'} - \frac{3z''^2}{4z'^2} - \frac{qz'^2}{z(1-z)^2} - \frac{(j_0^2 + p^2 - 1)z'^2}{4z^2} + \frac{z'^2}{2z^2}\frac{1+z}{1-z}pj_0. \quad (4.6)$$

The procedure to follow is entirely similar to that of §3. It is convenient to make the change of notation

$$j_0 = \frac{1}{2}(\sqrt{v} + \sqrt{w})$$

$$p = \frac{1}{2}(\sqrt{v} - \sqrt{w}) \quad (4.7)$$

to make easier a rather long (but trivial) algebraic manipulation that leads to

$$q = \frac{1}{4}(h_1 - c_1 E), \quad (4.8a)$$
$$v = 1 - aE + f, \quad (4.8b)$$
$$w = 1 - c_0 E + h_0. \quad (4.8c)$$

The square root of each of the last two relations gives a linear system for j_0 and p as defined in (4.7). One can eliminate p and obtain an expression for j_0 which has to be identified with (4.5). Combining these relations (4.8a) becomes

$$2v + 1 + \sqrt{1 - c_1 E + h_1} = \sqrt{1 - aE + f} - \sqrt{1 - c_0 E + h_0} \quad (4.9)$$

which is the equation that solves the bound state eigenvalue problem for the whole set of hypergeometric Natanzon potentials. It coincides of course with the expression given in ref.11.

5. SUPERSYMMETRIC SEQUENCES

In SUSYQM it is possible to associate to any Hamiltonian H_0 a sequence of Hamiltonians $H_k = p^2 + V_k(r)$ through the following mechanism. If the eigenfunctions $\Psi_\nu^{(k)}(r)$ of H_k are known, define the *superpotential*

$$W_k(r) = -(\ln \Psi_0^{(k)})'. \quad (5.1)$$

It is easy to see that $V_k = W_k^2 - W_k' + E_0^{(k)}$. The potential for the next Hamiltonian H_{k+1} in the supersymmetric sequence is defined by

$$V_{k+1}(r) = W_k^2 + W_k' + E_0^{(k)}. \quad (5.2)$$

The importance of this construction stems from the following properties. Defining the operators,

$$\hat{A}_k = \frac{d}{dr} + W_k, \qquad \hat{A}_k^\dagger = -\frac{d}{dr} + W_k, \quad (5.3)$$

it can be seen that the eigenfunctions of the Hamiltonians of the sequence are related:

$$\Psi_\nu^{(k+1)} \propto \hat{A}_k \Psi_{\nu+1}^{(k)}, \qquad \Psi_{\nu+1}^{(k)} \propto \hat{A}_k^\dagger \Psi_\nu^{(k+1)} \tag{5.4}$$

and it is also true that

$$E_\nu^{(k+1)} = E_{\nu+1}^{(k)} \tag{5.5}$$

Which shows that given the solution to the bound state problem for H_0, the solution to all the other Hamiltonians in the sequence is known through (5.4) and (5.5).

TABLE 5.1

The simplest Natanzon confluent potentials and associated SGA and SUSY parameters.

$V(r)$	$\frac{\omega^2 r^2}{4} + \frac{\ell(\ell+1)}{r^2}$ $+\omega(\ell - \frac{3}{2})$	$-\frac{e^2}{r} + \frac{\ell(\ell+1)}{r^2}$	$B^2 e^{-2\alpha r} - \alpha B(2K+1)e^{-\alpha r}$
$h(r)$	r^2	$2r$	$\exp[-\alpha r]$
g_1	$\omega(\ell - \frac{3}{2})$	$-2e^2$	$-\frac{4\alpha B(K+1)}{\alpha}$
g_2	$\frac{1}{4}\omega^2$	0	$\frac{4B^2}{\alpha^2}$
η	$(\ell + \frac{1}{2})^2$	$(2\ell+1)^2$	0
σ_1	1	0	0
σ_2	0	1	0
c	0	0	$\frac{4}{\alpha^2}$
β_ν	$\frac{1}{4}\omega^2$	$\frac{e^4}{4(\ell+\nu+1)^2}$	$\frac{4B^2}{\alpha^2}$
γ_ν	$\ell + \frac{3}{2}$	$2\ell+2$	$2(K-\nu)+1$
q_ν	$(2\ell+1)^2 - \frac{1}{4}$	$\ell(\ell+1)$	$(K-\nu)^2 - \frac{1}{4}$
δ_ν	$\omega(\frac{3}{2} + \ell + 2\nu)$	$2e^2$	$\frac{4B(2K+1)}{\alpha}$
E_ν	$2\omega(\ell + \nu)$	$-\frac{e^4}{4(\ell+\nu+1)^2}$	$-\alpha^2(K-\nu)^2$
$W(r)$	$\frac{\omega r}{2} - \frac{\ell+1}{r}$	$\frac{e^2}{2(\ell+1)} - \frac{\ell+1}{r}$	$\alpha K - B\exp[-\alpha r]$

In the case of the confluent Natanzon potentials the superpotential can be written explicitly after making use of the wave functions given in (3.7). W takes the form

$$W(r) = \frac{1}{2}\left(\frac{h''}{h'} - \gamma_0 \frac{h'}{h} + \sqrt{\beta_0} h'\right). \tag{5.6}$$

The subindex zero refers to the ground state ($\nu = 0$); see the note after (2.5). The last equation shows a first link between the SGA scheme and SUSYQM. The function h is used to express the superpotential, the generic ground state function (3.7) and it is the basic function to express the generators of the $SO(2, 1)$ algebra (2.4).

Of particular interest is the notion of *shape invariant potentials* [13]. In general the V_k are a sequence of different potentials. Shape invariance means that there is a unique function $V(r; k)$ giving all the potentials of the sequence. For example, there is a SUSY sequence of Coulomb potentials. They differ in the value of ℓ in the centrifugal barrier term. Gendenshtein [13] and Dutt et al [14] proved that wave functions associated to sequences of shape invariant potentials share interesting algebraic properties which allow to solve the bound state problem in a purely algebraic way.

In table 5.1 we give three examples of supersymmetric sequences as Natanzon confluent potentials: the harmonic oscillator, the Coulomb case and the Morse case. It is easy to see that (5.5) is satisfied in all three examples. In the first two examples the index k characterizing the SUSY sequence is the angular momentum number $\ell = 0, 1, \dots$.

In the Morse case the index that characterizes the SUSY sequence is $K = K_0, K_0 - 1, K_0 - 2, \dots$ and the inequality in (2.3) requires that $K \geq 0$, namely $K_0 \geq 0$ and the sequence stops at $K = K_0 - n$ where n is the integer part of K_0.

6. THE MORSE POTENTIAL SUSY SEQUENCE

In this section we give some details for the case of the Morse supersymmetric sequence that we have found to be interesting since a connection between the SUSY operators and the SGA operators is found.

In §5 and table 5.1 there are already many details concerning the Morse case. The Morse supersymmetric sequence is described by $K, K - 1 \dots$. We add that the maximum value for ν typical of a short range potential comes from requiring that $\gamma_\nu \geq 1$, conditions mentioned after (1.4). From table 5.1 it is seen that it yields $\nu_{max} \leq K$. In the nice paper by Dabrowska et al [9] they use a parameter A (for us $A = \alpha K$) to distinguish the different members of the Morse supersymmetric sequence, we have preferred to use the index K. B and α are common to all members of the Morse supersymmetric sequence.

Since the operators $\hat{A}_K$ of (5.3) are built with W_K they carry an index K. The action (5.4) of $\hat{A}_K^\dagger$ is

$$\Psi_{\nu+1}^{(K)} \propto \hat{A}_K^\dagger \Psi_\nu^{(K-1)}. \tag{6.1}$$

The effect of $\hat{A}_K^\dagger$ is to raise the values of ν and K by one unit.

For every value of K there is also a set of SGA generators (2.4) with the Casimir for which γ

$$\gamma = 2(K - \nu) + 1 \tag{6.2}$$

is invariant under the simultaneous change: $\nu \rightarrow \nu + 1$ and $K \rightarrow K + 1$.

We are ready to see the action of the SGA generators on the wave functions, and relate it with the action of the factorizing SUSY operators $\hat{A}_K^\dagger$ and $\hat{A}_K$.

In §2 the wave functions were called $\Psi_{\nu,\gamma}$. Now that we have the explicit form (6.2) for γ the notation $\Psi_\nu^{(K)}$ for the wave functions can be used with no ambiguity.

The action of the $\tilde{J}_+$ on $\Psi_\nu^{(K-1)}$ was given in (2.9). The effect is to increase the value of ν keeping γ (and therefore the Casimir eigenvalue q) invariant. The value of K then changes to $K + 1$ yielding $\Psi_{\nu+1}^{(K)}$ namely,

$$\Psi_{\nu+1}^K \propto \tilde{J}_+(q)\Psi_\nu^{K-1}. \tag{6.3}$$

The argument on both wave functions is the same $f(r)$. In fact, from (2.8) and the value of δ it is seen that K does not enter into the definition of f. This is a property of the Morse potential not shared by the Coulomb potential [7].

Equations (6.1) and (6.3) look much alike, but there is an important difference besides the fact that one is a first order operator and the other one is of second order. A fixed operator $\hat{A}_K^\dagger$ relates all couples of wave functions – having the same energy – associated to two supersymmetric partner Hamiltonians $[H^{(K)}, H^{(K-1)}]$. On the other hand $\tilde{J}_+(q)$ with fixed q relates all the wave functions of the supersymmetric sequence with fixed energy.

Although the original works on SGA [2] mention it explicitly that the SGA has one different irreducible representation for every energy now it is possible to see that the representation space contains wave functions of the whole supersymmetric sequence with a fixed energy.

To better appreciate the relation between these two operators consider the general result for shape invariant potentials [9]

$$\Psi_\nu(r, a_0) \propto \hat{A}^\dagger(r, a_0)\hat{A}^\dagger(r, a_1) \ldots \hat{A}^\dagger(r, a_{\nu-1})\Psi_0(r, a_\nu) \tag{6.4}$$

which now — in the case of the Morse potential — can be written as

$$\Psi_\nu^{(0)} \propto [\tilde{J}_+(q)]^\nu \Psi_0^{(\nu)} \tag{6.5}$$

that is, instead of acting with a string of ν different operators $\hat{A}_k^\dagger$, it is enough to act with the νth power of $\tilde{J}_+$.

60

REFERENCES

1) Natanzon G A 1979 Teor. Mat. Fiz. **38**, 146
2) Cordero P 1970 Lett. Nuovo Cimento **4**, 164
   Cordero P and Ghirardi G C 1971 Nuovo Cimento **2A**, 217
   Cordero P and Hojman S 1970 Lett. Nuovo Cimento **4**, 1123
   Cordero P and Ghirardi G C 1972 Forts. der Phys. **20**, 105
   Cordero P and Salamó S 1975 Int. Jour. Theor. Phys. **13**, 265
3) Cordero P, Hojman S, Furlan P and Ghirardi G C 1971 Nuovo Cimento **3A**, 807
4) Cordero P, Salamó S, to appear in Journ. Phys. A: Math. Ger.
5) Brajamani S and Singh C A, J. 1990 Phys. A: Math. Gen. **23** 3421
6) Ghirardi G C 1972 Nuovo Cimento **10A**, 97
   Ghirardi G C 1973 Forts. der Phys. **21**, 653
   Villasante M 1980, MSc thesis, unpublished
7) Cordero P, Salamó S, in preparation
8) Witten E 1981 Nucl. Phys. **B 185**, 513
   Salomonson P and van Holten J W 1982 Nucl. Phys. **B 196**, 509
   Adhikari R, Dutt R and Varshni Y P 1989 Phys Lett. **A 141, 1**
9) Dabrowska J W, Khare A and Sukhatme U P 1988 J. Phys. A **21**, L125
10) Lévai G, 1989 J. Phys. A: Math. Gen. **22**, 689
11) Cooper F, Ginocchio J and Kahare A 1987 Phys. Rev D **36**, 2458
12) Sinha A, Roychoudhury R 1991 J Phys A: Math Gen **23**, 3869
13) Gendenshtein L 1983 JEPT Lett. **38**, 356
14) Dutt R, Khare A and Sukhatme U P 1986 Phys. Lett **181 B**, 295
15) Alhassid Y, Gursey F and Iachello F 1983, Phys Rev Lett **50** 873
    1983 Ann. Phys., NY **148** 346
16) Pieite B and Vinet L 1987 Phys. Lett. **A 125** 380
17) Cooper F, Ginocchio J N and Wipf A 1989 J. Phys A: Math Gen **22** 3707
18) Kais S and Levine R D 1990 Phys Rev A **41** 2301
19) Wu J and Alhassid Y 1990 J. Math. Phys. **31** 557
20) Wybourne B G 1974 *Classical Groups for Physicists* (New York, Wiley)

GROUND STATES OF FINITE SYSTEMS

AND INFORMATION THEORY

L. Arrachea, N. Canosa, A. Plastino, M. Portesi, R. Rossignoli

Departamento de Física, Universidad Nacional de La Plata
C.C.67, 1900 La Plata, Argentina

Abstract

The features and properties of the quantal entropy are examined within the context of pure states of quantum systems with a finite number of particles. Accurate predictions of the behavior of the ground state distribution over a given basis are obtained.

Introduction

The possibility of reconstructing the ground state (GS) of a quantum system on the basis of a limited amount of information has been discused in Refs.[1]–[3] , within the context of a special version of the maximum entropy principle (MEP) of information theory [4],[5]. A suitably defined 'quantal' entropy was seen to constitute a useful tool for discussing some aspects of the many-body problem.

The ground state of a quantum system may undergo qualitative changes when the interaction parameters which appear in the Hamiltonian are varied. From a practical point of view the question that arises is to what an extent it is possible to make predictions, based on a limited amount of prior information, about the location of regions where the wave function exhibits significant changes. The aim of this work is to examine this problem in the context of systems with a finite number of particles on the basis of the behavior of the quantal entropy.

The work is organized as follows: first, the MEP formalism with its application to pure states is described, followed by the study of the behavior of the quantal entropy with respect to changes in the control parameters. Then, the formalism is applied to a particular many fermion model and typical results are shown. Finally, some conclusions are drawn.

Condensed Matter Theories, Vol. 7, Edited by A.N. Proto
and J.L. Aliaga, Plenum Press, New York, 1992

Maximum Quantal Entropy Approach to Pure States

Let us consider a quantum system described by a pure state $|\psi\rangle$ about which the available information consists of a set of expectation values $O_i \equiv \langle \hat{O}_i \rangle$ of n linearly independent observables $\hat{O}_i$. We shall suppose that the operators $\hat{O}_i$ commute with each other, so that there exists a common basis $\{|j\rangle, \; j = 1, \ldots, d\}$ in which they possess a diagonal representation, with elements $O_i(j) \equiv \langle j|\hat{O}_i|j\rangle$.

Let us consider now the expansion of the state $|\psi\rangle$ over this common basis,

$$|\psi\rangle = \sum_j C_{ex}(j) \, |j\rangle. \tag{1}$$

The ensuing quantal entropy is [6]

$$S = -\sum_j |C_{ex}(j)|^2 \ln(|C_{ex}(j)|^2) \tag{2}$$

$$= -\mathrm{Tr}\hat{\rho}_d \ln \hat{\rho}_d, \tag{3}$$

where $\hat{\rho}_d$ is a diagonal density satisfying $\langle j|\hat{\rho}_d|k\rangle = \delta_{jk}|C_{ex}(j)|^2$.

In order to univocally infer $\hat{\rho}_d$ from the given data, the operators $\hat{O}_i$ whose mean values are available should form a complete diagonal set. In the case of incomplete information, following the general prescription of information theory, we shall choose that $\hat{\rho}_d$ which maximizes the entropy (3) subject to the constraints

$$O_i = \mathrm{Tr}\hat{\rho}_d\hat{O}_i. \tag{4}$$

This leads to the well-known result [4],[5]

$$\hat{\rho}_d = \exp(-\lambda_0 - \sum_{i=1}^n \lambda_i\hat{O}_i), \tag{5}$$

where the λ_i $(i = 1, \ldots, n)$ are Lagrange parameters to be determined from (4), and λ_0 is the normalization constant, which satisfies

$$\frac{\partial \lambda_0}{\partial \lambda_i} = -O_i, \quad i = 1, \ldots, n. \tag{6}$$

The ensuing maximum entropy acquires the form

$$S = \lambda_0 + \sum_{i=1}^n \lambda_i O_i, \tag{7}$$

and satisfies the relationships

$$\frac{\partial S}{\partial O_i} = \lambda_i, \quad i = 1, \ldots, n. \tag{8}$$

We note that the quantal entropy (2) measures the missing information associated with the probability distribution of the state over the considered basis. In the case where the operators $\hat{O}_i$ form a complete basis for the expansion of any other *commuting* operator, (7) reduces to the exact quantal entropy (2). Otherwise, (7) will obviously be an upper bound to the exact quantal entropy.

Quantal Entropy Behavior

A very accurate prediction of the quantal entropy (2) in the common basis can be in general achieved with just a few expectation values of relevant operators. Thus, concerning the response of the system to changes in external parameters, the behavior of the inferred quantal entropy (7) will serve to indicate just where the exact distribution over the common basis changes significantly. For this purpose, let us assume that the state of the system, and thus the available information, depends upon the set of parameters $\{z_\alpha, \ \alpha = 1, \ldots, r\}$. For a particular parameter z_α, intervals of strongest variation of the inferred entropy, identified by maxima of $|\partial S/\partial z_\alpha|$ with (see (8))

$$\frac{\partial S}{\partial z_\alpha} = \sum_{i=1}^{n} \lambda_i [\frac{\partial O_i}{\partial z_\alpha} - \langle \frac{\partial \hat{O}_i}{\partial z_\alpha} \rangle], \tag{9}$$

will indicate 'transitional regions' for the distribution of the state wave function over the given basis. For the case in which the operators $\hat{O}_i$ do not explicitly depend on the parameters, the location of 'critical' values of z_α for finite systems is thus obtained from the necessary condition

$$\frac{\partial^2 S}{\partial z_\alpha^2} = \sum_{i=1}^{n} [\lambda_i \frac{\partial^2 O_i}{\partial z_\alpha^2} - \sum_{j=1}^{n} (A^{-1})_{ij} \frac{\partial O_j}{\partial z_\alpha} \frac{\partial O_i}{\partial z_\alpha}] = 0, \tag{10}$$

where A denotes the (non singular [5]) fluctuation matrix given by

$$A_{ij} = -\frac{\partial O_i}{\partial \lambda_j} = \langle \hat{O}_i \hat{O}_j \rangle - O_i O_j. \tag{11}$$

For instance, it is feasible to take as 'common basis' that spanned by the eigenstates of the unperturbed part $\hat{H}_0$ of the Hamiltonian of the system, by selecting observables $\hat{O}_i$ which commute with $\hat{H}_0$. The quantal entropy will obviously vanish for the unperturbed system (assuming a non degenerate ground state) but as the interaction is switched on, it will acquire a finite value. Thus, taking as control parameter the coupling constant of the interaction, the scheme developed gives a measure of the intensity of the transition from the weak to the strong interacting regime.

Description of the Model

This model [7],[8] is also known as the extended Lipkin-Meshkov-Glick (LMG) model [9] and consists of $N = 2\Omega$ interacting fermions distributed among n 2Ω–fold degenerate single particle (SP) levels denoted by $|p, i\rangle$, $p = 1, \ldots, 2\Omega$, $i = 1, \ldots, n$, which interact through a monopole interaction. The corresponding Hamiltonian reads

$$\hat{H} = \sum_{i=1}^{n} \epsilon_i \hat{G}_{ii} + \frac{1}{2} \sum_{i<j}^{n} V_{ij} (\hat{G}_{ij}^2 + \hat{G}_{ji}^2), \tag{12}$$

with $\epsilon_i \leq \epsilon_j$ for $i < j$. The collective operators

$$\hat{G}_{ij} = \sum_{p=1}^{2\Omega} c_{pi}^\dagger c_{pj} \tag{13}$$

satisfy an $U(n)$ algebra under commutation.

The ground state of the Hamiltonian (12) belongs to the completely symmetric representation of $U(n)$, $(N, 0, \ldots, 0)$ [2] and is an eigenstate of the n 'parities' $\hat{P}_i = \exp(i\pi \hat{G}_{ii})$, with eigenvalue $+1$. It can thus be expanded as

$$|\psi_0\rangle = \sum_{n_i \text{ even}} C(n_2, \ldots, n_n) |n_2, \ldots, n_n\rangle, \tag{14}$$

where the states [2]

$$|n_2, \ldots, n_n\rangle = (n_1!/N!)^{1/2} \prod_{i \geq 2} (n_i!)^{-1/2} \hat{G}_{i1}^{n_i} |0\rangle, \quad 0 \leq \sum_{i \geq 2} n_i \leq N, \tag{15}$$

constitute the complete orthonormal unperturbed basis of dimension $d = \binom{N+n-1}{n-1}$ within the symmetric representation. The quantity n_i denotes the number of particles in the ith level and $|0\rangle$ the unperturbed ground state.

Within the Hartree-Fock (HF) picture, this model exhibits second order ground state shape transitions in the classical limit ($N \to \infty$) [8] as the coupling parameters vary. This approach predicts $n - 1$ transitions at the critical values [10]

$$v_c^{(i)} = i\varepsilon_i - \sum_{j=1}^{i} \varepsilon_j, \quad i = 2, \ldots, n, \tag{16}$$

for the case $v_{ij} \equiv V_{ij}(N-1) = -v(1 - \delta_{ij})$, $v > 0$.

Results

The quantal entropy associated with an incomplete set of expectation values of commuting operators, which we choose here as functions of the collective operators, will be now investigated. We assume that the observer knows with certainty that the state to be inferred lies within the symmetric subspace , i.e., the expansion (15) for the ground state is to be considered. In addition to this, the parity of the state is supposed to be known, so that the traces will be restricted to the corresponding eigenspace.

We shall consider the description constructed on the basis of the information given by expectation values of one-body operators, $\hat{G}_{ii}$, as well as by two-body operators, $\hat{G}_{ii}\hat{G}_{jj}$, representing information about the diagonal elements of the SP density matrix and the corresponding fluctuation or covariance matrix $A_{ij} = \langle \hat{G}_{ii}\hat{G}_{jj} \rangle - \langle \hat{G}_{ii} \rangle \langle \hat{G}_{jj} \rangle$. The eigenvalues of any set of these operators will just be functions of the labels n_i, so that they completely identify a many-body state. Otherwise, a multiplicity factor should be included in the computation of the pertinent traces.

The corresponding statistical operator is

$$\hat{\rho} = \exp(-\lambda_0 - \sum_{i \geq 2} \lambda_i \hat{G}_{ii} - \sum_{i \geq j \geq 2} \lambda_{ij} \hat{G}_{ii}\hat{G}_{jj}), \tag{17}$$

where

$$\lambda_0 = \ln \sum_{n_2, \ldots, n_n} \exp(-\sum_{i \geq 2} \lambda_i n_i - \sum_{i \geq j \geq 2} \lambda_{ij} n_i n_j) \tag{18}$$

and the Lagrange parameters are determined from the available data by

$$\frac{\partial \lambda_0}{\partial \lambda_i} = -\langle \hat{G}_{ii} \rangle, \quad \frac{\partial \lambda_0}{\partial \lambda_{ij}} = -\langle \hat{G}_{ii}\hat{G}_{jj} \rangle. \tag{19}$$

The available information becomes complete in this context when $\binom{N+n-1}{n-1}$ independent expectation values of commuting collective operators are given. Then the inferred entropy will coincide with the quantal entropy of the ground state (2), now written as

$$S_{ex} = - \sum_{n_i even} |C(n_2,\ldots,n_n)|^2 \ln(|C(n_2,\ldots,n_n)|^2) \tag{20}$$

Numerical calculations have been performed for $n = 3$, assuming $\varepsilon_i = (i-1)\varepsilon$ and $V_{ij} = -v(1 - \delta_{ij})/(N-1)$. Within the HF picture, the ground state undergoes a transition at $v_c^{(1)}/\varepsilon = 1$ and $v_c^{(2)}/\varepsilon = 3$. Accordingly, as control parameter we choose the scaled coupling parameter v. The corresponding quantal entropy is depicted in

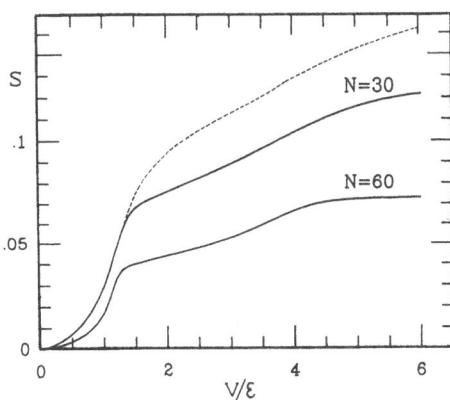

Figure 1. The scaled quantal entropy $s = S/N$ as a function of the dimensionless coupling parameter v/ε for different particle numbers. The dashed line corresponds to the one-body approximation, while the full line to the one and two-body approximation together with the exact results (indistinguishable in this scale).

Figure 1 as a function of this parameter. Three different regimes for the distribution of the wave function over the unperturbed basis, characterized by smooth variations of the slope of the exact quantal entropy, can clearly be distinguished. Between them, the regions where the distribution undergoes strong changes lie in the neighborhood of the HF critical points. Near these values, the first derivative of the quantal entropy exhibits a maximum, indicating transitional regions. The changes become sharper and the critical regions narrower as the number of particles increases, thus approaching the abrupt changes of the ground state distribution as the system goes closer to the classical limit ($N \to \infty$), in which a discontinuity of the slope is expected.

Finally, it can also be seen that the inference with one-body mean values gives a good qualitative description of the behavior of the entropy, while the inference with the addition of two-body information is extremely accurate, overlapping with the exact results within the scale of the figure.

Conclusions

Our goal was to give a new tool to study the main features of the ground state distribution without embarking oneself in an expensive and laborious many-body calculation. We have shown that a new theoretical construct, the quantal entropy [1], allows one to gain revealing insight into the complex nature of a quantum many-body system.

In the example considered, it is seen that just with the expectation values of a few relevant commuting observables, the exact quantal entropy is very accurately reproduced, and that adequate prescriptions about the behavior of the actual wave function can be extracted from analizing this quantity.

We conclude that the quantal entropy, besides playing the role of a variational functional within information theoretic formalisms, is by itself a meaningful physical quantity, reflecting changes of the wave function related with the critical behavior of the system.

Acknowledgments

L.A., N.C., A.P. and M.P. acknowledge support from Consejo Nacional de Investigaciones Científicas y Técnicas de la República Argentina. R.R. acknowledges support from Comisión de Investigaciones Científicas de la Provincia de Buenos Aires.

References

[1] N.Canosa, A.Plastino and R.Rossignoli, *Phys.Rev. A* <u>40</u>, *519 (1989).*

[2] N.Canosa, R.Rossignoli and A.Plastino, *Nucl.Phys. A* <u>512</u>, *492 (1990).*

[3] N.Canosa, R.Rossignoli and A.Plastino, *Phys.Rev. A* <u>43</u>, *1445 (1991).*

[4] J. Jaynes, *Phys.Rev.* <u>106</u> *(1957) 620;* <u>108</u> *(1957) 171.*

[5] A.Katz, *Principles of Statistical Mechanics* (Freeman, San Francisco, 1967).

[6] N.Canosa, R.Rossignoli and A.Plastino, *this volume.*

[7] N. Meshkov, *Phys.Rev. C* <u>3</u>, *2214 (1971).*

[8] R.Gilmore, *Catastrophe Theory for Scientists and Engineers* (Wiley, New York, 1981).

[9] H.J.Lipkin, N.Meshkov and A.J.Glick, *Nucl.Phys.* <u>62</u>. *188 (1965);* D.Agassi, H.J.Lipkin and N. Meshkov, *Nucl.Phys.* <u>86</u>, *321 (1966).*

[10] N.Canosa, A.Plastino and R.Rossignoli, *Nucl.Phys. A* <u>453</u>, *417 (1986).*

INFORMATION THEORY

AND QUANTUM WAVE FUNCTIONS

N. Canosa, R. Rossignoli, A. Plastino

Departamento de Física, Universidad Nacional de La Plata
1900 La Plata, Argentina

Abstract

A new scheme for the reconstruction and approximation of quantum wave functions is derived within the context of Information Theory. The method is applied to the inference of the energy spectra and the pertinent eigenstates of special hamiltonians from incomplete information concerning the system's ground state, and to the development of an approximation for the ground state of a superconducting many-fermion model. A substantial improvement over standard and projected BCS approximations is obtained, especially in transitional regions.

Introduction

Information Theory [1] provides the way for developing powerful inference methods for reconstructing probability distributions on the basis of a limited amount of information. In its conventional application to statistical mechanics and quantum physics, it is used for inferring a density or statistical operator from the knowledge of an incomplete set of expectation values of physical observables, via the maximum entropy principle. A quite general and already well known reformulation of statistical mechanics ensues [2]–[5].

However, in trying to apply the previous scheme to the inference of pure quantum states, one is faced with the obvious fact that pure states are characterized by density operators $\hat{\rho}$ which are idempotent ($\hat{\rho}^2 = \hat{\rho}$), implying thus a zero value of the entropy $S = -\text{Tr}\hat{\rho}\ln\hat{\rho}$. In other words, conventional inference schemes lead to density operators with a non vanishing entropy, which thus do not actually represent pure quantum states.

In order to overcome this difficulty, we discuss here a new scheme [6]–[8] for dealing with the inference of pure states, based on the concept of a suitably defined 'quantal' entropy [9], which is related with the probability distribution of a pure state on a given basis. A tractable method ensues, which allows for both an inference approach, based on the knowledge of a limited amount of information about the system, and a variational scheme based on a small set of variables of a very specific character. In particular, the variational approach is able to provide a very accurate description of the ground state of a many-body system, yielding a substantial improvement over conventional and projected mean field (i.e., Hartree-Fock and BCS) results [10], especially in transitional regions, where it is known that mean field methods fail to provide a good description in finite systems [11].

First, the maximum quantal entropy approach is developed. Then we discuss an alternative hamiltonian variational formulation. Finally, two different applications to particular many-fermion models are made.

Maximum Quantal Entropy Scheme

Let us consider a state $|\psi\rangle$ describing a quantum system, which can be expanded in a complete orthonormal basis $\{|j\rangle, j = 1, \ldots, L\}$ as

$$|\psi\rangle = \sum_j C_j |j\rangle. \tag{1}$$

This basis (for simplicity we take it to be discrete) is assumed known and can be chosen for instance as the eigenstates of an unperturbed Hamiltonian $\hat{H}_0$. The corresponding exact density matrix or statistical operator associated with this quantum state is given by $\hat{\rho} = |\psi\rangle\langle\psi|$. Instead of this, we shall employ a *diagonal* density operator $\hat{\rho}_d$ defined as

$$\hat{\rho}_d = \sum_j |j\rangle\langle j|\hat{\rho}|j\rangle\langle j|, \tag{2}$$

such that

$$\langle j|\hat{\rho}_d|k\rangle = \delta_{jk}\langle j|\hat{\rho}|j\rangle = \delta_{jk}|C_j|^2. \tag{3}$$

Thus, $\hat{\rho}_d$ contains information only about the probability distribution over the given unperturbed basis, corresponding to the state $|\psi\rangle$.

Associated with $\hat{\rho}_d$ we introduce a 'quantal' entropy defined by

$$\begin{aligned} S &= -\text{Tr}\hat{\rho}_d \ln \hat{\rho}_d \\ &= -\sum_j |C_j|^2 \ln |C_j|^2, \end{aligned} \tag{4}$$

which measures the lack of information related with the probability distribution over the unperturbed basis (PDUB). In this way (4) measures the information concerning the diagonal elements of $\hat{\rho}$ and vanishes only if $|\psi\rangle$ coincides with one of the unperturbed states. Its maximum value is acquired for a uniform distribution $|C_j|^2 = \frac{1}{L}$, $\forall j$, in which case $S = \ln(L)$. On the other hand, it is obvious that the conventional entropy associated with the exact density operator, $S = -\text{Tr}\hat{\rho}\ln\hat{\rho}$, vanishes for a pure state. This prevents us from using conventional maximum entropy techniques [2]–[3] for dealing with the inference of pure states.

Thus, for a non interacting system, the quantal entropy (4) associated with an energy eigenstate (assumed non degenerate for simplicity) vanishes. As the interaction is switched on, (4) acquires a finite value, measuring the departure of the PDUB from a δ distribution.

Assume now that the only available information concerning the system, supposed to be in the pure state $|\psi\rangle$, deals with a set of expectation values O_i of n linearly independent observables $\hat{O}_i$,

$$O_i \equiv \text{Tr}\hat{\rho}\hat{O}_i = \sum_{j,k} C_k^* \langle k|\hat{O}_i|j\rangle C_j, \quad i = 1, \ldots, n. \tag{5}$$

Unless the operators $\hat{O}_i$ constitute a complete set, this information does not suffice, in general, to univocally determine $\hat{\rho}$. Hence, many states $|\psi\rangle$ may exist which comply with the constraints (5), but which will predict different expectation values for the observables not included in the original set. In this situation, we shall choose (and this is the *central* idea of this work), according to the standard prescription of information theory, the least biased state $|\psi\rangle$ compatible with the available data. This state will be taken as that which maximizes the entropy (4) subject to the constraints (5). This goal can be accomplished by introducing n Lagrange multipliers λ_i and extremalizing the quantity $S' = S - \sum_i \lambda_i O_i$. The ensuing L equations are

$$\frac{\partial S'}{\partial C_j} \propto C_j[2\ln(|C_j|) + 1] + \sum_{i,k} \lambda_i C_k^* \langle k|\hat{O}_i|j\rangle = 0. \tag{6}$$

In the rest of this work we shall examine that situation in which the observables $\hat{O}_i$ are diagonal in the unperturbed basis, constituting thus an abelian set (alternatively, for commuting operators $\hat{O}_i$, we can choose as unperturbed basis the common basis in which they are diagonal). This is not, within the context of an inference procedure, a restrictive situation. We shall show indeed that very accurate predictions can be obtained just with this type of information, which leads to a simple and tractable scheme. In this case, we obtain from (6)

$$|C_j|^2 \propto \exp\{-\sum_i \lambda_i O_i(j)\}, \tag{7}$$

where $O_i(j) = \langle j|\hat{O}_i|j\rangle$. The normalization of $|\psi\rangle$ (and thus $\hat{\rho}_d$) can be tackled as a constraint (5) by including an additional Lagrange parameter λ_0 associated with the identity operator $\hat{O}_0 \equiv \hat{I}$, with $\langle \hat{O}_0 \rangle = 1$. The result (7) can thus be also written as

$$\hat{\rho}_d = \exp(-\lambda_0 - \sum_{i=1}^{n} \lambda_i \hat{O}_i), \tag{8}$$

with

$$\lambda_0 = \ln \operatorname{Tr} \exp(-\sum_{i=1}^{n} \lambda_i \hat{O}_i), \tag{9}$$

which obviously coincides with the well known expression for the statistical operator in the information theory approach to statistical mechanics [2]–[3]. Nevertheless, we would like to remark that (8) is just an inference of the *diagonal* part of the actual density, and its non vanishing entropy is hence to be interpreted as an estimate of the exact quantal entropy (4). The parameters λ_i can be explicitly obtained by solving the equations

$$\frac{\partial \lambda_0}{\partial \lambda_i} = -O_i. \tag{10}$$

The solution of equations (10) can be proved to exist and to be unique if the mean values O_i are independent and physically meaningful (i.e., if a density operator yielding these mean values exists). This expression provides us with a maximum of S for fixed mean values O_i, and a maximum of S' for fixed parameters λ_i.

It is important to remark that in the diagonal case, if $n = L - 1$, the independent observables $\hat{O}_i$ form a complete basis for the expansion of any *diagonal* operator. In this situation all $|C_j|$'s can be uniquely determined and the available information can be regarded as complete in this context. The values of $|C_j|$ and λ_i can be directly obtained in this case by inverting the expansions (5) and (7). This can be accomplished by means of the $L \times L$ matrix $A_{ij} = \langle j|\hat{O}_i|j\rangle$, leading to

$$|C_j|^2 = \sum_i A_{ji}^{-1} O_i, \tag{11}$$

$$\lambda_i = -\sum_j 2A_{ji}^{-1} \ln|C_j|. \tag{12}$$

In the case where an expectation value O_i is unknown, it is clear that we cannot employ (11), so that following the maximum entropy prescription, we set the corresponding multiplier $\lambda_i = 0$ and end up with (7). Thus, the maximum entropy prescription can be viewed as an inversion method for the equations (11) when not all O_i's are known, leading to the least biased distribution compatible with the available information. The maximum quantal entropy, which acquires the expression $S = \sum_{i=0}^n \lambda_i O_i$, provides obviously an upper bound to the 'exact' quantal entropy, obtained when all O_i's corresponding to a complete set are known, and which coincides obviously with the quantal entropy of the exact distribution.

We can assert that an information saturation has already occurred if, by adding new information, the predictions of interest do not differ (within our desired precision) from the former ones. It is apparent that this stability will depend on the particular predictions desired. Many quantities turn out to be very accurately inferred just with a small amount of information.

Finally, the formalism can be easily extended to the case where a specific unequal weight or multiplicity factor p_j (i.e., an approximate estimate or, alternatively, a given degeneracy) is a priori assigned to the unperturbed states $|j\rangle$. In this case we should maximize, instead of (4), the *relative* quantal entropy

$$S = -\sum_j |C_j|^2 \ln(|C_j|^2/p_j). \tag{13}$$

This leads to

$$|C_j|^2 = p_j \exp\{-\lambda_0 - \sum_i \lambda_i O_i(j)\}, \tag{14}$$

so that in case of no available information, we have just $|C_j|^2 \propto p_j$ instead of a uniform distribution.

Hamiltonian Variational Approach

With the solution (7) (or either (14)) for $|C_j|^2$, we are in a position to predict the mean value of any diagonal observable. Besides, in those cases where the phase of the coefficients is 'a priori' known (for instance those 'coherent' like cases where all coefficients possess the same phase) we are able to predict C_j and hence the full density $\hat{\rho}$.

The present formalism, being basically a smoothness criterion, is especially suited for the reconstruction and approximation of ground states, i.e., states characterized by smooth distributions over a given unperturbed basis. Due to this fact, it is also feasible to consider the state constructed with the coefficients (7) as a trial wave function for approximating the ground state of a specific (assumed known) Hamiltonian $\hat{H}$, with the Lagrange parameters (or at least some of them) obtained by minimization of the ensuing energy.

In this situation we can consider a mixed description, where in addition to the n operators $\hat{O}_i$ whose mean values are known, we include in the exponent of $\hat{\rho}_d$, m

extra diagonal operators $\hat{O}_k$, with $k = n+1, \ldots, m+n$, whose associated multipliers λ_k are to be determined by recourse to the minimization of $H = \langle\psi|\hat{H}|\psi\rangle$. The corresponding $n + m$ equations can be written as

$$\tfrac{1}{2}\langle\hat{H}\hat{O}_k + \hat{O}_k\hat{H}\rangle + \tfrac{1}{2}\sum_{i=0}^{n}\langle\hat{H}\hat{O}_i + \hat{O}_i\hat{H}\rangle D_{ik} = 0, \quad k = n+1, \ldots n+m, \qquad (15)$$

$$\langle\hat{O}_i\rangle - O_i = 0, \quad i = 0, \ldots, n, \qquad (16)$$

where

$$D_{ik} = \frac{\partial\lambda_i}{\partial\lambda_k} = -\sum_{l=0}^{n}(B^{-1})_{il}\langle\hat{O}_l\hat{O}_k\rangle, \qquad (17)$$

with $B_{il} = \langle\hat{O}_i\hat{O}_l\rangle$.

In case no a priori information (except normalization) about the system is considered, equations (15) and (16) reduce to

$$\tfrac{1}{2}\langle\hat{H}\hat{O}_k + \hat{O}_k\hat{H}\rangle - \langle\hat{O}_k\rangle\langle\hat{H}\rangle = 0, \quad k = 1, \ldots, m, \qquad (18)$$

with λ_0 given by (9). These equations imply simply the vanishing of the covariance of $\hat{H}$ with all the operators $\hat{O}_k$.

As we shall see in a specific example, the present formalism is able to provide a very accurate description of ground states of two-body hamiltonians with a small set of variational parameters λ_k associated with *diagonal* one and two-body observables.

Application

a) Inference of Energy Spectra

As an illustration of the broad possibilities of the present formalism, we shall first employ the inference scheme for reconstructing the unknown matrix elements and the corresponding spectrum of a special hamiltonian $\hat{H}$ which is only partially known [8]. Let us suppose that the available information about the system consists of a set of ground state expectation values. According to our prescription, we shall determine this ground state $|\psi\rangle$ by recourse to the the maximum quantal entropy principle, and will then consider the Schrödinger equation $\hat{H}|\psi\rangle = E|\psi\rangle$ as an equation giving information about E and $\hat{H}$.

For instance, let us suppose that the hamiltonian can be cast as $\hat{H} = \hat{H}_0 + \hat{V}$, where $\hat{H}_0$ is the unperturbed diagonal part assumed known, $\hat{H}_0|j\rangle = \epsilon_j|j\rangle$, whereas $\hat{V}$ represents the unknown interaction between unperturbed levels, which is assumed here to couple only nearest neighbors, i.e.,

$$\langle j|\hat{V}|k\rangle = \delta_{k,j+1}V_j + \delta_{k,j-1}V_{j-1}^*, \qquad (19)$$

with $\epsilon_k \leq \epsilon_j$ if $k < j$.

In this way, having obtained the coefficients C_j from the available information by means of (7), a complete prediction of the energy E and the $L-1$ matrix elements V_j can be obtained by solving the system of L linear complex equations

$$\epsilon_j C_j + V_j C_{j+1} + V_{j-1}^* C_{j-1} = E C_j, \quad j = 1, \ldots, L, \qquad (20)$$

i.e., we force the inferred wave function to fulfill the eigenvalue equation (we set $V_L = V_0 = 0$). The corresponding expression for the energy is given by

$$E = \sum_{j=1}^{L}(-1)^j \varepsilon_j |C_j|^2 / \sum_{j=1}^{L}(-1)^j |C_j|^2$$
$$= \langle \hat{H}_0 \hat{P}\rangle / \langle \hat{P}\rangle, \tag{21}$$

where $\hat{P} = \exp(i\pi\hat{n})$ with $\hat{n}|j\rangle = j|j\rangle$. The inferred matrix elements are given by

$$V_j = \sum_{k=j+1}^{L}(-)^{j+k}(\varepsilon_k - E)|C_k|^2/(C_j^* C_{j+1}), \tag{22}$$

for $j = 1,\ldots,L-1$. Note that the energy (21) is independent of the phase of C_j, so that the coefficients can be taken as positive in this system. This only fixes the phase of the matrix elements V_j.

Now, it is also possible to consider the case in which the energy E is an available datum, *without* knowing at the same time the exact matrix elements. In this situation (21) becomes a constraint which can be tackled in the conventional diagonal way by introducing, within the set of relevant operators $\hat{O}_i$, the diagonal operator

$$\hat{E}' = (\hat{H}_0 - E)\hat{P} \tag{23}$$

and setting

$$\langle \hat{E}'\rangle = 0. \tag{24}$$

One of the equations of the set (20) is automatically fulfilled in this way and (22) gives us again the inferred values of V_j.

It can be explicitly shown [8] that a very accurate inference of the matrix elements, and with them, of the remaining energy eigenstates, can be obtained with only a very few ground state expectation values (such as just the relevant diagonal one and two body observables) and the energy.

b) Maximum Entropy Variational Description of a Pairing Hamiltonian

We shall now apply the formalism to the development of an approximation for the ground state of a superconducting system. We shall consider a fermion two-level pairing model much employed in nuclear physics [12], described by the Hamiltonian

$$\hat{H} = \sum_{p,i} \varepsilon_i(c_{pi}^\dagger c_{pi} + c_{pi'}^\dagger c_{pi'}) - G \sum_{p,i,q,j} c_{pi}^\dagger c_{pi'}^\dagger c_{qj'} c_{qj}, \tag{25}$$

where $p = 1,\ldots,\Omega$, $i,j = 1,2$ and the prime denotes the time reversed state. We are dealing thus with two 2Ω fold degenerate single particle levels coupled by a pairing interaction. It is possible now to define the collective operators

$$\hat{Q}_{zi} = \tfrac{1}{2}\sum_{p,i}(c_{pi}^\dagger c_{pi} + c_{pi'}^\dagger c_{pi'}) - \Omega, \quad \hat{Q}_{+i} = \sum_{p,i} c_{pi}^\dagger c_{pi'}^\dagger = \hat{Q}_{-i}^\dagger, \quad i = 1,2 \tag{26}$$

which satisfy an $SU(2) \times SU(2)$ algebra.

We shall consider the number of particles $N = 2\Omega$, so that in this situation, $\hat{Q}_{z1} + \hat{Q}_{z2} = 0$ in a number projected calculation. In this case, it is obvious that the ground state belongs to the completely symmetric representation ($Q = \Omega$), which is spanned by the states $|M\rangle$, eigenvectors of $\hat{Q}_{z1}$, with $-\Omega \le M \le \Omega$. Hence, the ground state of (25) can be written as $|\psi\rangle = \sum_M C_M|M\rangle$, with all coefficients C_M possessing obviously the same phase for $G > 0$, so that they can be taken as positive.

We shall consider now a maximum entropy approximation based on (linearly independent) *diagonal* one and two-body collective observables, which reduce here to $\hat{Q}_{z1}$ and $\hat{Q}_{z1}^2$. The ensuing coefficients (7) can thus be cast as

$$C_M = \exp(-\tfrac{1}{2} \sum_{i=0}^{2} \lambda_i M^i). \tag{27}$$

As a matter of fact, the BCS wave function can be written, according to Thouless theorem [13], as

$$|\psi_{BCS}\rangle = C \exp(h_d \hat{Q}_{d+} + h_u \hat{Q}_{u+})|0\rangle, \tag{28}$$

where $|0\rangle$ represents the full vacuum and C a normalization constant. After projection onto good particle number, the projected BCS state becomes

$$|\psi_{PBCS}\rangle = \sum_M \sqrt{p_M} \exp[-\tfrac{1}{2}(\lambda_0 + \lambda_1 M)]|M\rangle, \tag{29}$$

where $\lambda_1 = -2\ln(h_u/h_d)$, and $p_M = \{\Omega!/[(\tfrac{1}{2}\Omega + M)!(\tfrac{1}{2}\Omega - M)!]\}^2$. The state (29) represents a maximum entropy wave function constructed with one body operators, but in the complete space (and not just the symmetric representation), corresponding to the expression (14). The multiplicity p_M is just the total number of states with a given value of M in the complete space of dimension $4\Omega!/(2\Omega!)^2$. Thus, we recover the particle number projected BCS approximation as a very particular case of our formalism. It is also possible to recover in a similar fashion the conventional (non projected) BCS wave function.

The result for the mean value of the interaction is shown in figure 1, as a function of the coupling constant. The conventional BCS approach exhibits as G increases, a transition from the normal to the superconducting state at $g = \Omega G/\epsilon = 1$, where $\epsilon = \tfrac{1}{2}(\epsilon_2 - \epsilon_1)$. On the other hand, the exact solution is obviously smooth for finite values of N, evolving gradually from the normal to the superconducting state. The same behaviour is obtained in the maximum entropy approach based on (27). It is seen that a substantial improvement over the conventional and even over the particle number projected BCS approximations is obtained, especially in the transitional region. The same conclusion holds for other quantities such as the overlap with the exact wave function and other expectation values [14].

We remark that the quality of our approximation is independent of the accessible space considered as far as one and two body diagonal observables are used in the exponent of (7). High quality results are also obtained for the fluctuation of relevant quantities, for which both projected and conventional BCS fail to give even an adequate estimate [14].

Conclusions

The maximum entropy principle derived from Information Theory is a powerful formalism for dealing with a wide variety of physical problems. The conventional application of Information Theory to Quantum Mechanics is based on the maximization of the thermodynamic or statistical entropy subject to appropriate constraints. Here we employ instead a diagonal or quantal entropy (expression (4)), which measures the information content of a quantum wave function in a given unperturbed basis, or alternatively, in a basis defined by the operators whose expectation values are available. The ensuing formalism provides us with a new and useful tool for developing various approximate techniques in quantum mechanics.

First, an inference scheme for reconstructing a quantum wave function from the knowledge of a limited amount of information is developed. In particular, we have shown that it is possible to infer in this way some information about the energy spectrum of the system even in cases where the pertinent Hamiltonian is not completely known. Secondly, the approach is also suitable for a hamiltonian variational

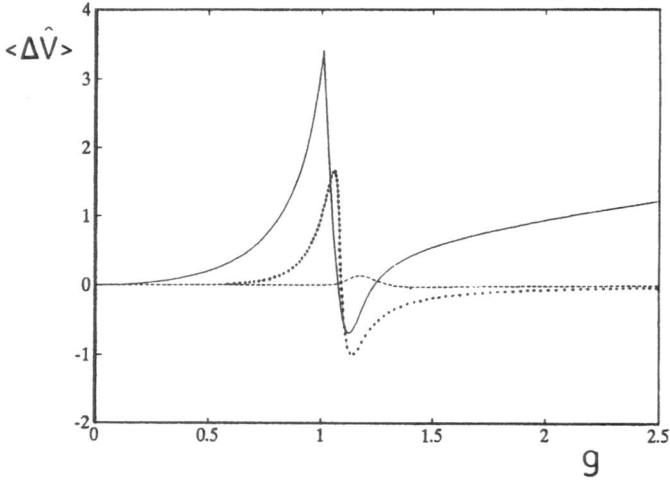

Figure 1. Difference between the approximate and the exact ground state mean value of the pairing interaction (in units of ε) as a function of $g = \Omega G/\varepsilon$, for $N = 2\Omega = 100$, according to standard BCS (solid line), particle number projected BCS (dotted line) and the present maximum entropy approach (dashed line).

approximation for the ground state, based on a limited set of parameters, when the Hamiltonian is known. Both, the inference and the variational schemes are able to yield a quite accurate description of relevant physical quantities employing just a few relevant mean values or parameters respectively.

In the examples considered, it can be seen that just with the expectation values or parameters associated with one and two-body *diagonal* operators an excellent agreement with the exact results can be obtained in systems described by two-body hamiltonians. The approximation is quite accurate in the prediction of the fluctuation of relevant expectation values and also in the so called transitional regions, where it is well known that mean field approaches fail to provide an adequate description of the pertinent physics in finite systems.

Acknowledgment

N.C. and A.P. are supported by Consejo Nacional de Investigaciones Científicas y Técnicas de Argentina (CONICET). R.R. acknowledges support from Comisión de Investigaciones Científicas de la Provincia de Buenos Aires (CIC).

References

[1] C. Shannon, *Bell Syst. Tech. J.* 27 (1948) 379, 623.

[2] J. Jaynes, *Phys.Rev.*106 (1957) 620; 108 (1957) 171.

[3] A.Katz, *Principles of Statistical Mechanics* (Freeman, San Francisco, 1967).

[4] R.Balian, Y.Alhassid and H.Reinhardt, *Phys.Rep.* 131,1 (1986).

[5] Y.Alhassid and R.D.Levine, *Phys.Rev. A* 18, 89 (1978); *Phys.Rev. C* 20, 1775 (1979).

[6] N.Canosa, A.Plastino and R.Rossignoli, *Phys.Rev. A* 40, 519 (1989).

[7] N.Canosa, R.Rossignoli and A.Plastino, *Nucl.Phys. A* 512, 492 (1990).

[8] N.Canosa, R.Rossignoli and A.Plastino, *Phys.Rev. A* 43, 1145 (1991).

[9] N.Canosa, A.Plastino and R.Rossignoli, *Nucl.Phys. A* 453, 417 (1986).

[10] P.Ring and P.Schuck, *The Nuclear Many Body Problem* (Springer, Berlin, 1980).

[11] R.Gilmore, *Catastrophe Theory for Scientists and Engineers* (Wiley, New York, 1981).

[12] J.Krumlinde and Z. Szymanski, *Ann. of Phys.* 79, 201 (1973).

[13] D.J.Thouless, *The Quantum Mechanics of Many Body Systems* (Academic, New York, 1972).

[14] N.Canosa, R.Rossignoli, A.Plastino, and H.G. Miller, *to be published.*

DECOMPOSITION OF SIGNALS AND INFORMATION THEORY

A. Plastino[*], L. Rebollo Neira[**], F. Zyserman[=]

Centro de Física Teórica – Universidad Nacional de La Plata
C.C. 67, 1900 La Plata, Argentina
* Member of CONICET
** Member of CICPBA

ABSTRACT

We discuss a method that is able to determine, with the help of the Maximum Entropy Principle, the detailed nature of a complex signal that can be conceived as a linear superposition of independent, elementary signals. The physical relevance of this decomposition is discussed.

1. INTRODUCTION

Techniques for the decomposition of signals are of physical interest because they can be used to study systems that react, after being impinged upon by an external probe, by producing an output signal that is composed by a linear superposition of independent, "elementary" signals The analysis of the nature of this superposition allows one to gain revealing insights concerning the structure of the system.

Signals are preconcerted signs that convey information We envisage a situation in which a well-known probe, (for example, electromagnetic radiation) impinges upon a physical system, interacts with it and is afterward analyzed by an appropriate detection procedure.

Shannon's[1] vectorial representation of signals independizes the pertinent considerations from the specific details characterizing the detection procedure. In a previous work[2], we have found it very convenient to proceed as follows: to any signal f a vector ket $|f\rangle$ is attributed and measurements performed upon f are described by linear functionals $\mathcal{L}_i$ that map $|f\rangle$ upon the set of the real numbers.

The process of decomposition of signals is thus applied with the idea of studying systems that react with the input probe producing a response signal that is a superposition of independent signals. "Statistical weights" appear as a coefficient in this superposition and they contain information about the (statistical) nature of the physical system. Illustrations of these ideas are given in Ref.2. The Maximum Entropy Principle (MEP)[3] is there employed and shown to provide one with a powerful algorithm that allows for successfully tackling this type of problems.

However, whenever recourse to the MEP was made, we have tacitly assumed in our previous work[2], that a positive-definite quantity, given as the exponential of a suitable linear form, is the protagonist of the concomitant algorithm. Finding it is the final goal to be achieved, that will provide the information one is searching for.

The purpose of the present effort is that of overcoming this restriction, at least with reference to the problem outlined in the first paragraphs above.

2. FORMALISM

We shall assume that our response signal $|f\rangle$, to be analyzed by a convenient detection procedure, belongs to a vector subspace U_M (of a suitable vector space) that is spanned by a basis $|n\rangle$ ($n = 1,\ldots,M$). Thus $|f\rangle$ acquires the form

$$|f\rangle = \sum_{n=1}^{M} C_n \, |n\rangle \qquad (2.1)$$

The decomposition of $|f\rangle$ is the procedure that allows one to find out the coefficients C_n. To this end, $|f\rangle$ is to be subject of a finite number, N, of independent measurements that will allow for a numerical representation of the response signal[2], in the form of a set of numbers f_i ($i = 1, \ldots, N$), where

$$f_i = \mathcal{L}_i \,|f\rangle \qquad\qquad i = 1,\ldots,N \qquad\qquad (2.2)$$

which under the assumption of linearity can be rewritten as

$$f_i = \sum_{n=1}^{M} C_n \, \ell_{ni} \qquad\qquad (2.3)$$

if ℓ_{ni} stands for the measurement, represented by $\mathcal{L}_i^{\,2}$, performed upon the "elementary" output signal $|n\rangle$

$$\ell_{ni} = \mathcal{L}_i \,|n\rangle \qquad\qquad n = 1,\ldots,M \qquad\qquad (2.4)$$

The new idea to be discussed here is that of allowing for non-positive definite (i.e., negative) coefficients C_n. This is achieved by writting them down as

$$C_n = p_n - B \qquad\qquad n = 1,\ldots,M \qquad\qquad (2.5)$$

where the p_n are positive-definite figures and B is an unknown constant, to be self-consistently determined from the data (measurements upon the response signal) according to the algorithm to be developed below. With (2.5), (2.3) acquires the form

$$f_i = \sum_{n=1}^{M} \{ p_n \, \ell_{ni} - B \, \ell_{ni} \} \qquad\qquad i = 1,\ldots,N \qquad\qquad (2.6)$$

Select one of these N equations, say the r-th one, so as to fix B

$$B = [\sum_{n=1}^{M} p_n \, \ell_{nr} - f_r] \,/\, [\sum_{n=1}^{M} \ell_{nr}] \qquad\qquad (2.7)$$

and introduce the two definitions

$$F_i = f_i \sum_{n=1}^{M} \ell_{nr} - f_r \sum_{n=1}^{M} \ell_{ni} \qquad\qquad (2.8)$$

$$O_{ni} = \ell_{ni} \sum_{n=1}^{M} \ell_{nr} - \ell_{nr} \sum_{n=1}^{M} \ell_{ni} \qquad\qquad (2.9)$$

It becomes apparent that the system (2.6) can be recast as

$$F_i = \sum_{n=1}^{M} p_n \, O_{ni} \qquad\qquad i \neq r; \; i = 1,\ldots,N \qquad\qquad (2.10)$$

The set of positive numbers $\{p_n\}$ can be thought of as representing a non-normalized probability distribution, whose informational entropy is[4]

$$S = -\sum_{n=1}^{M} p_n \ln p_n + \sum_{n=1}^{M} p_n - 1 \tag{2.11}$$

Each equation of the system (2.10) tells us that the datum F_i is proportional to the mean value of a random variable whose values are given by the O_{ni} $(n = 1,\ldots,M)$ and "weighted" by the p_n.

The idea is now to solve the system (2.10) by recourse to the MEP. An iterative procedure will be followed in which an "optimal conjecture" is successively improved according to the MEP.

We start with a zeroth-order guess, in which a set $p_n^{(0)}$ is obtained by requiring that the form (2.11) be maximized (that is, $p_n = 1$, independent of n). This approximation provides our zeroth-order estimate for B, to be called $B^{(0)}$

$$B^{(0)} = 1 - f_r \Big/ \sum_{n=1}^{M} \ell_{nr} \tag{2.12}$$

with which we can predict for the result of the remaining measurements the values

$$f_i^{(0)} = \sum_{n=1}^{M} p_n^{(0)} \ell_{ni} - B^{(0)} \sum_{n=1}^{M} \ell_{ni} \qquad i \neq r \; ; \; i = 1,\ldots,N \tag{2.13}$$

The quality of this conjecture can be measured by defining the "predictive error" ε_i (for the i-th measurement)

$$\varepsilon_i = |f_i - f_i^{(0)}| / |f_i| \qquad i = 1,\ldots,N \tag{2.14}$$

In order to improve upon the zeroth-order guess and construct a first-order estimate we select, among the members of the set $\{\varepsilon_i\}$, its largest one, to be called ε_{k1}. The first-order weights $p_n^{(1)}$ are chosen so as to maximize S subject to the constraint

$$F_{k1} = \sum_{n=1}^{M} p_n^{(1)} O_{nk1} \tag{2.15}$$

which is tantamount to enforce the fulfillment of the k1-th equation in the system (2.10). According to Jaynes' MaxEnt approach this leads to[3]

$$p_n^{(1)} = \exp(-\lambda_1 O_{nk1}) \tag{2.16}$$

where the Lagrange multiplier λ_1 is constructively obtained by solving (2.15) for it.

With the $p_n^{(1)}$ we can build up the "predictions" $f_i^{(1)}$ ($i = 1,\ldots,N$) and the concomitant (new) set of ε_i. After selection of the largest element of this new $\{\varepsilon_i\}$-set, let us call it ε_{k2}, we obtain the $p_n^{(2)}$ by maximizing S subject to two constraints, namely, the fulfillment of the equations in the set (2.10) corresponding to both $i = k1$ and $i = k2$.

In general, the J-th order estimate is

$$p_n^{(J)} = \exp \{ - \sum_{i=1}^{J} \lambda_i O_{nki} \} \qquad n = 1,\ldots,M \qquad (2.17)$$

where the Lagrange multipliers λ_i ($i = 1,\ldots,J$) are obtained by solving the J equations

$$F_{ki} = \sum_{n=1}^{M} p_n^{(J)} O_{nki} \qquad i = 1,\ldots,J \qquad (2.18)$$

The iterative process is to be ended when we reach the situation

$$\varepsilon_i \le \Delta f_i \qquad i = 1,\ldots,N \qquad (2.19)$$

where Δf_i are the errors (of whatever origin) that character_ze the experimental data f_i. Let us suppose that this happy circumstance occurs when we reach the L-th iteration. Our final results will be

$$B^{(L)} = [\sum_{n=1}^{M} p_n^{(L)} \ell_{nr} - f_r] / [\sum_{n=1}^{M} \ell_{nr}] \qquad (2.20)$$

$$C_n^{(L)} = p_n^{(L)} - B^{(L)} \qquad n = 1,\ldots,M \qquad (2.21)$$

and they allow for the <u>prediction</u> of any subsequent measurement performed upon $|f\rangle$, f_{N+1}, $f_{N+2}, \ldots, f_{N+k}, \ldots$ These predictions read

$$f_{N+k}^{(L)} = \sum_{n=1}^{M} C_n^{(L)} \ell_{n(N+k)} \qquad k = 1,2,3,\ldots \qquad (2.22)$$

3. A NUMERICAL ILLUSTRATION

As an application of the formalism expounded in the preceding paragraphs we shall consider that situation in which the measurements performed upon the signal $|f\rangle$ are obtained as a function of some appropriate parameter t_i ($i = 1,\ldots,N$). This is a common occurrence indeed. We assume of course, that we deal with N independent measurements, so that each value t_i can be regarded as defining an (orthogonal) direction $|t_i\rangle$ in an appropriate N-

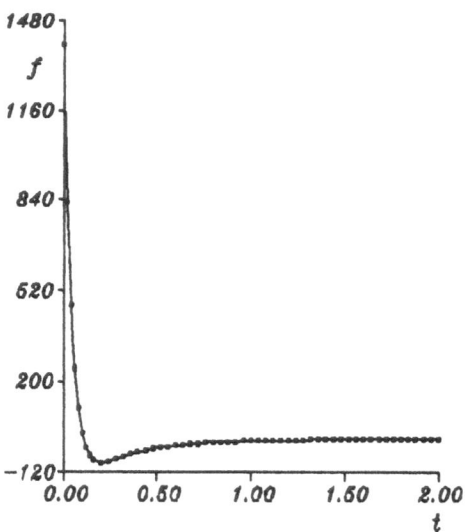

Fig.1 The continuous curve constitutes the theoretical predictions
of the data (squares).

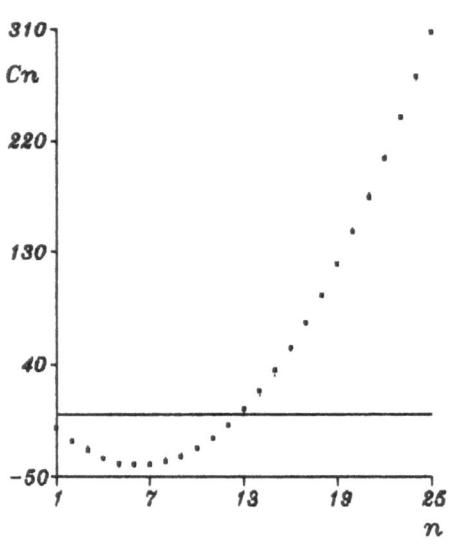

Fig.2 The exact Cn coefficients
(squares) are compared with the
results of the seventh order
iterative process (points).

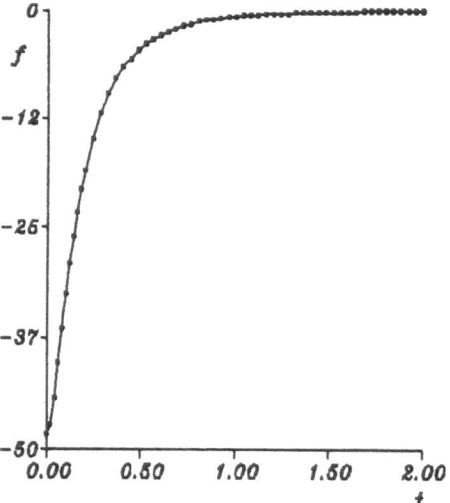

Fig.3 The curve is the theoretical
prediction, the squares the numerical representation of the signal.

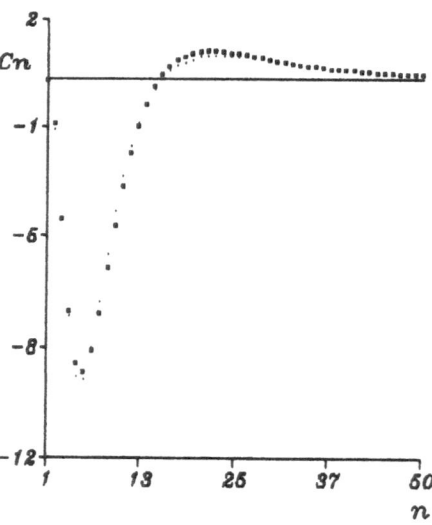

Fig.4 The exact Cn coefficients
(squares) are compared with the
results of the fifth order
iterative process (points).

dimensional space. The figures f_i that result after performing each measurements can now be regarded as "projections" of $|f\rangle$ upon the "direction" defined by t_i

$$f_i = \langle t_i | f \rangle \qquad\qquad i = 1,\ldots,N \qquad\qquad (3.1)$$

and, in an analogous fashion, we set

$$\ell_{ni} = \langle t_i | n \rangle \qquad\qquad n = 1,\ldots,M \qquad\qquad (3.2)$$

We shall consider two situations. In both of them we take

$$\langle t_i | n \rangle = \exp(-nt_i) \qquad\qquad (3.3)$$

and we simulate the data in two different manners. Those of Fig.1 (squares) arise from the expression

$$f_i = \sum_{n=1}^{M} n(n - 12.5)\,\exp(-nt_i) \qquad\qquad (3.4)$$

with $t_i = (j - 1)0.02$ $(j = 1,\ldots,100)$. The iterative process of section 2 converges after 7 iterations for an error $\Delta f_i = 0.01$ for all i. The coefficients $C_n^{(7)}$ are displayed in Fig.2 (points), while the exact ones are represented by squares. Coming back to Fig.1, the continuous curve represents the predictions afforded by the $C_n^{(7)}$. In a second example we simulate the data (squares) of Fig.3 by means of a set of coefficients C_n

$$C_n = (n - 15.6)\,\exp\{-(\ln n - \ln 7)^2/(\ln 1.8)^2\} \qquad\qquad (3.5)$$

Convergence, after 5 iterations and an error equal to that of Example 1 yields the points of Fig.4 which are to be compared to the exact ones (squares). The continuous curve in Fig.3 represents the predictions calculated by recourse to the $C_n^{(5)}$.

4.CONCLUSION

We have presented a method that is able to conveniently decompose a complex signal as a linear superposition of independent, elementary signals. With the help of the Maximum Entropy Principle, we have devised a practical algorithm that allows for the determination of the coefficients of that linear superposition.

The method can be of relevance for the study of systems that, upon interacting with well known probes, react producing complex output signals of the type described above.

Application to two simple examples allows one to appreciate the fact that this algorithm produces excellent results.

REFERENCES

1 - C.E. Shannon, Proc. IRE $\underline{37}$ (1949) 1

2 - A. Plastino, L. Rebollo Neira, A.G. Alvarez, Phys. Rev. $\underline{A40}$ (1989) 1644

3 - E.T. Jaynes, "The Maximum Entropy Formalism" edited by R.D. Levine and M. Tribus (MIT, Boston, 1979)

4 - R. Balian, Y. Alhassid, H. Reinhardt, Phys. Rep. Vol 131 (1986)

FINITE TEMPERATURE CORRELATED MEAN FIELD

TREATMENTS AND INFORMATION THEORY

R. Rossignoli, A. Plastino

Departamento de Física, Universidad Nacional de La Plata
1900 La Plata, Argentina

Abstract

A general self-consistent scheme for approximating statistical operators is discussed within the context of Information Theory. As an application, a special correlated finite temperature mean field approximation is derived and applied to many-fermion systems. A substantial improvement over conventional approaches such as finite temperature Hartree-Fock and finite temperature BCS is obtained in finite systems.

Introduction

Finite temperature (FT) mean field approaches such as FT Hartree-Fock (FTHF) [1], FTBCS and FT Hartree-Fock-Bogoliubov (FTHFB) [2], constitute the basic microscopic methods for dealing with many-body systems at finite temperature. They provide the first step upon which further approximations and also higher order treatments (such as FTRPA [3]) are constructed. The essential ingredient in these theories is the replacement of the full hamiltonian by an effective temperature dependent single particle (sp) or single quasiparticle Hamiltonian in the exponent of the statistical operator and the use of a Grand Canonical (GC) ensemble for calculating the pertinent traces. As a consequence of these approximations, the Fermi-Dirac expression for the average sp occupation number and the finite temperature version of Wick's theorem for calculating averages of m-body operators, hold.

However, important shortcomings are exhibited by these theories, especially in finite systems and in the so called transitional regions. In particular, the vanishing of the order parameter, such as the quadrupole moment or the pairing gap in systems like finite nuclei, at the corresponding mean field critical temperature, is not actually seen in exact canonical calculations [4]–[6].

In the present contribution our aim is first to make a short review of general self consistent approximations for statistical operators [7]–[8], within the general framework of Information Theory [9]–[11]. As a particular case, the usual FT mean field approximations are derived. Based on this general formalism, we derive next a correlated mean-field approximation, including a Correlated FTHF approach [12,14] (suitable also for *canonical* and projected statistics calculations), and also a general correlated FTHFB scheme [13] suitable for systems exhibiting pairing-like interactions in addition to long range forces. Finally, a particular application in a finite pairing model is developed.

General Statistical Formalism

Let us consider a quantum system about which the only available information consists of the expectation values O_i of n linearly independent observables $\hat{O}_i$. According to Information Theory, the least biased statistical operator $\hat{\rho}$ describing the system is that which maximizes the entropy (we set Boltzmann constant $k = 1$)

$$S = -\text{Tr}\hat{\rho}\ln\hat{\rho}, \tag{1}$$

subject to the constraints

$$\text{Tr}\hat{\rho}\hat{O}_i = O_i, \quad i = 1,\ldots,n. \tag{2}$$

The result can be cast as

$$\hat{\rho} = \exp\{-\lambda_0 - \sum_i \lambda_i \hat{O}_i\}, \tag{3}$$

where λ_0 is a normalization constant ($\text{Tr}\hat{\rho} = 1$) and λ_i a set of Lagrange parameters to be determined by means of the constraints (2).

Ordinary equilibrium finite temperature descriptions are obtained when one of the operators $\hat{O}_i$ is the Hamiltonian $\hat{H}$ of the system and the remaining ones commute with $\hat{H}$ (for instance, one of the $\hat{O}_i$'s is the particle number operator in GC treatments). Nevertheless, the present formalism is completely arbitrary and allows for general non-commuting operators $\hat{O}_i$, being thus suitable for off equilibrium statistical descriptions.

The ensuing maximum entropy reads

$$S = \lambda_0 + \sum_i \lambda_i O_i. \tag{4}$$

It is easy to show that S and λ_0, as functions of O_i and λ_i respectively, satisfy the conjugate relationships

$$\partial S/\partial O_i = \lambda_i, \quad \partial\lambda_0/\partial\lambda_i = -O_i. \tag{5}$$

Generalized Self-Consistent Approximation

In many-body systems, the previous exact statistical operator (3) is not tractable in general and one is forced to employ approximate schemes for effectively computing the pertinent traces. We shall now describe a general self-consistent approximation for the density (3) [7]–[8]. Starting with a set of (in principle) m arbitrary operators $\hat{P}_j$, $m \geq n$, we set up a trial approximate statistical operator possessing the form

$$\hat{\rho}_{ap} = \exp\{-\lambda_0 - \sum_j \lambda_j \hat{P}_j\}, \tag{6}$$

where now λ_j are a set of variational parameters to be determined by maximizing the approximate entropy

$$S_{ap} = -\mathrm{Tr}\hat{\rho}_{ap}\ln\hat{\rho}_{ap}, \tag{7}$$

subject to the constraints

$$\mathrm{Tr}\hat{\rho}_{ap}\hat{O}_i = O_i. \tag{8}$$

In other words, the exponent of the statistical operator is approximately expanded in the (undercomplete) space spanned by the operators $\hat{P}_j$.

The optimum $\hat{\rho}_{ap}$ can be obtained by introducing n additional Lagrange parameters β_i and maximizing the quantity

$$S' = S_{ap} - \sum_i \beta_i O_i, \tag{9}$$

with respect to the yet unknown expectation values $P_j = \mathrm{Tr}\hat{\rho}_{ap}\hat{P}_j$ Using equations (5) [applied to the operator (6)], we obtain the fundamental relations

$$\lambda_j = \sum_i \beta_i \partial O_i/\partial P_j, \tag{10}$$

which represent a *self-consistent* set of equations for the parameters λ_j (they are given by a function of the expectation values they determine). By recourse to the Kubo transforms [15] of the operators $\hat{P}_j$, defined here as

$$\hat{P}_j^{\,*} = \int_0^1 \hat{\rho}_{ap}^{-u}\hat{P}_j\hat{\rho}_{ap}^u du - \mathrm{Tr}(\hat{\rho}_{ap}\hat{P}_j), \tag{11}$$

it is possible to explicitly express (10) as

$$\lambda_j = \sum_{i,l} \beta_i A_{il} B_{lj}^{-1}, \tag{12}$$

where

$$A_{il} = \partial O_i/\partial\lambda_l = -\langle\hat{P}_l^{\,*}\hat{O}_i\rangle = -\langle\hat{O}_i^*\hat{P}_l\rangle, \tag{13}$$

and B is the $m \times m$ generalized covariance matrix

$$B_{jl} = \partial P_j/\partial\lambda_l = -\partial^2\lambda_0/\partial\lambda_j\partial\lambda_l = -\langle\hat{P}_l^{\,*}\hat{P}_j\rangle. \tag{14}$$

The approximate density operator can finally be written as

$$\hat{\rho}_{ap} = \exp\{-\lambda_0 - \sum_i \beta_i\hat{o}_i\}, \tag{15}$$

with

$$\hat{o}_i = \sum_j (\partial O_i/\partial P_j)\hat{P}_j. \tag{16}$$

The effective operator (16) can be considered as the (density dependent) *projection* of $\hat{O}_i$ onto the space spanned by the operators $\hat{P}_j$.

We are thus led to a formal self consistent solution for the density operator. In case all operators $\hat{O}_i$ are linearly related to the $\hat{P}_j$'s we obviously recover from (10) the exact statistical operator. Otherwise, the operator (15) provides obviously a lower bound to the exact entropy (4).

We can distinguish now two different applications: a) Those in which the expectation values of the operators $\hat{O}_i$ are actually given; b) Those in which the Lagrange parameters β_i are supposed to be known and we are interested in the behavior of the approximation as a function of the β_i's. In case a), it is obvious that situations may exist in which we cannot fulfill the constraints (8) (i.e. we cannot obtain a solution for the β_i's), since the range of expectation values spanned by the approximate density may be smaller than the exact range.

We shall consider in the present work case b) with the aim of applying the formalism to finite temperature problems. In this case, we shall take the operator $\hat{O}_1$ as the Hamiltonian $\hat{H}$ of the system and the associated parameter $\beta \equiv \lambda_1$ will be the inverse of the temperature T. We can write in this situation (15) in the more familiar form

$$\hat{\rho}_{ap} = \exp(-\lambda_0' - \beta \hat{h}'), \tag{17}$$

where

$$\hat{h}' = \hat{h} - \sum_{i=2}^{n} \mu_i \hat{o}_i, \tag{18}$$

with $\hat{h}$ the effective Hamiltonian constructed as in (16) and $\mu_i = -\lambda_i/\beta$ the chemical potentials.

As we shall see, the present formalism contains as a very particular case the usual FT mean field approaches. We shall proceed now to a rigorous and quite general rederivation of microscopic FT and statistical mean field theories.

Extended FTHF Formalism

In order to extract from the previous formalism a useful approximation for many-body systems, it is necessary to suitably select a set of operators $\hat{P}_j$ such that the ensuing density operator becomes tractable. The most obvious choice is to restrict the $\hat{P}_j$'s to *single particle* operators of the type $c_i^\dagger c_j$ in which case the density becomes

$$\begin{aligned} \hat{\rho}_{ap} &= \exp\{-\lambda_0 - \sum_{i,j} \lambda_{ij} c_i^\dagger c_j\} \\ &= \exp\{-\lambda_0 - \sum_i \lambda_i a_i^\dagger a_i\}, \end{aligned} \tag{19}$$

where $a_i^\dagger$, a_i are related to the original operators by means of a unitary (HF-like [16]) transformation

$$c_i = \sum_j U_{ij} a_j, \quad UU^\dagger = I, \tag{20}$$

such that $U^\dagger \Lambda U = \Lambda'$, with Λ and Λ' the matrices of elements λ_{ij}, $\lambda_i \delta_{ij}$ respectively. In what follows we shall consider the fermion case [7], so that $[c_i^\dagger, c_j]_+ = \delta_{ij}$ (for the boson case see [8]).

The formalism becomes in this situation equivalent to a *generalized* statistical HF approximation. The ensuing variational equations can be cast in two different pieces: one involving the parameters λ_i and the other the transformation matrix U. We obtain, using (10),

$$\lambda_i = \text{Tr} \hat{\rho}_{ap} \sum_j [(a_j^\dagger a_j - f_j) \hat{O}'] B_{ji}^{-1}, \tag{21}$$

with $\hat{O}' = \sum_i \beta_i \hat{O}_i$, and where the covariance matrix B reduces now to

$$B_{ij} = f_{ij} - f_i(f_j - \delta_{ij}), \tag{22}$$

with f_j and f_{ij} the generalized average one and two-body occupation numbers,

$$f_i = \mathrm{Tr}\hat{\rho}_{ap}a_i^\dagger a_i, \tag{23}$$

$$f_{ij} = \mathrm{Tr}\hat{\rho}_{ap}a_i^\dagger a_j^\dagger a_j a_i, \tag{24}$$

such that

$$\mathrm{Tr}\hat{\rho}_{ap}a_i^\dagger a_j = \delta_{ij}f_i, \quad \mathrm{Tr}\hat{\rho}_{ap}a_i^\dagger a_k^\dagger a_j a_l = f_{ik}(\delta_{kj}\delta_{il} - \delta_{kl}\delta_{iz}). \tag{25}$$

Minimization with respect to the HF transformation leads to the fundamental equation

$$\mathrm{Tr}\hat{\rho}_{ap}[a_i^\dagger a_j, \hat{O}'] = 0, \quad i \neq j. \tag{26}$$

Equations (21)–(26) are completely general and hold in *any* type of ensemble. The usual FTHF equations are recovered if the pertinent traces are taken in a *Grand Canonical* ensemble. Only in this case the familiar Fermi-Dirac expression for the one-body occupation numbers,

$$f_j = [1 + e^{\lambda_j}]^{-1}, \tag{27}$$

and the *finite temperature Wick's theorem* [17] for calculating averages of m-body operators ($m > 2$) hold. This implies in the present context

$$f_{ij} = f_i f_j(1 - \delta_{ij}). \tag{28}$$

In this case, $B_{ij} = f_i(1 - f_i)\delta_{ij}$, corresponding to a fully uncorrelated picture, and equations (21)–(26) reduce to the usual HF equations (see next section). We remark however that in other ensembles, expressions (27) and (28) do not necessarily hold and one should calculate them using the diagonal density operator (19).

Hence, the previous formalism allows us to extend the usual GC FTHF equations to any type of ensemble and to a general statistical context. We are thus in a position to perform *Canonical* FTHF calculations [12,14] (conserving exactly the number of particles). The formalism is also suitable for *projected* statistics [12] (ensembles which are microcanonical with respect to some set of operators), as for instance, angular momentum projected statistics, of much interest in finite temperature Nuclear Physics.

Nevertheless, the two-body occupation numbers are obviously still dependent on the one-body mean values, and do not constitute new degrees of freedom in the present formalism. Up to now, we have considered thus the possibility of introducing correlations just in the trace, but not in the density operator itself. We will now attempt to go beyond the statistical HF approximations (whether GC or not) by introducing special correlations in the density operator but preserving at the same time its tractability.

Correlated FTHF

We shall now consider the addition of *diagonal two-body* terms in the exponent of the density operator, such that

$$\hat{\rho}_{ap} = \exp\{-\lambda_0 - \sum_i \lambda_i a_i^\dagger a_i - \sum_{i,j} \lambda_{ij} a_i^\dagger a_j^\dagger a_j a_i\}, \tag{29}$$

with the operators $a_i^\dagger$, a_i given again by (20). In this way, (29) remains diagonal in an independent particle basis (i.e., consisting of Slater Determinants), so that no diagonalization is required for calculating traces.

The essential difference which is introduced now is that the two-body occupation numbers (24) are now *independent quantities*. In other words, they constitute new degrees of freedom. As we shall explicitly see, the density operator (29) is able to provide one with an essentially different probability distribution at finite temperature, in comparison with that given by ordinary or extended FTHF.

In order to determine the parameters λ_i, λ_{ij}, we employ now equations (10). We shall restrict ourselves to one and two-body operators $\hat{O}_i$, so that $\hat{O}'$ can be written as

$$\hat{O}' = \sum_{i,j} \varepsilon_{ij} c_i^\dagger c_j + \sum_{i,j,k,l} \tfrac{1}{4} V_{ijkl} c_i^\dagger c_j^\dagger c_l c_k \tag{30}$$

$$= \sum_{i,j} \varepsilon'_{ij} a_i^\dagger a_j + \tfrac{1}{4} \sum_{i,j,k,l} V'_{ijkl} a_i^\dagger a_j^\dagger a_l a_k, \tag{31}$$

with

$$\varepsilon'_{ij} = [U^\dagger \varepsilon U]_{ij}, \quad V'_{ijkl} = [U^\dagger U^\dagger V U U]_{ijkl}, \tag{32}$$

in obvious notation.

In this case, the expectation value of $\hat{O}'$ with respect to (29) will be a *linear* function of the elements f_i, f_{ij}. Hence, eqs. (10) lead to

$$\lambda_i = \varepsilon'_i, \quad \lambda_{ij} = \tfrac{1}{2} V'_{ijij}, \tag{33}$$

so that we obtain the expected result

$$\hat{\rho}_{ap} = \exp(-\lambda_0 - \hat{O}'_d), \tag{34}$$

where

$$\hat{O}'_d = \sum_i \varepsilon'_i a_i^\dagger a_i + \tfrac{1}{2} \sum_{i,j} V'_{ijij} a_i^\dagger a_j^\dagger a_j a_i \tag{35}$$

is just the diagonal part of $\hat{O}'$ in the independent particle basis.

The general equation determining the HF matrix U is still given by (26). For the case of a two-body $\hat{O}'$, it can be cast as

$$\varepsilon'_{ij}(f_j - f_i) + \sum_k V'_{ikjk}(f_{jk} - f_{ik}) = 0, \tag{36}$$

which is the fundamental equation of the CFTHF approach. We note that if Wick's theorem is applied in eqs. (36) [i.e. eq. (28)], we recover the ordinary FTHF equations

$$\varepsilon'_{ij} + \sum_k V'_{ikjk} f_k = 0, \tag{37}$$

with f_k given by expression (27). However, this does not obviously hold any longer in the correlated approach (whatever the ensemble) and equations (36) must be solved self-consistently with respect to the (U-dependent) diagonal operator (34).

We should bear in mind that in the statistical HF approximation, recovered in the present scheme for $\lambda_{ij} = 0$, we have in addition to solve the non linear eqs. (21) (i.e., we must solve for the sp energies), which represent the *projection* of $\hat{O}'_d$ onto the space spanned by the diagonal sp operators. On the other hand, in the present CFTHF treatment, eqs. (36) are the *only* ones to be solved, since the λ_i's and the

λ_{ij}'s are already known [eqs. (33)]. In this sense, the CFTHF eqs. are structurally simpler than the ordinary FTHF equations.

Hence, the central point we have considered is the addition of diagonal two-body correlations in the density operator. Equation (36) can be considered now as the extension of the usual FTHF eqs. to density operators *diagonal* in an independent particle basis. In some sense, this CFTHF treatment can be regarded as the 'true' or correct statistical extension of HF. For the operator (29) is, for one and two-body operators $\hat{O}_i$, the *best* statistical operator diagonal in this type of basis, that can be constructed. The extension of the present scheme to arbitrary m-body operators $\hat{O}_i$ is straightforward (one should just include in (29) diagonal m-body terms).

Finally, for thermal applications, it seems more familiar to cast the density (34) as

$$\hat{\rho}_{ap} = \exp(-\lambda_0 - \beta \hat{H}'_d), \tag{38}$$

where $\hat{H}'_d$ is the *diagonal* part of the generalized Hamiltonian [see (18)] $\hat{H}' = \hat{H} - \sum_{i=2}^{n} \mu_i \hat{O}_i$.

Extension to Pairing Interactions; Correlated FTHFB

The extension of the approximations of the previous sections to systems containing pairing-like interactions is straightforward. The standard FTBCS and FTHFB theories can be derived by considering just the set $\hat{P}_j = \{c_i^\dagger c_j, c_i^\dagger c_j^\dagger, c_i c_j\}$, i.e., by restricting the exponent of $\hat{\rho}_{ap}$ to a general quadratic function of fermion operators. The ensuing pairing extension of the correlated FTHF, leading in general to a correlated FTHFB (CFTHFB) [13] approximation, can be obtained again from a trial density of the form (29) but with the operators $a_i^\dagger$ a_i related to the original operators by means of a Bogoliubov transformation

$$c_i = \sum_j (U_{ij} a_j + V_{ij} a_j^\dagger), \tag{39}$$

with

$$U V^{tr} + V U^{tr} = 0, \quad U U^\dagger + V V^\dagger = I, \tag{40}$$

in order to ensure fermion commutation relationships for the operators a_i, c_i. In this way, the density operator (29) is diagonal now in an *independent quasiparticle* basis. The formalism obviously reduces to a correlated FTBCS (CFTBCS) approach if (39) is restricted to a BCS transformation [13]. We recover the ordinary FTBCS and FTHFB approaches by setting $\lambda_{ij} = 0$ in (29).

We again obtain for $\hat{\rho}_{ap}$ an expression similar to (34), where $\hat{O}'_d$ is now the full diagonal part of $\hat{O}'$ in the quasiparticle representation. In order to perform simple calculations, it is necessary to take the traces now in a GC ensemble. The fundamental equations that determine the transformation (39) are

$$\text{Tr}\hat{\rho}_{ap}[a_i^\dagger a_j, \hat{O}'] = 0, \quad \text{Tr}\hat{\rho}_{ap}[a_i^\dagger a_j^\dagger, \hat{O}'] = 0, \quad i \neq j, \tag{41}$$

which, for two-body operators $\hat{O}_i$, can be cast as the equation (36) plus

$$\varepsilon_{ij}^{02}(f_i + f_j - 1) + \sum_k V_{kkij}^{13}(f_{ik} + f_{jk} - f_k) = 0, \tag{42}$$

where now ε'_{ij}, V'_{ijkl}, ε_{ij}^{02} and V_{kkij}^{13} represent, respectively, the (suitably antisymmetrized) coefficients of $a_i^\dagger a_j$, $a_i^\dagger a_j^\dagger a_l a_k$, $a_j a_i$ and $a_k^\dagger a_k a_j a_i$ in the quasiparticle expansion of $\hat{O}'$, and are obviously transformation dependent. Thus, equations (36) and

(42) must be solved self-consistently. Again, we can recover the conventional FTBCS and FTHFB equations (or more adequately, the general statistical extensions of these approximations) if Wick's theorem and the Fermi-Dirac expression are employed in (36) and (42).

Application

As a particular illustration of the previous formalism, we shall consider a schematic two-level pairing model. We shall label the sp states as $|p, \nu\rangle$, $p = 1, \ldots, \Omega$, $\nu = \pm 1$. We shall examine the quasispin pairing Hamiltonian [18]

$$\hat{H} = \varepsilon \hat{J}_z - \tfrac{1}{2} G \hat{Q}_+ \hat{Q}_-, \tag{43}$$

where

$$\hat{Q}_+ = \sum_p c^\dagger_{p+} c^\dagger_{p-}, \quad \hat{J}_z = \tfrac{1}{2} \sum_{p,\nu} \nu c^\dagger_{p\nu} c_{p\nu}, \tag{44}$$

are collective operators satisfying, together with $\hat{Q}_z = \tfrac{1}{2}(\hat{N} - \Omega)$ and $\hat{J}_\nu = \sum_p c^\dagger_{p\nu} c_{p-\nu}$ an $SU(2) \times SU(2)$ algebra [18]. The present Hamiltonian represents a pairing interaction between levels separated by an energy ε, and reduces to the degenerate pairing model [2] for $\varepsilon = 0$. We shall consider the case $N = \Omega$.

The relevant Bogoliubov transformation reduces here to the BCS transformation

$$c_{p\nu} = a_{p\nu} \cos \tfrac{1}{2}\theta + \nu e^{i\phi} a^\dagger_{p-\nu} \sin \tfrac{1}{2}\theta, \tag{45}$$

which gives rise, in the conventional FTBCS approach [5,6], to a non vanishing pairing tensor represented here by

$$|\langle \hat{Q}_+ \rangle| = \Omega q \sin \theta = 2\Delta/G, \tag{46}$$

where $q = \tfrac{1}{2}(1 - f_+ - f_-)$, with $f_\nu = \langle a^\dagger_{p\nu} a_{p\nu} \rangle$, and Δ the FTBCS pairing gap. This particle number symmetry breaking tensor is the *essential* ingredient in the FTBCS theory of superconductivity. The mean value of the interaction becomes $-2\Delta^2/G$ in this approximation (neglecting small contributions of relative order $1/\Omega$ to the HF potential). For $G\Omega > 4\varepsilon$, the systems starts at $T = 0$ in a superconducting state characterized by $q = \tfrac{1}{2}$, $\theta = \tfrac{1}{2}\pi$, and, as T increases, the gap decreases approaching 0 at the critical temperature determined by $8T_c \cosh^2(\varepsilon/4T_c) = \Omega G$ [5], where there is a transition to the 'normal' state characterized by $q = 0$. For $T > T_c$, no effect of the interaction is seen in FTBCS ($\Delta = 0$). On the other hand, the exact canonical results exhibit no sharp transition for finite values of Ω, and the expectation value of the interaction decreases smoothly with temperature but does not vanish.

When considering the correlated FTBCS treatment based on the operator (29) and using the transformation (45) [13], the thermal description of the pairing interaction is substantially different from that given by ordinary FTBCS. One finds in CFTBCS a smooth thermal evolution of the system, with no sharp transitions and a smooth decrease of the mean value of the interaction as T increases, in agreement with exact results. Moreover, one finds also $\langle \hat{Q}_+ \rangle = 0$, implying thus a *vanishing* pairing tensor and hence an *approximate symmetry restoration* [13] (actually the particle number fluctuations are still non vanishing).

This behaviour is depicted in figure 1, for a typical situation. Since the gap is essentially a mean field parameter, we have defined the gap for the exact and

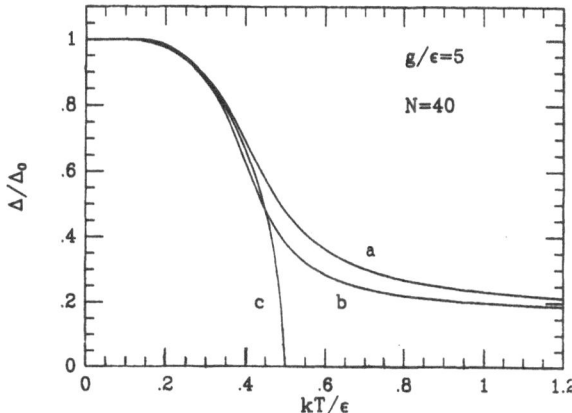

Figure 1. The pairing gap Δ as a function of the temperature T in the quasispin pairing model (see text). Δ_0 denotes the gap at $T = 0$, N is the number of particles and $g = \Omega G$. (a) denotes exact results, (b) results corresponding to the present correlated FTBCS treatment and (c) the conventional FTBCS results.

correlated treatment as $\Delta = \frac{1}{2}G\sqrt{\langle\hat{Q}_+\hat{Q}_-\rangle}$, which coincides with the gap defined in (46) in case we use the FTBCS expectation value. It is clearly seen that a substantial improvement over FTBCS results is obtained in the correlated approach for finite values of Ω.

Results in other finite systems containing different interactions, including realistic nuclei such as ^{20}Ne, are qualitatively similar (see [12]–[14]). A much better qualitative description of the pertinent physics is found in the correlated approach, in comparison with that given by ordinary FT mean field treatments.

Conclusions

Based on a general self-consistent approximation for statistical operators derived within the framework of Information Theory, we have developed a general correlated finite temperature mean field approximation. The method is based on the inclusion of special diagonal correlations in the density operator, and can be easily applied in arbitrary ensembles and general statistical contexts.

It is seen that the method provides a substantial improvement over the conventional FT mean field treatments, especially in finite systems and in transitional regions. Moreover, it is able to approximate restore some of the symmetries broken in mean field calculations, providing one with a much better global description of the pertinent physics.

In particular, we have seen in the example considered that a correct description of the pairing interaction is obtained in the correlated FTBCS treatment with a *vanishing* value of the pairing tensor. Furthermore, the interaction effects do not completely vanish at high temperatures, in contrast with ordinary FTBCS results. No sharp transition is obtained at finite temperatures for a finite value of the number of particles, in agreement with exact canonical results.

Acknowledgment

R.R. is supported by Comisión de Investigaciones Científicas de la Provincia de Buenos Aires (CIC) and A.P. by CONICET (Argentina).

References

[1] D.J.Thouless, *The Quantum Mechanics of Many Body Systems* (Academic, New York, 1972).

[2] A.L.Goodman, Nucl. Phys. **A352** (1981) 30; **A352** (1981) 45.

[3] P.Ring, L.Robledo, J.L.Egido and M.Faber, Nucl. Phys. **A419** (1984) 261; J.L.Egido, P.Ring and H.J.Mang, Nucl.Phys. **A451** (1986) 77.

[4] H.G.Miller, R.M.Quick and B.J.Cole, Phys.Rev. **C39** (1989) 1599.

[5] R.Rossignoli, and A.Plastino, Phys. Rev. **C32** (1985) 1040.

[6] R.Rossignoli, A.Plastino and J.Vary, Phys. Rev. **C37** (1988) 314.

[7] R.Rossignoli, and A.Plastino, Phys. Rev. **C40** (1989) 1798.

[8] R.Rossignoli, and A.Plastino, Phys. Rev. **A42** (1990) 2065.

[9] E.T.Jaynes, Phys. Rev **106** (1957) 620; **108** (1957) 171.

[10] R.Balian, Y.Alhassid and H.Reinhardt, Phys.Rep. **131** (1986) 1.

[11] R.Balian and M. Veneroni, Ann. of Phys. **164** (1985) 334; **174** (1987) 229.

[12] R.Rossignoli, H.G.Miller and A.Plastino, Phys. Rev. **C42** (1991) 314; Journal of Physics G, (1991), (in press).

[13] R.Rossignoli, Nucl. Phys. A (1991) (in press).

[14] R.Rossignoli, R.M.Quick and H.G.Miller, to be published.

[15] R. Kubo, J.Phys. Soc. Jpn. **12** (1957) 570; Rep. Pogr. Phys. **29** (1966) 225.

[16] P.Ring and P.Schuck, *The Nuclear Many Body Problem* (Springer-Verlag, New York, 1980).

[17] H.Reinhardt, R.Balian and Y.Alhassid, Nucl. Phys. **A422** (1984) 349.

[18] M.C.Cambiaggio and A.Plastino, Zeit. Phys. **A288** (1978) 153.

DYNAMICS OF QUANTUM INHIBITED PROCESSES

THROUGH REPEATED MEASUREMENTS

G. Crespo, Hilda A. Cerdeira* and A. N. Proto

Grupo de Sistemas Dinámicos, Regional Norte
Universidad de Buenos Aires, C.C. 2
(1638) Vicente López, Argentina

Usually in quantum mechanics, the computation of the wave function is achieved via the Schrödinger equation from t = 0 to a final time T, using the Born interpretation to predict the result of the measurement. In 1977, Misra and Sudarshan[1] pointed out that repeated projections lead to a new effect not included in the Schrödinger equation such as inhibitions of transitions by frequent measurements (in fact, they called it Zeno paradox). The possibility of this effect has attracted much interest over the last years.[2-5] More recently, Itano, Heinzen, Bollinger and Wineland[6] have claimed they succeeded in observing this effect. We can summarize the experiment by Itano et al (which is in fact based on a proposed experiment by Cook[5]) as follows: The experiment involves $^9Be^+$ ions. The energies of the excited levels 2 and 3 are higher than that of the ground level 1 by ω_2 and ω_3, which correspond respectively to rf and optical frequencies. It is important to notice that $1\leftrightarrow2$ and $1\leftrightarrow3$ transitions are allowed while $2\leftrightarrow3$ transitions are forbidden. They send an on-resonance rf field, which creates a coherent superposition state of levels 1 and 2 oscillating at the Rabi frequency Ω. If the rf field is applied during $T = \pi/\Omega$, all the ions that were initially prepared in level 1 will be brought into the excited level 2 (this is called a π-pulse) During the time T, they apply N short on-resonance optical pulses (the so-called "measurement" pulses) to perform quantum mechanical measurement of level 1, by promoting the ion two level 3 (from where it returns quickly to level 1 by spontaneous emission) in case it was in level 1 at the

* International Centre for Theoretical Physics (ICTP), P.O. Box 586, 34100 Trieste, Italy; and Instituto de Fisica, Universidade Estadual de Campinas (UNICAMP), C.P. 6165, 13081 Campinas, São Paulo, Brazil.

moment of "measuring", or leaving the ion in level 2 in case the measurement process finds the ion there.

The analysis made by Itano et al[6] and Cook[5] in order to obtain the expression for the probability of finding an ion into level 1 or level 2 does not take into account the presence of the level 3. This probability is derived by considering the time evolution of the density matrix ρ describing a two level system. These authors claim that the consequence of the measurement pulse is to project an ion into level 1 or level 2, thus destroying the superposition existing between these two states. This fact in turn corresponds to setting the coherences (ρ_{12} and ρ_{21}) of the density matrix to zero.

When one is dealing with a dynamical description of a two level system achieved by the temporal evolution of the expectation values of the relevant operators imposed by the dynamics, it is pertinent to find the dynamical behavior that proves to be tantamount to that of setting the coherences of the density matrix to zero.

A Hamiltonian describing a two level system can be written as

$$\hat{H} = \omega_1 \, \hat{a}_1^+ \, \hat{a}_1 + \omega_2 \, \hat{a}_2^+ \, \hat{a}_2 + \gamma_{12} \, \hat{a}_1 \, \hat{a}_2^+ + \gamma_{12}^* \, \hat{a}_2 \, \hat{a}_1^+ \tag{1}$$

where $\hbar = 1$.

As it was shown in Ref.(7), using the generalized Ehrenfest theorem, if we define the following set of relevant operators

$$\hat{N}_1 = \hat{a}_1^+ \, \hat{a}_1 \tag{2a}$$

$$\hat{N}_2 = \hat{a}_2^+ \, \hat{a}_2 \tag{2b}$$

$$\hat{N}_3 = i \, (\gamma_{12} \, \hat{a}_1 \, \hat{a}_2^+ - \gamma_{12}^* \, \hat{a}_2 \, \hat{a}_1^+) \tag{2c}$$

$$\hat{N}_4 = \gamma_{12} \, \hat{a}_1 \, \hat{a}_2^+ + \gamma_{12}^* \, \hat{a}_2 \, \hat{a}_1^+ \tag{2d}$$

the temporal evolution of the expectation values of these operators is governed by:

$$\frac{d < \hat{N}_1 >}{d t} = < \hat{N}_3 > \tag{3a}$$

$$\frac{d < \hat{N}_2 >}{d t} = - < \hat{N}_3 > \tag{3b}$$

$$\frac{d<\hat{N}_3>}{dt} = 2|\gamma_{12}|^2(<\hat{N}_2>-<\hat{N}_1>) - (\omega_2-\omega_1)<\hat{N}_4> \qquad (3c)$$

$$\frac{d<\hat{N}_4>}{dt} = (\omega_2-\omega_1)<\hat{N}_3> \qquad (3d)$$

Because of the very definition of these relevant operators, $\hat{N}_1$ and $\hat{N}_2$ are the number of particles or populations of level 1 and 2, respectively, $\hat{N}_3$ is related to the current of particles between these two levels, and $\hat{N}_4$ to the interaction energy.

In order to compare this approach to that used in Refs. (5) and (6), it is convenient to normalize the expectation value of the total number of particles of the system to one, i.e. $<\hat{N}_1> + <\hat{N}_2> = 1$, so that $<\hat{N}_1>$ and $<\hat{N}_2>$ thus represent the probabilities of finding the particle in level 1 or in level 2, respectively. The dynamical equivalent of setting the coherences in the density matrix to zero in order to provide the so-called "collapse" of the wave function due to the measurement performed on the system, is to set the expectation value of the current of particles, i.e. $<\hat{N}_3>$, to zero. In an attempt to reproduce the dynamical behavior expected, we can modulate a strong exponential decay by narrow gaussian functions. These gaussian functions would be present only at the measurement pulses, and their width (namely σ) related to the length of the measurement pulse. Thus, the Eq. (3c) would turn into

$$\frac{d<\hat{N}_3>}{dt} = 2|\gamma_{12}|^2(<\hat{N}_2>-<\hat{N}_1>) - (\omega_2-\omega_1)<\hat{N}_4> -$$

$$-\frac{<\hat{N}_3>}{\sigma}\left(\sum_{n=1}^{N}\exp\{-(t-nT/N-\Delta)^2/\sigma^2\}\right) \qquad (3c')$$

where N is the number of measurements made during the period T. In order to reproduce the experimental results, we set $T = 0.26$ sec, $\sigma = 0.001$ sec, and $\Delta = 0.00013$ sec on the one hand, and letting $\omega_2 = \omega_1 = 0$ and $\gamma_{12} = 6.14$ sec^{-1}, so as to adjust the π pulse. The time evolution of the number of particles of levels 1 and 2 and the current of particles between these two levels for the particular case of $N = 4$, which were obtained through numerical integration of Eqs. (3), Cast the final expectation values $<\hat{N}_1>(T) = 0.63631$ and $<\hat{N}_2>(T) = 0.36369$.

When considering carefully the details of the experiment proposed by Itano et al[6], it is necessary, in order to try to obtain a complete dynamical description of the quantum process involved, to include the third atomic level. To achieve this purpose, a suitable Hamiltonian would be:

$$\hat{H} = \omega_1\, \hat{a}_1^+ \hat{a}_1 + \omega_2\, \hat{a}_2^+ \hat{a}_2 + \omega_3\, \hat{a}_3^+ \hat{a}_3 + \gamma_{12}\, \hat{a}_1 \hat{a}_2^+ + \gamma_{12}^*\, \hat{a}_2 \hat{a}_1^+ +$$

$$+ f(t)\,(\gamma_{13}\, \hat{a}_1 \hat{a}_3^+ + \gamma_{13}^*\, \hat{a}_3 \hat{a}_1^+) \qquad\qquad (4a)$$

where we can define $f(t)$ as a succession of very narrow gaussian functions centred around the measurement times. Thus,

$$f(t) = \sum_{n=1}^{N} \exp\{-(t - nT/N - \Delta)^2/\sigma^2\} \qquad\qquad (4b)$$

with N, T, Δ, and σ defined as before. At this stage, it is pertinent to notice that the following set of operators:

$$\hat{O}_1 = \hat{a}_1^+ \hat{a}_1 \qquad\qquad (5a)$$

$$\hat{O}_2 = \hat{a}_2^+ \hat{a}_2 \qquad\qquad (5b)$$

$$\hat{O}_3 = \hat{a}_3^+ \hat{a}_3 \qquad\qquad (5c)$$

$$\hat{O}_4 = i(\gamma_{12}\, \hat{a}_1 \hat{a}_2^+ - \gamma_{12}^*\, \hat{a}_2 \hat{a}_1^+) \qquad\qquad (5d)$$

$$\hat{O}_5 = \gamma_{12}\, \hat{a}_1 \hat{a}_2^+ + \gamma_{12}^*\, \hat{a}_2 \hat{a}_1^+ \qquad\qquad (5e)$$

$$\hat{O}_6 = i(\gamma_{13}\, \hat{a}_1 \hat{a}_3^+ - \gamma_{13}^*\, \hat{a}_3 \hat{a}_1^+) \qquad\qquad (5f)$$

$$\hat{O}_7 = \gamma_{13}\, \hat{a}_1 \hat{a}_3^+ + \gamma_{13}^*\, \hat{a}_3 \hat{a}_1^+ \qquad\qquad (5g)$$

$$\hat{O}_8 = i(\gamma_{13}\, \gamma_{12}^*\, \hat{a}_2 \hat{a}_3^+ - \gamma_{13}^*\, \gamma_{12}\, \hat{a}_3 \hat{a}_2^+) \qquad\qquad (5h)$$

$$\hat{O}_9 = \gamma_{13}\, \gamma_{12}^*\, \hat{a}_2 \hat{a}_3^+ + \gamma_{13}^*\, \gamma_{12}\, \hat{a}_3 \hat{a}_2^+ \qquad\qquad (5i)$$

closes a semialgebra under commutation with the Hamiltonian (4), which in turn defines a closed system of differential equations for the expectation values of these operators. Due to the presence of the function $f(t)$, this system has time dependent coefficients, which does not allow for a direct analytical integration.

Nevertheless, the Hamiltonian given by Eqs. (4) does not yet take into account all the physical characteristics involved in the experiment of Ref. (6), since it does not include the spontaneous emission that occurs when an atom decays from level 3 to level 1, owing to the very short lifetime in the upper level. Thus, the Hamiltonian should be modified into

$$\hat{H} = \omega_1 \hat{a}_1^+ \hat{a}_1 + \omega_2 \hat{a}_2^+ \hat{a}_2 + \omega_3 \hat{a}_3^+ \hat{a}_3 + \gamma_{12} \hat{a}_1 \hat{a}_2^+ + \gamma_{12}^* \hat{a}_2 \hat{a}_1^+ +$$

$$+ f(t) (\gamma_{13} \hat{a}_1 \hat{a}_3^+ + \gamma_{13}^* \hat{a}_3 \hat{a}_1^+) + \sum_{k=1}^{\infty} \hat{b}_\kappa^+ \hat{b}_\kappa +$$

$$+ \sum_{k=1}^{\infty} (\gamma_\kappa \hat{a}_3 \hat{a}_1^+ \hat{b}_\kappa^+ + \gamma_\kappa^* \hat{a}_1 \hat{a}_3^+ \hat{b}_\kappa) \qquad (6)$$

where the rotation wave approximation is used.

The first summation in this Hamiltonian accounts for the quantized electromagnetic field, while the second one describes the interaction that allows for the destruction of a particle in level 3, its creation in level 1 and the creation of a photon in the electromagnetic field (plus H.C.).

The difficulty that arises when dealing with the Hamiltonian given by Eq. (6) is that it is no longer possible to close a finite semialgebra of operators under commutation with the Hamiltonian. This is due to the presence of trilinear operator terms appearing in the last summation of Eq. (6), whose effect is to introduce new operators of higher order than trilinear. Therefore we can ascertain that it is not possible to obtain an exact analytical resolution of the dynamical equations for the expectation values of the relevant operators related to the proper Hamiltonian for the physical system.

However, one can modify the system of equations related to Hamiltonian (4) in order to obtain the desired dynamical behavior and taking advantage of the finite semialgebra associated with it. The modifications introduced are indicated in bold characters in:

$$\frac{d<\hat{O}_1>}{dt} = <\hat{O}_4> + f(t)<\hat{O}_6> + \alpha <\hat{O}_3> \qquad (7a)$$

$$\frac{d<\hat{O}_2>}{dt} = -<\hat{O}_4> \qquad (7b)$$

$$\frac{d<\hat{O}_3>}{dt} = -f(t)<\hat{O}_6> - \alpha <\hat{O}_3> \qquad (7c)$$

$$\frac{d<\hat{O}_4>}{dt} = -(\omega_2-\omega_1)<\hat{O}_5> - 2 |\gamma_{12}|^2(<\hat{O}_1> - <\hat{O}_2>)$$
$$+ f(t)<\hat{O}_9> \qquad (7d)$$

$$\frac{d<\hat{O}_5>}{dt} = (\omega_2-\omega_1)<\hat{O}_4> + f(t)<\hat{O}_8> \qquad (7e)$$

$$\frac{d<\hat{O}_6>}{dt} = -(\omega_3-\omega_1)<\hat{O}_7> - f(t)2|\gamma_{13}|^2(<\hat{O}_1> - <\hat{O}_3>)$$
$$+ <\hat{O}_9> - \beta<\hat{O}_6> \tag{7f}$$

$$\frac{d<\hat{O}_7>}{dt} = (\omega_3-\omega_1)<\hat{O}_6> - <\hat{O}_8> - \beta<\hat{O}_7> \tag{7g}$$

$$\frac{d<\hat{O}_8>}{dt} = -(\omega_3-\omega_2)<\hat{O}_9> + |\gamma_{12}|^2<\hat{O}_7> - f(t)|\gamma_{13}|^2<\hat{O}_5>$$
$$- \beta<\hat{O}_8> \tag{7h}$$

$$\frac{d<\hat{O}_9>}{dt} = (\omega_3-\omega_2)<\hat{O}_8> - |\gamma_{12}|^2<\hat{O}_6> - f(t)|\gamma_{13}|^2<\hat{O}_4>$$
$$- \beta<\hat{O}_9> \tag{7i}$$

The modifications in Eq.(7a) and (7c) are made in order to introduce the spontaneous emission caused by the decay from level 3 to level 1 in the dynamics and the conservation of the number of particles. The magnitude of α would entail the strength of that decay. The modifications in Eqs.(7f-i) allows to set the current of particles between levels 1 and 2 to zero after each measurement as required. The parameter β should be adjusted numerically depending on the number of measurements (i.e. N) that are performed during the experiment.

The numerical integration of the final system of differential equations was obtained using the Runge-Kutta method. In order to compare our numerical results through this approach with the experimental ones, we set the parameters T, σ, Δ and γ_{12} as before, and $\omega_3 = \omega_2 = \omega_1 = 0$ and $\gamma_{13} = (800000)^{1/2}$ sec^{-1}, so as to adjust the corresponding π pulses. This fact enables us to uncouple the original set of differential equations (7) retaining only Eqs. (7a), (7b), (7c), (7d), (7f) and (7i). To provide the strong spontaneous decay from level 3 to 1, we set $\alpha = 2000$ sec^{-1}, and β is to be adjusted numerically depending on N. In Figs. 3 and 4, we depict the dynamical behavior of $<\hat{O}_1>$, $<\hat{O}_2>$, $<\hat{O}_3>$, and $<\hat{O}_4>$ for N = 4 and 16, respectively, indicating in each figure the choice for β. If one compares the results obtained through this dynamical approach to the experimental ones exhibited in Ref.(6), the agreement is almost complete for all values of N.

At this point, we would like to mention the differences between the present dynamical description to that presented by T. Petrosky, S. Tasaki and I. Prigogine[8]. There, the authors make a description in terms of wave functions and the evolution operator, restoring to a Hamiltonian split into two in order to introduce the Wigner-Weisskopf approximation to deal with the spontaneous emission. Thus, the dynamical evolution is to be obtained from the successive application of two different Hamiltonians, while a strong approximation is used.

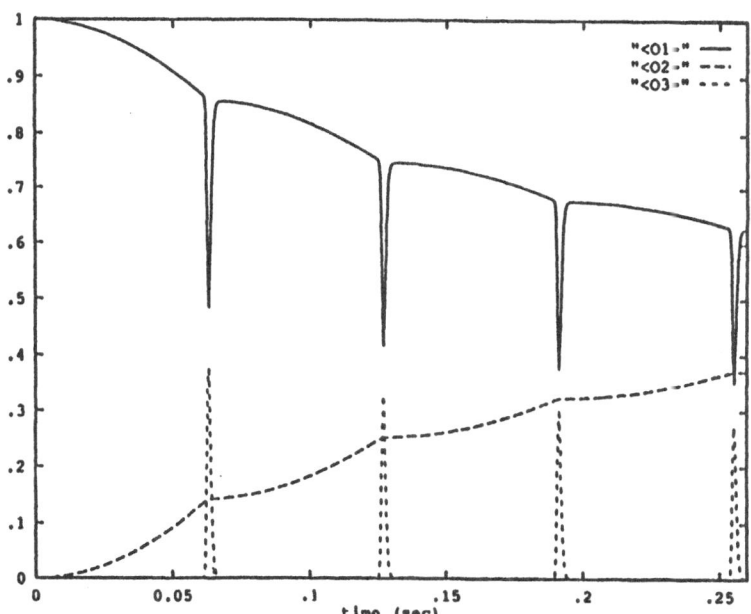

Fig. 1.   Expectation values $< \hat{O}_1 >$, $< \hat{O}_2 >$ and $< \hat{O}_3 >$
versus time, for $N = 4$ and $\beta = 67.5$. Final
values $< \hat{O}_1 >_T = 0.6279$, $< \hat{O}_2 >_T = 0.3720$

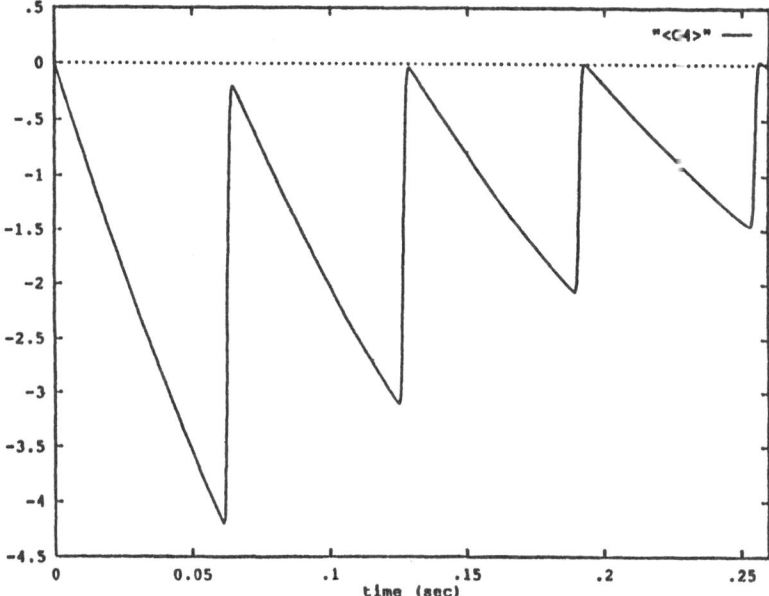

Fig. 2.   Expectation value of $< \hat{O}_4 >$ versus time, for
$N = 4$ and $\beta = 67.5$.

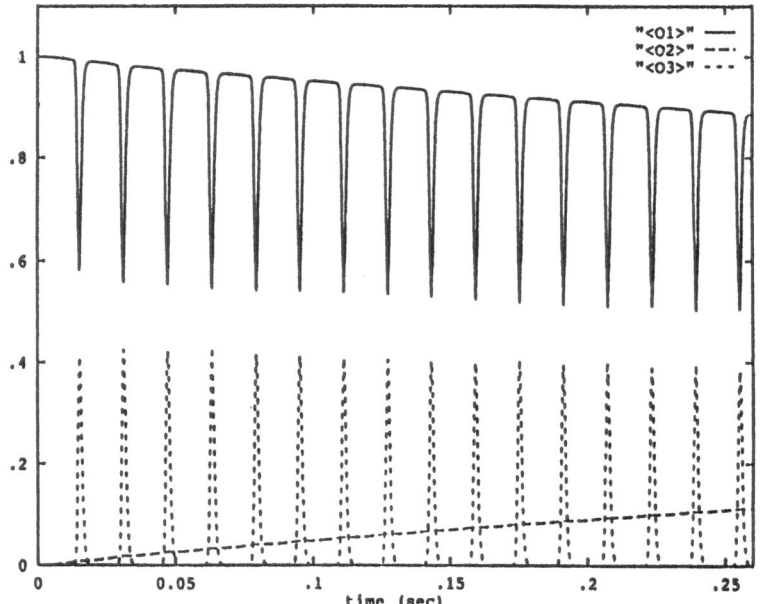

Fig. 3.   Expectation values $<\hat{O}_1>$, $<\hat{O}_2>$ and $<\hat{O}_3>$ versus time, for $N = 4$ and $\beta = 160$. Final values $<\hat{O}_1>_T = 0.8872$, $<\hat{O}_2>_T = 0.1127$.

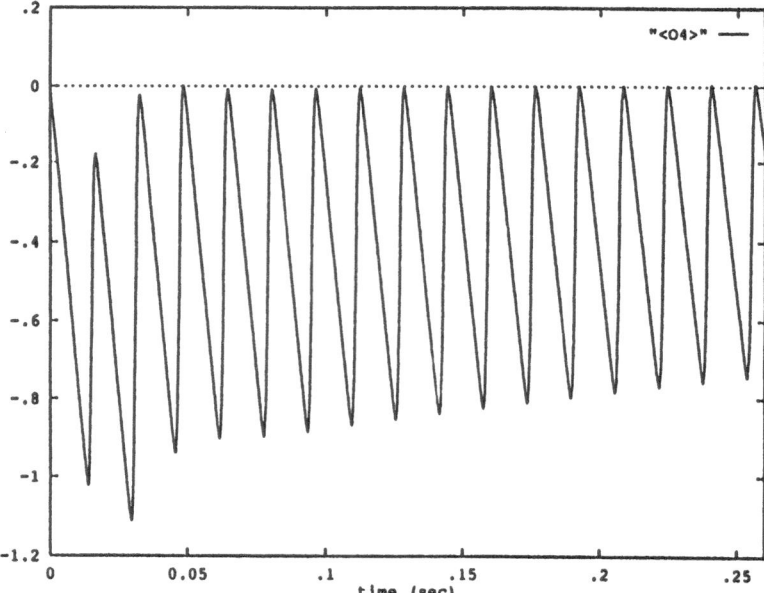

Fig. 4.   Expectation value of $<\hat{O}_4>$ versus time, for $N = 16$ and $\beta = 160$.

As a conclusion, we would like to stress the main features of Eqs. (7). First of all, we should mention that they were obtained starting from the Hamiltonian given by Eqs. (4). Although this Hamiltonian does not represent the physical system meant to be described (in fact, the correct Hamiltonian is given by Eq. (6)), it has the advantage of generating a dynamics related to a finite set of operators as the one given by Eqs. (5) (which is not the case for the Hamiltonian (6) that has a necessarily infinite semialgebra of operators related to it). As a consequence, the system of differential equations for the expectation values of the relevant operators were modified in order to introduce phenomenologically the dynamical behavior that was not present in the Hamiltonian itself. As it was stated before, a necessary condition so as to mimic the setting to zero of the coherences between levels 1 and 2 in the density matrix is to obtain the setting to zero of the expectation value of the operator representing the current of particles between these two levels (i.e. $< \hat{O}_4 >$). This dynamical character is provided by the inclusion of the terms containing the parameter β, and it should be noticed that these terms do not appear in the equations governing the dynamical evolution of the expectation values of the energy of interaction between levels 1 and 2 (i.e. $< \hat{O}_5 >$) and $< \hat{O}_4 >$ itself. Finally, it should be mentioned that the magnitude of the parameter β must be adjusted numerically depending on the number of measurement pulses in order to obtain the expected dynamical evolution for $< \hat{O}_4 >$.

ACKNOWLEDGEMENTS

Two of the authors (G. C. and A. N. P.) gratefully acknowledge the Comisión de Investigaciones Científicas de la Provincia de Buenos Aires for its support. Also two of the authors (G. C. and H. A. C.) gratefully acknowledge support by the International Centre for Theoretical Physics, Trieste, Italy.

REFERENCES

1. B. Misra and E.C.G. Sudarshan, J. Math. Phys. **18**, 756 (1977).
2. A. Peres, Am. J. Phys. **48**(11), 931 (1980).
3. H. Dehmelt, Proc. Natl. Acad. Sci. U.S.A. **83**, 2291 (1986); **83**, 3074 (1986).
4. M. Porrati and S. Putterman, Phys. Rev. A **36**, 929 (1987).
5. R.J. Cook, Phys. Scr. **T21**, 49 (1988).2) Cook
6. W.M. Itano, D.J. Heinzen, J.J. Bollinger and D.J. Wineland, Phys. Rev. A **41**, 2295 (1990).
7. A.N. Proto, in *Proceedings of the XIII International Workshop on Condensed Matter Theories*, edited by V. Aguilera-Navarro (Plenum, New York, 1989); J. Aliaga, G. Crespo, and A.N. Proto, Phys. Rev. A 42, 618 (1990); J. Aliaga, G. Crespo, and A.N. Proto, Phys. Rev. A 44, (1991), (to be published).
8. T. Petrosky, S. Tasaki and I. Prigogine, Phys. Lett. A **151**, 109 (1990).

FEATURES OF HNC/0 SOLUTIONS FOR SYMMETRIC FILMS

OF LIQUID ^4He AT $T = 0$ K

L. Szybisz and R.O. Vallejos

Departamento de Física, Comisión Nacional de Energía Atómica
Avda. del Libertador 8250
1429 Buenos Aires, Argentina

INTRODUCTION

Much work has been devoted in recent years[1-7] to microscopc c calculations of the ground state of inhomogeneous liquid ^4He at $T = 0$ K. The Hamiltonian of an interacting N-body quantum system immersed in an external one-body field is

$$H = -\frac{\hbar^2}{2m}\sum_{i=1}^{N}\nabla_i^2 + \sum_{i<j=1}^{N} v(r_{ij}) + \sum_{i=1}^{N}U_{ext}(\mathbf{r}_i) , \qquad (1)$$

where $v(r_{ij} = |\mathbf{r}_i - \mathbf{r}_j|)$ is a two-body potential and $U_{ext}(\mathbf{r}_i)$ an external one-body potential. In the present work we shall focus our attention on some features of variational solutions obtained when the the ground state of a Bose system is described by an N-body wave function of the Feenberg type[8]

$$\Psi_0(\mathbf{r}_1, ..., \mathbf{r}_N) = \exp[\frac{1}{2}\sum_{i=1}^{N} u_1(\mathbf{r}_i) + \frac{1}{2}\sum_{i<j=1}^{N} u_2(\mathbf{r}_i, \mathbf{r}_j) . \qquad (2)$$

Here $u_1(\mathbf{r}_i)$ and $u_2(\mathbf{r}_i, \mathbf{r}_j)$ are, respectively, the one- and two-body correlation factors. The factor $u_1(\mathbf{r}_i)$ accounts for the inhomogeneity of the system. On the other hand, the factor $u_2(\mathbf{r}_i, \mathbf{r}_j)$ exhibits some degree of anisotropy depending, especially near a surface, on the positions $\mathbf{r}_i$ and $\mathbf{r}_j$ of both particles. The optimal values of the correlation factors should be determined from the Euler-Lagrange (hereafter denoted as EL) equations derived by minimizing the energy expectation value

Condensed Matter Theories, Vol. 7, Edited by A.N. Proto
and J.L. Aliaga, Plenum Press, New York, 1992

$$H_{00}[u_1, u_2] = \frac{\int d^3 r_1 ... d^3 r_N \Psi_0(r_1, ..., r_N) H \Psi_0(r_1, ..., r_N)}{\int d^3 r_1 ... d^3 r_N \Psi_0^2(r_1, ..., r_N)} = E \qquad (3)$$

with respect to these one- and two-body quantities, i.e.,

$$\frac{\delta H_{00}[u_1, u_2]}{\delta u_n} = 0 \qquad (4)$$

for n=1 and 2.

Early calculations of the ground state of non-uniform helium liquid always invoked some sort of local-density approximation (LDA), see for instance ref. 1. A solvable set of coupled EL equations has been derived from eq.(4) by Krotscheck *et al.*[2,3] within the framework of the paired-phonon analysis (PPA). The solution of these coupled equations for films of planar symmetry yielded the first self-consistent variational calculation for an inhomogeneous system. Krotscheck and co-workers solved the PPA integro-differential equation by introducing a normal-mode decomposition of the density fluctuations reducing the initial form to an eigenvalue equation. In ref. 6 it has been developed an alternative numerical procedure for solving the two-body PPA equation by implementing a finite-difference relaxation (FDR) method. In that work[6] and in a subsequent one reported in ref. 7, the effort has been mainly devoted to study the convergence of the novel method. In order to continue the project, it is the main purpose of the present investigation to search for the extent to which the two-body functions obtained at the central part of a symmetric film agree with results evaluated for an homogeneous system of density equal to the central density ρ_c. To this end we will apply the method described in ref. 6 to solve the EL equations corresponding to symmetric films of liquid ^4He exposed to external potentials; these potentials will be adjusted in order to get systems with a central density approximately equal to the experimental equilibrium density of bulk liquid ^4He, i.e., $\rho_{exp} = 0.0218$ Å^{-3}. This work will be closed with a brief discussion on the most important normal-modes.

RELEVANT FORMULAS

In this section we will give all necessary formulas as well as a summary of the algorithm. In the case of a planar geometry, the system is translationally invariant in the x-y plane and symmetry is broken in the z direction giving rise to a surface structure. Accordingly, the external potential and the one-body density are reduced to one-dimensional functions depending only upon z, i.e., $U_{ext}(r) = U_{ext}(z)$ and $\rho(r) = \rho(z)$. Furthermore, all two-body quantities depend only on three variables, the distance between the two particles parallel to the surface

$$\eta = |\eta_{12}| = |\eta_2 - \eta_1| = [(x_2 - x_1)^2 + (y_2 - y_1)^2]^{\frac{1}{2}} \tag{5}$$

and the z coordinate of each of the two particles, i.e., z_1 and z_2. Therefore, the functional dependence of any two-body quantity, from now on we drop the corresponding subindex 2, can be indicated by $F(\eta, z_1, z_2)$.

The EL equation corresponding to $n = 1$ is derived from eq. (4) by imposing the constraint of keeping constant the particle number N. Its final expression has the form of a Hartree equation for the square root of the one-body density, $\sqrt{\rho(\mathbf{r})}$, which in the case of planar symmetry reads

$$[-\frac{\hbar^2}{2m}\frac{d^2}{dz^2} + U_{ext}(z) + V_{\mathrm{H}}(z)]\sqrt{\rho(z)} = \mu\sqrt{\rho(z)} , \tag{6}$$

where μ is the chemical potential, and $V_{\mathrm{H}}(z)$ a generalized Hartree potential

$$V_{\mathrm{H}}(z_1) = \int_{-\infty}^{\infty} dz_2 \rho(z_2) W_{\mathrm{H}}(z_1, z_2) . \tag{7}$$

The constraint of a fixed particle number can be expressed as

$$n_z = \int_0^{\infty} dz \rho(z) . \tag{8}$$

For the calculations carried out in the present work we adopted for $W_{\mathrm{H}}(z_1, z_2)$ the formula published by Saarela et al.[1]

$$W_{\mathrm{H}}(z_1, z_2) = 2\pi \int_0^{\infty} \eta d\eta (g(\eta, z_1, z_2) v(r_{12})$$

$$+ \frac{\hbar^2}{2m}[|\nabla_1\sqrt{g(\eta, z_1, z_2)}|^2 + |\nabla_2\sqrt{g(\eta, z_1, z_2)}|^2]$$

$$- \frac{\hbar^2}{8m}[\nabla_1 g(\eta, z_1, z_2) \cdot \nabla_1 N(\eta, z_1, z_2) + \nabla_2 g(\eta, z_1, z_2) \cdot \nabla_2 N(\eta, z_1, z_2)]$$

$$- \frac{\hbar^2}{8m}\nabla_2 N(\eta, z_1, z_2) \cdot \nabla_2 X(\eta, z_1, z_2)) \tag{9}$$

with

$$r_{12} = \sqrt{\eta^2 + (z_2 - z_1)^2} . \tag{10}$$

The relation between the two-body distribution function $g(\eta, z_1, z_2)$ and $u(\eta, z_1, z_2)$ is provided by the hypernetted equation

$$g(\eta, z_1, z_2) = \exp[u(\eta, z_1, z_2) + N(\eta, z_1, z_2) + E(\eta, z_1, z_2)] . \tag{11}$$

Here the quantities $N(\eta, z_1, z_2)$ and $E(\eta, z_1, z_2)$ are sums of nodal (N) and elementary (E) diagrams. On the other hand, $g(\eta, z_1, z_2)$ is related to $N(\eta, z_1, z_2)$ by the chain equation, for instance see eq.(10) in ref. 1. However, for the planar symmetry one can take advantage of the translational invariance performing a two-dimensional Fourier transform in the plane x-y (usually denoted Hankel transform) of the original chain equation obtaining

$$N(q, z_1, z_2) = \int_{-\infty}^{\infty} dz_3 [X(q, z_1, z_2) + N(q, z_1, z_2)] X(q, z_3, z_2) , \tag{12}$$

where the sum of non-nodal diagrams $X(\eta, z_1, z_2)$ is defined as

$$X(\eta, z_1, z_2) \equiv g(\eta, z_1, z_2) - 1 - N(\eta, z_1, z_2) . \tag{13}$$

This procedure decouples the chain equation for each momentum q parallel to the surface. In this representation all two-body functions $F(q, z_1, z_2)$ are the Hankel transforms of $F(\eta, z_1, z_2)$, which can be evaluated according to

$$F(q, z_1, z_2) = 2\pi \sqrt{\rho(z_1)\rho(z_2)} \int_0^{\infty} \eta d\eta J_0(\eta q) F(\eta, z_1, z_2) , \tag{14}$$

where $J_0(\eta q)$ is the cylindrical Bessel function of the first kind and zero order. The set of coupled eqs.(11)-(13) with $E(\eta, z_1, z_2) = 0$ is known as the hypernetted-chain/0 (HNC/0) approximation. In order to solve the Hartree eq.(6) the Newton-Raphson method has been applied.

The EL equation corresponding to $n = 2$ leads to a PPA equation which can be formulated in terms of the sum of non-nodal diagrams $X(\mathbf{r}_1, \mathbf{r}_2)$. For the planar symmetry considered in this paper the PPA equation can be also decoupled for each momentum q parallel to the surface

$$-\frac{1}{2}[H(q, z_1) + H(q, z_2)] X(q, z_1, z_2)$$

$$+ \frac{1}{2} \int_{-\infty}^{\infty} dz_3 X(q, z_1, z_3) H(q, z_3) X(q, z_3, z_2) = V_{p-h}(q, z_1, z_2) . \tag{15}$$

The quantity $H(q, z)$ is the symmetric one-body operator

$$H(q, z) = \frac{\hbar^2}{2m}[q^2 - \frac{1}{\sqrt{\rho(z)}}\frac{d}{dz}\rho(z)\frac{d}{dz}\frac{1}{\sqrt{\rho(z)}}] \,. \tag{16}$$

Finally, the "particle-hole" $(p - h)$ interaction is

$$V_{p-h}(\eta, z_1, z_2) = g(\eta, z_1, z_2)v(r_{12})$$

$$+ \frac{\hbar^2}{2m}[|\nabla_1\sqrt{g(\eta, z_1, z_2)}|^2 + |\nabla_2\sqrt{g(\eta, z_1, z_2)}|^2]$$

$$+ [g(\eta, z_1, z_2) - 1][\frac{\hbar^2}{4m}(D(1) + D(2))N(\eta, z_1, z_2) + V_C(\eta, z_1, z_2)] \,. \tag{17}$$

Here $D(i)$ is the abbreviated derivative

$$D(i) = \nabla_\eta^2 + \frac{1}{\rho(z_i)}\frac{d}{dz_i}\rho(z_i)\frac{d}{dz_i} \,, \tag{18}$$

where ∇_η^2 is the Laplace operator in the x-y plane. The quantity $V_C(\eta, z_1, z_2)$ is a convolution type contribution which can be conveniently expressed in momentum space as

$$V_C(q, z_1, z_2) = -\frac{1}{2}\int_{-\infty}^{\infty} dz_3 X(q, z_1, z_3)H(q, z_3)X(q, z_3, z_2) \,. \tag{19}$$

Of course, any function $F(\eta, z_1, z_2)$ can be obtained from the inverse Hankel transform of $F(q, z_1, z_2)$.

The FDR algorithm to solve the PPA equation has been already described in detail in ref. 6. It follows the main idea of the iterative procedure devised by Chang and Campbell[9] for solving bulk liquid. The crucial assumption of this method is that the two-body correlation factor is split into two parts

$$u(\eta, z_1, z_2) = u_{SR}(\eta, z_1, z_2) + \Delta u(\eta, z_1, z_2) \,, \tag{20}$$

where $u_{SR}(\eta, z_1, z_2)$ is an analytic function which accounts for the short-range correlations, and $\Delta u(\eta, z_1, z_2)$ is the corrective Jastrow function which is determined numerically and includes the middle- and long-range correlations. Within this framework the sum of non-nodal diagrams can be written as

$$X(\eta, z_1, z_2) = X_{\mathrm{SR}}(\eta, z_1, z_2) + \Delta u(\eta, z_1, z_2). \qquad (21)$$

The optimization of the two-body correlation factor is performed as follows. (i) The algorithm starts adopting an initial guess for the Jastrow function and solving self-consistently the chain eq. (12) and the Hartree eq. (6). For the initial short-range two-body correlation factor a generalization of the expression used by McMillan[10] and Schiff and Verlet[11] is adopted. (ii) Next, a corrective Jastrow function expressed in momentum space, i.e., $\Delta u(q, z_1, z_2)$, is obtained from the solution of the PPA eq. (11) by applying the FDR method. (iii) Next, eqs. (12) and (6) are solved using the new Jastrow function in order to obtain an improved one-body density, the corresponding long-range two-body quantities, and a lower total energy per unit area. The steps (ii) and (iii) of the algorithm are iterated until convergence is achieved.

RESULTS AND DISCUSSION

For the calculations reported in the present work we assumed that helium atoms interact via the improved Hartree-Fock dispersion potential (dubbed HFDHE2) with the parameters determined by Aziz et al.[12] In addition, the atoms are immersed in an external field $U_{ext}(z)$. This external potential is meant to model stable films which should exhibit both following features: (i) a central density near to the experimental equilibrium density of bulk liquid ^4He and (ii) a so wide as possible central region with almost constant density. In order to fulfil these requirements, a somewhat more complicated form for $U_{ext}(z)$ than that used in ref. 6 was adopted

$$U_{ext}(z) = -U_2 \exp[-z^2/(2s^2)] - U_4 \exp[-z^4/(4s^4)] \qquad (22)$$

We shall describe the results obtained for films of $n_z = 0.22$, 0.30 and 0.40 Å^{-2}. The parameters of the potential adopted in each case are listed in table 1. The numerical task was performed on the VAX 11/780 and μVAX-II computers of the Laboratorio TANDAR. Figure 1 shows the one-body densities $\rho(z)$ determined for the different n_z. The corresponding central densities ρ_c, chemical potentials μ, and binding energies per surface area E/L^2 are included in table 1.

Table 1. The parameters U_2, U_4 and s of the external potential $U_{ext}(z)$ adopted to confine the symmetric films with $n_s = 0.22$, 0.30 and 0.40 Å^{-2}. The central density ρ_c, the total energy per surface area E/L^2 and the chemical potential μ obtained for the optimal solution corresponding to each coverage n_s.

	$n_s = 0.22$ Å^{-2}	$n_s = 0.30$ Å^{-2}	$n_s = 0.40$ Å^{-2}
U_2 (K)	16.0	7.0	4.0
U_4 (K)	0.0	7.0	8.0
s (Å)	3.3	5.0	6.0
ρ_c (Å^{-3})	0.02159	0.02155	0.02157
E/L^2 (K/Å^2)	-3.0854	-4.3523	-5.1874
μ (K)	-9.8607	-10.1118	-8.1911

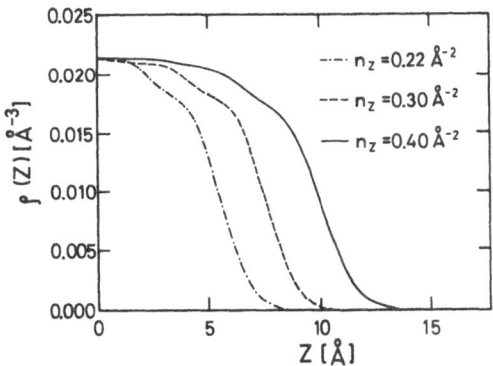

Figure 1. Density profile of the ^4He films for the coverages $n_s = 0.22$, 0.30, and 0.40 Å^{-2}. The density profile is symmetric at $z = 0$.

The central densities are almost the same for all the cases, being approximately equal to $\rho_c = 0.0216$ Å^{-3}. In addition, as expected, for increasing coverages the smooth central part of the film becomes broader. The changes of curvature exhibited by all three profiles just before the fall off are caused by the large strength of the external potential required to model films with ρ_c near to the experimental equilibrium density $\rho_{exp} = 0.0218$ Å^{-3}. It happens that the potentials are just strong enough to begin to build up a layer structure of the profiles. In what follows we shall analyze the solutions obtained for the two-body distribution function.

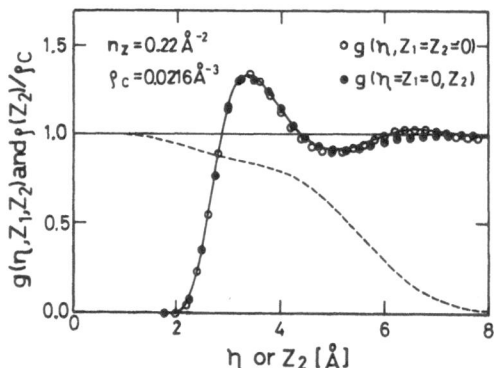

Figure 2. Two-body distribution function for $n_s = 0.22$ Å^{-2}. One particle is located at the center of the film, $z_1 = 0$ Å, and other removed in the direction parallel to the surface (open circles) and orthogonal to the surface (filled circles). The solid line is $g(r_{12})$ for bulk liquid ^4He of density $\rho = 0.0216$ Å^{-3}. The dashed line represents $\rho(z_2)/\rho_c$.

The Two-Body Distribution Function

Let us now focus our attention on the results for $g(\eta, z_1, z_2)$. It is known that LDA predicts anisotropy of two-body functions. In this approach, one supposes that any two-body quantity $F(\eta, z_1, z_2)$ corresponding to an inhomogeneous fluid can be approximated by that calculated for a uniform system of a hypothetical average density, which is usually assumed to be equal to the arithmetic or the geometric mean of the local densities at z_1 and z_2. However, there is no solution of the HNC/0 equations for bulk ^4He of density below the spinodal value $\rho_{sp} \approx 0.016$ Å^{-3}. This fact leads to an important limitation for the LDA, since it cannot be applied to treat a large part of the fall off of profiles unless one adopts a further approximation as in ref. 1.

Figures 2 and 3 show the behavior of $g(\eta, z_1 = z_2 = 0)$ and $g(\eta = z_1 = 0, z_2)$ for the coverages $n_z = 0.22$ and 0.30 Å^{-2}. To make possible a direct visual comparison, we also display in all these graphs the bulk $g(r_{12})$ calculated at density $\rho = \rho_c = 0.0216$ Å^{-3}. From these figures it is apparent that the function $g(\eta, z_1 = z_2 = 0)$ always agrees with that evaluated for the uniform system. This result is to be expected according to LDA, since in this case both particles are always located at places of density ρ_c. Moreover, by studying $g(\eta, z_1 = z_2 = z)$ as a function of η at several fixed values of z, it was also found a good agreement with results obtained for bulk helium at $\rho = \rho(z)$. The nearest-neighbor peak exhibited by these functions is weaker and shifted towards larger η for decreasing density, a feature that has already been shown by Krotscheck et al.[2] in their fig. 3.

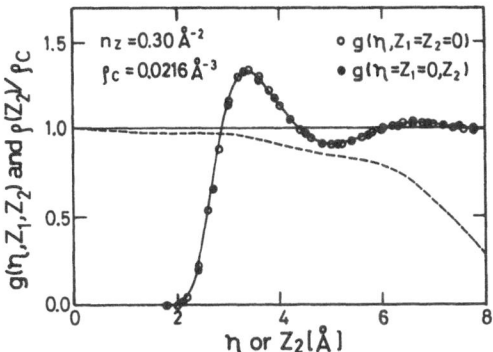

Figure 3. Same as fig. 2 for $n_s = 0.30$ Å^{-2}.

Returning to figs. 2 and 3 one can see that $g(\eta = z_1 = 0, z_2)$ does not always coincide with the bulk values. In this respect, fig. 1 shows that the distributions as a function of η and z_2 are not equal. The difference being noticeable for distances greater than 4 Å. This is a case of anisotropy, which was not discussed by Krotscheck and co-workers because it is not clearly exhibited in fig. 5(c) of ref. 2. In fact, this behavior is to be expected from the LDA and in the following lines we shall trace a qualitative interpretation. In the case of $g(\eta = z_1 = 0, z_2)$ one particle is situated in the region of density ρ_c, while the other is located at sites of lower density leading to mean densities smaller than the central one. In order to facilitate the explanation of this anisotropy, we displayed in figs. 2 and 3 the ratio $\rho(z)/\rho_c$. The first overshoot of the two-body distribution function occurs at about $z_2 = 3.4$ Å, where the density $\rho(z_2)$ takes the values 0.0182 (84%), 0.0205 (95%) and 0.0211 Å^{-3} (98% of the central density) for $n_z = 0.22$, 0.30 and 0.40 Å^{-2}, respectively. Hence the geometric mean density

$$\bar{\rho} = \sqrt{\rho_c \rho(z_2)} \tag{23}$$

is, for the same sequence of coverages, equal to 0.0198 (92%), 0.0210 (97%) and 0.0213 Å^{-3} (99% of the value at the center). Due to the fact that $\bar{\rho}$ and ρ_c are very similar it is hard to see the anisotropy on the scale of the figures even in fig. 2, where the difference is the largest one. Of course, the effect becomes greater for larger z_2. For instance, the first minimum lies at about $z_2 = 5.2$ Å, where the densities $\rho(z_2)$ for increasing n_z are 0.0116 (54%), 0.0181 (84%) and 0.0204 Å^{-3} (95% of the central density), yielding geometric mean densities $\bar{\rho} = 0.0158$ (73%), 0.0198 (92%) and 0.0210 Å^{-3} (97%), respectively. Therefore, according to the LDA, in this region it is reasonable to expect an importantant deviation from the results of the bulk helium for $n_z = 0.22$ Å^{-2}, a much smaller anisotropy for $n_z = 0.30$ Å^{-2} and a negligible one for $n_z = 0.40$ Å^{-2}. The data depicted in figs. 2 and 3 are in accordance with these qualitative predictions.

On the other hand, for slices parallel to the surface, where the density is lower than the spinodal point, $g(\eta, z_1 = z_2 = z)$ as a function of η, for fixed z, exhibits a broad maximum instead of the usual nearest-neighbor peak. In addition, similar anisotropy to that shown in fig. 5(a) of ref. 2 is also observed.

The Normal-Modes

In this section we shall look at some properties of the normal-modes of the density fluctuations. The PPA eq.(15) can be rewritten as an eigenvalue problem

$$H^2(q, z_1)\psi_l(q, z_1) + 2\int_{-\infty}^{\infty} dz_2 V_{p-h}(q, z_1, z_2) H(q, z_2)\psi_l(q, z_2) = \hbar^2 \omega_l^2 \psi_l(q, z_1) . \tag{24}$$

The eigenstates of this equation are related to the normal-modes. It has been demonstrated in ref. 7 that for the optimal two-body functions eq.(24) provides the same solutions as

$$H(q, z_1)\psi_l(q, z_1) - \int_{-\infty}^{\infty} dz_2 X(q, z_1, z_2) H(q, z_2)\psi_l(q, z_2) = \hbar \omega_l \psi_l(q, z_1) . \tag{25}$$

The eigenfunctions $\psi_l(q, z)$ are real and can be normalized according to

$$< \psi_l(q, z)|\psi_l(q, z) >= \int_{-\infty}^{\infty} dz \psi_l^2(q, z) = 1 . \tag{26}$$

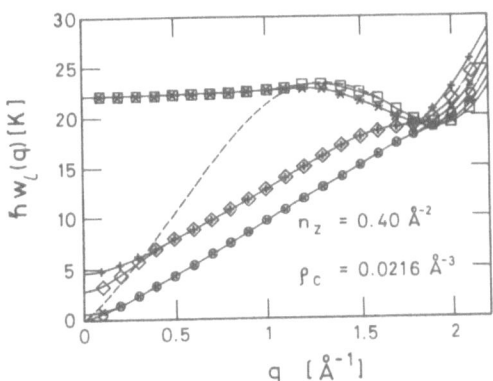

Figure 4. Eigenvalues $\hbar\omega_l(q)$ for $n_z = 0.40$ Å^{-2}. These results correspond essentially to surface modes (circles, crosses, rombs and pluses) and volume modes (squares and stars). The dashed line indicates the excitations calculated with eq.(27).

They form a complete set of functions, but they are not orthogonal in the usual way (see refs. 2 and 6). In fig. 4 we plotted some representative eigenvalues as a function of q together with the excitation energies for a uniform system calculated with the Feynman's[13,14] formula

$$\epsilon_F(q) = \frac{\hbar^2 q^2}{2mS(q)}, \tag{27}$$

where $S(q)$ is the static form factor. One sees that there is a level crossing at about the roton minimum. Below $q = 1.9$ Å^{-1} the two lowest-lying eigenstates are essentially surface modes and above this momentum value the lowest-lying eigenstates are volume modes as it was clearly established in refs. 3 and 7. In the present work it was possible, for the first time, to identify the bulk mode for momenta below the roton minimum, some examples of this kind of eigenfunctions are displayed in fig. 5. It is interesting to note the fact that in fig. 4 the eigenvalues corresponding to the even volume mode coincide with the results provided by eq.(27) for $q > 1.1$ Å^{-1}.

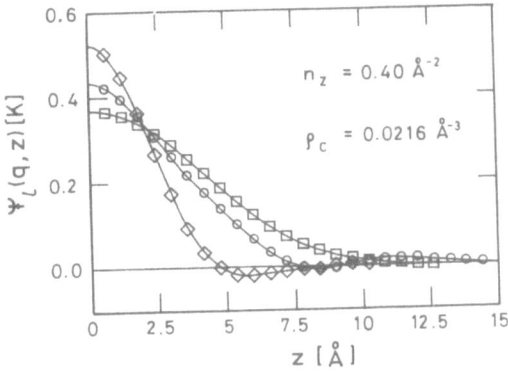

Figure 5. Eigenfunctions $\psi_l(q,z)$ for the bulk mode for $q = 1.2$ Å^{-1} (rombs), $q = 1.7$ Å^{-1} (circles) and $q = 2.2$ Å^{-1} (squares).

SUMMARY

In conclusion we can state that the analysis of two-body distribution function indicates that the results at the center of the film reproduce satisfactorily well those evaluated for homogeneous helium. In addition, both kind of anisotropy, i.e., that found by Krotscheck *et al.*[2] at density lower than the spinodal point and that predicted by the LDA, are also obtained. On the other hand, it was possible to follow the volume mode for momenta bellow the roton minimum.

ACKNOWLEDGEMENTS

One of the authors (LS) is Member of the Carrera del Investigador Científico of the Consejo Nacional de Investigaciones Científicas y Técnicas. This work has been supported in part by Fundación Antorchas, Argentina.

REFERENCES

1. M. Saarela, P. Pietiläinen, and A. Kallio, Phys. Rev. $B27$:231 (1983).

2. E. Krotscheck, G.-X. Qian, and W. Kohn, Phys. Rev. $B31$:4245 (1985).

3. E. Krotscheck, Phys. Rev. $B31$:4258 (1985).

4. E. Krotscheck, Phys. Rev. $B32$:5713 (1985).

5. J.L. Epstein and E. Krotscheck, Phys. Rev. $B37$:1666 (1988).

6. L. Szybisz and M.L. Ristig, Phys. Rev. $B40$:4391 (1989).

7. L. Szybisz, Phys. Rev. $B41$:11282 (1990).

8. E. Feenberg, Theory of Quantum Fluids (Academic, New York, 1969).

9. C.C. Chang and C.E. Campbell, Phys. Rev. $B15$:4238 (1977).

10. W.L. McMillan, Phys. Rev. 138:A442 (1965).

11. D. Schiff and L. Verlet, Phys. Rev. 160:208 (1967).

12. R.A. Aziz, V.P.S. Nain, J.S. Carley, W.L. Taylor, and G.T. McConville, J. Chem. Phys. 70:4330 (1979).

13. R.P. Feynman, Phys. Rev. 94:262 (1954).

14. R.P. Feynman and M. Cohen, Phys. Rev. 102:1189 (1956).

COMPRESSIBLE RAYLEIGH-BENARD CONVECTION

IN A HARD DISKS SYSTEM

Dino Risso* and Patricio Cordero†

Departamento de Física
Facultad de Ciencias Físicas y Matemáticas
Universidad de Chile, Casilla 487, Santiago 3, Chile

1. SUMMARY

Rayleigh-Bénard convection in systems of a few thousand hard disks was obtained for the first time in [MK86, MK87]. This stimulating development has been followed by further results on the same line: [Ra88, MMPK88, PMM89] .

In the present work we report our own study of the behavior of a hard disk systems. In our computer experiments a two dimensional fluid is simulated through the dynamics of 1521 hard disks that move inside a square box under the influence of gravity and a temperature gradient pointing upwards. Under appropriate conditions Rayleigh-Bénard convection is clearly observed. See figure 1.

We have observed convection even for Rayleigh numbers well below the critical values predicted by the Oberbeck-Bussinesq approximation of the Navier-Stokes equation suggesting that this effect is due to the compressibility of the system.

In the simulations the disks collide elastically and when they hit a wall, the tangential component of the velocity is preserved. Namely, we have set stress-free boundary conditions in all four walls. The collisions with the lateral (vertical) walls are perfectly elastic, while the collision with the floor and roof simulate contact with heat baths at temperatures T_b at the bottom and T_t at the top. We have been able to achieve our simulations thanks to the efficient *event list scheduling* algorithm developed by Rapaport [Ra80].

*Partially supported by grant *FONDECYT* 90-0005 and Project Fundación Andes C-10003.
†Partially supported by grant *FONDECYT* 90-1240.

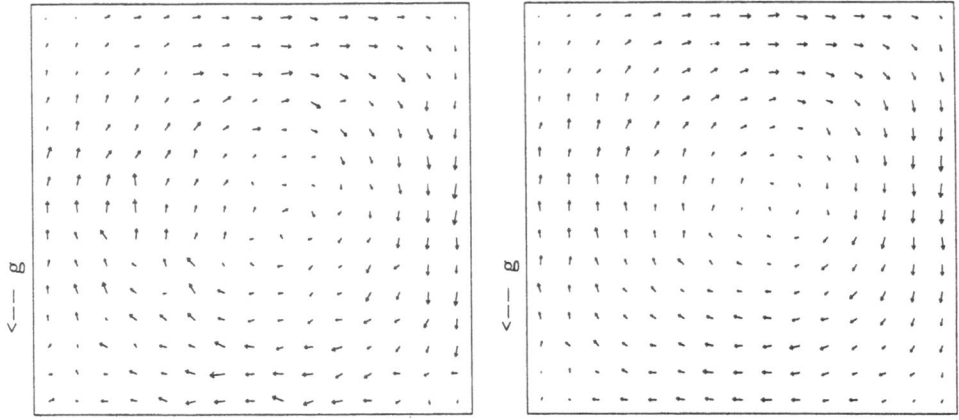

Figure 1. A typical velocity field obtained with the simulations. The right figure corresponds to an average over the last 10 million collisions of a total of 20 million while the left figure corresponds to an average over the last million collisions.

After an introductory §2, section 3 is devoted to describe with some details the conditions under which we performed our simulations. §4 summarizes our results.

2 INTRODUCTION

The study of the fluid state of matter has a long and rich history, from both the experimental and theoretical standpoints. But in the last decades it has also been possible to make computational studies: numerical integrations of the Navier-Stokes equation [Fl87, PT90] and references therein as well as microscopic studies through simulation using Monte Carlo techniques or molecular dynamics [CFMc87, AT87].

2.1 Rayleigh-Bénard convection. Bénard convection occurs when there is a temperature gradient pointing against gravity. If the size of the temperature gradient is above a certain threshold the upwards buoyancy force acting on the lower (hot and less dense) layers overcomes the drag force due to viscosity. Simultaneously the upper cooler and denser layers are subject to the reciprocal effect. Under favorable circumstances a non-trivial stationary situation may be reached in the sense that macroscopic nonuniform velocity, temperature, density fields set up.

Bénard made the first quantitative experiments in 1900 and Rayleigh in 1916 formulated the theory of convective instabilities . In his original formulation Rayleigh was able to prove that convection exists only if a certain dimensionless parameter called the *Rayleigh number* has a value above a critical threshold. In his derivation Rayleigh applied the Bussinesq approximation (due independently to Oberbeck (1879) and Bussinesq(1903). An account of the history of this early discoveries can be seen in [NP77]. In this approximation it is assumed that the temperature and the density vary little, yet in which the buoyancy drives the motion. The density variation is neglected everywhere except in the buoyancy term of the Navier-Stokes equations.

On convection see the important book by Chandrasekhar [Ch61] as well as [NP77] and [SG85]. In particular see the article by Busse in the last reference.

2.2 Fluid description.

There are two rich and complementary ways to picture fluid motion. The kinetic picture (many simple atomic elements colliding rapidly with simple interactions) [RL69, M76] and the Navier-Stokes partial differential equations (which relies on the idea of the continuum) [LL59].

2.2.1 Continuous description.

Convective flow has traditionally been studied using the Navier-Stokes equations. The description is precise but incomplete. Constitutive relations for stress tensor, heat flux, etc. must be supplied [LL59]. When Newtonian fluids are considered transport coefficients like viscosity η, bulk viscosity ξ and thermal conductivity κ, are incorporated using linear constitutive relations [LL59, M76]. These transport coefficients must be derived from a different theory or supplied as an empirical relation. No information about this can be deduced from the continuous approximation itself.

There is no analytic method of integrating Navier-Stokes equations for the convective flow. Single component fluids and the Oberbeck-Boussinesq approximation have been introduced to simplify the study of convection in Newtonian fluids [SV60]. Here the range of problems studied goes from Rayleigh-Bénard convection with stress-free and rigid boundary conditions [Cha61, NP77] to time-dependent [SH88] and corrugated geometry boundary conditions [B91]. Maxwellian fluid convection which is of interest in geology has recently been studied [H90].

2.2.2 Molecular Dynamics description.

For about three decades, most applications of molecular dynamics focused in the study of equilibrium properties of fluids, solids, etc. In the pioneer work of Alder and Wainwright [AW57, AW59] they worked with a system of hard spheres. In their simulation the spheres move at constant speed between perfectly elastic collisions. It took some time before a successful attempt was made to tackle the problem of a set of points interacting with a Lennard-Jones potential between each other [R64]. From then on a large sequence of important contributions has been made for increasingly complicated molecular systems. For a complete list of references see [AT87].

In present days, thanks to the availability of powerful computers, it has been possible to begin simulations of inhomogeneous and/or far from equilibrium systems.

The first large size molecular dynamic studies (of the order of 10^5 particles) were [RC86, M86] who studied the flow past an obstacle.

The first molecular dynamics simulation where Rayleigh-Bénard convection was achieved is in [MK86, MK87]. They studied a two-dimensional system of a few thousand hard disks. Soon later a similar result was obtained in [Ra88] (see also [MK89]).

The problem of boundary macroscopic conditions which arise from the characteristics of the interaction of the fluid with: (a) the walls that surround the system or (b) a different fluid, has been studied also using molecular dynamics [TR89, KBW88, KBW89].

3 THE SIMULATIONS

In the following paragraphs we describe with some detail the boundary and initial conditions under which we carried out our simulations. Essentially these are: isolating boundary conditions on the vertical walls and fixed temperature on the horizontal walls. In all the walls there are stress-free boundary conditions. Furthermore the initial state is chosen so that the stationary situation is reached in a shorter time. The parameters describing the system are also given.

3.1 Boundary conditions. In the simulations reported here the collisions at lateral boundaries are specularly reflecting. Hence no heat can be exchanged through lateral walls and no forces tangential to the wall are exerted to the fluid. This collision rule corresponds to macroscopic stress-free and thermal isolating boundary conditions.

To impose a vertical temperature gradient the collisions at the upper and lower boundaries simulate contact with a heat bath at temperatures T_t and T_b. In these collisions particles bounce preserving the horizontal component of their velocity but the new vertical component is taken from a Maxwellian distribution.

3.2 The initial state. The disks are initially distributed uniformly across a rectangular region with some initial velocities. Due to the large thermal diffusion time the initial disk velocities are chosen such that the horizontal average kinetic energy profile is linear between the upper and lower temperatures as in a purely conductive regime.

3.3 Units and parameters. The type of simulations that we are describing are performed with N hard disks of mass m and diameter D. They collide elastically with each other and move under the effect of gravity. The N disks are in a rectangular box of width L_X and height L_Z. The temperatures at the upper and lower walls are T_t and T_b.

The units of length, mass, temperature and time are chosen so that the disk diameter, the disk mass, the Boltzmann's constant and the temperature at the bottom horizontal boundary are set equal to 1.

With these parameters it is useful to define the density number $n = \frac{N}{L_X L_Z}$ and the aspect ratio $\lambda = \frac{L_Z}{L_X}$.

In our simulations we have fixed the number of disks to be $N = 1521$, $\lambda = 1$ and $n = 0.25$. This leaves at our disposal the gravitational acceleration g and the top temperature T_t.

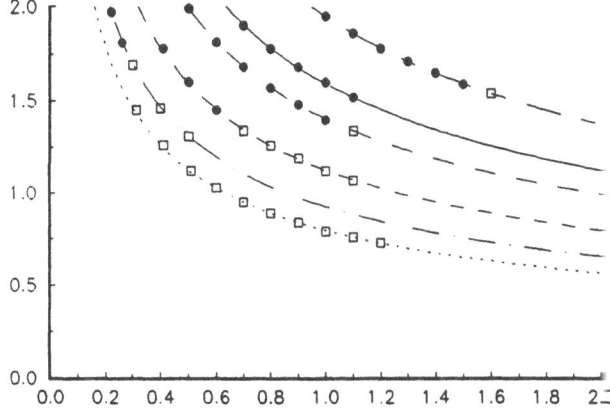

Figure 2. Summary of experimental observations. The vertical axis is the reduced temperature gradient $\frac{\Delta T}{T}$ and in horizontal axis is the Froude number. The lines are curves of equal Rayleigh number (from bottom to top we have Rayleigh number curves of 250, 338, 500, 780, 1000 and 1500). Circles are points where developed convection is observed. Boxes represent points where convection is not developed.

Further we define the 'experimental' Rayleigh number Ra

$$Ra = (\frac{\Delta T}{T})^2 \frac{N\,Fr}{\lambda}\,J(n). \tag{1}$$

$J(n)$ is a function of the density number and involves the viscosity and the thermal conductivity transport coefficients (see the Appendix). Transport coefficients are evaluated using Enskog's hard disk theory [DG71]. T is the mean temperature $\frac{1}{2}(T_t + T_b)$ and $\Delta T = (T_b - T_t)$ is the imposed temperature gradient.

The Froude number Fr — which measures the ratio between the potential energy gained by a particle raised from the bottom to the top of the box over the kinetic energy release — is defined by

$$Fr = \frac{g\,L_z}{\Delta T}. \tag{2}$$

To distinguish our different experiments, then it suffices to give the values of (Ra, Fr).

The gravity g and the top temperature T_t are fixed through (1) and (2) choosing different values for the Rayleigh and Froude numbers. The Rayleigh and Froude numbers considered in our simulations are (250, 338, 500, 780, 1000, 1500) and (0.2, 0.3, 0.4, ... 1.6) respectively. A Froude number of about 1.56 corresponds to the Bussinesq approximation as it can be seen in the Appendix. For each Rayleigh number considered there is a lowest Froude number that can be reached which corresponds to the maximum possible temperature gradient ($\Delta T = 1$).

3.4 Simulation algorithm. The hard particle molecular dynamics simulations are *event-driven* in the sense that the particles move free of interparticle interaction except for the moments when particles experience an impulse interaction due to their having become in contact either with another particle or with a wall. The movement of each particle between events is dictated by Newton's equation with whatever external force (e.g., gravity) may exist. The time intervals by which the simulation proceeds are the increments between events, and these are determined by the system itself.

The strategy to make efficient computer programs is relatively new. Verlet [Ve67] suggested a technique for improving the speed of a program by maintaining a list of neighbors of each molecule which is updated at intervals. Later it was suggested [EW77] maintaining as efficiently as possible a table of future collisions between pairs of particles. But the most important improvement was given by Rapaport [Ra80]. He designed a highly efficient *event list scheduling* on which we have based the code for our simulations.

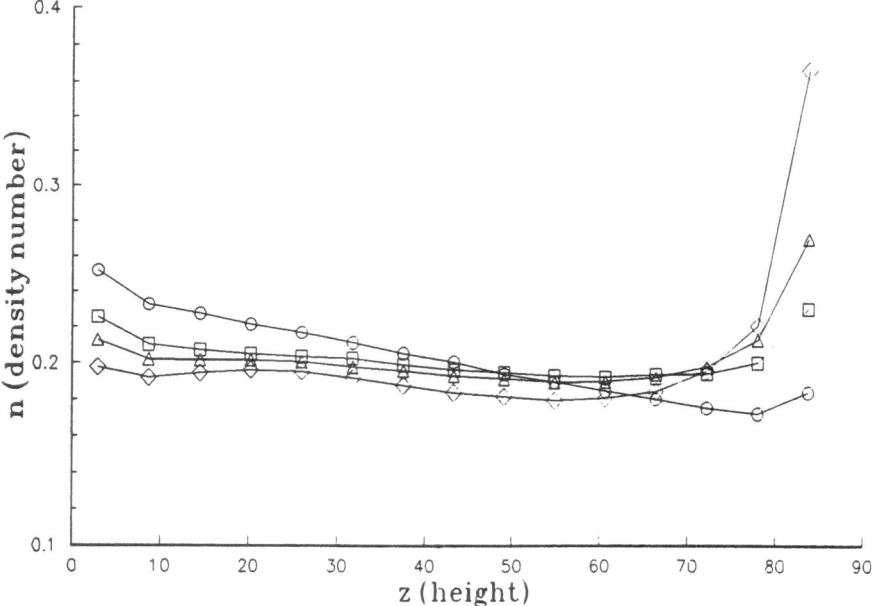

Figure 3. Horizontal averaged density-number profile for different Froude numbers. Diamonds, triangles, boxes and circles correspond to Froude numbers 1.0, 1.1, 1.2 and 1.6 respectively. The Rayleigh number is 1500. In the cases with Froude numbers of 1.0, 1.1 and 1.2 there is convection and a pronounced top boundary layer is observed.

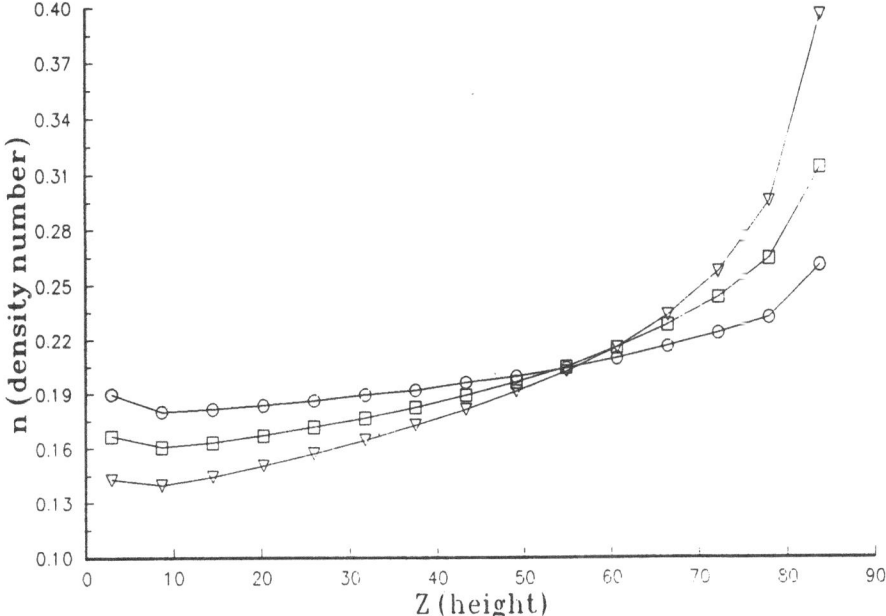

Figure 4. Horizontal averaged density-number profile for different Froude numbers. Triangles boxes and circles correspond to Froude numbers 0.25, 0.40 and 0.50. The Rayleigh number is 338. The case with Froude number of 0.25 correspond to a convective situation.

The code was run in an AViiON 200 Work Station and a DEC 5400 computer. The collision rate of our code being 3.8×10^6 and 5.1×10^6 collisions/hour of central processor respectively.

3.5 Measurements. Because the observables we want to see are macroscopic fields as in Navier-Stokes equations, local-spatial and temporal macroscopic averages must be done.

With this aim the system is divided in 15×15 cells which means about 6 disks per cell. A disk number large enough to define spatial-averages within each cell and small enough to have sufficiently many cells to be able to detect eventual pattern formation.

Spatial measurements of density, velocity, temperature, and pressure in each cell are taken at regular intervals of about 20 collisions per particle. These spatial measurements are time-averaged every one million collisions (1300 collisions per particle). This gives about 70 spatial measurements to be averaged.

Roughly speaking there are two time scales to be considered: flight time between collisions, and hydrodynamics time. Hydrodynamics time being of the order of the thermal diffusion time

$$\tau_d = \frac{L_Z^2}{\kappa} = \frac{N}{\kappa n}.$$

Here κ is the thermal diffusivity (see Appendix). In our simulations this means about one million total collisions.

In every simulation the system was left to evolve during a total of 20 million collisions.

4. THE RESULTS

In order to control the pattern evolution, for each one-million-collision space-time-average of the velocities, we evaluated the *circulation of the velocity field* in concentric square paths. A graphical representation of the velocity field was also generated at these intervals (as in figure 1).

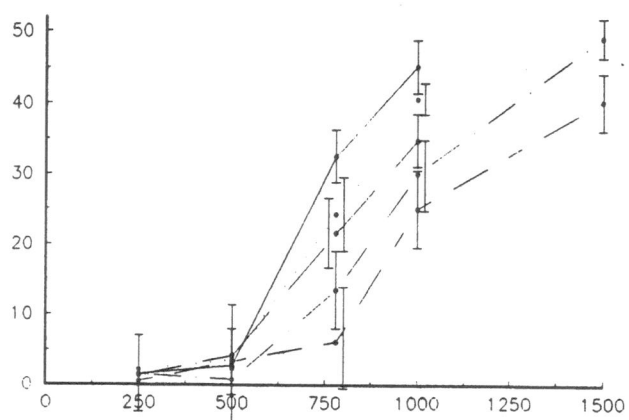

Figure 5. Averaged accumulated circulation versus Rayleigh number for different Froude numbers. From bottom to top and from right to left the curves correspond to Froude numbers 0.70, 0.80, 0.90, 1.00 and 1.10. Error bars correspond to r.m.s. time fluctuations.

We have added the circulation of the different paths and averaged them over the last 10 million collisions. The values for these averaged accumulated circulations obtained for most of the cases that we have studied in this way can be seen in the Table. To decide whether there was a stable convective pattern we observed the circulation values before averaging. The cases when the circulation changed sign (or did not change sign) were classified as non-convective (or convective). In figure 2 these two situations are shown in the Froude-Rayleigh plane including the $Ra = 338$ case which we did not include in the table.

Figures 3 and 4 show the density change with height in several cases. In the cases that appear in figure 3 the density in the bulk varies little and one could say that the Oberbeck-Bussinesq approximation applies while in figure 4 the density changes appreciably with height. In both figures there are convective situations but in the

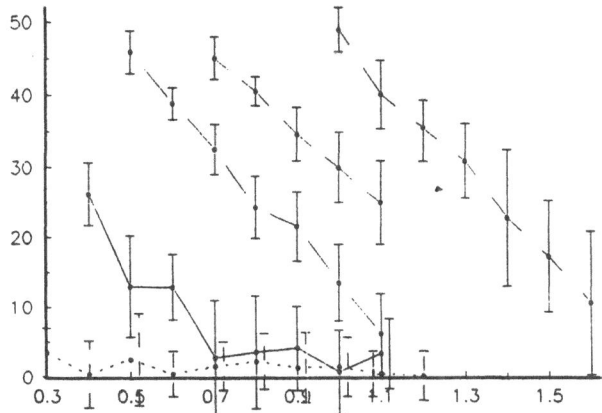

Figure 6. Average accumulated circulation versus Froude number for different Rayleigh numbers (from bottom to top and from left to right we have curves of Rayleigh number 250, 500, 780, 1000 and 1500). Error bars correspond to r.m.s. time fluctuations.

cases of figure 4 (rather small Rayleigh number) convection is seen only when the Froude number is quite small. See also figure 2.

In figure 5 the averaged accumulated circulation is plotted against Ra for different Froude numbers and it is possible to see that there is a larger circulation for smaller Froude number. In figure 6 we present the same information but plotting the averaged accumulated circulation against Froude number.

Table 1

Averaged accumulated circulation for different values of the Froude and Rayleigh numbers.

Fr	$Ra = 250$	$Ra = 500$	$Ra = 780$	$Ra = 1000$	$Ra = 1500$
0.3	3.48 ± 3.42				
0.4	0.40 ± 5.47	26.17 ± 4.17			
0.5	2.54 ± 5.25	12.96 ± 7.95	45.91 ± 2.25		
0.6	0.46 ± 3.06	12.84 ± 3.97	38.81 ± 1.93		
0.7	1.55 ± 5.17	2.76 ± 8.12	32.49 ± 3.12	45.07 ± 2.78	
0.8	2.22 ± 3.83	3.53 ± 7.73	24.28 ± 4.07	40.51 ± 1.70	
0.9	1.35 ± 3.77	4.15 ± 5.62	21.61 ± 4.51	34.68 ± 3.06	
1.0	1.43 ± 3.87	0.72 ± 5.97	13.46 ± 4.56	29.99 ± 4.05	49.02 ± 2.02
1.1	0.45 ± 3.12	3.34 ± 4.41	6.17 ± 6.59	24.99 ± 5.16	40.15 ± 3.87
1.2	0.18 ± 3.00				35.57 ± 3.09
1.3					30.91 ± 4.31
1.4					22.79 ± 8.79
1.5					17.25 ± 6.77
1.6					10.57 ± 10.35

APPENDIX

In what follows we summarize the expressions for the hydrodynamics parameters in terms of the basic parameters defined in §3.3

The Rayleigh number Ra [VN80] is defined through:

$$Ra = \frac{g \, n \, L_z^3 \Delta T \alpha}{\nu \kappa} \tag{A1}$$

where ν is the kinematic viscosity, κ is the thermal diffusivity, and α the thermal expansion constant.

Rigid disk fluid transport coefficients were studied by [DG71] using Enskog theory [M76, RL69]. When specialized to unitary disk mass, disk diameter and Boltzmann's constant, the kinematic viscosity ν and thermal diffusivity κ are

$$\nu = \frac{\eta_0 \sqrt{T}}{n} = 0.2555 \sqrt{\pi} (1 + \frac{2}{\pi n \chi} + 0.4364 \pi n \chi) \sqrt{T} \tag{A2}$$

$$\kappa = \frac{\kappa_0 \sqrt{T}}{n} = 1.029 \sqrt{\pi} [\frac{3}{2} + \frac{2}{\pi n \chi} + 0.4355 \pi n \chi] \frac{1}{c_v} \sqrt{T} \tag{A3}$$

here we have taken a constant volume process, $\chi(n)$ is the Enskog factor and the specific heat capacity in the units defined in §3.3 is $c_v = 1$ [M76].

The thermal expansion coefficient α and the Enskog factor χ may be evaluated using the empirical formula of Henderson [H75] for the state equation:

$$\frac{p}{nT} = H(n) = \frac{1 + \frac{(\pi n)^2}{128}}{(1 - \frac{\pi n}{4})^2} \tag{A4}$$

the thermal expansion coefficient and the Enskog factor are (the prime denotes derivation respect to the argument):

$$\alpha = \frac{\alpha_0(n)}{T} = \frac{1}{T} \frac{1}{1 + nH'(n)/H(n)} \tag{A5}$$

$$\chi(n) = \frac{1 - 7\frac{\pi n}{64}}{1 - \frac{\pi n}{4}}. \tag{A6}$$

Using relations (A1) to (A6), $n = \frac{N}{L_x L_z}$ and $\lambda = \frac{L_z}{L_x}$ the Rayleigh number can be written as in (1). Here we have introduced the Froude number $Fr = \frac{gL_z}{\Delta T}$ and the function [PMM89]

$$J(n) = \frac{n\alpha_0(n)}{\eta_0(n)\kappa_0(n)}. \tag{A7}$$

Top temperature T_t then is

$$T_t = T_b \frac{\sqrt{Fr} - \sqrt{Fr_{min}}}{\sqrt{Fr} + \sqrt{Fr_{min}}} \qquad (A8)$$

where $Fr_{min} = \frac{Ra\lambda}{4NJ(n)}$ is the minimum Froude number.

Boussinesq approximation corresponds to taking Froude number $Fr \approx H(n)$ as follows from (A4) and the hydrostatic equilibrium equation $\nabla p = -\mathbf{n} g \hat{z}$ (which holds near the onset of convection). We get

$$\frac{\partial n}{\partial z} = -n \left(\frac{\partial T}{\partial n} + \frac{g}{H(n)} \right) \frac{\alpha_0(n)}{T}$$

in Boussinesq approximation density variations are considered small, $n \approx \bar{n}$ (the averaged density) and $\frac{\partial n}{\partial z} \approx 0$ hence

$$\frac{\partial T}{\partial z} \approx -\frac{g}{H(\bar{n})}$$

and $Fr = \frac{gL\mathbf{z}}{\Delta T} \approx H(n)$. For the density number of our experiments this corresponds to a Froude number of 1.56.

REFERENCES

AT87 M P Allen and D J Tildesley *Computer Simulation of Liquids* Oxford Science Pub, 1987

AW57 B J Alder and T E Wainright, J Chem Phys 27 (1957) 1208

AW59 B J Alder and T E Wainright, J Chem Phys 31 (1959) 459

B91 J K Bhattacharjee, Phys. Rev. A, 43 (1991) 819

CFMc87 G Ciccotti, D Frenkel, I R McDonald, *Simulation of Liquids and Solids* North-Holland, 1987

Cha61 S Chandrasekhar, *Hydrodynamic and Hydromagnetic stability* Clarendon Press, Oxford, 1961.

DG71 D Gass, J. Chem. Phys., 54 (1971) 1898

EW77 J J Erpenbeck and W W Wood in *Statistical mechanics B. Modern Theoretical Chemistry.* B J Berne ed., Vol 1. Plenum NY 1977

Fl87 C A J Fletcher *Computational Techniques for Fluid Dynamics* Vols. 1 & 2, Springer-Verlag, 1987

H75 D Henderson Mol. Phys. 30 (1975) 971

H90 H Harder, J. Non-Newtonian F Mech, 39 (1991) 67

KBW88 J Koplick, J R Banavar, J F Willemsen, Phys. Rev.Lett. 60 (1988) 1282

LL59 L D Landau and E M Lifshitz *Fluid Mechanics* Pergamon, 1959

M76 D A McQuarrie (1976) *Statistical Mechanics*, Harper & Row, 1976

M86 E Meilburg, Phys. Fluids 29 (10), October 1986

MK84 M Mareschal and E Kestemont, Phys Rev A 30 (1984) 1158

MK86	M Mareschal and E Kestemont, Nature 329 (1986) 427
MK87	M Mareschal, E Kestemont, J. Stat. Phys., 48, (1987) 1187
MK89	M Mareschal and E Kestemont, Phys Rev Lett 62 (1989) 691; D C Rapaport, Phys Rev Lett 62 (1989) 692
MMPK88	M Mareschal, M Malek-Mansour, A Puhl and E Kestemont, Phys. Rev. Lett., 61 (1988) 2550
NP77	C Normand and Y Pomeau, Rev. Mod. Phys. 49, 3 (1977) 581
PMM89	A Puhl, M Malek-Mansour and M Mareschal, Phys. Rev. A, 40 (1989) 1999
PT90	R Peyret and T D Taylor *Computational Methods for Fluid Flow* Springer-Verlag, 1990
R64	A Rahman, Phys Rev 136A (1964) 405
Ra80	D C Rapaport, J Comp Phys 34 (1980) 184
Ra88	D C Rapaport, Phys. Rev. Lett., 60 (1988) 2480
RC86	D C Rapaport and E Clementi, Phys. Rev. Lett. 57 (1986) 695
RL69	L Liboff *Introduction to the Therory of Kinetic Equations* Wiley NY (1969)
SG85	H L Swinney and J P Gollub eds. *Hydrodynamic Instabilities and the Transition to Turbulence* Springer-Verlag, 1985
SH88	J B Swift and P C Hohenberg, Phys. Rev. A, 39 (1988) 4132
SV60	E A Spiegel and G Veronis, Astrophys. J. 132 (1960) 716
TR89	P A Thompson and M O Robbins, Phys.Rev.Lett. 63 (1989) 766
Ve67	L Verlet Phys Rev 159 98 (1967)
VN80	M G Velarde and C Normand Sci. Am. 243, (1980) 78

POSITRON ANNIHILATION AS A PROBE OF LOCALIZED STATES IN FLUIDS

Bruce N. Miller

Department of Physics, Box 32915
Texas Christian University
Fort Worth, Texas 76129, U. S. A.

ABSTRACT

This paper concerns the properties of localized states of a light particle (e.g. electron, positron or positronium atom) thermalized in a fluid. In contrast with the more familiar Anderson localization which occurs in solids, in a fluid the atoms are free to rearrange their positions in the vicinity of the light particle and therefore participate in the formation of the average quantum state. Because electron localization affects the mobility of the charge carrier more than any other single factor, it has important consequences for charge transfer. In the following I will describe the experimental evidence for the existence of localized states and how positron annihilation provides a useful window for observing their properties. I will then introduce the theoretical models which have proven helpful for predicting the equilibrium structures and compare their relative merits. As an example, I will conclude with new results obtained from an application of the path integral to the behavior of a positron in Xenon.

INTRODUCTION

Experimental Evidence for Localization

The mobility of a charged particle in a fluid determines its mean drift in an electric field. It is the essential parameter governing the transport of electric charge. Experimental studies of electrons in fluids show that moderate changes in the mean density or temperature of the fluid can result in order of magnitude variations in the electron's mobility for certain ranges of these parameters, usually in a broad region of the liquid-vapor critical point.[1] The accepted explanation for the attenuation in mobility is that, for a particular range of <u>average</u> fluid density and

Condensed Matter Theories, Vol. 7, Edited by A.N. Proto
and J.L. Aliaga, Plenum Press, New York, 1992

temperature, the electron becomes localized or trapped in a region of altered local density. Upon altering the density, the electron sits in the ground state of the potential well it has created. This type of localization is called self-trapping because the electron plays a role in determining the environment it experiences. When self-trapping occurs the electron's effective mass increases because it has to remain in the altered environment even in the presence of the field. It is remarkable that the electron is able to influence the distribution of the relatively massive fluid atoms. The change in the electron's environment is strictly a quantum effect and has no classical analogue. It occurs because the electron deBroglie wavelength is on the order of 50Å at room temperature. Thus it can interact with many fluid atoms simultaneously. If the electron-fluid interaction is repulsive, it stabilizes in a region of reduced fluid density, whereas if it is attractive it collects additional atoms. Since there is a competition between the attraction produced by polarization of the atom, and Fermionic repulsion between the free electron and the atomic electrons in filled shells, the dominant force will depend on the specific fluid.

Clearly any other low mass particle (lp generically for light particle) can also localize. Two examples are the positron (e+) and the positronium atom (Ps).[2] Radioactive positron sources such as ^{22}Na are readily available. When a positron from such a source enters a fluid it rapidly loses energy and thermalizes as a consequence of inelastic ionizing collisions with the fluid atoms. The thermalized positron can annihilate with an electron associated with the fluid, or it can form positronium which will, in turn, decay. In each case the annihilation process provides an additional indicator of the state of the combined particle-fluid system. For e+, the annihilation rate is simply proportional to the available electron density provided by the fluid atoms. Since the dominant e+-atom interaction is due to polarization the mean force is attractive and the fluid density near the localized particle increases, resulting in a shorter mean lifetime than that which would result if the positron simply experienced the ambient fluid density.

Positronium comes in two varieties, ortho (spin triplet) and para (spin singlet). The vacuum lifetime of pPs is 1.23×10^{-10} seconds. When pPs forms (25% of the time) the positron annihilates with it's partner electron via a two photon process. If the pPs atom is not localized, it's mean kinetic energy is (3/2)kT. If it is localized, then the uncertainty principle requires that the translational energy is greater. In fact, the momentum distribution of the center of mass of pPs can be determined experimentally by measuring the angular correlation of the annihilation photons. The shape of the distribution provides detailed information concerning its state. There is experimental evidence that localization can take place on this short time scale![3] A consequence of angular mementum conservation is that the 2γ decay process is not available to oPs in the vacuum. Its lifetime is 1.47×10^{-7} sec., three orders of magnitude greater than that of pPs, and it decays via a 3γ process. The long vacuum lifetime permits an alternative decay scheme for oPs in which the e+ annihilates with an atomic electron in the fluid. This mechanism, for obvious reasons called pick-off decay, is sensitive to the local environment of the oPs and provides a further probe of localization. The Ps-atom interaction is dominated by the Fermionic repulsion of the electron in positronium and the atomic electrons of the fluid atoms. Here the signature of self-trapping is a decrease in annihilation rate resulting from the relative dearth of atoms near the localized oPs.

As an example consider Figure 1, which reproduces experimental measurements of the decay rate of oPs in ethane.[4] The decay rate is plotted as a function of average fluid density at four different temperatures. We see that, at the highest temperature, the plot is nearly linear, indicating the natural increase in the pick-off decay rate due to the increase in available electrons. As the temperature is lowered, increasingly stronger deviations from linearity occur,

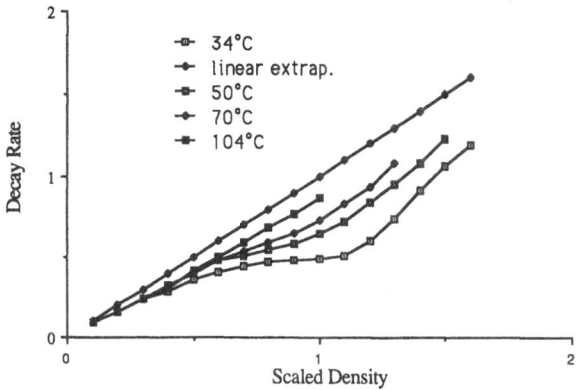

Figure 1. Decay rate of ortho-positronium versus density in ethane.

indicating that the oPs has reduced the fluid density in its environment. The existence of a plateau, which occurs just above the critical temperature of ethane ($33°C$), shows that the oPs environment is hardly changing in response to the increased fluid density. This strongly suggests that here the particle is localized. Notice that the deviations from linearity are greatest in a broad vicinity of the fluid's critical point where the large compressibility makes it easier for the Ps to displace the fluid atoms and form a trap. Localization of $e+$ and/or Ps has been observed in Ar [5], N_2 [6], CH_4 [7], C_2H_6 [4], Ne [8], and He [9].

Theoretical Models

The first theoretical models of trapped particles assumed that the fluid density varied abruptly near the lp, confining it in the ground state of a finite square well.[2,10] The fluid density in the well was either greater or less than the background, depending on whether the lp-atom interaction was attractive or repulsive. The criteria for the existence of the trapped state is a reduction of the free energy of the confined particle-fluid system below that of the extended (propagating) mode. The model was generalized to account for a smoothly varying local fluid density. It represents the fluid and the lp by a semi-macroscopic density profile $\rho(R)$ and wavefunction $\psi(r)$. The lp experiences the average potential $W_L(r)$ and, conversely, a fluid atom experiences the additional mean potential $W_F(R)$:

$$W_L(r) = \int w(r\text{-}R') \, \rho(R')dR' , \quad W_F(R) = \int w(R\text{-}r') \, |\psi(r')|^2 dr'. \quad (1)$$

Coupled equations for ρ and ψ are provided by the Schrodinger equation and the condition for macroscopic equilibrium,

$$-(\hbar^2/2m)\Delta\psi(r) + \psi(r)W_L(r) = E\psi(r) , \quad W_F(R) + \mu[\rho(R)] = \mu_0 \quad (2)$$

where μ is the chemical potential of the fluid, w is the effective lp-atom interaction, and

$\mu_0 = \mu_0(\rho_0)$ is the value of μ far from the lp, where the density is ρ_0. They can be derived from a variational principle by minimizing a free energy density functional which depends on both $\rho(\mathbf{R})$ and $\psi(\mathbf{r})$.[11,12] Since μ is determined by the equation of state, this is the only information required of the fluid.[12] Straightforward numerical algorithms have been constructed for solving the coupled pair when w is represented by a Fermi pseudo-potential.[11,12] This approximation is consistent with the assumption that ψ and ρ are slowly varying on an atomic scale. While propagating solutions always exist, bound states occur in specific regions of fluid density and temperature. In earlier work I showed that, in their simplest form, density functional theories (DFTs) satisfy a universality principle when the thermodynamic variables are scaled by their values at the liquid-vapor critical point: Quantitative differences in localization among fluids depend on a single dimensionless, system specific, parameter which is determined by the thermal wavelength of the lp, the mean separation between fluid atoms at the critical point, and the scattering length of the lp-atom interaction potential.[12,13]

About a decade ago density functional theories were extended in two directions: Since the size of the trapped region is typically about 20Å, density variations occur over distances on the order of the range of inter-atomic correlation within the fluid. Thus the condition for macroscopic equilibrium employed for the closure (see Eq. 1) needs to be modified by accounting for the presence of non-local correlations.[14,15] Second, the true wavefunction will vary rapidly near a fluid atom, whereas the macroscopic $\psi(\mathbf{r})$ is slowly varying by hypothesis (and construction). A Wigner-Seitz approach was used to take into account the rapid short range variation, in which the role of $\psi(\mathbf{r})$ is reduced to an envelope function which still controls the degree of localization by determining the average potential experienced by the fluid.[16] All of these variants are mean field theories which can be derived from an appropriate density functional. Although they are used to model phenomena which occur on a "mesoscopic" scale, or smaller, they ignore fluctuations. DFT has proved useful for modeling localization in He at low temperatures where fluctuations are small, but has been less successful at higher temperatures, in spite of various efforts to massage them with further heuristic assumptions. A common feature of DFT is that the existence of localized solutions abruptly turn on and off at specific values of the density, whereas experimental data at ordinary temperatures shows a continuous transition from extended to localized behavior. Another difficulty is that, by ignoring fluctuations, the mean field theories overemphasize the degree of localization, resulting in decay rate plots that have the wrong shape. Reese and I have demonstrated that a mild improvement in agreement can be obtained by explicitly considering transitions between localized and extended states.[18]

In order to improve on DFT it is necessary to construct a microscopic model which accounts for the interaction between the lp and the fluid atoms. In the simplest version, which is appropriate when the fluid is a Noble Gas, the translational degrees of freedom of the fluid atoms are assumed to obey classical mechanics. This is justified because, in the temperature range of interest, the mean deBroglie wavelength of the fluid atoms is less then an angstrom. When combined with the additional approximations that the internal degrees of freedom of the fluid atoms only enter implicitly in the determination of the effective scalar potential w(r) and the inter-atomic potential u(r), and that the lp is annealed by the fluid, the adiabatic model is realized.

The adiabatic model can be considered the starting point for DFT, i.e. the system which DFT attempts to model accurately. In fact, we have shown that DFT follows from this model with the further approximation that the Jensen inequality is an identity, i.e. that $\langle \varphi | \exp(-\beta H) | \varphi \rangle = \exp[-\beta \langle \varphi | H | \varphi \rangle]$ for any lp microstate $|\varphi\rangle$.[19] With this assumption, it can be shown that the wavefunction φ induces the "external" potential $W_F(R)$ at every point in the classical fluid. The free energy density functional in the canonical ensemble is then simply

$$DF[\varphi(r), \rho(R)] = \langle \varphi | \hat{p}^2/2m | \varphi \rangle - (1/\beta) \ln Q_N, \tag{3}$$

where $Q_N = Q_N(\varphi)$ is the canonical partition function of the inhomogeneous classical fluid experiencing the external potential $W_F[\varphi]$. By minimizing DF with respect to variations in φ and ρ which respect the appropriate constraints ($\int |\varphi|^2 dr = 1$; $\int \rho(R)dR = N$, the number of fluid atoms), we obtain the optimal (most probable) system "meso" state, ψ, ρ as defined by Eq. 2 above.

In the last decade much work has been focused on surpassing DFT by considering the complete adiabatic model without further approximation. The most useful tool to date for accomplishing this is Feynman's imaginary time path integral representation of the lp density matrix.[20] It has been used extensively to explain qualitative features of the localized electron.[21,22] In the discretized version of Feynman's path integral, the quantum mechanical electron is represented by a <u>classical</u> closed chain of p interacting harmonic oscillators, i.e. a ring polymer with p sites. In the limit $p \to \infty$, the discretized path integral rigorously converges to the adiabatic model[23] and is frequently referred to in the literature as the ring-polymer iso-morphism (RPI). At the present time it's implementation is more difficult than DFT because it requires large blocks of computer time to carry out computations of mean values using "smart" Monte Carlo algorithms.[22] An analytic approximation to the continuous path integral based on Feynman's construction of the polaron has been developed by Chandler et. al. using integral equation methods. While less CPU intensive than the Monte Carlo techniques, the RISM (reference interaction site model)-polaron theory is also difficult to implement numerically for realistic interaction potentials.[21] It is our impression that we are the first group to use both RPI and RISM-polaron theory to model the equilibrium behavior of positrons and positronium in fluids.[19,25]

In the remainder of the paper I will show how the RPI can be used to characterize localized states and compute experimental observables such as the e+ annihilation rate and the momentum distribution of the lp. I will conclude with an unpublished application of RPI to the unusual e+-Xenon system recently carried out with Greg Worrell which demonstrates the power of the method and provides new understanding of the highly non-linear decay rate observed at low density.

THE DISCRETIZED PATH INTEGRAL

The microscopic Hamiltonian for the lp-fluid system is

$$H = \hat{p}^2/2m + \sum_i w(r-R_i) + \sum_{i<j} u(R_i-R_j) + \sum_i P_i^2/2M, \tag{4}$$

where the first term is the kinetic energy of the lp, the second, $W(r, \underline{R})$ is the total lp-fluid interaction energy, and the third, $U(\underline{R})$, and fourth are the total internal potential energy and kinetic energy of the fluid. R_i, P_i and M are, respectively, the position, momentum and mass of a fluid atom and u is the interatomic pair potential (e.g. a Lennard-Jones potential). As explained in the introduction, this is a hybrid representation in which the lp is treated via quantum mechanics and the fluid atoms are treated classically. Important issues concerning its use are the computation of observables and the identification of signatures of localization.

In thermal equilibrium, the mean value of a physical observable, O, which may depend on both the lp and fluid coordinates, but _not_ the fluid momenta, is given by

$$\langle O \rangle = (1/Z) \int d\mathbf{R} \exp[-\beta U(\mathbf{R})] \, \text{Tr} \, [O \exp(-\beta H')]. \tag{5}$$

The trace refers the quantum part, so $H' = \mathbf{p}^2/2m + W(\mathbf{r}, \mathbf{R})$, and the classical part consists of the integral over the fluid translational coordinates ($\int d\mathbf{R}_1 \cdots \int d\mathbf{R}_N \equiv \int d\mathbf{R}$) weighted by the classical Gibb's factor $\exp(-\beta U)$. The partition function, Z, is obtained from the numerator by letting $O=1$ and, as usual, $\beta = 1/kT$.

Calculations using (5) directly are nearly impossible with present methods. For example, in a representation where H' is diagonal, it would be necessary to compute the complete set of eigenstates of H' for every realization of the collection of atomic positions $\mathbf{R}$, and then perform the classical average. Feynman showed that the calculation could be closely approximated with an equivalent classical system by introducing the discretized path integral (RPI).[20] Consider a closed chain of harmonic oscillators (ring polymer) consisting of p elements with positions x_i. Each element interacts harmonically with its nearest neighbors on the chain with force constant ($4mp/\hbar^2\beta^2$) and with all of the fluid atoms within the interaction range through the potential W/p. Thus the potential energy of the polymer is

$$\Phi(\mathbf{x}, \mathbf{R}) \equiv \sum_{1 \leq i \leq p} [(2mp/\beta^2\hbar^2)|x_{i+1} - x_i|^2 + (1/p)W(x_i, \mathbf{R})] \tag{6}$$

where $x_{p+1} = x_1$. The mean value of the configuration dependent operator $O(\mathbf{r}, \mathbf{R})$ for this system is

$$\langle O \rangle = (1/Z) \int d\mathbf{R} \exp[-\beta U(\mathbf{R})] \int d\mathbf{x} \exp[-\beta \Phi(\mathbf{x}, \mathbf{R})] O(x_1, \mathbf{R}) \tag{7a}$$

and converges to (5) in the limit $p \to \infty$. Reduction of the quantum average to a many body classical average reduces the calculation to a level that is currently feasible using modern computers.

The Annihilation Rate and its Variance

Physically the annihilation rate is proportional to the overlap of the positron and the electron density of the fluid. If $f(\mathbf{r})$ is the quantum averaged electron density of an atom fixed at the origin, then the quantum state $|\psi\rangle$ and molecule i contribute $\int |\psi(\mathbf{r})|^2 f(\mathbf{r}-\mathbf{R}_i)d\mathbf{r}$ to the decay rate, λ. Thus the decay rate is represented by the operator $\hat{\lambda} = \sum_{1 \leq i \leq N} f(\mathbf{r}-\mathbf{R}_i)$.[19]

Because all of the labelled sites of the ring polymer are equivalent, the thermally averaged annihilation rate may be written as

$$\langle \hat{\lambda} \rangle = \langle \sum_{1 \leq j \leq N} f(\mathbf{r}-\mathbf{R}_j) \rangle = \langle (1/p) \sum_{1 \leq \alpha \leq p} \sum_{1 \leq j \leq N} f(x_\alpha - \mathbf{R}_j) \rangle$$

$$= (1/p) \int d\mathbf{r} \int d\mathbf{R} \, f(\mathbf{r}-\mathbf{R}) \, n^{(2)}_{LF}(\mathbf{r}-\mathbf{R}) = \rho_0 \int d\mathbf{r} \, f(\mathbf{r}) \, g_{LF}(\mathbf{r}), \tag{7b}$$

where, in (7),

$$n^{(2)}{}_{LF}(r-R) = \langle \sum_{1 \leq \alpha \leq p} \sum_{1 \leq j \leq N} \delta(r-x_\alpha)\delta(R-R_j) \rangle = \rho_0 \, \rho_{pol} \, g_{LF}(r-R) \qquad (8)$$

is the two point lp-atom distribution function of statistical mechanics and $\rho_{pol} = p/V$ is the mean site density of the polymer. Finally, express the radial distribution in terms of the pair correlation function, $g_{LF}(r) = 1 + h_{LF}(r)$, to arrive at

$$\langle \hat{\lambda} \rangle = \rho_0 \int dr \; f(r) \; (1 + h_{LF}(r)). \qquad (9)$$

This simple looking result supports our intuition that, in the absence of correlations, the mean decay rate is directly proportional to the average fluid density. All of the effects of both localization and fluctuations enter in the computation of the pair correlation function between polymer sites and the elements of the fluid. Because f has a short range (typically less than one Å) the experimental deviations from linearity displayed in Fig. 1 in the critical region of the fluid arise from the reduction in the <u>short range</u> behaviour of h_{LF} (or g_{LF} resulting from the large compressibility.

It is also useful to have information concerning the variance of the decay rate. Since the environment of the lp is not static, annihilation phenomena cannot be completely characterized by a single decay rate. The statistical variance of $\hat{\lambda}$ provides limited information concerning the distribution of decay rates. Like $\langle \hat{\lambda} \rangle$ it may be expressed in terms of polymer-fluid distribution functions in the RPI. The use of similar methods yields[20]

$$\sigma^2{}_\lambda = \langle (\hat{\lambda} - \langle \hat{\lambda} \rangle)^2 \rangle = \rho_0 \int dr \; f(r)^2 \; g_{LF}(r) +$$

$$\rho_0{}^2 \int dr \int dr' \; f(r) \; f(r') \; g_{LF}(r) \; g_{LF}(r') \; h^{(3)}{}_{LFF}(r, r') . \qquad (10)$$

In (10), $h^{(3)}{}_{LFF}(r, r')$ is the three point correlation function and vanishes unless both r and r' are "small". If $\sigma_\lambda / \langle \hat{\lambda} \rangle \ll 1$ then the lp sees roughly the same fluid environment and the experimental measurements are successfully represented by a single decay rate. Conversely values of order one indicate that either the lp is undergoing transitions between a few well characterized states, or that the environment is changing continuously. Either way, the variance provides a hint of possible dynamic complexity.

<u>The Momentum Distribution</u>

Modern studies of positron annihilation frequently obtain information concerning the momentum of the annihilating pair immediately prior to their demise from either Doppler broadening techniques or the direct measurement of the photon angular correlation (LCAR).[2] These measurements demand a theory that provides the momentum distribution of the lp. This is easily obtained from a mild variation on RPI.

The probability distribution for the momentum, p, of the lp is the expectation of $\delta(\hat{p} - p)$:

$$P(p) = \langle \delta(\hat{p} - p) \rangle = \int dx \int dx' (2\pi\hbar)^{-3} \exp[ip \cdot r/\hbar] \langle \rho_d(x, x') \rangle$$

$$= V \int dr \, (2\pi\hbar)^{-3} \exp[i\mathbf{p}\cdot\mathbf{r}/\hbar] \langle \rho_d(\mathbf{r}) \rangle = V \int d\mathbf{r} \, (2\pi\hbar)^{-3} \langle \exp[i\mathbf{p}\cdot\mathbf{r}/\hbar] \rangle \qquad (11)$$

where V is the system volume, $\mathbf{r}=\mathbf{x}-\mathbf{x}'$, and the Fourier representation of $\delta(\hat{\mathbf{p}}-\mathbf{p})$ was used to obtain the result.[19] Thus the momentum distribution is simply the Fourier transform of $\langle \rho_d(\mathbf{x},\mathbf{x}') \rangle$ the <u>thermal average</u> of the lp density matrix, $\langle \mathbf{x}'|\exp(-\beta H')|\mathbf{x} \rangle / \mathrm{Tr}(\exp(-\beta H'))$. The development of a discrete path integral representation for $\langle \rho_d(\mathbf{x},\mathbf{x}') \rangle$ is analogous to that for $\langle \rho_d(\mathbf{x},\mathbf{x}) \rangle$ implicitly carried out in (16). The sole, and important, difference is that in (22) the endpoints of the polymer, or the smooth path in the limit $p\to\infty$, are separated by $\mathbf{r}$, so that the polymer is no longer ring-shaped, but rather open ended.[19]

<u>Quantifying Localization within the Framework of RPI</u>

There are two central characteristics of the localized state in RPI. First, the fluid influences the state of the lp. If the interaction potential w is essentially repulsive (has positive scattering length) the fluid literally squeezes the localized lp. The chief consequence is the contraction of the polymer. This can be quantified by constructing the distribution function, $P_\alpha(r)$, for the distance between pairs of polymer sites separated by α-1 intervening sites, and it's variance:

$$P_\alpha(r)=(1/p)\langle \sum_{1\leq j\leq p} \delta[(x_{j+\alpha}-x_j)-r] \rangle, \quad \sigma^2_\alpha = \int dr \, r^2 P_\alpha(r). \qquad (12)$$

If the lp is in an extended state, σ_α has values characteristic of a free polymer, e.g. it has a strong maximum at $\alpha=p/2$.[20] If it is localized, the polymer is highly compressed and σ_α is approximately independent of the index α (except near the end points $\alpha\approx 1,p$). Chandler refers to this configuration as ground state dominance[21] and argues convincingly that the lp is in a single quantum state. Evidence of polymer compression should be observable from photon correlation experiments of pPs decay : It will broaden the momentum distribution of the pPs atom. It should be kept in mind, however, that the momentum distribution itself is related to the broken polymer, not the ring. Intuition suggests that the two are closely related with the size of the broken polymer on the order of $\sigma_{p/2}$.

The second signature of localization is the alteration of the local atomic density in the neighborhood of the lp. If a deep trap is created, a semi-macroscopic region of altered density (about 20 Å in diameter) surrounds the lp. Similar behavior has been found in Monte Carlo simulations of an electron solvated in a fluid of hard spheres.[2] Two quantitative measures of this change are (1) the distribution of molecules about the polymer center of mass and (2) the lp-atom radial distribution function. The former corresponds to our usual intuitive ideas of localization and the local density profiles which are produced by mean field theory. However, I have shown above (Eqs. 8,9) that it is the latter that determines the mean annihilation rate for either positrons or oPs.[19]

In an ordinary fluid with interaction potential u(r), in the limit of low density or high temperature, the radial distribution is approximately $\exp(-\beta u)$. Collective effects of longer range appear for higher densities and lower temperatures and are most pronounced near the liquid-vapor critical point. In the case of the polymer, in the limit of high temperature and low fluid density, $g_{LF}(r)\approx\exp[-\beta w(r)]$. It is the deviation from this behavior as the temperature is lowered that signifies the onset of quantum behavior.

Positronium in Argon

Fan and I applied the RISM-polaron technique to a fluid interacting via the Lennard-Jones 6-12 potential $(u(r)=\varepsilon[(\sigma/r)^{12}-(\sigma/r)^{6}])$.[25] We selected values of the potential parameters which provide a good fit of pressure measurements on Argon to the low density virial series. We represented the lp-atom interaction $w(r)$ by a hard sphere potential of diameter d and computed both $g_{LF}(r)$ and σ_{τ} for all combinations of the densities $\rho_0 c^3 = 0.1, 0.3, 0.5$ and temperatures $T/T_c = 1.0, 2.0, 20.0$. (Here the path is continuous so the imaginary time, τ,

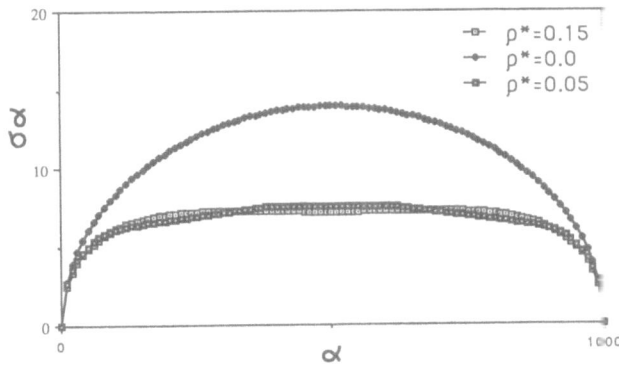

Figure 2. Distance between a pair of polymer sites versus number of intervening sites.

labels a point on the path rather than α.) Molecular dynamics calculations indicate that for the pure fluid the critical density is 0.285 in these units and $kT_c/\varepsilon = 1.3$. We found that:

1. Polymer compression increased with density and hard sphere diameter d at all temperatures, and decreased with increasing temperature. Compression was negligible at the highest temperature.
2. Fluid displacement is strongest at the critical point where there is a very large, long ranged deviation of g_{LF} from the step function $\Theta(r-d)$.

This is the first work which systematically investigates the influence of the critical region on localization. Qualitatively it shows that, due to the large isothermal compressibility, in the critical region the lp can have a significant influence on its surroundings. In contrast, polymer compression is more sensitive to the total pressure. This work points out that compression and displacement are not simply related. Consequently, at present, there is no single, simple criteria for localization.

Positron Annihilation in Xenon: An Application of RPI

Worrell and I recently decided to investigate the annihilation rate and local structure of a positron in Xenon. We chose this system because experiments have demonstrated atypical nonlinearity in plots of decay rate versus density on isotherms.[26] In contrast, plots for Neon are nearly linear.[8] The difference arises because the polarizability of Xe is an order of magnitude greater than that for Ne. We constructed a pseudopotential based on the computations of Schrader,[27] and Lane and Geltman,[28] who have optimized particular functional forms to fit e^+-Xe scattering cross sections. As for Argon, the fluid interaction was modelled by a Lennard-Jones interaction with $\sigma = 4.0551$ Å and $\varepsilon = 229K$. While final calculations of the annihilation rate over large ranges of density have not been completed, preliminary results for $\rho\sigma^3 = 0.05, 0.15$ at $T = 340°K$ will be discussed here.

We applied the "smart" Metropolis algorithms developed by Pollock and Ceperley,[24] and also used by Coker, Berne and Thirumalai[22] to study an excess electron in Xe, to the case of a positron. The basic method has two essential features: First, sample the free (noninteracting)

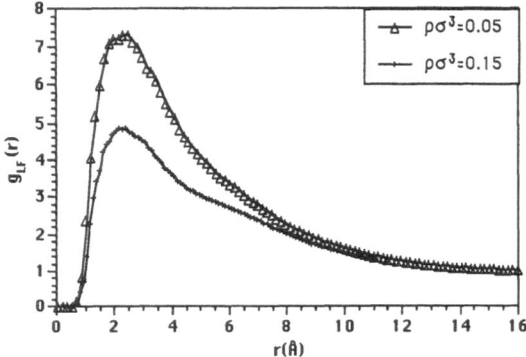

Figure 3. Polymer site-atom radial distribution functional.

polymer distribution to avoid a large percentage of rejections; Second, displace a segment of the polymer sites at each move to avoid being "trapped" in a local potential minimum, i.e. to guarantee sufficiently ergodic behavior of the sampling process. We obtained good convergence for the mean decay rate and interaction energy with p=1000 by making 2,000 passes through the system for each data point (density and temperature).

In Figure 2, σ_α is plotted for each density. The upper curve corresponds to a free polymer, whereas $\rho\sigma^3 = 0.05$ and 0.15 correspond to the lower curves. It is clear that the positron is strongly localized (compressed) at each density. In Figure 3 we plot the radial distribution function $g_{LF}(r)$. The results are surprising: They show very strong clustering of Xe atoms around the positron. In each case the cluster consists of about 20 atoms. In dimensionless units, the decay rate at the lower density is 1.35 and 2.52 at the higher density. Also, in each case, the variance is about 0.001, supporting the claim that the positron is localized.

The values of λ are consistent with the highly nonlinear experimental measurements. It is clear that the polarization potential induces clustering about the positron which results in the large decay rate. The repulsive short range interaction between the Xenon atoms saturates the cluster density, resulting in the nonlinearity of plots of λ versus ρ on isotherms. This concluding example demonstrates the power of RPI for localization studies in fluids. The large clusters could not be predicted by a virial type expansion event though the density (.05) is low. An interesting feature of the cluster not illustrated here is the exclusion of atoms from a small region about the positron center of mass: i.e. the cluster has structure. It is unlikely that such a rapidly changing fluid density could be modeled by DFT.

CONCLUDING REMARKS

The strength of the discrete ring-polymer approximation to the path integral is that it converges rigorously in the limit of an infinite number of polymer sites. This convergence can be followed numerically. Consequently RPI, coupled with state of the art computing, provides the opportunity to study the predictions of the assumed microscopic model with satisfactory precision. In contrast with additional approximations, such as density functional theory,[11-16] or the RISM-polaron approximation,[20] any inconsistency between RPI and experimental measurements is a consequence of the model, and not the method of approximation. The disadvantage of RPI is that the computations are expensive (for someone) and time consuming. In practice, this simply limits the number of data points which one would like to investigate. With the rapid changes in computing technology, these comments will no doubt be obsolete in a few years.

In the final example considered above of a positron annihilating in Xe, RPI has recovered the nonlinearity of the experimental data directly from the microscopic model without introducing further assumptions or heuristic, density dependent, nuclear charges. As in experiments, it demonstrates that trapping does not turn on and off suddenly, but rather varies continuously with the thermodynamic parameters. For the two densities studied here, there is convincing evidence for a strongly localized, stable e^+-fluid configuration.

ACKNOWLEDGEMENTS

I would first like to recognize my students and co-workers, Terrence Reese, Yzhong Fan and Greg Worrell who assisted in this work and performed most of the numerical computations, as well as David Schrader who provided us with information concerning the Xe charge distribution and e+-Xe interaction, and Suresh Sharma for providing the cPs lifetime data for ethane. I am also grateful for the support of the Robert Welch Foundation of Houston Texas, the Research Foundation of Texas Christian University, the Pittsburgh Supercomputer Center. In addition, travel support was provided by the U. S. Army Research Office.

REFERENCES

1. N. Gee and G. R. Freeman, Can. J. Chem. **64**, 1810 (1986).
2. For an excellent review of the behavior of positrons in gases see
   I. T. Iakubov and A. G. Khrapak, Prog. Phys. **45**, 697 (1982). For a review of the behavior of electrons in fluids see J. Hernandez, Rev. Mod. Phys. **63** 675, (1991).

3. C. V. Briscoe, S. I. Choi, and A. T. Stewart, Phys. Rev. Lett. **20**, 493, 1968.

4. S. C. Sharma, R. H. Arganbright, and M. H. Ward, J. Phys. B **20**, 867 (1987);
   S. C. Sharma and E. H. Juenguman, Phys. Lett. A **144**, 47 (1986).

5. M. Tuomisaari, K. Rytsola, and P. Hautojarvi, Phys. Lett. **112A**, 279 (1988).

6. K. Rytsola, K. Rantapuska, and P. Hautojarvi, J. Phys. B **17**, 347 (1984).

7. J. D. McNutt and S. C. Sharma, J. Chem. Phys. **68**, 130 (1978).

8. K. F. Canter and L. O. Roellig, Phys. Rev. A **12**, 386 (1975).

9. T. B. Daniel and R. Stump, Phys. Rev. **115**, 1599 (1959).

10. J. L. Levine and T. M. Sanders, Phys. Rev. **154**, 138 (1967).

11. R. L. Moore, C. L. Cleveland, and H. A. Gersch, Phys. Rev. B **18**, 1183 (1978).

12. B. N. Miller and T. Reese, Phys. Rev. A **39**, 4735 (1989).

13. B. N. Miller, in *Positron Annihilation Studies of Fluids*,
    ed. S. Sharma (World Scientific, Singapore, 1988) p. 81.

14. C. Ebner and C. Punyanita, Phys. Rev. A **19**, 856 (1979).

15. M. J. Stott and E. Zaremba, Phys. Rev. Lett. **38**, 1493 (1977).

16. R. M. Nieminen, M. Manninen, I. Vÿalimaa and P. Hautojarvi,
    Phys. Rev. A **21**, 1677 (1980).

17. M. Tuomisaari, K. Rytsola, R. M. Niemenen and P. Hautojarvi,
    J. Phys. B **19**, 2667 (1986).

18. T. Reese and B. N. Miller, Phys. Rev. A **42**, 6068 (1990).

19. B. N. Miller and Y. Fan, Phys. Rev. A **42**, 2228 (1990).

20. R. P. Feynman and A. R. Hibbs, "Quantum Mechanics and Path Integrals" (McGraw-Hill,
    New York, 1965); R. P. Feynman, "Statistical Mechanics", (Benjamin, Reading, Mass.
    1972); R. P. Feynman, Phys. Rev. **97**, 660 (1955).

21. D. Chandler, Y. Singh, and D. M. Richardson, J. Chem. Phys. **81**, 1975 (1984);
    A.L. Nichols III and D. Chandler, J. Chem. Phys. **81**, 5109 (1984).

22. D. F. Coker, B. J. Berne, and D. Thirumalai, J. Chem. Phys. **86**, 5689 (1987);
    D. F. Coker and B. J. Berne, J. Chem. Phys. **86**, 5689 (1987);
    M. F. Herman, E. J. Bruskin, and B. J. Berne, J. Chem. Phys. **76**, 6150 (1982);
    D. Thirumalai, R. W. Hall and B. J. Berne, J. Chem. Phys. **81**, 2523 (1984).

23. E. H. Lieb, J. Math. Phys **8**, 43 (1967).

24. E. L. Pollock and D. M. Ceperley, Phys. Rev. B **30**, 2555 (1984).

25. Y. Fan and B. N. Miller, J. Chem. Phys. **93**, 4322 (1990).

26. M. Tuomisaari, K. Rytsola, and P Hautojarvi in "Positron Annihilation Studies of Fluids",
    ed.S. C.Sharma (World Scientific, Singapore, 1988).

27. D. Schrader, Phys. Rev. A **20**, 918 (1978); **34**, 1810 (1986).

28. N. F. Lane and Geltman, Phys. Rev. **160**, 53 (1967); **173**, 183 (1968).

LOWER BOUND ASPECTS OF

FERMION DENSITY FUNCTIONALS*

J. K. Percus

Courant Institute of Mathematical Sciences and
Physics Department, New York University
251 Mercer Street, New York, NY 10012

ABSTRACT

The density functional formalism for N-fermion ground state energies is reviewed, stressing the role of the grand ensemble and the associated minimum principle. Piecewise minimization is shown to provide a variety of lower bounds; these are preserved under model estimates of the kinetic energy functional. Consistency conditions on this functional are derived, and a WKB expansion carried out. The interaction energy functional for coulomb forces is bounded by a modified mean field approach which picks up the correlation hole, and a more systematic technique is introduced but not evaluated.

INTRODUCTION

N-fermion systems are surely the most common microscopic entities found on earth, and an enormous amount of effort has gone into elucidating their properties. Exact solutions of even model systems are very few and far between, and anyway it is the electron fluid in the field of nuclei — and not a model system — that provides the physical importance. Thus, in addition to massive numerical computations, ranging from multiconfiguration variational to in principle exact Monte Carlo, attention has been focussed on more approximate but correspondingly more intuitive and computationally simpler approaches. One of the most popular of these is that of density functional estimation,[1] in which an appropriate thermodynamic energy is expressed with tolerable accuracy in terms of the electron density alone (or simple extensions in which e.g. spin is identified), which is then determined by minimization of the energy. Here, we would like to comment upon some aspects of this activity, incorporating concepts and estimates taken over from the theory of reduced density matrices,[2] which on its own has done little to fulfill its original promise.

To keep the discussion at the most elementary but meaningful level, we will consider N identical spinless fermions in the system ground state (which should

---

* Supported in part by a U. S. Army Research Office travel grant and by the National Science Foundation.

be generalized to $T \rightarrow 0°$ in case of ground state degeneracy). Including only nonrelativistic kinetic energy, pair interaction, and external field, we then have the Hamiltonian

$$\hat{H} = \hat{T} + \hat{\phi} + \hat{U}, \tag{1.1}$$
$$\hat{T} = \sum_i p_i^2/2m , \quad \hat{\phi} = \tfrac{1}{2}\sum_{i \neq j} \phi(r_i - r_j) , \quad \hat{U} = \sum_i u(r_i).$$

We will adopt the notation

$$Q[\psi] \equiv \langle \psi | \hat{Q} | \psi \rangle \tag{1.2}$$

where $\psi(r_1, \ldots, r_N)$ is a normalized antisymmetric wave function, so that the Rayleigh-Ritz principle for the ground state energy becomes

$$E_0 = \min_\psi (T[\psi] + \phi[\psi] + U[\psi]) \tag{1.3}$$

The minimization is now carried out in two steps[3] by introducing the microscopic density

$$\hat{n}(r) = \sum_i \delta(r - r_i) \tag{1.4}$$

and rewriting (1.3) as

$$E_0 = \min_{n:\ \int n(r)\, d^3 r = N} \min_{\psi:\ \hat{n}(r)[\psi] = n(r)} (T[\psi] + \phi[\psi] + U[\psi]). \tag{1.5}$$

If, with slight abuse of notation, we set

$$Q[n] = \min_{\psi:\ \hat{n}(r)[\psi] = n(r)} Q[\psi] \tag{1.6}$$

$n(r)$ again denoting the expectation of the particle density $\hat{n}(r)$, (1.5) translates to

$$E_0 = \min_{n:\ \int n(r)\, d^3 r = N} [(T + \phi)[n] + \int u(r)\, n(r)\, d^3 r], \tag{1.7}$$

where we have used the fact that $U[\psi]$ depends only upon $n(r)$. Finally, introducing the "residual" energy functional

$$\overline{E}_0[n] \equiv (T + \phi)[n], \tag{1.8}$$

that in which the external field energy is subtracted out, we have the basic

$$E_0 = \min_{n:\ \int n(r)\, d^3 r = N} (\overline{E}_0[n] + \int u(r)\, n(r)\, d^3 r) \tag{1.9}$$

MINIMUM PROPERTIES

The final minimization, (1.9), can be carried out in routine fashion by introducing the chemical potential or Fermi energy μ as Lagrange parameter to automatically incorporate the restriction $\int n(r)\, d^3 r = N$:

$$\frac{\delta}{\delta n(r)}\left(\overline{E}_0[n] - \int \mu(r)\, n(r)\, d^3 r\right) = 0 \tag{2.1}$$
$$\text{where} \quad \mu(r) \equiv \mu - u(r),$$

leading to the profile equation in inverse form

$$\mu(r) = \frac{\delta \overline{E}_0[n]}{\delta n(r)}. \tag{2.2}$$

The quantity

$$\Omega_0 = \overline{E}_0[n] - \int \mu(r)\, n(r)\, d^3 r \tag{2.3}$$

satisfies $\delta\Omega_0 = \int \delta\overline{E}_0[n]/\delta n(r)\, \delta n(r)\, d^3 r - \int (\mu(r)\, \delta n(r) + \delta\mu(r)\, n(r)\,)\, d^3 r$, or

$$n(r) = -\frac{\delta\Omega_0[\mu]}{\delta\mu(r)}. \tag{2.4}$$

It is hence recognized as the (second quantized) ground state grand potential, a functional of $\mu(r)$:

$$\Omega_0[\mu] = \langle \hat{H} - \hat{N}\mu \rangle_0, \tag{2.5}$$

since both (2.3) and (2.5) satisfy (2.4) (Ehrenfest, Hellman-Feynman[4]) as well as vanishing when $\mu \to -\infty$. Making the N-dependence explicit, we have

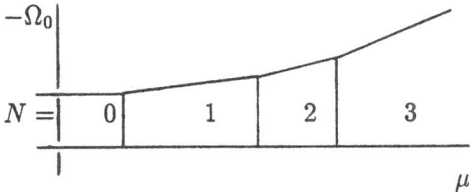

$$\Omega_0(\mu) = \min_N(E_0^N - N\mu), \tag{2.6}$$

so that Ω_0 is piecewise linear in the chemical potential μ, as shown, incurring a "phase transition", and resulting ensemble mixture of N-particle and $N+1$-particle states at a sequence of discrete values of μ. The ensemble nature is the reason that $N = \int n(r)\,d^3r$ in (2.1) is not restricted to integers, but also shows that the analytic properties of $\overline{E}_0[n]$ in terms of $n(r)$ must be intrinsically quite complex. But we do observe[5] that except at the singular values of μ, the "ensemble" provides a unique integer N, and this fact will be used routinely.

How does one determine $\overline{E}_0[n]$ of (1.8)? A versatile format for this purpose is that in which the interaction is broken up into a reference interaction and a remainder

$$\phi = \phi_0 + \Delta\phi. \tag{2.7}$$

Now quite generally,

$$(A + B)[n] \geq A[n] + B[n], \tag{2.8}$$

since the ψ which achieves the minimum $(A+B)[n]$ will do so for neither $A[n]$ nor $B[n]$, resulting at once in the *lower bound principle*

$$\overline{E}_0[n] \geq (T + \phi_0)[n] + \Delta\phi[n], \tag{2.9}$$

familiar from the reduced density matrix field.

It is also to be noted from the grand canonical Ehrenfest (or Helman-Feynman) theorem

$$\langle \hat{Q} \rangle_0 = \tfrac{\partial}{\partial\lambda}\Omega_0[\hat{H} - \hat{N}\mu + \lambda\hat{Q}]\big|_{\lambda=0}, \tag{2.10}$$

together with the convexity of Ω_0 in λ, that expectations are bounded from both sides: if $\lambda \geq 0$, then[6]

$$\tfrac{1}{\lambda}(\Omega_0[\hat{H} - \hat{N}\mu] - \Omega_0[\hat{H} - \hat{N}\mu - \lambda\hat{Q}])$$

$$\geq \langle \hat{Q} \rangle_0 \geq \tfrac{1}{\lambda}(\Omega_0[\hat{H} + \hat{N}\mu + \lambda\hat{Q}] - \Omega_0[\hat{H} - \hat{N}\mu]). \tag{2.11}$$

Thus, a lower bound for $\Omega_0[\hat{H} - \hat{N}\mu + \lambda\hat{Q}]$, available from (2.9), together with any traditional upper bound for $\Omega_0[\hat{H} - \hat{N}\mu]$ will give a guaranteed lower bound for $\langle Q \rangle_0$ — and similarly for upper bounds.

The most primitive use of (2.9) is that in which the reference system has $\phi_0 = 0$, independent fermions:

$$\overline{E}_0[n] \geq T[n] + \phi[n]. \tag{3.1}$$

The effect of (3.1) on exchange-correlation contributions will be referred to later. But the problem of determining $T[n]$ is not trivial; indeed, the answer is not known. There are several ways of proceeding, all depending upon the fact that minimizing $\langle \widehat{T} \rangle$ at given $\langle \hat{n}(r) \rangle$, on imposing the restriction with a Lagrange parameter function $u(r)$, is equivalent to minimizing $\langle \widehat{T} + \int \hat{n}(r) u(r) dr \rangle$, i.e. to solving the independent-fermion problem with arbitrary external potential $u(r)$. In the Kohn-Sham approach,[7] this problem is "solved" by translating to the resulting orthonormal orbitals $\{\phi_\alpha(r)\}$, in terms of which

$$n(r) = \sum_\alpha |\phi_\alpha(r)|^2 , \quad \langle \widehat{T} \rangle = (\hbar^2/2m) \int \sum_\alpha |\nabla \phi_\alpha(r)|^2 d^3r, \tag{3.2}$$

and minimizing with respect to the set $\{\phi_\alpha\}$. Taking $\phi[n]$ in (3.1) in mean field approximation would then yield precisely the Hartree approximation to E_0, but much more intelligent hybrid procedures are available.

We will aim directly at the functional $T[n]$, to start with by finding lower bounds from physical considerations, and thereby of course preserving the inequality (3.1). In essence, we just parametrize the set $\{\phi_\alpha\}$ to be used in the minimization associated with (3.2), but a convenient interpretation is as follows. Suppose that the independent N-fermion ground state energy is known to be E_{0M}^N for a "model" potential $u_M(r)$. Then for any valid independent-fermion one-body density matrix γ, i.e. one of the form $\gamma(r, r') = \sum_\alpha \phi_\alpha(r) \phi_\alpha^*(r')$, the Rayleigh-Ritz bound

$$\text{Tr}(\widehat{T} + \sum u_M(r_i))\gamma \geq E_{0M}^N \tag{3.3}$$

holds. In particular, selecting γ as the minimizing density matrix for $\text{Tr}\,\widehat{T}\gamma$ at given $n(r)$, then

$$T[n] \geq E_{0M}^N - \int n(r) u_M(r) d^3r. \tag{3.4}$$

The art now consists of choosing a solvable u_M, or better a set u_M^λ, which is expected to reasonably bracket the effective potential of the system under consideration. Then (3.4) generalizes at once[3] to

$$T[n] \geq \max_\lambda(E_{0M\lambda}^N - \int n(r) u_{M\lambda}(r) d^3r) \tag{3.5}$$

For example, for an N-electron atom, a reasonable choice might be to take $u_{M\lambda}(r)$ as coulomb, but of unknown effective charge λ. The expression (3.5) is then readily computed. For large N, one finds[3]

$$T[n] \geq \tfrac{\hbar^2}{2m}(\int n(r) d^3r/r)^2 / (12 \int n(r) d^3r)^{1/3}; \tag{3.6}$$

for smaller N, (3.6) is more complicated — but still very explicit — as details of shell structure enter.

CONSISTENCY CONDITIONS

Another way of proceeding — essentially Galerkin[8] in spirit — is to construct parametric forms for $T[n]$ and then determine the parameters by using exact conditions that $T[n]$ must satisfy. As we have indicated, the most subtle structural effects are those associated with the integer particle number aspect of the system. To get a feeling for the nature of the ensuing restrictions, let us confine our attention to one-dimensional fermions, in which case explicit results are available through the use of what are generally termed supersymmetric partners.[9]

Let us set $\hbar^2/2m = 1$, so that we are concerned with the bound state Schrodinger equation (we suppose that $u \to 0$ as $|x| \to \infty$)

$$\psi'' + E - u)\psi = 0. \tag{4.1}$$

If ψ_0 is the nodeless ground state and E_0 its energy, then in terms of

$$v = -\psi_0'/\psi_0, \tag{4.2}$$

we have

$$E_0 - u = v' - v^2, \tag{4.3}$$

so that

$$\left(\tfrac{d}{dx} - v\right)\left(\tfrac{d}{dx} + v\right)\psi = -\epsilon\psi \tag{4.4}$$
$$\epsilon = E - E_0.$$

Now defining

$$\bar{\psi} = \left(\tfrac{d}{dt} + v\right)\psi, \tag{4.5}$$

it follows from (4.4) that

$$\left(\tfrac{d}{dx} + v\right)\left(\tfrac{d}{dr} - v\right)\bar{\psi} = -\epsilon\bar{\psi}, \tag{4.6}$$

and that $\langle\psi|\psi\rangle = 1$ implies $\langle\bar{\psi}|\bar{\psi}\rangle = |\epsilon|$. In particular, $\bar{\psi}_0 = 0$. Further, from (4.6), $\bar{\psi}$ now satisfies the Schrodinger equation in which

$$E_0 - \bar{u} = -v' - v^2, \tag{4.7}$$

with $\bar{u} \to 0$ as $|x| \to \infty$, i.e. that for which

$$\bar{v} = -v. \tag{4.8}$$

Thus, a state sequence $\psi_0, \ldots, \psi_N$ determined by $E_0 + \epsilon_N = \mu$ is carried over into a sequence $\bar{\psi}_1, \ldots, \bar{\psi}_N$, again with $E_0 + \epsilon = \mu$. Now suppose any state occupation $f(\epsilon)$, e.g. $\theta(\mu - E_0 - \epsilon)$ for a fermi sea, where θ is the Heaviside step function. Clearly

$$n(x) = \langle x|f((v - \tfrac{d}{dx})(v + \tfrac{d}{dx}))|x\rangle. \tag{4.9}$$

According to (4.6), the corresponding transformed density will be given by

$$\bar{n}(x) = \langle x|f((v + \tfrac{d}{dx})(v - \tfrac{d}{dx}))|x\rangle. \tag{4.10}$$

But from the general

$$\tfrac{d}{dx}\langle x|g|x\rangle = \langle x|[\tfrac{d}{dx}, g]|x\rangle, \tag{4.11}$$

it readily follows that if

$$q(x) = n'(x) + 2v(x)n(x), \tag{4.12}$$

then

$$q(x) = \langle x|(v + \tfrac{d}{dx})f((v - \tfrac{d}{dx})(v + \tfrac{d}{dx}))|x\rangle \tag{4.13}$$
$$+ \langle x|f((v - \tfrac{d}{dx})(v + \tfrac{d}{dx}))(v - \tfrac{d}{dx})|x\rangle = -\bar{q}(x),$$

to complement (4.8).

$q(x)$ and $v(x)$ are in fact conjugates, i.e. from $E_0 = -v(\infty)^2$, then $\delta E_0/\delta v(x) = 0$, and so $\delta\Omega_0/\delta v(x) = \int \delta\Omega_0/\delta u(y)\,\delta u(y)\,/\,\delta v(x)\,dy = \int n(y)\,(\delta'(x-y) + 2v(y)\,\delta(x-y)\,)\,dy$, or

$$q(x) = \delta\Omega_0/\delta v(x). \tag{4.14}$$

We conclude from (4.8) and (4.13) that

$$\Omega_0[E_0 - v' + v^2] \text{ is odd in } v, \tag{4.15}$$

a fairly strong condition with a somewhat more complicated consequence for $T[n]$.

WKP APPROXIMATION

The functional $T[n]$, or more generally $\overline{E}_0[n]$, can also be developed sequentially through a number of different expansions, ranging from path integral discretization to linear response expansions. A more traditional approach in the face of external potentials which do not vary rapidly is something along the lines of a WKB approximation. As in Section 4, it is easier to formulate this in terms of a $\mu(r)$ expansion, and we shall do so. Confining our attention to the independent particle system in D dimensions appropriate to $T[n]$, we can choose to examine the profile equation

$$n(r) = \langle r|\theta(\mu(r) - p^2/2m)|r\rangle, \tag{5.1}$$

the kinetic energy functional

$$T[\mu] = \text{Tr}(p^2/2m\,\theta(\mu(r) - p^2/2m)\,), \tag{5.2}$$

or via $\Omega_0 = T - \int \mu(r)\,n(r)\,dr^D$, the grand potential

$$\Omega_0[\mu] = \text{Tr}(p^2/2m - \mu(r)\,)\,\theta(\mu(r) - p^2/2m)\,). \tag{5.3}$$

Direct approximations to (5.1) too often fail to satisfy the required integrability condition $\delta n(r)/\delta\mu(r') = \delta n(r')/\delta\mu(r)$, and so we will consider Ω_0, in the convenient form ($' = \partial/\partial\mu$, $\mu(r) = \mu - u(r)$).

$$\Omega_0''[\mu] = -\,\text{Tr}\,\delta(\mu(r) - p^2/2m), \tag{5.4}$$

whose double integral over μ recovers (5.3).

On Fourier representing the δ-function in (5.4), we must evaluate $\exp iz(\mu(r) - p^2/2m)$. But

$$\langle r'|e^{iz(\mu(r)-p^2(2m)}|r\rangle = e^{iz(\mu(r)+\hbar^2/2m\,\nabla^2)}\,\delta(r-r')$$

$$= \int e^{iz(\mu(r)+\hbar^2/2m\,\nabla^2)}\,e^{i(r-r')\cdot p/\hbar}\,(dp/h)^D$$

$$= \int e^{i(r-r')\cdot p/\hbar}\,e^{iz[\mu(r)-p^2/2m+i\hbar/m\,p\cdot\nabla+\hbar^2/2m\,\nabla^2]}\cdot 1\,(dp/h)^D,$$

so that

$$\text{Tr}\,e^{iz(\mu(r)-p^2/2m)} = \int\int e^{iz[\mu(r)-p^2/2m+i\hbar/mp\cdot\nabla+\hbar^2/2m\nabla^2]}\cdot 1\,(dr\,dp/h)^D. \tag{5.5}$$

Now applying the standard Campbell-Baker-Hausdorf expansion

$$e^{A+B} = e^A\,e^{\frac{1}{2}[B,A]+\cdots}\,e^B, \tag{5.6}$$

the operator integrand becomes

$$e^{iz[\mu(r)-p^2/2m+i\hbar/mp\cdot\nabla]}\,e^{-\frac{1}{2}z^2(\hbar^2/2m)(\nabla\mu(r)\cdot\nabla+\nabla^2\mu(r)+\cdots)}\,e^{iz\hbar^2/2m\nabla^2}\cdot 1 \tag{5.7}$$

To leading order in this development, we now retain only the undifferentiated $\mu(r)$:

$$\Omega_0''[\mu] = (1/2\pi) \int \int \int e^{iz[\mu(r)-p^2/2m+i\hbar/mp\cdot\nabla]} \cdot 1 \,(dr\,dp/h)^D \,dz. \tag{5.8}$$

Evaluation of (5.8) is routine. We have quite generally

$$e^{a(r)+b\cdot\nabla} \cdot 1 = e^{-(b\cdot\nabla)^{-1}a(r)} \, e^{b\cdot\nabla} \, e^{(b\cdot\nabla)^{-1}a(r)} \cdot 1$$

$$= e^{(b\cdot\nabla)^{-1}a(r+b)-(b\cdot\nabla)^{-1}a(r)} = e^{\int_0^1 a(r+\lambda b)\,d\lambda},$$

so (5.8) becomes

$$\Omega_0''[\mu] = \tag{5.9}$$

$$(1/2\pi) \int_{-\infty}^{\infty} \int \int \exp\left\{ iz[\int_{-1/2}^{1/2} \mu(r - \lambda z\hbar p/m)\,d\lambda - p^2/2m] \right\} (d\,dp/h)^D \,dz,$$

or via the transformation $p \to p/z$,

$$\Omega_0''[\mu] = \tag{5.10}$$

$$\frac{1}{\pi}\mathrm{Re} \int_0^{\infty} \int \int \exp i\left[z \int_{-1/2}^{1/2} \mu(r - \lambda\hbar p/m)\,d\lambda - z^{-1}p^2/2m\right] (dr\,dp'h)^D \,dz/z^D.$$

The z-integral is wel known[10]

$$\int_0^{\infty} \exp i(az - b/z)\,dz/z^D = 2(-a/b)^{(1/2)(D-1)} K_{D-1}(2\sqrt{ab}) \tag{5.11}$$

$$\mathrm{Im}\,z > 0\,, \qquad \mathrm{Im}\,b < 0.$$

Here, $\mathrm{Im}\,a$ and $\mathrm{Im}\,b$ act as convergence factors and we must imagine branch cuts for $\mathrm{Im}\,a \leq 0$, $\mathrm{Im}\,b \geq 0$. Hence

$$\Omega_0''[\mu] = (2/\pi)\,\mathrm{Re} \int \int \left(-2m/p^2 \int_{-1/2}^{1/2} \mu(r - \lambda\hbar p/m)\,d\lambda \right)^{(1/2)(D-1)} \tag{5.12}$$

$$K_{D-1}(2(p^2/2m \int_{-1/2}^{1/2} \mu(r - \lambda\hbar p/m)\,d\lambda)^{1/2}) (dr\,dp/i)^D.$$

To recover $\Omega_0[\mu]$, we integrate twice from $-\infty$ to μ, using[11]

$$\int_{-\infty}^{x} \left(\sqrt{\tfrac{x}{y}}\right)^{V-1} K_{V-1}(2\sqrt{xy})\,dx = -\left(\sqrt{\tfrac{x}{y}}\right)^{V} K_V(2\sqrt{xy}), \tag{5.13}$$

to obtain the desired result

$$\Omega_0[\mu] = (2/\pi)\,\mathrm{Re} \int \int \left(-2m/p^2 \int_{-1/2}^{1/2} \mu(r - \lambda\hbar p/m)\,d\lambda \right)^{(1/2)(D+1)} \tag{5.14}$$

$$K_{D+1}(2(p^2/2m \int_{-1/2}^{1/2} \mu(r - \lambda\hbar p/m)\,d\lambda)^{1/2}) (dr\,dp/h)^D$$

again with appropriate imaginary parts understood.

In this approximation, the external potential suffers a quantum smearing, absent if $h \to 0$. The profile equation can of course now be computed from $n(r) = -\delta\Omega_0[\mu]/\delta\mu(r)$, and the kinetic energy from $T[\mu] = (1 - \int \mu(r)\delta/\delta\mu(r)\,dr^D)\Omega_0[\mu]$, but the task of combining them to find $T[n]$ does not seem trivial.

POTENTIAL ENERGY BOUNDS

Following the formulation of (3.1),now that we have examined bounds, restrictions, and approximations to the kinetic energy functional $T[n]$, we must do the same for the interaction energy functional $\Phi[n]$. The difference between $\Phi[n]$ and the Hartree or mean field estimate

$$\Phi_{MF}[n] = \tfrac{1}{2} \int \int n(r)\,n(r')\,\phi(r,r')\,dr^D\,dr'^D \tag{6.1}$$

is referred to as the exchange-correlation correction. However, it must be pointed out that there is a distinct difference between the associated correlation holes in their classical and quantum forms: since only $|\psi|^2$ is employed in computing $\langle\Phi\rangle$, the anti-symmetry of the wave function plays no role at all — there is no required exchange hole — and can only be picked up by a modified $\phi_0 \neq 0$ or by including aspects of momentum density or positional pair density in $T[n]$.

Subject to the above caveat, we proceed. Keeping in mind the example of a coulomb potential, in which $\phi(r-r') \geq 0$, we observe first that one way of expressing the existence of a correlation hole is by a reduction in the effective potential,

$$\phi(r - r') = \bar{\phi}(r - r') + \Delta\phi(r - r')$$
$$\Delta\phi(r - r') \geq 0. \tag{6.2}$$

Indeed, if (6.2) holds, then certainly

$$\Phi[n] \geq \tfrac{1}{2} \int \int \bar{\phi}(r - r')\, n_2(r, r')\, dr^D\, dr'^D \tag{6.3}$$

where $\quad n_2(r, r') = \langle \sum_{i \neq j} \delta(r - r_i)\, \delta(r' - r_j) \rangle.$

But now quite generally, $n_2(r, r')$ is bounded by its no-fluctuation part, i.e. $\langle (\hat{n}(r) - n(r))(\hat{n}(r') - n(r')) \rangle = n_2(r, r') - n(r)\, n(r') + n(r)\, \delta(r - r')$ is positive definite as a matrix, s that if

$$\bar{\phi}(r - r') \text{ is positive definite} \tag{6.4}$$

then[6]

$$\Phi[n] \geq \tfrac{1}{2} \int \int \bar{\phi}(r - r')\,(n(r)\, n(r') - n(r)\, \delta(r - r'))\, dr^D\, dr'^D \tag{6.5}$$
$$= \Phi_{MF}[n] - \tfrac{1}{2}\bar{\phi}(0) \int n(r)\, dr^D.$$

For example, take ϕ as coulomb, so that it and its Fourier transform are given in three dimensions by

$$\phi(r) = e^2/r\,, \qquad \phi_k = 4\pi\, e^2/k^2. \tag{6.6}$$

To avoid the singularity $\phi(0) = \infty$, take $\Delta\phi$ as Yukawa:

$$\Delta\phi(r) = e^2/r\, e^{-\gamma r}\,, \qquad \Delta\phi_k = 4\pi\, e^2/(k^2 + \gamma^2); \tag{6.7}$$

then indeed

$$\bar{\phi}_k = 4\pi\, e^2 \gamma^2/k^2(k^2 + \gamma^2) \tag{6.8}$$

is positive definite, and

$$\bar{\phi}(0) = \gamma e^2. \tag{6.9}$$

Adjusting γ for the strongest restriction, we conclude from (6.5) that

$$\Phi[n] \geq \max_\gamma \left[\tfrac{1}{2} \int \int \tfrac{e^2}{|r_{12}|}(1 - e^{-\gamma|r_{12}|}) n(r_1)\, n(r_2)\, d^3 r_1\, d^3 r_2 - \tfrac{1}{2}\gamma e^2 N \right], \tag{6.10}$$

with γ thereby determined by

$$\int \int e^{-\gamma|r_{12}|}\, n(r_1)\, n(r_2)\, d^3 r_1\, d^3 r_2 = N, \tag{6.11}$$

and the associated profile by (2.2) and (3.1) with any suitable $T[n]$.

It is of course not necessary to guess the form of $\Delta\phi$. For a more systematic procedure which again focuses on the pair distribution n_2 (and therefore does not produce quite as tight a lower bound as consideration of the full underlying N-body distribution would) we can proceed as follows, taking advantage of our ability to assume a fixed N system. The problem is now to minimize $\int \int n_2(r, r') \phi(r - r') d^3r \, d^3r'$, subject to the restriction

$$\int n_2(r, r') d^3r' = (N - 1) n(r). \tag{6.12}$$

For this purpose, we start with a correlated guess for $n_2(r, r')$, say $\bar{}(r, r') \geq 0$, and try to determine $w(r)$ so that

$$n_2(r, r') = w(r) f(r, r') w(r') \tag{6.13}$$

satisfies (6.12), finally minimizing with respect to $f(r, r')$. The difficulty lies in that of finding the correction factor $w(r)$. However, (6.12) and (6.13) can be cast in the iterative form

$$w(r) = \frac{1}{N - 1} \frac{n(r)}{\int f(r, r') w(r') d^3r'}, \tag{6.14}$$

starting e.g. with $w(r) = 1$ on the right hand side. This iteration converges and can be used as an approximation after a few passes.

REMARKS

We have seen that density functional theory can be cast in a form in which guaranteed ground state energy lower bounds are obtained, depending upon the fashion in which the system is developed into reference plus perturbation. The simplest version, in which the reference is nonuniform but noninteracting, accounts for correlation effects but not exchange effects in the interaction energy. This version has been examined in some detail, with kinetic energy and interaction components giving rise both to exact bounds and to approximations which should be very effective for coulomb interactions. The possibility of using reference systems which incorporate preliminary interactions as well should provide improvement at low computational cost, and is now under study.

REFERENCES

1. R. G. Parr and W. Yang, "Density Functional Theory of Atoms and Molecules," Oxford University Press, New York (1989).
2. E. R. Davidson, "Reduced Density Matrices in Quantum Chemistry," Academic Press, New York (1976).
3. J. K. Percus, *Int. J. Quantum Chem.* XIII:89 (1978).
4. R. P. Feynman, *Phys. Rev.* 56:340 (1939).
5. J. K. Percus, *Ann. N. Y. Acad. Sci.* 491:36 (1987).
6. J. K. Percus, *in:* "Statistical Mechanics," Rice, Freed, and Light, Eds., University of Chicago Press, Chicago (1972).
7. W. Kohn and L. J. Sham, *Phys. Rev.* 140:A1133 (1965).
8. See e.g. F. B. Hildebrand "Methods of Applied Mathematics," p. 451, Prentice-Hall, New York (1952).
9. E. Witten, *Nucl. Phys.* B185:513 (1981).
10. A. Erdelyi et al., "Tables of Integral Transforms, Vol. I," p. 143, McGraw-Hill, New York (1954).
11. A. Erdelyi et al, "Higher Transcendental Functions, Vol. II," p. 79, McGraw-Hill, New York (1953).

SOLUTION OF THE ORNSTEIN-ZERNIKE EQUATION FOR A MIXTURE OF STICKY
HARD SPHERES AND YUKAWA CLOSURE

J.N. Herrera

Escuela de Ciencias Físico-Matemáticas
Universidad Autónoma de Puebla
Apdo. Postal 1152
C. P. 72001 México

L. Blum

Department of Physics, P.O. Box 23343, University of
Puerto Rico, Río Piedras, PR 00931-3343
and

Fernando Vericat

Instituto de Física de Líquidos y Sistemas
Biológicos (IFLYSIB),c.c 565 (1900), and
Departamento de Físicomatematicas, Facultad de
Ingeniería, Universidad Nacional de La Plata, La
Plata, Argentina

ABSTRACT

We consider the solution of the Ornstein-Zernike equation for the
most general closure consisting of a sum of M Yukawa type
exponentials.

$$c_{ij}(r) = \sum_{n=1}^{M} \hat{K}_{ij}^{(n)} e^{-z_n(r-\sigma_{ij})}/r$$

A formal solution was found for an arbitrary mixture of hard
spheres in previous work. We study here the limiting case when one of
the exponentials, labelled s becomes infinitely attractive with zero
range: $\hat{K}_{ij}^{(s)} \to \infty$ and $z_s \to \infty$, but

$$\hat{K}_{ij}^{(s)}/z_s = \theta_{ij}^{(s)}$$

In this limit the s potential is equal to Baxter's sticky
potential, as was shown Mier y Terán and co-workers.[1] We obtain formal
equations for the sticky plus Yukawa potential.

Condensed Matter Theories, Vol. 7, Edited by A.N. Proto
and J.L. Aliaga, Plenum Press, New York, 1992

1. INTRODUCTION

Analytical solutions of equations for fluid mixtures are necessary to study phase transitions. Since the early work of Lebowitz,[2] much progress has been achieved. In particular, the solution of the Yukawa closure of the Ornstein- Zernike (OZ) equation by Waisman[3] has made possible a number of extensions and generalizations to rather general closures of arbitrary mixtures of spherical objects.[4-11]   There has been a number of very interesting calculations using these solutions.[12-15]

The solution of the general closure of the hard core OZ equation

$$c_{ij}(r) = \sum_{n=1}^{M} \hat{K}_{ij}^{(n)} e^{-z_n(r-\sigma_{ij})}/r \tag{1}$$

was discussed in an earlier publication.[16]

A very interesting case is the sticky potential of Baxter,[17] which provides an approximate solution to the general Yukawa closure, and is a physically interesting model in itself.[18,19]

In a recent paper, Mier y Terán and co-workers[1] have shown that the limit which in our case corresponds to

$$\hat{K}_{ij}^{(s)} \to \infty$$

$$z_s \to \infty$$

so that

$$\hat{K}_{ij}^{(s)}/z_s = \theta_{ij}^{(s)}. \tag{2}$$

We remark that we have used a different definition for the interaction coefficient

$$\hat{K}_{ij}^{(s)} = K_{ij}^{(s)} e^{z_s \sigma_{ij}}. \tag{3}$$

We call this limit the sticky limit (SL) and use the superscript s throughout. In section 2 we review the solution of the general mixture with a M Yukawa closure. In section 3 we discuss the SL for a single Yukawa. Finally in section 4 we present the solution for the case of M Yukawas plus a sticky interaction in the MSA (Mean Spherical Approximation).

2. BASIC FORMALISM

We study the Ornstein-Zernike (OZ) equation

$$h_{ij}(12) = c_{ij}(12) + \sum_k \int d3 h_{ik}(13)\rho_k c_{kj}(32) \tag{4}$$

where $h_{ij}(12)$ is the molecular total correlation function and $c_{ij}(12)$ is the molecular direct correlation function, ρ_i is the number density of the molecules i, and $i = 1,2$ is the position $\vec{r}_i$, $r_{12} = |\vec{r}_1 - \vec{r}_2|$ and σ_{ij} is the distance of closest approach of two particles i,j. The direct correlation function is

$$c_{ij}(r) = \sum_{n=1}^{M} K_{ij}^{(n)} e^{-z_n r}/r, \qquad r > \sigma_{ij} \tag{5}$$

and the pair correlation function is

$$h_{ij}(r) = g_{ij}(r) - 1 = -1., \qquad r \leq \sigma_{ij} \tag{6}$$

We use the Baxter-Wertheim[17,20] (BW) factorization of the OZ equation

$$\left[\mathbf{I} + \rho\tilde{\mathbf{H}}(\mathbf{k})\right]\left[\mathbf{I} - \rho\check{\mathbf{C}}(k)\right] = \mathbf{I} \tag{7}$$

where $\mathbf{I}$ is the identity matrix, and we have used the notation

$$\tilde{\mathbf{H}}(k) = 2\int_0^\infty dr \cos(kr)\mathbf{J}(r) \tag{8}$$

$$\check{\mathbf{C}}(k) = 2\int_0^\infty dr \cos(kr)\mathbf{S}(r) \tag{9}$$

The matrices $\mathbf{J}$ and $\mathbf{S}$ have matrix elements

$$J_{ij}(r) = 2\pi\int_r^\infty ds\, s h_{ij}(s) \tag{10}$$

$$S_{ij}(r) = 2\pi\int_r^\infty ds\, s c_{ij}(s) \tag{11}$$

$$\left[\mathbf{I} - \rho\check{\mathbf{C}}(k)\right] = \left[\mathbf{I} - \rho\check{\mathbf{Q}}(k)\right]\left[\mathbf{I} - \rho\check{\mathbf{Q}}^T(k)\right] \tag{12}$$

where $\check{\mathbf{Q}}^T(-k)$ is the complex conjugate and transpose of $\check{\mathbf{Q}}(k)$. The first matrix is non-singular in the upper half complex k-plane, while the second is non-singular in the lower half complex k-plane.

It can be shown that the factored correlation functions must be of the form

$$\check{\mathbf{Q}}(k) = \mathbf{I} - \rho\int_{\lambda_{ji}}^\infty dr e^{ikr}\check{\mathbf{Q}}(r) \tag{13}$$

where we used the following definition

$$\lambda_{ji} = \frac{1}{2}(\sigma_j - \sigma_i) \tag{14}$$

$$\mathbf{S}(r) = \mathbf{Q}(r) - \int dr_1 \mathbf{Q}(r_1)\rho\mathbf{Q}^T(r_1 - r) \tag{15}$$

Similarly, from eq. (12) and eq. (7) we get, using the analytical properties of $\mathbf{Q}$ and Cauchy's theorem

$$\mathbf{J}(r) = \mathbf{Q}(r) + \int dr_1 \mathbf{J}(r - r_1)\rho\mathbf{Q}(r_1) \tag{16}$$

The general solution is discussed in,[9,11] and yields

$$q_{ij}(r) = q_{ij}^0(r) + \sum_{n=1}^{M} D_{ij}^{(n)} e^{-z_n r} \qquad \lambda_{ji} < r \tag{17}$$

$$q_{ij}^0(r) = (1/2)A_j[(r - \sigma_j/2)^2 - (\sigma_i/2)^2] + \beta_j[(r - \sigma_j/2) - (\sigma_i/2)]$$

$$+ \sum_{n=1}^{M} C_{ij}^{(n)} e^{-z_n \sigma_j/2} [e^{-z_n(r-\sigma_j/2)} - e^{-z_n \sigma_i/2}] \qquad \lambda_{ji} < r < \sigma_{ij}. \tag{18}$$

The solution of this system of equations leads to

$$\beta_j = \frac{\pi}{\Delta} \sigma_j + \frac{2\pi}{\Delta} \sum_n \mu_j^{(n)} \tag{19}$$

and

$$A_j = \frac{2\pi}{\Delta} \left[1 + (1/2)\zeta_2 \beta_j + \sum_n M_j^{(n)} \right]. \tag{20}$$

Furthermore the coefficients of all the exponentials must satisfy eq. (16)

$$C_{ij}^{(n)} + D_{ij}^{(n)} = \sum_k \gamma_{ik}^{(n)} D_{kj}^{(n)} \tag{21}$$

We have used

$$\zeta_n = \sum_k \rho_k \sigma_k^n \tag{22}$$

$$\Delta = 1 - \pi\zeta_3/6 \tag{23}$$

$$\gamma_{ij}^{(n)} = 2\pi \tilde{g}_{ij}(z_n) \rho_j/z_n \tag{24}$$

$$\mu_j^{(n)} = \sum_k \rho_k C_k^\mu(z_n) D_{kj}^{(n)} e^{-z_n \sigma_{kj}} \tag{25}$$

$$M_j^{(n)} = \sum_k \rho_k C_k^M(z_n) D_{kj}^{(n)} e^{-z_n \sigma_{kj}} \tag{26}$$

$$\tilde{g}_{ij}(s) = \int_0^\infty dr r g_{ij}(r) e^{-sr} \tag{27}$$

$$C_k^\mu(z_n) = \sum_\ell \sigma_\ell^2 e^{z_n \sigma_{k\ell}} \gamma_{k\ell}^{(n)} z_n \sigma_\ell \psi_1(z_n \sigma_\ell) + \frac{1 + z_n \sigma_k/2}{z_n^2} \tag{28}$$

$$C_k^M(z_n) = \sum_\ell \sigma_\ell e^{z_n \lambda_{k\ell}} \gamma_{k\ell}^{(n)} z_n \sigma_\ell \phi_1(-z_n \sigma_\ell) - \frac{1 + z_n \sigma_k}{z_n} \tag{29}$$

with

$$\psi_1(x) = [1 - x/2 - (1 + x/2)e^{-x}]/(x^3) \tag{30}$$

$$\phi_1(x) = [1 - x - e^{-x}]/(x^2) \tag{31}$$

$$\phi_0(x) = [1 - e^{-x}]/(x) \tag{32}$$

The following relation is also useful

$$x^2 \psi_1(x) = -1 + (1 + x/2)\phi_0(x) \tag{33}$$

Some quantities will be of interest:

$$q_{ij}(\lambda_{ji}) = -\sigma_i \beta_j + \sum_m \left[(C_{ij}^{(m)} + D_{ij}^{(m)}) e^{-z_m \lambda_{ji}} - C_{ij}^{(m)} e^{-z_m \sigma_{ji}} \right] \tag{34}$$

which can be expressed as

$$q_{ij}(\lambda_{ji}) = -\sigma_i \sigma_j \frac{\pi}{\Delta} - \sum_m F_{ij}^{(m)} \tag{35}$$

156

where we have defined the convenient quantity

$$F_{ij}^{(m)} = \frac{2\pi}{\Delta}\sigma_i\mu_j^{(m)} - (C_{ij}^{(m)} + D_{ij}^{(m)})e^{-z_m\lambda_{ji}} + C_{ij}^{(m)}e^{-z_m\sigma_i} \tag{36}$$

or

$$F_{ij}^{(m)} = \frac{2\pi}{\Delta}\sigma_i\mu_j^{(m)} - \left[(C_{ij}^{(m)} + D_{ij}^{(m)})(1 - e^{-z_m\sigma_i}) + D_{ij}^{(m)}e^{-z_m\sigma_i}\right]e^{-z_m\lambda_{ji}} \tag{37}$$

Because of the symmetry of the direct correlation function, we require from eq. (15)

$$q_{ij}(\lambda_{ji}) = q_{ji}(\lambda_{ij}) \tag{38}$$

which implies that $F_{ij}^{(m)}$ must satisfy the symmetry relation

$$F_{ij}^{(m)} = F_{ji}^{(m)} \tag{39}$$

Using the relations (25) and (26)

$$M_j^{(m)} + \mu_j^{(m)}(z_m + \frac{\pi}{\Delta}\zeta_2) = (1/2)S_j^{(m)} \tag{40}$$

with

$$S_j^{(m)} = \sum_\ell \rho_\ell\sigma_\ell F_{\ell j}^{(m)}. \tag{41}$$

We change eq. (20) to

$$A_j = A_j^0 + \frac{\pi}{\Delta}\sum_m \left[S_j^{(m)} - 2z_m\mu_j^{(m)}\right] \tag{42}$$

where

$$A_j^0 = \frac{2\pi}{\Delta}\left[1 + (1/2)\zeta_2\frac{\pi}{\Delta}\sigma_j\right]. \tag{43}$$

We obtain the contact pair correlation function from the discontinuity of the first derivative of the factor function $q_{ij}(r)$ eq. (17):

$$y_{ij}^{(0)} \equiv 2\pi\sigma_{ij}g_{ij}(\sigma_{ij}) = q_{ij}'(\sigma_{ji}^-) - q_{ij}'(\sigma_{ji}^+) = A_j(\sigma_i/2) + B_j - \sum_{m=1}^M z_m C_{ij}^{(m)}e^{-z_m\sigma_{ij}}. \tag{44}$$

Using the continuity relation[7,22]

$$q_{ij}'(\lambda_{ji}) + q_{ji}'(\lambda_{ij}) = -\sum_k \rho_k q_{ij}(\lambda_{ki})q_{ij}(\lambda_{kj}) \tag{45}$$

and

$$q_{ij}'(\lambda_{ji}) = -A_j(\sigma_i/2) + B_j - \sum_{m=1}^M z_m(C_{ij}^{(m)} + D_{ij}^{(m)})e^{-z_m\lambda_{ji}} \tag{46}$$

we get the following relation for the contact pair distribution function

$$2\pi\sigma_{ij}g_{ij}(\sigma_{ij}) - 2\pi\sigma_{ij}g_{ij}^0(\sigma_{ij}) = (1/2)\sum_m\left[z_m(F_{ij}^{(m)} + F_{ji}^{(m)}) - \sum_{k,n}\rho_k F_{kj}^{(m)}F_{ki}^{(n)}\right] \tag{47}$$

where $g_{ij}^0(\sigma_{ij})$ is the contact pair distribution function of the hard sphere reference system. From eq. (16) we can show that the

Laplace transform of the pair correlation function must satisfy the consistency relation

$$2\pi \sum_{\ell} \tilde{g}_{i\ell}(z_n)[\delta_{\ell j} - \rho_{\ell}\tilde{q}_{j\ell}(iz_n)] = \tilde{q}_{ij}^{0'}(iz_n) \tag{48}$$

where

$$\tilde{q}_{ij}^{0'}(iz_n) = \int_{\sigma_{ij}}^{\infty} dr e^{-z_n r}[q_{ij}^0(r)]'$$

$$= [(1 + z_n\sigma_i/2)A_j + z_n\beta_j]e^{-z_n\sigma_{ij}}/z_n^2 - \sum_m \frac{z_m}{z_n + z_m}e^{-(z_n+z_m)\sigma_{ij}}C_{ij}^{(m)}. \tag{49}$$

The Laplace transform of Eqs.(17) and (18) yields

$$e^{z_n\lambda_{ji}}\tilde{q}_{ij}(iz_n) = \sigma_i^3\psi_1(z_n\sigma_i)A_j + \sigma_i^2\phi_1(z_n\sigma_i)\beta_j$$

$$+ \sum_m \frac{1}{z_n + z_m}\left[(C_{ij}^{(m)} + D_{ij}^{(m)})e^{-z_m\lambda_{ji}} - C_{ij}^{(m)}e^{-z_m\sigma_{ji}} - z_m z_n\sigma_i\phi_0(z_n\sigma_i)C_{ij}^{(m)}e^{-z_m\sigma_{ji}}\right]. \tag{50}$$

Using eq.(44) we get

$$\tilde{q}_{ij}^{0'}(iz_n) = e^{-z_n\sigma_{ij}}/z_n^2[A_j + z_n y_{ij}^{(0)} + z_n \sum_m \frac{z_m^2}{z_n + z_m}e^{-z_m\sigma_{ij}}C_{ij}^{(m)}] \tag{51}$$

After some lengthy but straightforward algebra we find the following simplification of the result of Ginoza[11]

$$\Pi_{ij}^{(n)} = \sum_m \sum_t \frac{e^{-z_m\sigma_{tj}}}{z_n + z_m}D_{tj}^{(m)}\sum_{\ell}\rho_{\ell}\left[z_m\Omega_{i\ell}^{(n)}\Omega_{i\ell}^{(m)} - \Omega_{i\ell}^{(m)}\Pi_{i\ell}^{(n)} + \Omega_{i\ell}^{(n)}\Pi_{i\ell}^{(m)}\right] \tag{52}$$

where

$$\Omega_{ij}^{(m)} = C_i^{\mu}(z_m)\frac{2\pi}{\Delta}\rho_i\sigma_j - \gamma_{ji}^{(m)}z_m\sigma_j\phi_0(z_m\sigma_j)e^{z_m\sigma_{ij}} - \delta_{ij} \tag{53}$$

and

$$\Pi_{ij}^{(m)} = -C_i^{\mu}(z_m)\frac{2\pi}{\Delta}\rho_i(1 + z_m\sigma_j/2) + \gamma_{ji}^{(m)}z_m e^{z_m\sigma_{ji}} - \frac{\pi}{2\Delta}\sigma_j\sum_{\ell}\rho_{\ell}\sigma_{\ell}\Omega_{i\ell}^{(m)} \tag{54}$$

Notice that

$$F_{ij}^{(m)} = \sum_{\ell}\Omega_{\ell i}^{(m)}D_{\ell j}^{(m)}e^{-z_m\sigma_{ji}} \tag{55}$$

and also that $\Pi_{ij}^{(m)}$ and $\Omega_{ij}^{(m)}$ are functions of the same set of parameters, and therefore

$$\delta_{ij} + \Omega_{ij}^{(m)} + \sigma_j\phi_0(z_m\sigma_j)\Pi_{ij}^{(m)} = -\frac{2\pi}{\Delta}\left[C_i^{\mu}(z_m)\rho_i z_m^2\sigma_j^3\psi_1(z_m\sigma_j) + \sigma_j^2\phi_0(z_m\sigma_j)\sum_{\ell}\rho_{\ell}\sigma_{\ell}\Omega_{i\ell}^{(m)}\right] \tag{56}$$

The MSA closure condition (5) yields

$$2\pi K_{ij}^{(n)}/z_n = \sum_{\ell}D_{i\ell}^{(n)}[\delta_{\ell j} - \rho_{\ell}\tilde{q}_{j\ell}(iz_n)]. \tag{57}$$

This expression can be combined with (48) to obtain an expression for the excess MSA energy due to the interaction $'n'$

$$\sum_{\ell}\gamma_{j\ell}^{(n)}K_{i\ell}^{(n)} = \frac{1}{2\pi}\sum_{\ell}\rho_{\ell}D_{i\ell}^{(n)}\tilde{q}_{j\ell}^{0'}(iz_n) \tag{58}$$

158

Eqs. (52) and (58) are the full solution of the general problem, since we have a set of algebraic equations for $\gamma_{ji}^{(m)}$ as a function of $K_{ij}^{(m)}$: We must however first solve eq. (52) for the unknown $D_{ij}^{(m)}$.

3.   THE ONE YUKAWA CASE AND THE LIMIT OF STICKY HARD SPHERES MIXTURE

When the Yukawa closure has only one set of exponentials, that is

$$c_{ij}(r) = K_{ij}e^{-zr}/r \tag{59}$$

then the equations of the preceding section simplify considerably, if we take the strong screening limit, the range of the interaction is reduced to σ_{ij}, i.e. we get a potential of the form of a delta function $(\delta(r - \sigma_{ij}))$, which is precisely the sticky potential of Baxter[17](Also studied by Perram and Smith,[18] and Barboy and Tenne[19]).

We take the sticky limit (SL), of the Yukawa potential which corresponds to

$$\hat{K}_{ij}^{(s)} \to \infty$$

$$z_s \to \infty$$

so that

$$\hat{K}_{ij}^{(s)}/z_s = \theta_{ij}^{(s)}. \tag{60}$$

In this limit all the equations of the preceding section are simplified. The SL the asymptotic value of $\gamma_{ij}^{(s)}$ is:

$$\gamma_{ij}^{(s)} \to e^{-z_s\sigma_{ij}}/z_s^3 \tag{61}$$

Similarly,

$$\psi_1(z_s\sigma_i) \to -1/z_s^2,$$

$$\phi_1(z_s\sigma_i) \to -1/z_s,$$

and

$$\phi_0(z_s\sigma_i) \to 1/\sigma_i z_s. \tag{62}$$

Furthermore, from Eqs. (57) and (62) we must have

$$D_{ij}^{(s)} \to 2\pi\theta_{ij}^{(s)}e^{z_s\sigma_{ij}} \tag{63}$$

and from eq.   (21)

$$C_{ij}^{(s)} \to -2\pi\theta_{ij}^{(s)}e^{z_s\sigma_{ij}} \tag{64}$$

$$\mu_j^{(s)} \to (2\pi/z_s)\sum_k \rho_k\theta_{kj}^{(s)}e^{z_s\sigma_{kj}} \tag{65}$$

Now, we can simplified eq. (36), the result is

$$F_{ij}^{(s)} \to -2\pi\theta_{ij}^{(s)}. \tag{66}$$

It is evident that $F_{ij}^{(s)} = F_{ji}^{(s)}$. In same limit (SL) eq. (29) is given by

$$C_k^M(z_s) \to \sigma_k \tag{67}$$

then, it is easy to prove that

$$M_j^s \to (-2\pi) \sum_k \rho_k \theta_{kj}^{(s)} e^{z_s \sigma_{kj}} \tag{68}$$

the later results permits to obtain the factor correlation function in the SL, which in this case is

$$q_{ij}(r) = q_{ij}^0(r) + 2\pi \theta_{ij}^{(s)}. \tag{69}$$

Now, if we call $t_{ij} = 2\pi \theta_{ij}^{(s)}$, and we identify t_{kj} with the expression of Barboy-Tenne, then we have

$$t_{kj} = \lambda_{kj}^{(s)} \sigma_k \sigma_j / (12\Delta) \tag{70}$$

here $\lambda_{kj}^{(s)}$ is the sticky parameter, which is proportional to the probability of adherence. Using this equality, we get:

$$M_j^s = -\frac{\pi \sigma_j}{12\Delta} \sum_k \rho_j \sigma_k^2 \lambda_{kj}^{(s)}. \tag{71}$$

Substituting the values of the parameters in SL the coefficients of q_{ij}^0 are given by the following expressions

$$\beta_j^{(s)} = \frac{\pi}{\Delta} \sigma_j \tag{72}$$

and

$$A_j = \frac{2\pi}{\Delta} \left[1 + (1/2)\zeta_2 \beta_j - \frac{\pi \sigma_j}{12\Delta} \sum_k \rho_j \sigma_k^2 \lambda_{kj}^{(s)} \right]. \tag{73}$$

The expression that we obtain by the factor correlation function, in the SL is the same of Barboy-Tenne.[19]

4. THE CASE OF STICKY MIXTURE WITH M YUKAWA CLOSURE

We consider now the somewhat more complicated case of a sticky mixture with M Yukawa closure, i.e., the MSA for this new system is:

$$c_{ij}(r) = \sum_{n=1}^{M} \hat{K}_{ij}^{(n)} e^{-z_n(r-\sigma_{ij})}/r + c_{ij}^{(s)}(r) \tag{74}$$

where

$$c_{ij}^{(s)}(r) = \hat{K}_{ij}^{(s)} e^{-z_s(r-\sigma_{ij})}/r \tag{75}$$

and $\hat{K}_{ij}^{(n)}$ and z_s satisfies the (60). In the case of a mixture sticky spheres with M Yukawa closure, we obtain the factor correlation function, which is represented by the expression:

$$q_{ij}(r) = q_{ij}^{(Y)}(r) + q_{ij}^{(s)}(r) \qquad \lambda_{ji} < r \tag{76}$$

here $q_{ij}^{(Y)}(r)$ is the same as eq. (17), but the coefficients are now

$$\beta_j = \frac{\pi}{\Delta} \sigma_j + \frac{2\pi}{\Delta} \sum_n \mu_j^{(n)} \tag{77}$$

and

$$A_j = \frac{2\pi}{\Delta} \left[1 + (1/2)\zeta_2\beta_j + \sum_n M_j^{(n)} - \frac{\pi\sigma_j}{12\Delta} \sum_k \rho_j\sigma_k^2\lambda_{kj}^{(s)} \right]$$ (78)

and the sticky contribution can be written as:

$$q_{ij}^{(s)}(r) = 2\pi\theta_{ij}^{(s)}.$$ (79)

We can use a similar procedure to study more complicated systems. This will be done in future work.

ACKNOWLEDGMENT

We acknowledge support from the National Science Foundation through grants NSF-CHE-89-01597, NSF-INT-89-01291 and Epscor RII-86-10677. The authors are grateful to E. Cantoral and E. García Llanos for very helpful observations.

REFERENCES

[1] L. Mier y Terán, E. Corvera and A. E. Gonzáles, Phys. Rev. A39: 371 (1989).

[2] J.L.Lebowitz, Phys. Rev. A133: 895 (1964).

[3] E. Waisman Mol. Phys. 25: 45 (1973).

[4] J.S. Høye and G. Stell, Mol. Phys. 32: 195 (1976).

[5] E. Waisman, J.S. Høye and G. Stell Chem. Phys. Letters 40: 514 (1976).

[6] J.S. Høye, G. Stell and E. Waisman Mol. Phys. 32: 209 (1976).

[7] J. S. Høye and L. Blum, J. Stat. Phys. 16: 399 (1977).

[8] L. Blum and J. S. Høye, J. Stat. Phys. 19: 317 (1978).

[9] L. Blum , J. Stat. Phys. 22: 661 (1980).

[10] L. Blum, Mol. Phys. 30: 1529 (1975).

[11] M.Ginoza, J.Phys. Soc. Japan 55: 95 (1986).

[12] C.Jedrzejek, J. Konior and M. Streszewski, Phys. Rev.A35: 1226 (1987).

[13] E. Arrieta, C.Jedrzejek and K.N. Marsh, J. Chem.Phys. 86: 3607 (1987).

[14] G. Giunta, M. C. Abramo and C. Caccamo Mol. Phys. 56: 319 (1985),

[15] D. J. González, M. J. González and M.Silbert, Mol. Phys. 71: 157 (1990).

[16] L. Blum, F.Vericat, and J. N. Herrera, J. Stat. Phys.   (in press) (19xx).

[17] R. J. Baxter, J.Chem. Phys. 49: 2770 (1968).

[18] J. W. Perram and E. R. Smith, Chem. Phys. Letters 35, 138 (1975).

[19] B. Barboy and R. Tenne, Chem. Phys. 38, 369 (1979).

[20] M.S.Wertheim, J.Math. Phys. 5: 643 (1964).

[21] E. Waisman and J. L. Lebowitz, J. Chem. Phys. 52: 4307 (1970).

[22] M.S.Wertheim, J.Chem.Phys. 88: 1214 (1988).

DYNAMICAL PROPERTIES OF

STRONGLY COUPLED COULOMB SYSTEMS

G. Kalman

Department of Physics
Boston College
Chestnut Hill, MA 02167

I. INTRODUCTION

Coulomb systems (i.e. many particle systems where the interaction be ween the particles is predominantly due to Coulomb forces) are ubiquitous in nature, although the guises in which they appear can be quite different. Laboratory plasmas, space plasmas, stellar interiors, solid and liquid metals, electron-hole liquids, ionic crystals and liquids, electrolytes, two dimensional electron films, semiconductor hetero-structures are the most immmediate examples. In some of these systems particle correlations do not play a significant role; in others correlations are important or even crucial. These latter are the ones of interest for the present discussion.

The measure of the correlations or coupling is the ratio of the potential and kinetic energies in the system. As long as the system can be regarded classical the latter can be characterized by the temperature, and if only one species of particles is of interest the coupling parameter is

$$\Gamma = Z^2 e^2 / akT$$

with a being the Wigner-Seitz radius, $\frac{4\pi}{3}a^3 n = 1$. In the other limit, for a zero temperature electron gas the kinetic energy is given by the Fermi energy, and the coupling parameter is

$$r_s = a/a_B$$

where $a_B = \dfrac{h^2}{e^2 m}$ is the Bohr radius. Even though classical and quantum systems are in many ways different, inference can be drawn for one system from information pertaining to the other: in order to do this the useful correspondence relation (note that $Z=1$)

$$\Gamma \rightarrow 1.36\, r_s$$

based on the comparison of the potential energy/kinetic energy ratios, can be established. The strong coupling domain is now characterized by

$$\Gamma \gtrsim 1 \text{ or } r_s \gtrsim 1$$

Condensed Matter Theories, Vol. 7, Edited by A.N. Proto
and J.L. Aliaga, Plenum Press, New York, 1992

The questions that can be posed concerning the physical behavior of these systems pertain partly to the static (time- or frequency- independent) or to the dynamic (time- or frequency-dependent) behavior of the system. Examples for the first category are problems relating to the pair correlation function, equation of state, phase transition, transport coefficients, etc. In the second category belong the related problems of response functions, collective modes, dynamical structure functions, etc. It is these dynamical problems that we wish to address in the present paper.

The simplest model for a Coulomb system is the one component plasma (OCP), consisting of one single species of particles dispersed in a neutralizing background. In addition to the normal three dimensional (3-d) OCP, a two-dimensional (2-d) OCP can serve as a physical model for 2-d electron films and layers. With two different ion species dispersed in a neutralizing background, one of faced with the more involved model of a binary ionic mixture (BIM). These are the systems we consider in the present paper. The more realistic models with two oppositely charged species present additional problems (bound states, etc.) in the strong coupling domain, which are outside the scope of this description.

When correlations are negligible (formally $\Gamma = 0$), the Vlasov or Random Phase Approximation (RPA) is appropriate. In this approximation the longitudal dielectric response functions $\varepsilon_L(k\omega)$ and $\varepsilon_T(k\omega)$ are given by

$$\varepsilon_L(k\omega) = 1 - \varphi(k) \, \chi_{L0}(k\omega)$$

$$\varepsilon_T(k\omega) = 1 + \varphi(k)\frac{k^2 c^2}{\omega^2} \chi_{T0}(k\omega)$$

where

$$\varphi(k) = \frac{4\pi e^2 Z^2}{k^2}$$

represents the Coulomb potential and $\chi_{L0}(k\omega)$ and $\chi_{T0}(k\omega)$ are the density and current response functions of the non-interacting (ideal) gas. The salient features of the collective mode spectrum in this approximation are the following:

(i)   The longitudinal plasmon mode $\omega_L(k)$ for 3-d systems starts, even for

multicomponent plasmas, at the plasma frequency $\omega_p = \left\{ \sum\limits_A \frac{4\pi Z_A^2 e^2 n_A}{m_A} \right\}^{1/2}$ and

for finite $\underline{k}$-values develops a positive dispersion $\left(\frac{d\omega}{dk} > 0\right)$; for 2-d systems $\omega_L(k) \sim \sqrt{k}$.

(ii)   The only damping mechanism is the collisionless Landau damping, which vanishes exponentially for $k \to 0$.

(iii)   There is no genuine shear generated transverse mode (shear mode) in the system: this is manifested by the fact that in the $c \to \infty$ limit the existing transverse photon mode doesn't survive.

These features may be contrasted with the behavior in the very strong coupling ($\Gamma \gg 1$) limit. It is known that the 3-d OCP crystallizes into a bcc lattice around $\Gamma = \Gamma_m = 178$ and the 2-d OCP into a hexagonal lattice around $\Gamma = \Gamma_m = 137$. In the lattice the collective excitations are the optical and acoustic plasmons, whose relevant features now can be listed as follows [1,2]:

(i)   The equivalent of the plasmon mode, the longitudinal optical phonon $\omega_L(k)$ for the 3-d OCP starts at the plasma frequency and for finite $\underline{k}$ develops a negative dispersion; for the 2-d OCP $\omega_L(k) \sim \sqrt{k}$ only for small $\underline{k}$, for higher $\underline{k}$-values a maximum develops.

(ii)   There exist two transverse shear modes $\omega_{T1}(k)$, $\omega_{T2}(k)$ which for small k-values have acoustic type $\omega_T(k) \sim k$ dispersion both in the 3-d and in the 2-d system; in addition in the 3-d lattice the phonons satisfy the Kohn sum rule: $\Sigma\omega_i^2(k) = \omega_p^2$.

(iii)   In the linear approximation the phonon modes are undamped.

The theoretical challenge is to understand the development of the dynamical properties of Coulomb systems as correlations become more important from $\Gamma=0$ through $\Gamma >> 1$.

II. APPROXIMATION METHODS

The general correlational problem can be approached from different angles. Starting from the RPA, a systematic perturbation technique of the BBGKY hierarchy can be worked out [3,4,5]. This technique has been around for a long time, but by its very nature, perturbation technique is limited to the $\Gamma < 1$ domain. From the strong coupling side one can generate an approximation [6] which exploits the main feature of the strongly coupled system, namely the quasi-localization of the particles (Quasilocalized Charges-QLC-model). A quite general method, valid in principle for arbitrary values of the coupling, is the Mean Field Theory [7], (MFT) which represents the dielectric function as

$$\varepsilon(k\omega)=1 - \varphi(k)\frac{\chi_{L0}(k\omega)}{1+\varphi(k)G(k\omega)\chi_{L0}(k\omega)}$$

$$\varepsilon_T(k\omega) = 1 + \varphi(k)\frac{k^2c^2}{\omega^2}\frac{\chi_{T0}(k\omega)}{1-\varphi(k)\frac{k^2c^2}{\omega^2}H(k\omega)\chi_{T0}(k\omega)}$$

(1)

and collects [8] all correlational effects in the frequency and wavenumber dependent mean fields $G(k\omega)$ and $H(k\omega)$. We have detailed experience with the longitudinal $G(k\omega)$ only: the frequency dependence of this latter is crucial for the correct treatment of dynamical effects and its proper inclusion distinguishes the Dynamical Mean Field Theory (DMFT) from static mean field theories with frequency independent $G(k)$-s.

In this paper we discuss recently obtained results [6,8,9,10] through the application of QLC to various physical systems, such as the 3-d and 2-d OCP-s and the BIM. We show how the QLC description accounts for the profound modification of the longitudinal plasmon mode in the strong coupling domain and how the transverse shear mode appears as an immediate consequence of the quasi-localization. Next we combine it with the DMFT proposed by Golden and Kalman [11] and we show how a reliable $\varepsilon_L(k\omega)$ can be obtained for the 3-d OCP over a broad range of Γ-values. For the transverse $\varepsilon_T(k\omega)$ and for the properties of the 2-d system we have only the less accurate, but from the point of view of the dispersion still quite satisfactory static MFT.

III. QUASI LOCALIZED CHARGES

In this Section we review the results obtained through the Quasi Localized Charge model. [6,8,9,10] Those results include (i) plasmon dispersion for the 3-d OCP; (ii) shear mode dispersion for the 3-d OCP; (iii) plasmon dispersion for the 2-d OCP; (iv) shear mode dispersion for the 2-d OCP; (v) plasma frequency shift for the BIM.

As stated before, the principal idea of the QLC model is that for strong coupling ($\Gamma \geq 10$) the particles are trapped in local potential minima and oscillate around their quasi-equilibrium positions. Averaging over these quasi-sites the Hamiltonian for the particles can be written down in terms of a dynamical matrix $D_{\alpha\beta}^{AB}(\mathbf{k})$ (A,B, etc. are species indices, α, β, etc. designate Cartesian components; $\omega^A(\mathbf{q})$ is the plasma frequency of species A; $g_{AB}(\mathbf{k})$ is the pair correlation function between species A and B.)

$$D_{\alpha\beta}^{AB}(\mathbf{k}) = \frac{1}{V} \sum_q \frac{q_\alpha q_\beta}{q^2} \omega^A(\mathbf{q})\omega^B(\mathbf{q})$$

$$\times \left[g^{AB}(\mathbf{k}\text{-}\mathbf{q})\text{-}\delta^{AB}\frac{N_{\bar{C}}\bar{Z}^{\bar{C}}}{N_A Z^A} g^{A\bar{C}}(\mathbf{q}) \right] \qquad (2a)$$

or for the 3-d OCP

$$D_{\alpha\beta}(\mathbf{k}) = \frac{1}{V} \sum_q \frac{q_\alpha q_\beta}{q^2} \omega_p^2 [g(\mathbf{k}\text{-}\mathbf{q}) - g(\mathbf{q})] \qquad (2b)$$

which then provides the Hamiltonian

$$H = \frac{1}{2} \sum_k \pi_{\mathbf{k},\alpha}^{\bar{A}} \pi_{-\mathbf{k},\alpha}^{\bar{A}}$$

$$+ \frac{1}{2} \sum_k \left[D_{\alpha\beta}^{\bar{A}\bar{B}}(\mathbf{k}) + \frac{k_\alpha k_\beta}{k^2} \omega^{\bar{A}}(\mathbf{k})\omega^{\bar{B}}(\mathbf{k}) \right] \xi_{\mathbf{k},\alpha}^{\bar{A}} \xi_{-\mathbf{k},\beta}^{\bar{B}}$$

$$- \frac{i}{V} \left[\frac{N_{\bar{A}}}{m_{\bar{A}}} \right]^{1/2} \sum_k k_\alpha \, \hat{\Phi}^{\bar{A}}(-\mathbf{k},t) \xi_{\mathbf{k}\alpha}^{\bar{A}}, \qquad (3)$$

(Barred indices are summed over; $\hat{\Phi}$ is an external perturbing potential, $\xi_{\mathbf{k},\alpha}$ and $\pi_{\mathbf{k},\alpha}$ are the dynamical coordinates and momenta).

Concentrating now on the OCP, the longitudinal dielectric function becomes

$$\varepsilon_L(\mathbf{k}\omega) = 1 - \frac{\omega_p^2}{\omega^2 - D_L(\mathbf{k})} \qquad (4a)$$

with

$$D_L(\mathbf{k}) = \frac{1}{k^2} k_\alpha D_{\alpha\beta}(\mathbf{k}) k_\beta \qquad (4b)$$

which then leads to the dispersion relation

$$\omega_L(\mathbf{k}) = \sqrt{\omega_p^2 + D_L(\mathbf{k})} \qquad (5)$$

The equilibrium pair correlation function is known from Monte Carlo (MC) simulations and Hypernetted Chain (HNC) calculations [12]. The calculated $\omega(\mathbf{k})$ curves for different Γ-values are given in Fig. 1.

For high Γ values one can observe the oscillating behavior, which reflects the short range order in the liquid phase. For $\Gamma > 100$ the oscillation period fairly well matches the Brillouin zone structure of the Wigner crystal; the first minimum of $\omega_L(\mathbf{k})$ (at ka = 4.2) almost coincides with the boundary of the first Brillouin zone in the [1,1,0] direction (ka = 4.44). In addition, it can be shown that for small ka and high Γ Eq. (5) becomes identical with the angle-averaged optical phonon dispersion of the Wigner crystal, as calculated by Caldwell-Horsefall and Maradudin [1].

In addition to the longitudinal plasmon mode the strongly coupled Coulomb liquid possesses transverse shear excitations. [10] These are analogous to the acoustic phonons in the Wigner crystal. The dispersion relation can be derived from the transverse dielectric function, which in the QLC becomes

$$\epsilon_T(k\omega) = 1 - \frac{\omega_\beta^2}{\omega^2 - D_T(k)} \tag{6}$$

where, from Eq. (2)

$$D_T(k\omega) = -\frac{1}{2} D_L(k\omega) \tag{7}$$

The transverse dispersion relation is

$$\frac{1}{\epsilon_T(k\omega)} = 0 \tag{8}$$

One obtains a doubly degenerate shear-mode

$$\omega_T(k) = \sqrt{D_T(k)} \tag{9}$$

which has, for small k an acoustic type dispersion relation with the phase velocity

$$V_0 = \sqrt{\frac{|E_{corr}(\Gamma)|}{\frac{15}{2}m}} \tag{10}$$

For larger k and high Γ values the dispersion curve emulates the behavior of the angle averaged acoustic phonon of the Wigner crystal. (See Fig. 2) The mode frequencies satisfy the Kohn sum rule in the form

$$\omega_L^2(k) + 2\omega_T^2(k) = \omega_\beta^2 \tag{11}$$

We note that as $k \to \infty$, both $\omega_L(k)$ and $\omega_T(k)$ go to the value $\frac{\omega_p}{\sqrt{3}} = 0.577\,\omega_p$. This can be interpreted as the frequency of individual particle motion in the background of a uniform distribution of opposite charges.

Turning now to the 2-d system, it is well-known that in the RPA description the longitudinal plasmon mode has an $\omega \sim \sqrt{k}$ dispersion. In the QLC description this feature is maintained for $k \to 0$, but strongly modified for higher k-values. The dispersion relation is similar to (3):

$$\omega_L(k) = \sqrt{\omega_p^2(k) + D_L(k)} \tag{12}$$

with the difference that $\omega_p(k)$ is the k-dependent RPA plasma frequency and $D_L(k)$ has its 2-d form [9]:

$$\omega_p(k) = \sqrt{2\pi e^2 nk/m} \tag{13}$$

$$D_L(k) = \omega_\beta^2(k)\frac{1}{A}\sum_q \frac{(k \cdot q)^2}{k^3 q} \{g(k-q) - g(q)\}$$

The correlation function g(k) can be obtained from Monte Carlo or HNC [13] data. The resulting dispersion curve is plotted in Fig. 3. Again, for $\Gamma > 100$ the dispersion curve matches the optical phonon branch of the 2-d Wigner lattice [2]. As to the transverse shear mode, its behavior is also very similar to that of its 3-d counterpart. (See Fig. 4) The

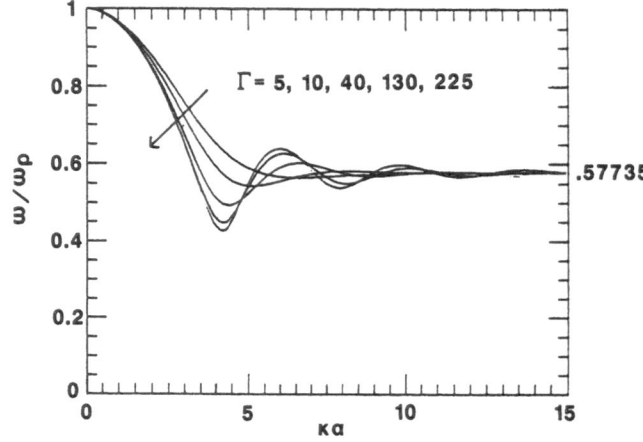

Fig. 1. Plasmon dispersion in the 3-d OCP, as calculated in the QLC
approximation; taken from Ref. [8].

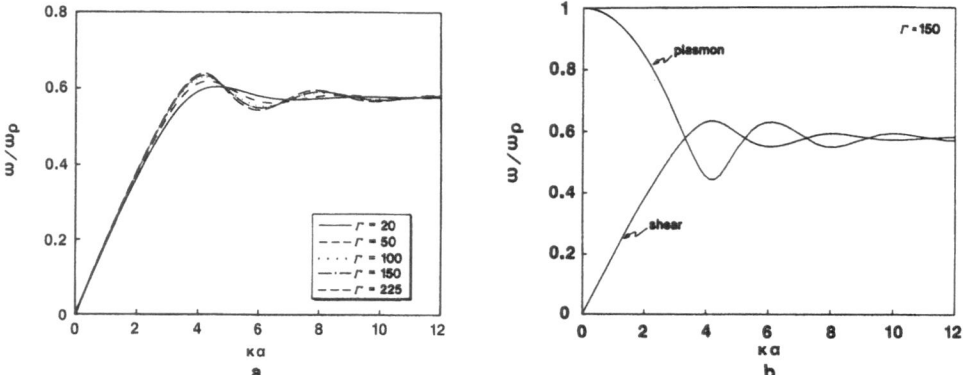

Fig. 2. Transverse shear mode dispersion in the 3-d OCP as calculated in the QLC
approximation; (a) shown the Γ-dependence, (b) the comparison of the
plasmon and shear modes: note the manifest satisfaction of the Kohn sum
rule; taken from Ref. [11b].

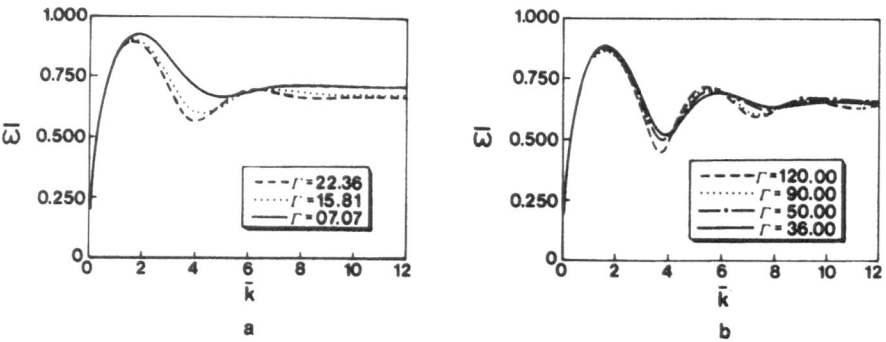

Fig. 3. Plasmon dispersion in the 2-d OCP, as calculated in the QLC
approximation; taken from Ref. [9].

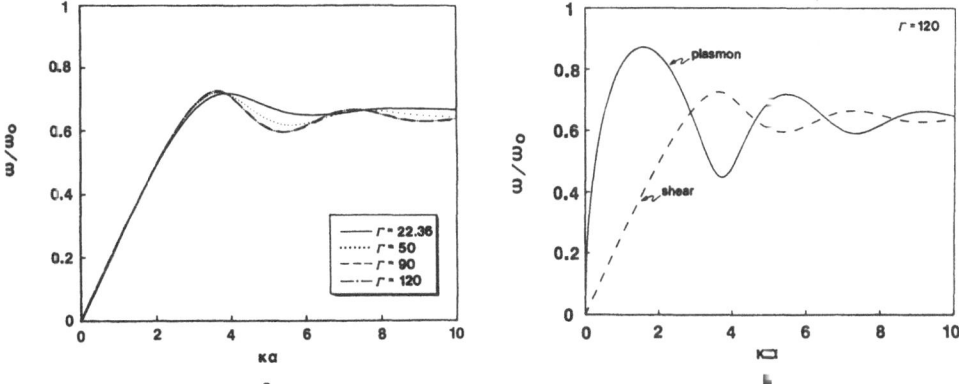

Fig. 4. Transverse shear mode dispersion in the 2-d OCP, as calculated in the QLC
approximation; (a) shows the Γ-dependence, (b) the comparison of
plasmon and shear modes: note that in contrast to the 3-d case, the modes
are not linked; taken from Ref. [10]

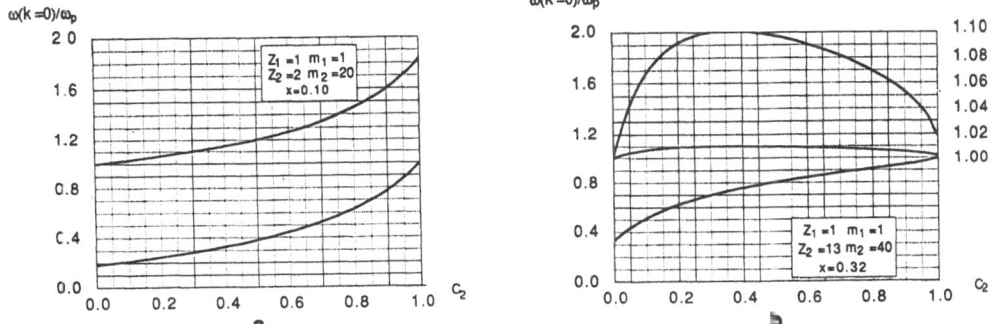

Fig. 5. Upper curve (to be read by the right-hand side scale): variation of the
plasma frequency at k=0 in terms of the RPA plasma frequency of the

mixture, $\omega_p = \left(\omega_1^2 + \omega_2^2 \right)^{1/2}$ vs. the concentration of the heavier species in
hydrogen. The lower curve (to be read by the left-hand side scale)
represents a second, low frequency longitudinal mode; taken from Ref. [6].

important difference is, however, that in 2-d the Kohn sum rule doesn't hold, and therefore $D_T(k)$ is independent of $D_L(k)$:

$$D_T(k) = \frac{1}{A}\sum_q \left\{\frac{q}{k} - \frac{(k\cdot q)^2}{k^3 q}\right\}\{g(k-q) - g(q)\}$$

(14)

The shear velocity is now

$$V_o = \sqrt{\frac{|E_{corr}(\Gamma)|}{8m}}$$

(15)

very close to the 3-d value. In the $k\rightarrow\infty$ limit both $\omega_L(k)$ approach the individual particle

oscillation frequency which now becomes $0.641 \times \left(\frac{2\pi e^2 n}{ma}\right)^{1/2}$ (a is the 2-d Wigner-Seitz radius, $\pi na^2 = 1$)

In the OCP momentum conservation precludes any deviation from the RPA behavior at $k=0$. In contrast, in the case of two or more components, both a collisional damping and a shift of the plasma frequency develop at $k=0$. The latter has been measured in Molecular Dynamics experiments and is easily calculable from the QLC formalism.[6] The result for the shifted plasma frequency in the case of the BIM is

$$\omega^2(k=0) = \frac{\omega_p^2}{2}\left\{1+p \pm[(1-p)^2+4d^2]^{1/2}\right\}$$

(16)

with

$$p = \frac{1}{3}\frac{x+y}{1+xy}$$

and

$$d = \sqrt{y/3}\,\frac{1-x}{1+xy}$$

$$x = \frac{Z_2 m_1}{Z_1 m_2},\ y = \frac{Z_2 c_2}{Z_1 c_1}$$

Fig. 5 shows the calculated variation of $\omega(k=0)$ as a function of concentration, for a few typical mixtures.

The calculated frequency shifts can be compared with the MD results of Hansen and his collaborators [14] for the 50% - 50% H^+ - He^{2+} BIM system. Results of the latter for $\Gamma = 40$ show a 3.9% upward shift of the plasma frequency, which is higher than the 1.9~2% shift predicted by a naive sum rule approximation [14,15]. Our calculated value at this point is 3.2%, quite close to the MD result.

The merits and the deficiencies of the QLC approach can be summarized as follows.

The good features are:
(i)   the smooth approach from the liquid state to the Wigner crystal behavior of the dispersion relation;
(ii)   the prediction of the shear mode;
(iii) the satisfaction of the Kohn and ω^{-4} third moment sum rules;
(iv)   the prediction of the shift of the plasma frequency for multi-component systems.

On the other hand the unsatisfactory features are:
(i)   the inability of the model to describe damping;
(ii)   the poor $\omega\rightarrow 0$ properties;
(iii) no transition to weak coupling regime.

In order to address these problems one has to turn to the mean field theory description. This is done in the next Section.

IV. DYNAMICAL MEAN FIELD THEORY FOR THE OCP

As discussed in the Introduction, in the DMFT approach one concentrates on the determination of the dynamical mean field $G(k\omega)$. There is no simple way to do this. A method that has proved to be quite successful [16] has been worked out by Golden and Kalman [11]. The general idea is to achieve self-consistency by simultaneously using kinetic equations and fluctuation-dissipation relations.

The central approximation used in the scheme is velocity average approximation (VAA): this consists of replacing the perturbed (symbol (1)) two-body function $G^{(1)}(x_1,v_1; x_2,v_2)$ by its average over the velocities v_1 and v_2, $G^{(1)}(x_1, x_2)$. As a result $X(k\omega)$ becomes a functional of the perturbed two point function $\langle n(x_1t_1) n(x_2t_2)\rangle^{(1)}$. At this point the quadratic Fluctuation-Dissipation theorem (QFDT) derived by Golden and Kalman [17] and by Kalman and Gu [18] comes to one's help; it relates $\langle n\, n\rangle^{(1)}$ to its three-point equilibrium (symbol (0)) counterpart; $\langle n\, n\rangle^{(1)} \leftrightarrow \langle n\, n\, n\rangle^{(0)}$. Further application of the QFDT allows one to express $\langle n\, n\, n\rangle^{(0)}$ in terms of the quadratic response function $X(k_1\omega_1; k_2\omega_2)$. Finally, an additional approximation helps to express $X(k_1\omega_1; k_2\omega_2)$ in terms of the linear X-s, $X(k_1\omega_1)$ and $X(k_2\omega_2)$. The final outcome of this procedure is a self-consistency relation (integral equation) for $X(k\omega)$ or for $G(k\omega)$

$$G(k) = - \{D_L(k) + F(k\omega)\} \tag{17}$$

$D_L(k)$ is given by Eq. (4b); for small k, $F(k\omega)$ is determined by the integral equation

$$F(k\omega) = - (ka)^2 \frac{1}{N}\sum_q (1-6\lambda^2 + 8\lambda^4)\, [\varphi(q)]^2 \int_-^- d\mu\, \delta\text{-}(\mu)\frac{\chi(q\mu)\,\chi(q\omega-\mu)}{\varepsilon(q\mu)\,\varepsilon(q\omega-\mu)} \tag{18}$$

where
$$\lambda = \frac{k.q}{kq}$$

Based on our experience with the QLC model, it is reasonable - although certainly not rigorously justifiable - to extrapolate the k-dependence so that

$$G(k) = - D_L(k)\, X(\omega) \tag{19}$$

where $X(\omega)$ is evaluated from (18). The calculation is made possible by an approximation on the integrand in (18): this is the "two-pole" approximation.[16] This approximation picks up the contribution from $\varepsilon^{-1}(\omega)$ in the vicinity of the pole; the position of the pole is taken as $\omega=\omega_p + A(ka)^2 + iB(ka)^2$, with floating $A=A(\Gamma)$ and $B=B(\Gamma)$ coefficients: the coefficients are ultimately determined by requiring the satisfaction of (18) (For details of the latter involved procedure the reader is referred to Ref. [8]). The resulting $G(k\omega)$ is depicted in Fig. 6. The main characteristics of this function are as follows:

    (i)   The relatively large positive Re $G(k\ \omega=0)$ has an important effect on the static isothermal compressibility.

    (ii)   Re $G(k\omega)$ develops a cusp-like singularity around $\omega=2\omega_p$; the Im $G(k\omega)$ develops a bump in the same region: these features are the consequences of the enhanced plasmon-plasmon interaction for $\Gamma\approx\Gamma_{crit}$ and $\Gamma>\Gamma_{crit}$.

(iii) As $\omega \to \infty$, Im $G(k\omega)$ decays slowly through a power-law behavior ($\sim \omega^{-5/2}$) The formation of this "algebraic tail" is a well established feature [19,20,21] of Coulomb systems.

(iv) For $\omega < 2\omega_p$ Im $G(k\omega)$ becomes significant only if $\Gamma > \Gamma_{crit}$.

With the aid of $G(k\omega)$ one can now construct $\varepsilon(k\omega)$. First we examine the static $\varepsilon(k0)$ (Fig. 7). One can observe the onset of the negative compressibility at $k=0$ for $\Gamma \gtrsim 3.8$. One can also see that there is a small $\underline{k}$ region, whose extension depends on Γ, where $\varepsilon(k0)$ enters the "forbidden" unphysical zone $0<\varepsilon<1$. This is a defect of the approximation. Otherwise the agreement with the known exact results [12] is quite good.

In particular the isothermal compressibility, K_T which can be derived from $\varepsilon(k0)$ (K_T^0 is the compressibility of the ideal gas) is

$$K_T = K_T^0 \left(1 + \frac{4}{15}\beta E_{corr} X(\omega=0)\right)$$

(20)

and is shown in Fig. 8. It is the appearance of $X(\omega=0)$ which is responsible for the dramatic improvement in $\varepsilon(k0)$ and in the compressibility over static MFT results.

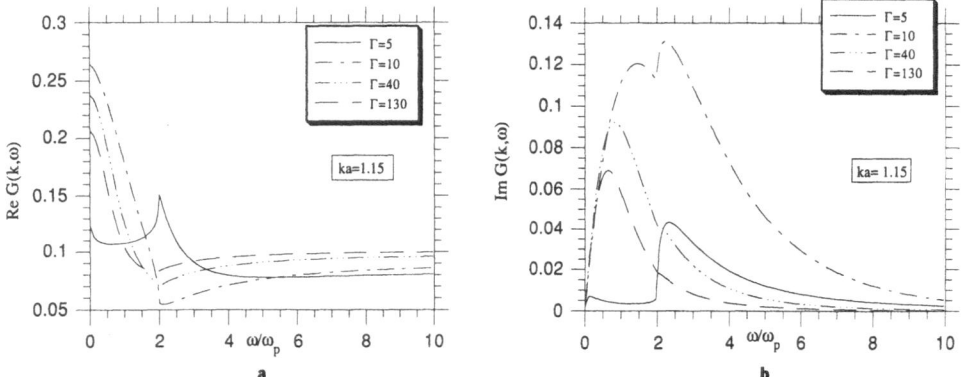

Fig. 6. Real (a) and imaginary (b) parts of the dynamical mean field, $G(k\omega)$. Note the singular behavior around $\omega \cong 2\omega_p$; taken from Ref. [8].

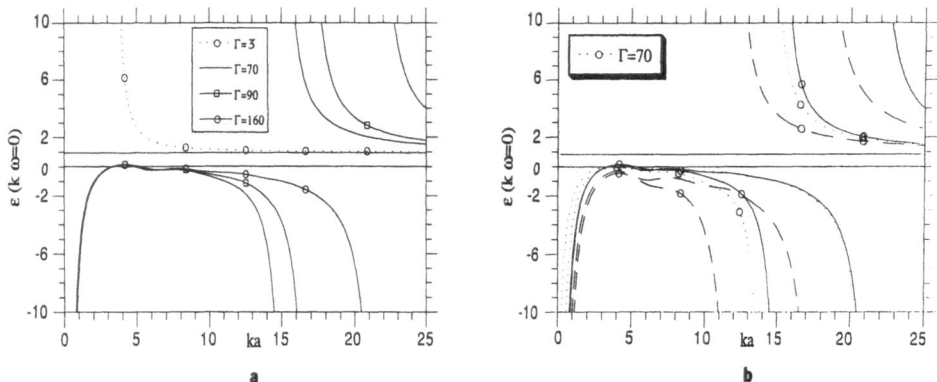

Fig. 7. The static dielectric function $\varepsilon_L(k0)$; (a) shows the variation with Γ; (b) shows the comparison of the results of this work (full lines) with the exact Monte Carlo - HNC data (dotted lines) and with the results of the a static mean field theory (dashed lines), for $\Gamma = 70$ (circles) and $\Gamma = 160$(unmarked). Note the small anomalous region near ka = 5; taken from Ref. [8].

The full Re $\varepsilon(k\omega)$ and Im $\varepsilon(k\omega)$ graphs are shown in Fig. 9. We have calculated the dynamical structure function $S(k\omega)$ from $\varepsilon(k\omega)$ via the fluctuation-dissipation theorem and compared it with the $S(k\omega)$ values obtained by Hansen, Pollock and McDonald [22] through MD calculations. Fig. 10 shows the results for a broad range of Γ values. The agreement is very satisfactory, especially in view of the fact that the derivation has been effected from first principles, without the introduction of any adjustable parameter.

Finally, Fig. 11 shows the dispersion and damping of plasma oscillations compared with MD data [22]. The agreement for the dispersion is very good and for the damping quite good. Further corroboration for the correctness of the theory is provided by comparing [23] its predictions for plasmon dispersion with recent experimental results [24] obtained on alkali metals. The coefficient $A(\Gamma)$ (or $A(r_s)$) has been extracted from the measured data and is compared with the calculated data. (Fig. 12) It is evident that the present theory is much superior to previous static [25,26] and semi-dynamic [27] theories. The reproducibility of the abrupt change in the vicinity of $r_{s,crit}$ is due to the central role played by the plasmon-plasmon interaction in the theory [23].

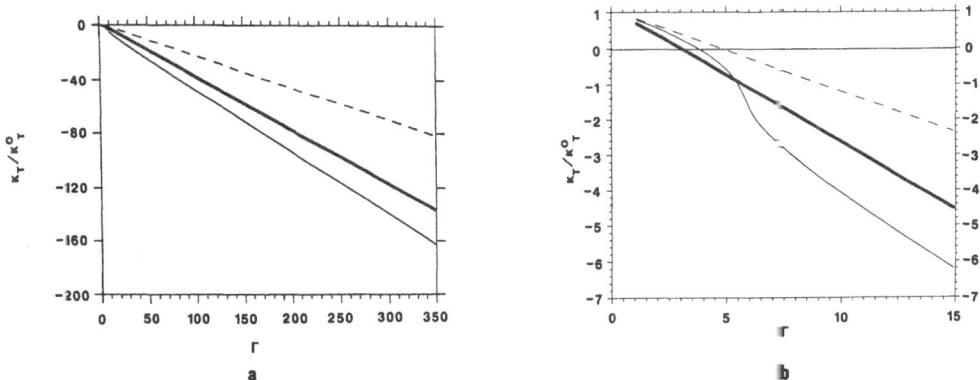

Fig. 8. The relative isothermal compressibility as calculated from this work (light line) compared with the exact result (heavy line) and with the results of a static mean field theory (dashed lines); taken from Ref. [8].

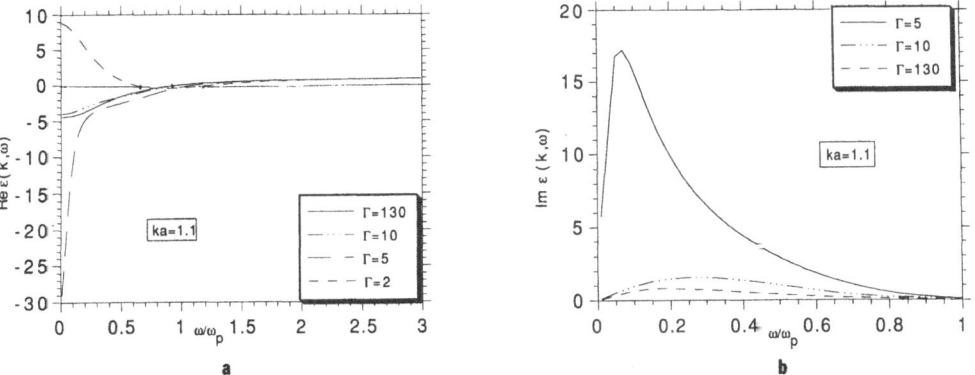

Fig. 9. Real (a) and imaginary (b) parts of the dielectric function $\varepsilon(k\omega)$; taken from Ref.[8].

173

V. STATIC MEAN FIELD THEORY

For the cases where the dynamical mean field theory is not available, a static mean field theory extension of the QLC can be generated [9]. For the longitudinal $\varepsilon_L(k\omega)$ this is done simply by omitting $F(k\omega)$ in Eq. (17). For the transverse $\varepsilon_T(k\omega)$ one replaces $D_L(k\omega)$ by $D_T(k)$, i.e. one sets $H(k\omega) = - D_T(k)$ in Eq. (1). The MFT generalization of the QLC is expected to account for most of the "indirect" thermal effects, ignored in the original QLC. These originate from the slow thermal migration and diffusion of the quasi-sites. To be more exact, one expects the MFT to work reasonably well for frequencies $\omega > t_D^{-1}$ where t_D is the (k-dependent) diffusion time. In particular, the MFT is not reliable for the acoustic shear mode below a k_{min} determined by t_D. The main effect of the introduction of the MFT description emerges in the higher $\underline{k}$ domain: here the onset of Landau damping and the disappearance of the modes beyond a k_{max} are the principal new features. In this Section we discuss

    (i)   the transverse shear mode in the 3-d OCP;
    (ii)   the longitudinal plasmon mode in the 2-d OCP;
    (iii)   and the transverse shear mode in the 2-d OCP.

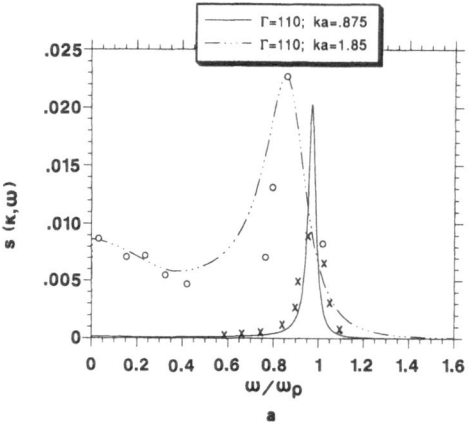 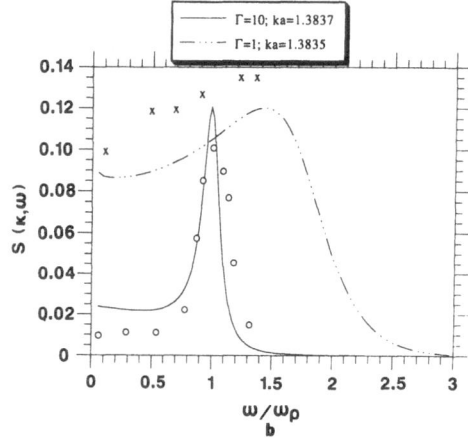

Fig. 10. Comparison of the calculated results for the dynamical structure function
$S(k\omega)$ given in this paper, with the MD results of Ref. [22] ((a) $\Gamma=110.4$,
ka = 0.875 (x) and ka = 1.85 (o); (b) $\Gamma= 9.7$ (o) and $\Gamma = 0.99$ (x)); taken
from Ref. [8].

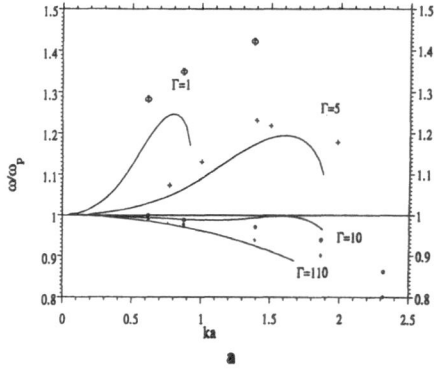 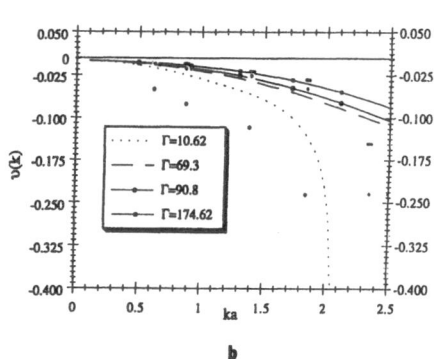

Fig. 11. Comparison of the calculated results for (a) plasmon dispersion and
(b) damping ($\nu = Im\omega/\omega_p$) as given in this paper, with the MD results of
Ref. [22] ($\Gamma= 0.99$ (Φ), $\Gamma= 2$ (+), $\Gamma= 9.7$ (o), $\Gamma= 110.4$ (*))
$\Gamma=152.4($) taken from Ref. [8].

The modification of the 3-d shear mode brought about by the MFT as compared to the QLC, is illustrated in Fig. 13. The main feature, namely the acoustic behavior up to about $ka \tilde{} 4$, and the shear velocity given by Eq. (10) are unchanged. The principal new feature is that the shear mode terminates at a maximum k value around $k_{max} a \tilde{} 12.5$. Landau damping (not shown in the graph) increases with increasing k, and by the time k_{max} is reached extinguishes the shear wave.

Fig. 13 also shows the comparison with the MD data of Hansen and collaborators [22]. The agreement for the shear velocity in the acoustic domain is very good. However, for higher k values the MD data indicate a splitting of the mode into a low-frequency and a high-frequency branch. No such effect is suggested by the calculated behavior: we believe that nonlinear shear-plasmon interaction (which is not a part of the theoretical model) may be responsible for this effect. For lower Γ values one finds that k_{max} decreases, roughly $k_{max} a \tilde{} \sqrt{\Gamma}$, and the shear mode completely ceases to exist for $\Gamma < 9.41$.

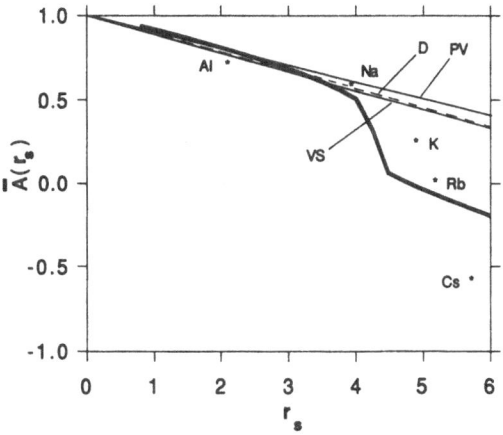

Fig. 12. Comparison of experimental results on plasmon group velocity with various theories. Experiments on A1 from Ref. [24a], on Na, K, Rb and Cs from Ref. [24b]. Thin solid lines refer to different approaches neglecting genuine dynamical correlations: Ref. [26a] (VS, Ref. [26b] (PV), Ref. [27] (D); heavy line refers to calculated values given in this paper (Ref. [8] and [16]); taken from Ref. [23].

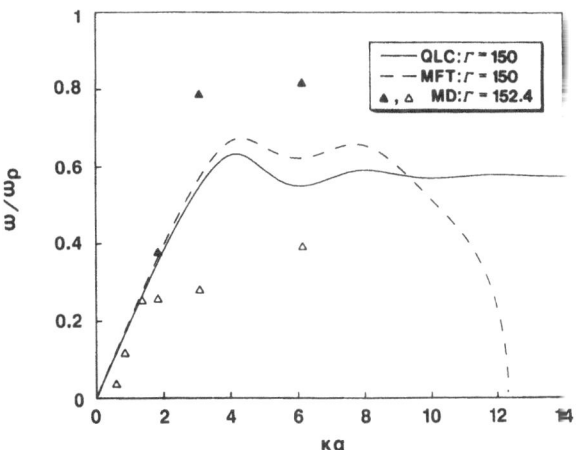

Fig. 13. Comparison of the calculated results for shear mode dispersion in the 3-d OCP as given in this paper, with the MD data of Ref. [14] (triangle points); the graph also shows the comparison between the QLC and MFT data; taken from Ref. [10].

Turning now to the analysis of the 2-d system, one finds that the introduction of the direct thermal effects through the MFT in the QLC has qualitatively the same effect as in the case of the 3-d shear mode. Fig. 14 shows the behavior of the plasmon mode [11]. One can observe the disappearance of the mode for $k>k_{max}(\Gamma)$ where $k_{max}a$ increases with Γ towards the value $k_{max}a \cong 3.65$. In the vicinity of k_{max} the mode heavily Landau damped. The comparison of our dispersion curve with the MD data of Totsuji and Kakeya. [28] The agreement, including the cut-off near k_{max}, is very good.

Fig. 15 shows the shear mode in the 2-d system for $\Gamma=50$. Most of the comments made for the 3-d shear wave apply in this situation as well. The shear mode exists now down to $\Gamma \cong 1.82$, a value substantially lower than its 3-d equivalent. The high-$\underline{k}$ cutoff is now given by $k_{max}a = 0.9\sqrt{\Gamma}$, but heavy Landau damping extinguishes propagation for somewhat lower values.

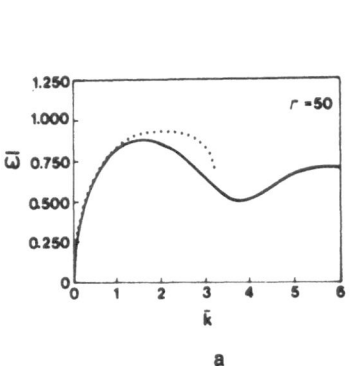 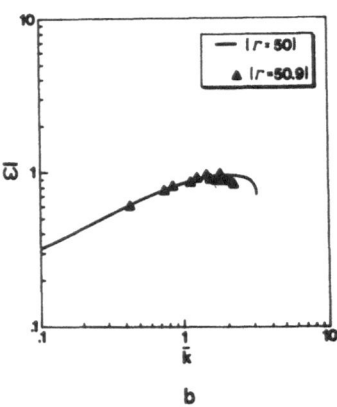

a   b

Fig. 14. Comparison of the plasmon dispersion in the 2-d OCP (a) as calculated in the QLC (full line) and in the MFT (dotted line); (b) as given by the MFT theory calculations and by the MD data of Ref. [28] (triangle points); taken from Ref.[9].

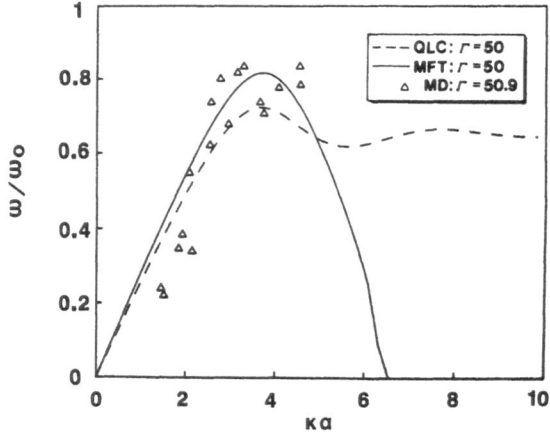

Fig. 15. Comparison of the calculated results for the shear mode dispersion in the 2-d OCP as given in this paper with the MD data of Ref. [28] (triangle points), the graph also shows the comparison between the QLC (dashed lines) and MFT data (full line); taken from Ref. [10].

Fig. 15 also shows the comparison with the MD data of Totsuji and Kakeya, [28]. Overall qualitative agreement with the theory is good. In contrast to the 3-d case, the MD data do not indicate any splitting of the shear mode. The observed k_{max} values are also in good agreement with theory, once allowance for the reduction of k_{max} because of Landau damping is made. On the other hand, the MD data clearly indicate the existence of a k_{min}, below which the shear mode doesn't propagate. This feature is not predicted by the theory: this is, however, not surprising in view of the limitations of the QLC in the small-k domain.

VI. CONCLUSIONS

In this paper we have shown that a theoretical framework based on the quasi-localization of the particles and combined with a mean field formalism gives a satisfactory description of most of the MD analyzed properties both of 3-d and 2-d plasmas. A rather sophisticated dynamical mean field theory with a frequency dependent mean field, based on the relationship between the linear and quadratic response function through the quadratic fluctuation-dissipation theorem, is capable to generate a detailed formulation of $\varepsilon_L(k\omega)$ and $S(k\omega)$ for the 3-d OCP and of the Γ-dependence of the plasmon dispersion and damping. The more modest static mean field theory with a frequency independent mean field provides a satisfactory description of the 2-d plasmon dispersion and of the transverse shear mode both in 3 and in 2 dimensions.

ACKNOWLEDGEMENTS

The works reported in this paper have been the result of a longstanding collaboration with Kenneth Golden and of outstanding contributions by a number of former graduate students: Paul Carini, Massimo Minella, Xiaoyue Gu, Zichi Tao, Hong Zhang and Philippe Wyns. This work has been partially supported by the National Science Foundations through Grant ECS 87-13337 and by the Army Research Office.

REFERENCES

[1]   R. A. Caldwell-Horsfall and A.A. Maradudin, J. Math. Phys. **1**, 395 (1960).

[2]   L. Bonsall and A.A. Maradudin, Phys. Rev. **B15**, 1959 (1977).

[3]   (a) V. I . Perel and G. M. Eliashberg, Zh. Eksperim. i Teor. Fiz **41**, 886(1961) [translation: Sov. Phys. JETP **14**, 633(1962)].
       (b) J. Dawson and C. Oberman, Phys Fluids **5**, 517 (1962); Phys. Fluids **6**, 394 (1962); C. Oberman, A. Ron and J. Dawson, Phys. Fluids **5**, 1514 (1962).

[4]   (a) J. Coste, Nuclear Fusion **5**, 284, 293 (1965).
       (b) M. G. Kivelson and D. F. DuBois, Phys. Fluids **7**, 1578 (1954).

[5]   P. Carini, G. Kalman and K. Golden, Phys. Rev. **A 26**, 1686 (1982).

[6]   G. Kalman and K. Golden, Phys. Rev. **A41**, 5516 (1990).

[7]   e.g. V. D. Gorbchenko and E. G. Maksimov, Usp. Fiz Nauk **130**, 65 (1980) [Soviet Phys - Usp. **23**, 35 (1980)].

[8]   M. Minella, G. Kalman, K.I. Golden, and P. Carini, submitted to Phys. Rev. **A**.

[9]   K.I. Golden, G. Kalman and P. Wyns, Phys. Rev. **A41**, 6940 (1990).

[10]   K. I. Golden, G. Kalman and Ph. Wyns, submitted to Phys. Rev A.

[11]   K. I. Golden and G. Kalman, Phys. Rev. A **19**, 2112 (1979).

[12]   (a) G. S. Stringfellow, H.DeWitt and W. Slattery, Phys Rev. A **41**, 1105 (1989).

    (b) F.J. Rogers, D. A.Young, H.E. DeWitt and M. Ross, Phys. Rev. A **28,** 2990 (1983).

[13]   (a) H. Totsuji and N. Kakeya, Phys. Rev. A **22,** 1220 (1980).

    (b) R. C. Gann, S. Chakravarty, and G. V. Chester, Phys. Rev. B **20,** 326 (1979).

[14]   I. R. McDonald, P. Vieillefosse, and J. P. Hansen, Phys. Rev. Lett **39**, 271 (1977); J. P. Hansen, I. R. Mc Donald, and P. Vieillefosse, Phys. Rev. A **20,** 2590 (1979); J. P. Hansen, F. Joly, and I. R. Mc Donald, Physica **132A,** 4721 (1985).

[15]   K. I. Golden, F. Green, and D. Neilson, Phys. Rev. A **31**, 3529 (1985); **32,** 1969 (1985).

[16]   P. Carini, G. Kalman, and K. I. Golden, Phys. Lett. **78A,** 450 (1980); P. Carini and G. Kalman, *ibid.* **105A,** 229 (1984).

[17]   K. I. Golden, G. Kalman and M. B. Silevitch, J. Stat. Phys. **6,** 87 (1972).

[18]   G. Kalman and Xiao-Yue Gu, Phys. Rev. A36, 3399 (1987).

[19]   A. J. Glick and W. F. Long, Phys. Rev. B **4**, 3455 (1971).

[20]   F. Family, Phys. Rev. Lett. **34,** 1375 (1975).

[21]   Z. C. Tao and G. Kalman, Phys. Rev. A **43,** 973 (1991); G. Kalman and Z. C. Tao, Phys. Rev. A **43,** 7073 (1991).

[22]   J. P. Hansen, E. L. Pollock, and I. R. McDonald, Phys. Rev. Lett. **32,** 277 (1974); J. P. Hansen, I.R. McDonald, and E. L. Pollock, Phys. Rev. A **11,** 1025 (1975).

[23]   G. Kalman, K. Kempa, and M. Minella Phys. Rev. B **43**, 14238 (1991).

[24]   (a) A. vom Felde, J. Sprosser-Prou, and J. Fink, Phys. Rev. B **40**, 10 181 (1989).

    (b) J. Sprosser-Prou, A. vom Felde, and J. Fink, Phys. Rev. B **40**, 5799 (1989).

[25]   P. Vashishta and K.S. Singwi, Phys. Rev. B **6**, 875 (1972).

[26]   K. N. Pathak and P. Vashishta, Phys. Rev. B **7**, 3649 (1973).

[27]   B. Dabrowski, Phys. Rev. B **34**, 4989 (1986).

Z-DEPENDENT PERTURBATION THEORY AND

COMPLEX ROTATION METHOD IN

AUTOIONIZING STATES OF ATOMS

Lonnie W. Manning and Frank C. Sanders

Department of Physics, Southern Illinois University-Carbondale
Carbondale, IL 62901-4401   USA

ABSTRACT

A combined Complex Rotation and Feshbach Projection method is implemented within Z-dependent perturbation theory to obtain the lowest-order contribution to the widths of autoionizing states of two- and three-electron atoms. This approach relies on the fact that in the complex rotation method, it is the open-channel part of the wave function that produces the imaginary component of the energy. Calculation of this part of the wave function involves the solution of a simple one-electron differential equation which, in lowest-order, can be obtained to any desired level of accuracy. These results represent the limiting values of the width for high Z values in the non-relativistic approximation, and are particularly useful for states with extremely narrow widths, as these are difficult to calculate accurately with other methods.

INTRODUCTION

Knowledge of the properties of autoionizing states of atoms is important for an understanding of atomic processes in high-temperature plasmas and in astrophysics. Such multiply-excited states of atoms present special theoretical and computational problems, however. Perturbation theory is perhaps computationally more naturally suited to the study of autoionizing states than approaches based on the variational method.[1] The presence of a degeneracy between each doubly-excited, zero-order wave function and a singly-excited, zero-order continuum function is one difficulty which is not present in a variational calculation but must be dealt with if perturbation theory is to be applied to these states. Application of the Feshbach projection method[2] to Z- dependent perturbation theory circumvents this problem by projecting out these continuum functions. This results in a variational perturbation calculation for the closed- channel part of the resonance wave function similar to that for an ordinary bound-state problem[1]. Calculating the width and shift of the autoionizing state, however, requires knowledge of the continuum-like part of the wave function that has been projected out, and this is computationally more difficult.

An alternative approach to the study of such states is the complex rotation method.

This method has proven to be very effective for the study of autoionizing states, yielding both resonance positions and widths in a single calculation. Much recent work applying this method to atomic systems has been reviewed by Ho,[3] who also provides much information on the computational aspects of implementing this method. The complex rotation method involves a complex scaling of all the radial coordinates of the Hamiltonian for the system. The resultant, non-hermitean Hamiltonian yields complex eigenvalues for the resonance states of the system. The real part of such an eigenvalue is the resonance position while the imaginary part is proportional to the width of the state. The complex scaling transformation removes the degeneracy with the continuum zero-order functions noted above, as the doubly-excited, zero-order, hydrogenic functions have real eigenvalues while the zero-order continuum energies are complex, having been rotated downward into the lower energy half plane by the complex scaling transformation.

Applications of Z-dependent perturbation theory to autoionizing states of atoms have been rather limited and most[1,4] have relied on use of the Feshbach projection method. Recently, we have used Z-dependent perturbation theory together with a combined complex rotation and Feshbach projection method to obtain very precise values of the lowest-order contribution to the width for the 50 lowest autoionizing P-states of two-electron atoms below the n=2 threshold.[5]

In this article we extend the application of the $1/Z$ expansion to autoionizing states of three-electron atoms. We present results for the lowest-order contribution to the widths for the rydberg series of autoionizing doublet S-states lying between the first and second ionization thresholds of three-electron atoms. To do this we make use of the fact that, in Z-dependent perturbation theory, results obtained for simple systems are exactly transferable into complex systems which incorporate the identical electronic configuration of the simple system. Hence, the first step in our procedure is to obtain the results for the appropriate doubly-excited singlet and triplet S-states of two-electron atoms. Results for those singly-excited states of two-electron atoms that are also subsystems of these three electron autoionizing states are readily available[6] but are not required here as they do not contribute to the lowest-order widths. Following the procedure of Ref. 5, lowest-order widths are obtained for the 40 lowest singlet and triplet autoionizing S-states of two-electron ions below the n=2 threshold. The lowest-order contributions to the widths of the corresponding autoionizing doublet S-states of three-electron atoms are then easily obtained from these two-electron widths as there are no three-electron contributions to the width at this order of the perturbation theory.

These doublet S-states are all four-fold degenerate. Classification of these states is complicated by the fact that the mixing coefficients of the individual components of the zero-order wave functions behave erratically for the first few members of this rydberg sequence. Similarly, the lowest-order widths also exhibit this erratic behavior. These states can be classified into four unambiguous series, however, based on their first-order energies. These first-order energy expansion coefficients exhibit very simple behavior with increasing degree of excitation. Examination of these first-order coefficients also reveals that two of the four series become nearly degenerate in first-order. This near degeneracy is the source of the erratic behavior of the mixing coefficients noted above.

Finally, it should be noted that there is no information available in the literature for the widths of most of these states. Hence these results, despite their low order, do give useful qualitative information about the total widths of these states.

THEORY

Complex Rotation Method

In the complex rotation method[3] the rotated Hamiltonian is obtained via a complex scaling transformation of the radial coordinates, $r = e^{i\theta}\rho$. For the case of a Coulomb potential, the Hamiltonian becomes:

$$H(\theta) = e^{-i2\theta}T + e^{-i\theta}V(\rho). \tag{1}$$

The transformed Hamiltonian is non-hermitean with complex eigenvalues, W. The discrete, real eigenvalues of the rotated Hamiltonian are independent of θ and are the discrete eigenvalues of the non-rotated Hamiltonian which, under this transformation, remain on the real axis. However, the continuous spectra will be rotated by an angle 2θ about their respective thresholds into the lower energy half-plane. The resonances are the discrete complex eigenvalues of the rotated Hamiltonian. For these states, the real part of W is the resonance position, E_r, while the imaginary part is the half-width, Γ_2. As increasing θ sweeps the associated continua past a resonance, the wave-function for that resonance becomes square-integrable. With the wavefunction square-integrable, calculation of the eigenvalues proceeds in the manner of a standard variational calculation for a bound state.

Adapting this method to Z-dependent perturbation theory results in few changes to the usual perturbation expressions. The rotated zero-order Hamiltonian, in charge-scaled atomic units, is now given by

$$H_0 = \sum_i h_i(\theta), \tag{2}$$

$$h_i(\theta) = \left[-e^{-i2\theta}\tfrac{1}{2}\Delta_i - e^{-i\theta}\tfrac{1}{\rho_i}\right], \tag{3}$$

while the perturbation becomes

$$\lambda H_1 = Z^{-1}\sum_{j>i}\frac{e^{-i\theta}}{\rho_{ij}}. \tag{4}$$

Thus Γ_2 is the leading contribution to the width and, in ordinary atomic units (obtained from the charge-scaled results by multiplying by Z^2), is the limiting value of Γ for high Z as well.

Under the complex rotation transformation, the resonant state has a complex energy, ω. This is expanded in a power series in $1/Z$,

$$\omega = \sum_n \omega_n Z^{-n}. \tag{5}$$

The ω_n can themselves be expressed in terms of the expansion coefficients of the resonance position and width,

$$\omega_n = \epsilon_n - i\frac{\Gamma_n}{2}. \tag{6}$$

Application of the complex rotation method removes the degeneracy with the continuum in zeroth order, resulting in a bound zero-order wave function. This in turn means that the lowest-order contributions to the total energy are purely real: $\omega_0 = \epsilon_0$ and $\omega_1 = \epsilon_1$.

Feshbach Projection Method

The Feshbach projection method divides the Hilbert space into two orthogonal subspaces with the help of projection operators, P and Q. These satisfy the relations: $P + Q = 1$, $PQ = QP = 0$, and the asymptotic conditions,

$$
\begin{aligned}
P\Psi &\sim \Psi, \\
Q\Psi &\sim 0.
\end{aligned}
\tag{7}
$$

Expressing the first-order wave function in terms of the Feshbach projection operators, P and Q, the expression for the second-order coefficient becomes,

$$
\omega_2 = <(P+Q)\psi_1^* | \tfrac{e^{-i\theta}}{\rho_{12}} - \epsilon_1 | (P+Q)\psi_0 >.
\tag{8}
$$

Complex conjugation of the bra indicates that, in accordance with the bi-orthogonality of the eigenfunctions in the complex rotation method, there is no complex conjugation of the radial part of this function.

For the particular case of S-state resonances of two-electron atoms below the n=2 threshold, P is chosen to be $P_1 + P_2 - P_1 P_2$ where $P_i = |1s(i)\rangle\langle 1s(i)|$ and $1s$ represents a hydrogenic, ground state orbital. This projector results in $P\psi_0 = 0$ and a $P\psi_1$ of the form,

$$
P\psi_1 = \tfrac{1}{\sqrt{2}} [1s(1)\phi_1(2) \pm \phi_1(1)1s(2)].
\tag{9}
$$

ω_2 can be divided into a purely real part,

$$
\omega_2^Q = < Q\psi_1^* | \tfrac{e^{-i\theta}}{\rho_{12}} - \epsilon_1 | Q\psi_0 >,
\tag{10}
$$

and a complex part containing the lowest-order contributions to the shift, Δ_2, and the width, Γ_2:

$$
\omega_2^P = < P\psi_1^* | \tfrac{e^{-i\theta}}{\rho_{12}} | Q\psi_0 > = \Delta_2 - i\tfrac{\Gamma_2}{2}.
\tag{11}
$$

The lowest-order contribution to the width and shift can thus be obtained if the one-electron function, ϕ_1, is known. Since the projector P commutes with H_0, the differential equation for ϕ_1 can be obtained from the first-order perturbation equation with the help of P, and can be written as

$$
g_0\phi_1(\rho) + Y_1(\rho) = 0,
\tag{12}
$$

with

$$
g_0 = h(\theta) - (\epsilon_0 - \epsilon_0^{1s}),
\tag{13}
$$

where $h(\theta)$ is given in Eq. ϵ_0^{1s} is (3), the energy for the $1s$ state of hydrogen, and

$$
Y_1(\rho_j) = \int d\vec{\rho_i}\, 1s(\rho_i)\frac{e^{-i\theta}}{\rho_{ij}}\psi_0(\rho_i, \rho_j)
\tag{14}
$$

is just the $1s$ component of $\frac{1}{\rho_{12}}\psi_0$. From Eq.(12) one obtains a variational functional for ϕ_1,

$$\omega_2^P \approx \langle \phi_1 | g_0 | \phi_1 \rangle + 2 \langle \phi_1 | Y_1 \rangle. \tag{15}$$

On variation of ϕ_1, this functional is stationary about ω_2^P.

RESULTS AND DISCUSSION

The ^{1,3}S Autoionizing States of Two-electron Atoms

With the hydrogenic H_0 utilized here, the zero-order wave functions of the ^{1,3}S resonances below the n=2 threshold are doubly degenerate. Thus, they are a mixture of the properly symmetrized products of hydrogenic orbitals $|nl \, n'l'\rangle$.

$$|\psi_0(N)\rangle = a_1 \, | \, (2sNs) \, ^{1,3}S \rangle + a_2 \, | \, (2pNp) \, ^{1,3}S \rangle. \tag{16}$$

The (a_1, a_2) are obtained in the usual manner of conventional perturbation theory for degenerate states and are the eigenvectors of the 2×2 secular determinant constructed from the perturbation matrix over the degenerate zero-order states. The eigenvalues of this secular determinant are of course just the first-order energies, ϵ_1, which remove the degeneracy. Table I contains the a_i, ϵ_1, and Γ_2 for these two sequences of states.

TABLE I.   Doubly excited states of He-like ions below the n=2 threshold.

^{1}S

N	c_1	c_2	ϵ_1(a)	Γ_2(a)	ϵ_1(b)	Γ_2(b)
2	0.87964	0.47564	0.1229524	0.008325303	0.2442351	0.0004182294
3	0.86510	0.50159	0.0778425	0.004251225	0.1324833	0.0001664848
4	0.85493	0.51874	0.0476280	0.001716481	0.0703601	0.0000545336
5	0.85005	0.52671	0.0322242	0.000861635	0.0438219	0.0000246012
6	0.84736	0.53101	0.0232310	0.000493363	0.0299332	0.0000132578
7	0.84574	0.53360	0.0175277	0.000308718	0.0217455	0.0000079911
8	0.84468	0.53528	0.0136880	0.000205967	0.0165125	0.0000052015
9	0.84395	0.53642	0.0109819	0.000144250	0.0129650	0.0000035812
10	0.84343	0.53724	0.0090040	0.000104947	0.0104495	0.0000025736
11	0.84304	0.53785	0.0075151	0.000078731	0.0086011	0.0000019131

^{3}S

N	c_1	c_2	ϵ_1(a)	Γ_2(a)	ϵ_1(b)	Γ_2(b)
3	0.82561	0.56425	0.065051	0.0000204587	0.101445	0.000002030933
4	0.81789	0.57537	0.043315	0.0000142197	0.060360	0.000001285357
5	0.81412	0.58070	0.030198	0.0000087279	0.039235	0.000000752411
6	0.81202	0.58363	0.022107	0.0000055202	0.027425	0.000000463346
7	0.81075	0.58539	0.016838	0.0000036571	0.020217	0.000000301972
8	0.80992	0.58654	0.013232	0.0000025293	0.015510	0.000000206607
9	0.80934	0.58733	0.010665	0.0000018147	0.012271	0.000000147143
10	0.80893	0.58790	0.008775	0.0000013430	0.009949	0.000000108315
11	0.80863	0.58832	0.007344	0.0000010202	0.008228	0.000000081955

The mixing coefficients and first-order energies were obtained exactly with the aid of the REDUCE programming language. Hence the number of significant figures reported here is much less than what was actually used in the calculation of the ω_2. The results of Ref. 5 indicate that the degree of accuracy possible with the present approach far exceeds any possible requirement for precision. For the $2s2p$ $^{1,3}P$ states, for example, the second-order widths had converged to more than 20 significant figures. These calculations for the corresponding P-states of two-electron atoms[5] have indicated that even these lowest-order results for the widths can give useful information on the total widths, particularly for sufficiently high values of Z and N. In the case of these P-states, at least, the second-order widths yield estimates for the total width good to at least one or more significant figures for $Z \geq 8$ and $N \geq 5$. In contrast to this, variational calculations have difficulty yielding reliable estimates of widths for the more highly excited members of these series, which have very narrow widths. We can expect the present results to be equally useful and of comparable accuracy.

The ^{2}S Autoionizing States of Three-electron Atoms

The lowest series of ^{2}S autoionizing states of three-electron atoms are four-fold degenerate:

$$|\psi_0^\alpha(N)\rangle = a_1(\alpha)\big|\, [1s,(2sNs)^1\text{S}]\ ^2\text{S}\big\rangle + a_2(\alpha)\big|\, [1s,(2sNs)^3\text{S}]\ ^2\text{S}\big\rangle$$

$$+ a_3(\alpha)\big|\, [1s,(2pNp)^1\text{S}]\ ^2\text{S}\big\rangle + a_4(\alpha)\big|\, [1s,(2pNp)^3\text{S}]\ ^2\text{S}\big\rangle, \tag{17}$$

where the label α distinguishes between the four degenerate states, here simply designated as (a,b,c,d). The $P\phi_1$ for these states can be written in terms of the two-electron $P\phi_1$. Hence, the second-order width can be expressed entirely in terms of the two-electron Γ_2:

$$\Gamma_2(\alpha) = \Big[\sqrt{\tfrac{1}{2}}\, a_1(\alpha)\, \mathcal{M}_2(2sNs\ ^1\text{S}) + \sqrt{\tfrac{3}{2}}\, a_2(\alpha)\, \mathcal{M}_2(2sNs\ ^3\text{S})$$

$$+ \sqrt{\tfrac{1}{2}}\, a_3(\alpha)\, \mathcal{M}_2(2pNp\ ^1\text{S}) + \sqrt{\tfrac{3}{2}}\, a_4(\alpha)\, \mathcal{M}_2(2pNp\ ^3\text{S})\Big]^2. \tag{18}$$

where $\mathcal{M}_2 = \sqrt{\Gamma_2}$.

Note that no three-body contributions appear in this second-order width (or, perhaps more correctly, the three-body contributions reduce to two-body contributions). Hence the calculation of these lowest-order widths is particularly simple. This characteristic of the second-order width extends to all multi-electron atoms and arises from the one-body character of $P\psi_1$. This, in turn, is due to the character of the zero-order Hamiltonian, Eq. (1), which ensures that in perturbation theory the projector P is always a symmetrized product of single-particle wave functions.[5] This is in contrast to the usual non-perturbative approach to Feshbach projection, where the projectors for an n-body resonant state are constructed from (n-1)-particle wave functions.

TABLE II.   First-order energies of doubly excited ^2S states of Li-like ions.

N	$\epsilon_1(a)$	$\epsilon_1(b)$	$\epsilon_1(c)$	$\epsilon_1(d)$
3	0.37180	0.44991	0.40306	0.47402
4	0.29753	0.35150	0.33665	0.37962
5	0.26038	0.30570	0.30194	0.33683
6	0.23923	0.28107	0.28186	0.31354
7	0.22611	0.26636	0.26921	0.29945
8	0.21743	0.25683	0.26080	0.29027
9	0.21140	0.25031	0.25493	0.28396
10	0.20704	0.24564	0.25067	0.27943
11	0.20379	0.24219	0.24749	0.27606
∞	$\epsilon_1(1s2s^3S)$	$\epsilon_1(1s2p^3P)$	$\epsilon_1(1s2s^1S)$	$\epsilon_1(1s2p^1P)$

TABLE III. Lowest-order widths of the doubly excited ^2S states of Li-like ions.

N	$\Gamma_2(a)$	$\Gamma_2(b)$	$\Gamma_2(c)$	$\Gamma_2(d)$
3	0.001165970	0.0000705036	0.0006859413	0.0003201745
4	0.000416395	0.0001059510	0.0002647682	0.0001216509
5	0.000186825	0.0000656192	0.0001335803	0.0000713134
6	0.000099511	0.0000747716	0.0000423495	0.0000456534
7	0.000059552	0.0000369254	0.0000372429	0.0000305728
8	0.000038634	0.0000245907	0.0000251946	0.0000212681
9	0.000026571	0.0000173506	0.0000176366	0.0000153004
10	0.000019097	0.0000127048	0.0000128027	0.0000113332
11	0.000014205	0.0000095765	0.0000095848	0.0000086089

The $\epsilon_1(\alpha)$ for these states are presented in Table II, while the corresponding $\Gamma_2(\alpha)$ are listed in Table III. The mixing coefficients, $a_i(\alpha)$, and the second-order widths exhibit somewhat erratic behavior with respect to N, the level of excitation of the state, for $N \leq 7$, so that it is difficult to classify these states on the basis of these parameters. However, examination of the first-order energies of Table II indicates that they behave smoothly over the entire range of N. Note in particular that, for large N, the ϵ_1 converge rapidly to the first-order energies of the singly-excited states of two-electron atoms indicated at the bottom of this table. Hence, as the level of excitation increases, a more appropriate coupling for the hydrogenic orbitals in the zero-order wave function would be

$$|\psi_0^a(N)\rangle = c_1(\alpha)\,\big|\,[2s,(1sNs)^3S]\,^2S\big\rangle + c_2(\alpha)\,\big|\,[2p,(1sNp\ ^3P]\,^2S\big\rangle$$

$$+c_3(\alpha)\,\big|\,[2s,(1sNs)^1S]\,^2S\big\rangle + c_4(\alpha)\,\big|\,[2p,(1sNp)^1P\ ^2S\big\rangle, \tag{19}$$

where the $c_i(\alpha)$ are obtained from the $a_i(\alpha)$ by a simple unitary transformation. The behavior of these new mixing coefficients with increasing N is presented in Fig. 1. This figure clearly demonstrates that the appropriate coupling scheme is that of Eq. (19), with the exception, of course, of $N = 2$ and, to a lesser degree, $N = 3$. These

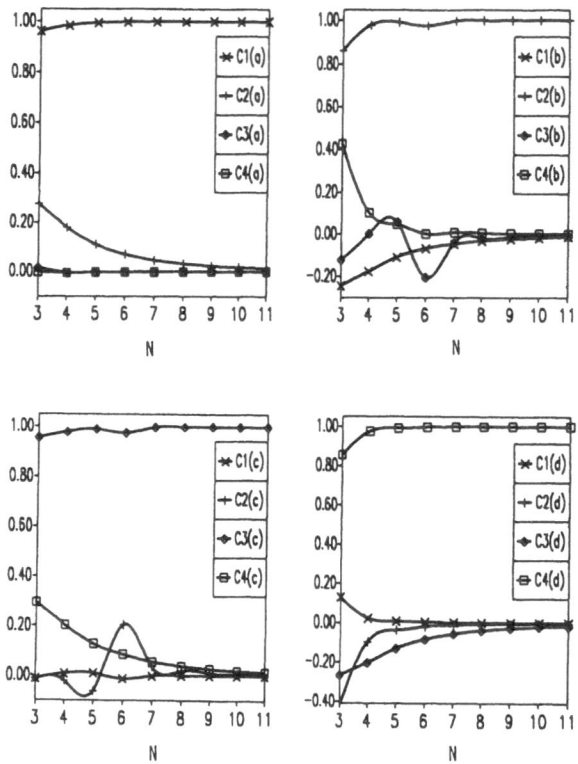

Fig. 1 The mixing coefficients of the four-fold degenerate zero-order wave functions of the ^{2}S states of lithium-like atoms. $c_i(\alpha)$ is the ith component of the state α.

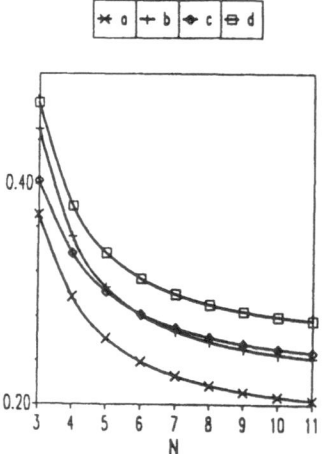

Fig. 2 The first-order energies of the doubly-excited, ^{2}S states of lithium-like atoms in $a.u.$.

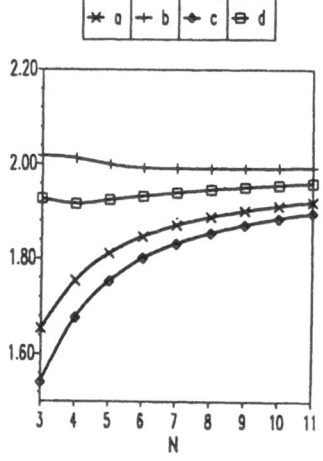

Fig. 3 The $\mathcal{E}_1$ (see Eq.(20) of the text) for the doubly-excited, ^{2}S states of lithium-like atoms in $a.u.$.

$c_i(\alpha)$ rapidly go to zero with the exception of the one coefficient that corresponds to the infinite-N limit for ϵ_1. Hence these states can be unambiguously classified by their first-order energies. Nevertheless, the coupling scheme of Eq. (17), based as it is on the doubly-excited, two-electron states, is the appropriate one for generating the $\Gamma_2(\alpha)$ of Eq. (18). Regardless of which coupling scheme is chosen to describe the zero-order wave function, the mixing coefficients for the (b) and (c) states exhibit erratic behavior in the vicinity of $N = 6$. These two states correspond to autoionizing states of the atom which, for high N, decay to the $(1s2p)^3P$ and $(1s2s)^1S$ states respectively. At $N = 6$ each of these two states shows a significant enhancement of the component of the other state, so that each state becomes less "pure" and more of a mixture of the two components. Examining the behavior of the first-order energies in Fig. 2, we see smooth behavior with respect to N throughout the range of N studied here with the ϵ_1 for the b and c states becoming very nearly degenerate at $N = 6$, thus explaining the enhanced mixing of these two states at this point in the series. Fig. 2 offers convincing evidence that the labeling of these two states has not been erroneously interchanged at $N = 6$. This last figure plots

$$\mathcal{E}_1 = N^2 [\, \epsilon_1(\alpha, N) - \epsilon_1(\alpha, N = \infty)] \tag{20}$$

so that what is seen in the figure is primarily the coefficient of the leading, $\frac{1}{N^2}$ term of ϵ_1. This can be seen to approach a value of 2 as N increases, as could have been anticipated on the basis of quantum defect theory.[7]

SUMMARY

In this paper, the complex rotation and Feshbach projection methods are combined within Z-dependent perturbation theory to obtain the lowest-order contribution to the width for autoionizing S-states of two- and three-electron atoms. Calculation of the two-electron $P\psi_1$ is the first step in obtaining the three-electron results of the present paper. The method presented here is applied to the doubly-excited S-states of two-electron atoms lying below the n=2 threshold. These two-electron widths are then utilized to obtain the second-order widths of the three-electron atoms exactly. There are no three-body contributions to these lowest-order widths.

These results represent the limiting values of the width and shift for large Z in the non-relativistic approximation. Comparison of earlier second-order widths for doubly-excited P-states of two-electron atoms with accurate total widths suggests that results such as these can give useful information even for low values of Z. Of particular interest are those states where the total widths are extremely small. Such widths have considerable interest but are extremely difficult to calculate with accuracy. Given that the total widths of the states considered here rapidly become very small as the level of excitation increases, it becomes very difficult to calculate these widths variationally with any precision. Since the extent to which the Γ_2 accurately represent Γ improves rapidly as Z and N increase, these second-order results can be quantitatively useful, especially when variational results are unavailable or unreliable. The ease with which results are obtained with the present approach makes it ideal for identifying such possible long-lived states.

ACKNOWLEDGEMENTS

One of the authors (F.C.S.) would like to express his gratitude to Professor A. N. Proto and her colleagues for organizing such a stimulating workshop, and to thank the U.S. Army Research Office for travel support in conjunction with this workshop.

REFERENCES

1. J. M. Seminario and F. C. Sanders, Phys. Rev. A**42**, 2562 (1990).

2. A. Temkin and A. K. Bhatia, in *Autoionization: Recent Developments and Applications*, edited by A. Temkin (Plenum Press, New York, 1985) describe recent applications of the Feshbach method to atoms.

3. Y. K. Ho, Phys. Repts. **99**, 1 (1983) and references therein.

4. G. W. F. Drake and A. Dalgarno, Proc. R. Soc. London Ser. A **320** , 549 (1971).

5. L. W. Manning and F. C. Sanders, Phys. Rev. A **43**, 272 (1991).

6. F. C. Sanders and R. E. Knight, Phys. Rev. A **27** , 1279 (1983).

7. P. Blanchard, Ph.D. dissertation, Harvard University, 1969.

SELFCONSISTENT SEMICLASSICAL MEAN FIELD

M. Casas[*], H. Krivine[+] and A. Puente[*]

[*] Departament de Física, Universitat de les Illes Balears
E-07071 Palma de Mallorca, Spain
[+] Division de Physique Théorique[1], Institut de Physique
Nucléaire- 91406 Orsay Cedex, France

ABSTRACT

We propose a method to obtain a selfconsistent semiclassical mean field for a given two body interaction. For the direct term the Thomas-Fermi method applied to the Coulomb potential is generalized to other potentials and we discuss the possibility to improve them going up to order $\hbar^2$. For the exchange term the Slater approximation is improved using the Density Matrix Expansion.

1. INTRODUCTION

As it is well known, the semiclassical expansions use explicitly the "smallness" of $\hbar$: These approximations are valid when the magnitude of the typical action of the problem is much larger than $\hbar$. In the case of one particle level we obtain the standard WKB approach. For the N fermion systems the "smallness" of $\hbar$ is also used to make the level density continuous, in order to smooth the oscillating contributions, and one reaches the Wigner-Kirkwood (WK) expansion of which the Thomas-Fermi (TF) is the first order in $\hbar$.

As it is shown in reference [1] following the treatment proposed by Dunham[2] in the thirties and using a power series expansion for the level density (approximating sums over discrete levels by integrals) it is possible to show the direct connection between WKB and WK expansion for N fermion systems (order by order in $\hbar$).

Developing the method proposed by Dagens[3] it is possible to compute the expectation value of an operator in the WKB approach without introducing wave functions. Using the development of the level density one recovers the expectation value of this operator in the Wigner Kirkwood approach (order $\hbar$, $\hbar^2$, $\hbar^4$...).

[1] Unité de Recherche des Universités Paris 11 et Paris 6, Associée au CNRS

In this context, given a two body interaction we derive the corresponding selfconsistent semiclassical mean field and we discuss in some cases the possibility to improve the exchange term using the Density Matrix Expansion.

The plan of the article is as follows. In Section 2 we briefly recall the connection between WKB and WK approach for the expectation values. We discuss explicitly the case of the diagonal density for one dimensional systems and the three dimensional case with spherical symmetry.

In section 3, given a two body interaction we derive the semiclassical expansion for the selfconsistent mean field. We discuss the direct term for the following examples: N particles interacting through a harmonic-oscillator force, N particles interacting via a zero range force and in the three dimensional case we discuss also the Coulomb potential.

In section 4, we discuss the exchange term in the Slater approximation, and the improvement of this term using the Campi Bouyssy[4] approach. We finally present the results for the harmonic oscillator force and the Coulomb potential.

Section 5 contains the summary and the conclusions.

2. CONNECTION BETWEEN WKB AND WK

We consider a system of non interacting fermions confined by an external potential $V(x)$. Here and in the following we take units such that $2m = 1$ (m is the particle mass). For the one dimensional case it is shown in reference [1] and [3] how it is possible to compute the expectation value of a given one body operator $F(x)$ without introducing wave functions. Applying first order perturbation theory, one obtains for a given level of energy ϵ

$$< F >_\epsilon = \lim_{\lambda \to 0} \frac{(\epsilon[V + \lambda F] - \epsilon[V])}{\lambda} \tag{2.1}$$

In this expression the ϵ represents successively the eigenenergies corresponding to the external potential $V(x) + \lambda F(x)$ and $V(x)$. We can now introduce the semiclassical expansion of (2.1) using the asymptotic WKB series for the determination of the eigenenergies.

For the ground state of N-fermion system the expectation value of an operator F is obtained summing $< F >_\epsilon$ over all occupied levels. The semiclassical expansion is then recovered replacing sums by integrals up to the Fermi energy ϵ_F and using the semiclassical level density $g_{sc}(\epsilon)$ as the weighting function

$$< F >_{sc} = \int_{-\infty}^{\epsilon_F} < F >_\epsilon g_{sc}(\epsilon) d\epsilon \tag{2.2}$$

The semiclassical level density g_{sc} is obtained after smoothing the oscillating contributions of the quantal level density using the Euler-Mc Laurin or the Poisson expansion as it is shown in section 2.2 of reference [1].

The Fermi energy is obtained by requiring that the integral of the semiclassical level density up to ϵ_F gives the correct number of particles N.

Going up to order $\hbar^2$ one finds in one dimensional case

$$< F >_{sc} = \frac{2}{h} \int F \sqrt{\epsilon_F - V} \theta(\epsilon_F - V) dx$$

$$-\frac{\hbar}{24\pi}[\frac{d}{d\epsilon_F}\int\frac{2FV''}{\sqrt{\epsilon_F-V}}\theta(\epsilon_F-V)dx - \frac{d^2}{d\epsilon_F^2}\int\frac{FV'^2}{\sqrt{\epsilon_F-V}}\theta(\epsilon_F-V)dx] + \Theta(\hbar^4) \quad (2.3)$$

where primes mean derivatives of the potential with respect to x.

We introduce the semiclassical one body density by requiring that for any arbitrary regular function F the semiclassical expansion of the expectation value should be given by

$$< F >_{sc} = \int \rho_{sc}(x)F(x)dx. \quad (2.4)$$

Using the expansion (2.3) one obtains the semiclassical expression of the matter density up to order $\hbar$, which corresponds to the Thomas-Fermi term

$$\rho_0(x) = \frac{1}{\pi\hbar}\sqrt{\epsilon_F - V}\theta(\epsilon_F - V) \quad (2.5a)$$

or up to order $\hbar^2$, which corresponds to the following term of the Wigner-Kirkwood expansion

$$\rho_1(x) = -\frac{1}{24\hbar\pi}(\frac{d}{d\epsilon_F}\frac{2V''}{\sqrt{\epsilon_F-V}} - \frac{d^2}{d\epsilon_F^2}\frac{V'^2}{\sqrt{\epsilon_F-V}})\theta(\epsilon_F - V) \quad (2.5b)$$

In the three dimensional case with spherical symmetry the treatment is analogous and one needs only to take carefully into account the contribution of the centrifugal term $\frac{\hbar^2(l+1/2)^2}{r^2}$ (Langer correction) in the derivation of the $\hbar$ expansion. The Thomas-Fermi diagonal density is then easily reproduced

$$\rho_0(\vec{r}) = \frac{1}{3\pi^2\hbar^3}(\epsilon_F - V(\vec{r}))^{3/2}\theta(\epsilon_F - V(\vec{r})) \quad (2.6a)$$

or the Wigner-Kirkwood diagonal density at order $\hbar^2$

$$\rho_1(\vec{r}) = -\frac{1}{24\pi^2\hbar^3}[-\frac{d}{d\epsilon_F}\frac{(\vec{\nabla}V(\vec{r}))^2}{2(\epsilon_F-V(\vec{r}))^{1/2}} + \frac{\Delta V(\vec{r})}{(\epsilon_F-V(\vec{r}))^{1/2}}]\theta(\epsilon_F - V(\vec{r})) \quad (2.6b)$$

It is clear that in the expansion $\rho = \rho_0 + \hbar^2\rho_1 + ...$ all the values of ρ_n with $n \geq 1$ diverge at the turning points and can be only used in the computation of expectation values; indeed ρ is no more a function but a distribution in the mathematical sense.

3. SELFCONSISTENT SEMICLASSICAL MEAN FIELD

Following the method described before for the expectation values of operators, one can solve a system of N interacting fermions within a selfconsistent semiclassical mean field approximation.

As it is well known, in coordinate space given a two body interaction $v(\vec{r}_i, \vec{r}_j)$ which does not depend on "spin" variables, the local Hartree potential is given by

$$V_H(\vec{r}_i) = \int d\vec{r}_j v(\vec{r}_i, \vec{r}_j)\rho(\vec{r}_j) \quad (3.1.a)$$

or

$$V_H(\vec{r}_i) = v * \rho \quad (3.1.b)$$

if $v(\vec{r}_i, \vec{r}_j) = v(\vec{r}_i - \vec{r}_j)$.

The averaged exchange potential can be written as

$$V_{ex}(\vec{r_i}) = -\int \frac{d\vec{r_j}\, v(\vec{r_i},\vec{r_j})\rho^2(\vec{r_i},\vec{r_j})}{\rho(\vec{r_i})} \tag{3.2}$$

where $\rho(\vec{r_i})$ and $\rho(\vec{r_i},\vec{r_j})$ are the diagonal and the off-diagonal parts of the density matrix.

Equation (3.1) can be solved in a selfconsistent way using the semiclassical expansion (2.5) or (2.6) for the diagonal density matrix. In the exchange term, for the non diagonal density we use some approaches that allow us to express $\rho(\vec{r_i},\vec{r_j})$ in terms of the diagonal density and its derivatives: the Density Matrix Expansion[5] of which the Slater approximation is the first order term.

3.1 One dimensional system

The equation (3.1.a) for the direct term can be solved by iteration

$$V_H^{n+1}(x_i) = \frac{\nu}{\pi\hbar}\int_{x_-^n}^{x_+^n} v(x_i,x_j)\sqrt{\epsilon_F^n - V_H^n(x_j)}\,dx_j$$

$$-\frac{\nu\hbar}{24\pi}[\frac{d}{d\epsilon_F}\int_{x_-^n}^{x_+^n} v(x_i,x_j)\frac{2V_H''^n(x_j)}{\sqrt{\epsilon_F^n - V_H^n(x_j)}}dx_j -$$

$$-\frac{d^2}{d\epsilon_F^2}\int_{x_-^n}^{x_+^n} v(x_i,x_j)\frac{V'^2(x_j)}{\sqrt{\epsilon_F^n - V_H^n(x_j)}}dx_j] + \Theta(\hbar^4) \tag{3.3}$$

where ν is the spin-isospin degeneration factor, the index n refers to the iteration number and $\epsilon_F^n, x_\pm^n$ are the Fermi Energy and the turning points computed at the same iteration. We start in eq (3.3) with some initial guess V_H^o and we iterate up to the selfconsistency.

To illustrate the accuracy of this approximation we shall consider now some simple examples.

3.1.1 Harmonic oscillator force

Let us consider a system of N fermions interacting through a harmonic oscillator force

$$v(x_i,x_j) = \frac{\omega^2}{4}(x_i - x_j)^2 \tag{3.4}$$

The selfconsistent Hartree mean field obtained by solving eq (3.3) is also a harmonic oscillator which differs from the quantal result[6] only by the zero point energy

$$V(x_j) = \frac{N\omega^2 x_j^2}{4} + \frac{N^{3/2}\hbar\omega}{4}(1 + \frac{N^{-2}}{12}) \tag{3.5}$$

The term in $N^{-2}/12$, comes from the inclusion of the $\hbar^2$ correction, and approaches numerically the semiclassical result to the quantal one.

This result can be easily generalized, if the two body interaction is

$$v(x_i,x_j) = |x_i - x_j|^{2n} \tag{3.6}$$

the direct term of the selfconsistent semiclassical mean field is given by

$$V(x_j) = Nx_j^{2n} + C_1 x_j^{2n-2} + ... + C_n \tag{3.7}$$

where $C_1, ..., C_n$ are constants. Namely, taking the $2n + 1$ derivative of the folding product (3.1.b) one gets

$$V(x_j)^{(2n+1)} = 0$$

3.1.2 Zero range force

If the N fermion system interacts via a zero range force with saturation term

$$v(x_i, x_j) = t_o \delta(x_i - x_j) + t_3 \rho(x_j) \delta(x_i - x_j) \tag{3.8}$$

going up to order $\hbar$ one recovers the same selfconsistent semiclassical mean field

$$V(x_j) = t_o \rho(x_j) + t_3 \rho^2(x_j) \tag{3.9}$$

that is given by the quantal Hartree calculation, but ρ is the semiclassical density and eq (3.9) is in principle valid between the turning points only.

In all cases in which the mean field is proportional to the density the improving of the direct term going up to order $\hbar^2$ is not possible, because both the matter density and Hartree mean field diverge at the turning points and can be only used in the computation of expectation values.

3.2 Three dimensional case

In the three dimensional case with spherical symmetry, equation (3.1) reads

$$V_H^{n+1}(\vec{r}_j) = \frac{\nu}{6\pi^2\hbar^3} \int v(\vec{r}_i, \vec{r}_j)(\epsilon_F^n - V_H^n(\vec{r}_j))^{3/2}\theta(\epsilon_F - V_H^n)d^3r_i$$

$$-\frac{\nu}{24\pi^2\hbar}[-\frac{d}{d\epsilon_F} \int v(\vec{r}_i, \vec{r}_j)\frac{(\vec{\nabla}V_H^n(\vec{r}_j))^2}{2(\epsilon_F - V_H^n(\vec{r}_j))^{1/2}}\theta(\epsilon_F - V_H^n)d^3r_i$$

$$+\int v(\vec{r}_i, \vec{r}_j)\frac{\Delta V_H^n(\vec{r}_j)}{(\epsilon_F - V_H^n(\vec{r}_j))^{1/2}}\theta(\epsilon_F - V_H^n)d^3r_i] + \Theta(\hbar^4) \tag{3.10}$$

3.2.1 Harmonic Oscillator force

As a test of our approximation we consider the case in which the two body interaction is a harmonic oscillator force.

$$V(\vec{r}_i, \vec{r}_j) = \frac{\omega^2}{4}|\vec{r}_i - \vec{r}_j|^2 \qquad . \tag{3.11}$$

the direct term of the selfconsistent mean field going up to order $\hbar$ in (3.10) is also a harmonic oscillator like the quantal result[7].

$$V(r_j) = \frac{N\omega^2 r_j^2}{4} + (3/2)^{4/3}N^{5/6}\frac{\hbar\omega}{4} \tag{3.12}$$

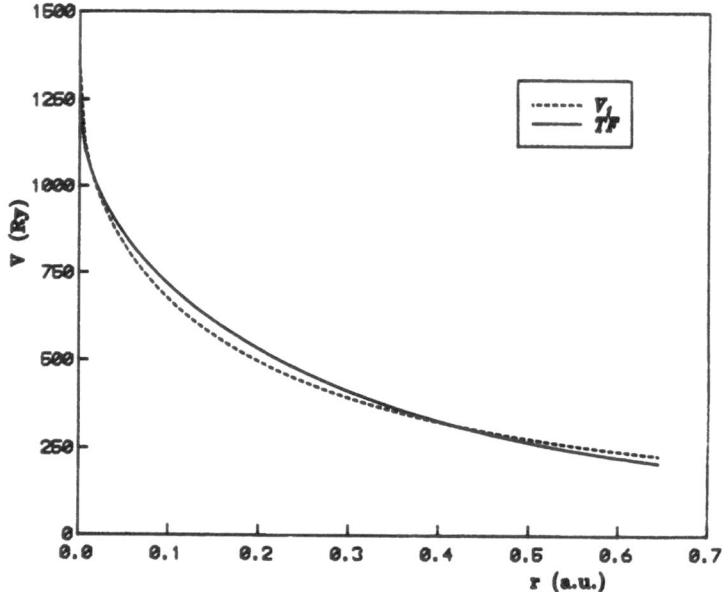

Fig 1- Comparison between the selfconsistent mean field in the Thomas-Fermi approach (TF) for the Rn atom ($Z = 86$) with the result (V_1) obtained at the first iteration of eq (3.3) using $\tilde{Z} = 68$.

3.2.2 Coulomb potential

If the two body interaction is the Coulomb potential $\frac{Ze^2}{|\vec{r}_i - \vec{r}_j|}$ it is easy to recover the Thomas-Fermi equation for neutral atoms. Namely, taking the laplacian of (3.1.b) one gets

$$\Delta V_H = \Delta v * \rho.$$

But $\Delta(1/r) = -4\pi\delta$, therefore ΔV_H is proportional to $\delta * \rho = \rho$ which is the Poisson equation. Now replacing ρ by ρ_{TF} one finds exactly

$$\frac{1}{r^2}\frac{d}{dr}(r^2 V'(r)) = -\frac{\nu}{3\pi\hbar^3}(\epsilon_F - V(r))^{3/2}. \tag{3.13}$$

We finally make the usual change

$$\phi = ar(\epsilon_F - V)$$

$$r = bx$$

with a and b constants, and we get the universal Thomas-Fermi equation

$$\phi'' = \frac{1}{\sqrt{x}}\phi^{3/2}$$

It should be noticed that the iterative procedure (3.10) converges quickly and using an input mean field of type $-\tilde{Z}/r$ (screened punctual Coulomb potential) the first iteration is analytical and can be a reasonable approximation even for small values of r as it is shown in fig [1] for the Rn atom ($Z = 86$).

As for the zero range force it is not possible to improve the direct term going up to the order $\hbar^2$.

4. EXCHANGE TERM

In the case of the zero range interaction the exchange and the direct terms of the mean field are proportional one to another and the selfconsistent semiclassical Hartree Fock calculations give the same formal result that the quantal ones.

In the case of a finite range interaction the computation of the exchange term for a two body interaction is a cumbersome process because of the angular dependence of the density matrix.

Many approximations have been used for the non-diagonal density matrix, one of the most accurate being the density matrix expansion proposed by Negele and Vautherin[5], which appears as a series expansion in the relative distance $\vec{s} = \vec{r}_i - \vec{r}_j$ followed by the averaging over the direction of s; keeping only the first two terms, one obtains:

$$\rho(\vec{R}, \vec{s}) = \rho(\vec{R})\hat{j}_1(k_F s) + 1/6 s^2 \hat{j}_3(k_F s)[1/4\Delta\rho(\vec{R}) - \tau(\vec{R}) + 3/5 k_F^2 \rho(\vec{R})] \qquad (4.1)$$

where

$$\vec{R} = \frac{\vec{r}_i + \vec{r}_j}{2}, \quad \hat{j}_l(x) = (2l + 1)!! \frac{j_l(x)}{x^l}$$

The value of $\rho(\vec{R}, \vec{s})$ depends on k_F because of the truncation of the expansion. The natural choice of k_F is that of the infinite system $k_F = (\frac{6\pi^2 \rho(R)}{\nu})^{1/3}$.

In the framework of semiclassical methods the most common approximation keeps only the first term (this is the Slater approach) using $\rho(R) = \frac{\nu}{6\pi^2\hbar^3}(\epsilon_F - V(r))^{3/2}$. This approach is generally used in the case of the Coulomb potential using the quantal density for $\rho(R)$.

Campi and Bouyssy[4] proposed to improve the approximation (4.1) by keeping only the first term but with k_F fixed by the condition that the second term vanishes identically and the expression is formally identical to the Slater approach. So that we have

$$\rho(\vec{R}, \vec{s}) = \rho(\vec{R})\hat{j}_1(\hat{k}_F s) \qquad (4.2)$$

with $\hat{k}_F(\vec{R}) = [\frac{5}{3\rho(\vec{R})}(\tau(\vec{R}) - 1/4\Delta\rho(\vec{R}))]^{1/2}$

Using for $\tau(\vec{R})$ the semiclassical expansion up to order $\hbar^2$

$$\tau(\vec{R}) = 3/5(\frac{6\pi^2}{\nu})^{2/3}\rho^{5/3}(\vec{R}) + \beta\frac{|\vec{\nabla}\rho(\vec{R})|^2}{\rho(\vec{R})} + \gamma\Delta\rho(\vec{R}) \qquad (4.3)$$

with $\beta = 1/36$ and $\gamma = 1/3$, it is possible to improve the semiclassical density matrix by replacing the first order $k_F(R)$ by the new value $\hat{k}_F(R)$.

We analyze the Campi-Bouyssy improvement using the semiclassical development (C.B.S.) for the case of the harmonic oscillator and the Coulomb potentials.

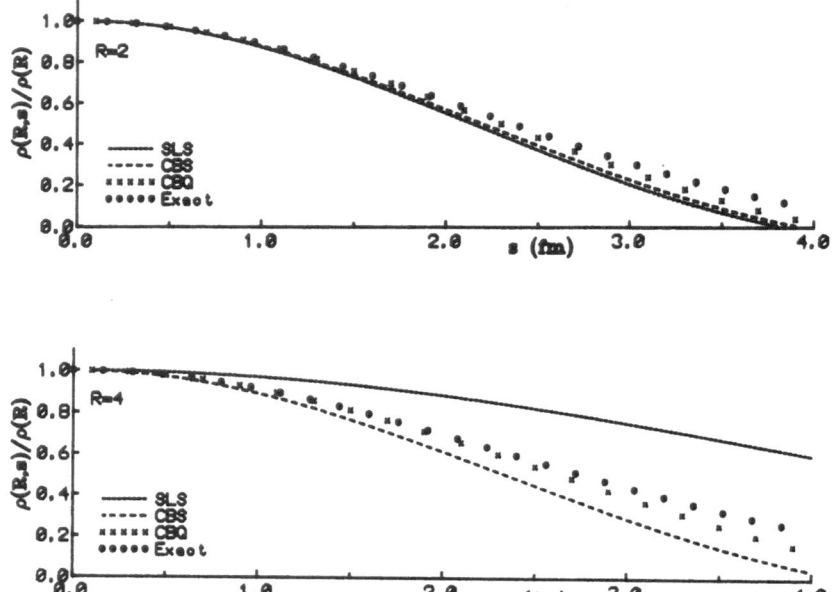

Fig 2- Comparison between the exact and semiclassical normalized density matrices $\rho(\vec{R}, \vec{s})/\rho(\vec{R})$ and $\rho_{sc}(\vec{R}, \vec{s})/\rho_{sc}(\vec{R})$ for the H. O. potential in the case $A = 16$ as a function of s at different values of R ($b = 1.83 fm$). The exact values are averaged over the direction of s. The different lines are explained in the text.

4.1 The Harmonic Oscillator

The three dimensional harmonic oscillator is used as a simple parametrization of the mean field in the ground state of some light nuclei .

As in the quantal case[4] the value of $\hat{k}_F(R)$ semiclassical is analytical

$$\hat{k}_F(R) = \{1/b^2[2(3/2N)^{1/3} - z^2] + \frac{5}{24}\frac{z^2}{b^2[(3/2N)^{1/3} - z^2/2]^2}$$

$$-\frac{5}{8}\frac{1}{b^2[(3/2N)^{1/3} - z^2/2]}\}^{1/2} \tag{4.4}$$

where N is the number of particles, $b = \sqrt{\hbar/m\omega}$ and $z = R/b$. Keeping only the first term, we obtain the semiclassical Slater approach. The difference between Slater and C.B.S. approaches becomes more important near the surface where the Slater approach is worse.

The importance of this improvement is shown in Fig. 2 by comparing the exact and semiclassical normalized density matrices ($\rho(\vec{R}, \vec{s})/\rho(\vec{R})$ and $\rho_{sc}(\vec{R}, \vec{s})/\rho_{sc}(\vec{R})$) as a function of s for two different values of R. We consider a pseudonucleus with $A/2 = N = Z = 8$ without Coulomb. The exact values are averaged over the direction of s. From the comparison we conclude that at the interior all the approximations considered (Campi Bouyssy quantal (CBQ), Slater Semiclassical (SLS) and Campi Bouyssy Semiclassical (CBS)) coincide. At the surface the Semiclassical Campi Bouyssy approach improves significantly the Semiclassical Slater approach.

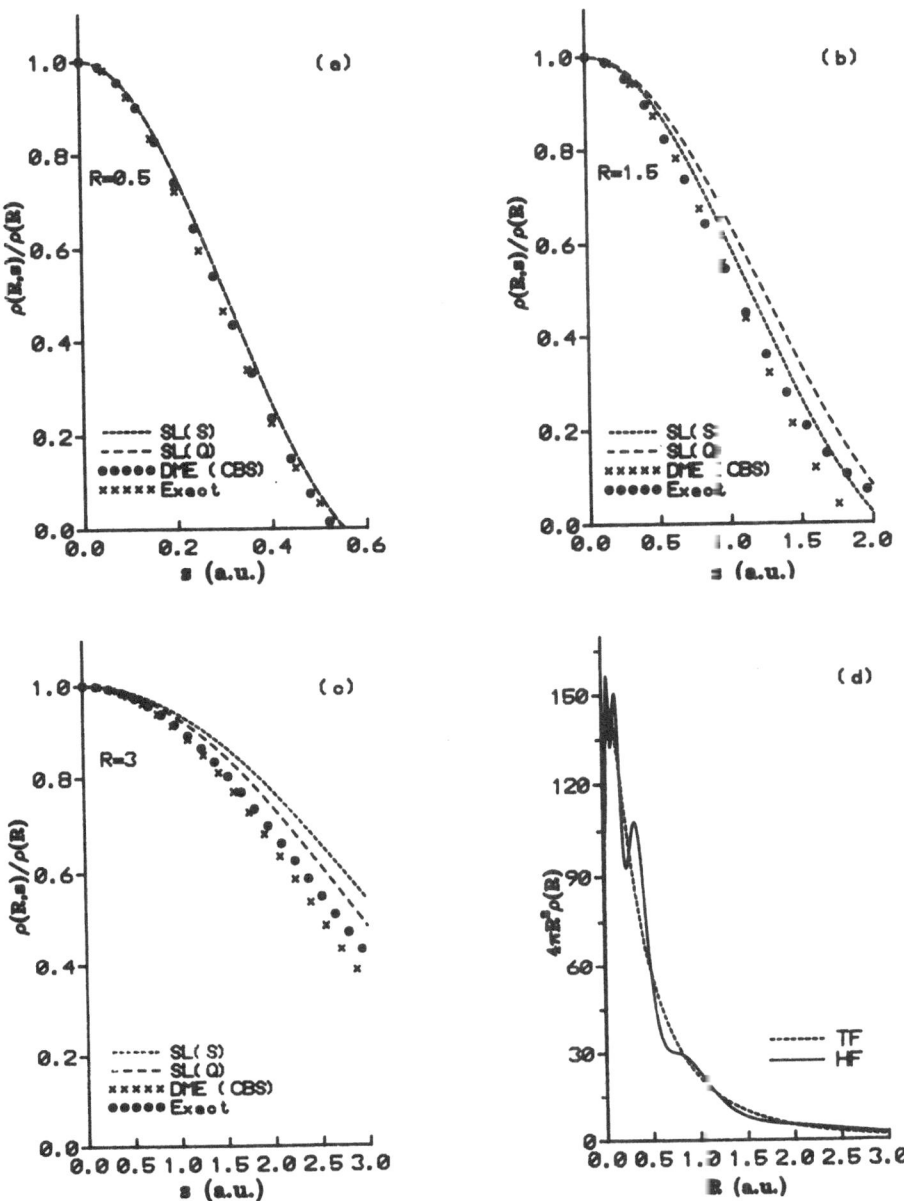

Fig 3 (a) (b) (c)- Same as Fig 2 for the Radon atom. SL (S) and SL (Q) are the semiclassical and quantal Slater approaches. DME (CBS) is the semiclassical Campi Bouyssy approach. The Hartree-Fock (exact) values are averaged over the direction of s. (d) Hartree-Fock and Thomas-Fermi diagonal densities for this atom.

The exchange term for the case of the Coulomb potential is usually computed in the plane wave approximation. The improvement of this term can be very important in some Energy Density Functionals used to solve the problem of metallic clusters where the exchange potential gives more than 70% of the total mean field[8].

In figures 3a, 3b and 3c we compare the quantal and the semiclassical normalized density matrices for the Radon ($Z = 86$) atom. The exact values, which are computed using the Hartree-Fock wave functions, are averaged over the direction of s. We conclude that at the interior all the approximations (Quantal Slater, Semiclassical Slater, and Semiclassical Campi Bouyssy) give reasonable results, but near the surface the CBS approach provides significantly better results, keeping the simplicity of the Slater approximation.

5. SUMMARY AND CONCLUSION

Starting from an $\hbar$ expansion of the expectation values of one body operator we propose a method to obtain a selfconsistent semiclassical mean field for a given two body interaction. In the particular case of the Coulomb potential the direct term up to order $\hbar$ is equivalent to the Thomas-Fermi equation for atoms.

In agreement with the quantal result, if the two body interaction is a harmonic oscillator force, the direct term of the selfconsistent mean field is also a harmonic oscillator.

The $\hbar^2$ corrections can be taken into account only if the direct term of the mean field is not proportional to the diagonal density otherwise the first term of the Wigner-Kirkwood approach diverges at the turning point.

The semiclassical Campi-Bouyssy approach improves in a significant way the off diagonal density matrix for the cases of the H. O. and the Coulomb potentials. This correction can be important in some Energy Density Functionals used to solve the problem of metallic clusters.

This approximation provides a simple way to obtain the Hartree-Fock semiclassical mean field with non-local exchange term even for the realistic two body forces used in nuclear and molecular physics problems. Work in this direction is in progress.

ACKNOWLEDGEMENTS

We are indebted to J. Martorell who suggested the subject of this work. We are also grateful to D.W.L. Sprung, M. de Llano and J. Treiner for discussions. This work has been supported by DGICYT (Spain) grant (PS88-0045).

REFERENCES

1- H. Krivine, M. Casas and J. Martorell, Ann. Phys. (N.Y.) 200 (1990) 304

2- J. L. Dunham, Phys. Rev. 41 (1932) 713

3- L. Dagens, J. Phys. (Paris) 30 (1969) 593

4- X. Campi and A. Bouyssy, Phys. Lett. 73B (1978) 263

5- J. W. Negele and D. Vautherin, Phys. Rev. $\underline{C5}$ (1972) 1472 and Phys. Rev $\underline{C11}$ (1975) 1031

6- M. Moshinsky, Am. J. of Phys. $\underline{36}$ (1968) 52

7- A. Calles and M. Moshinsky, Am. J. of Phys. $\underline{38}$ (1974) 456

8- Ll. Serra, Ph. D. Thesis, Universidad de les Illes Balears (1991)

A SIMPLE APPROACH TO SURFACE STATES IN COVALENT CRYSTALS

D. Mirabella, R. Deza[+] and C.M. Aldao[*]

Universidad Nacional de Mar del Plata
J.B. Alberdi 2695, 7600 Mar del Plata, Argentina

ABSTRACT

Within a simple model for a semiconductor (we study the energy spectrum of one electron in a one-dimensional crystal in the tight-binding approximation) and by imposing a minimum of boundary conditions, we address the following issues:
-Occurrence of surface Tamm's states
-States induced by chemisorption
-Pinning of the Fermi level
-Formation of Schottky barriers
Qualitative agreement with experimental results is obtained.

INTRODUCTION

Understanding the mechanism responsible for Schottky-barrier formation in metal-semiconductor interfaces is by now a fifty-year-old problem.[1,2] The origin and characterization of interface states have been the main issue to microscopically understand the barrier, and a number of first-line scientists have devoted to them, formulating very imaginative theories and models and applying (and even developing) very powerful theoretical tools.[3-7]

It has been systematically observed that whereas the barrier height is largely independent on the nature of the metal for a covalent semiconductor, it does strongly depend on the metal for an ionic one.[8] This experimental result remains to be properly explained and we believe it is of crucial importance. None of the theories in the literature seems to explain both kinds of behavior at the same time, or be free of contradictions with experimental data.

A metal-semiconductor contact is not a simple system and numerous phenomena capable of modifying the Schottky barrier can occur at the interface.[3,9] Because of this complexity, there is no consensus for an interpretation to explain all the experimental results. Although electrical properties of devices can be described with macroscopic models, the physical origin of the interface states responsible for establishing the barrier height is not clear. This puzzling situation challenged us to try to see how many fundamental facts could we grasp by starting from a simple-minded model, and indeed we are astonished that we could get so far.

In this paper we address, from a qualitative (even semiquantitative) viewpoint, the mechanism of generation of surface states in covalent crystals and its relevance to explain the independence of the Schottky barrier height on the metal. An associated phenomenon we concern with is Fermi level pinning at the surface in presence of a few absorbed metal atoms (Fig. 1 on next page shows Fermi level evolution as a function of coverage).

It is known that electron correlations do not play a significant role in these phenomena since the bands are either filled or empty, so we restrict for the moment to one-electron theory. In order to get a qualitative picture of what occurs at the interface we chose to keep the model as simple as possible. We then restrict for the moment to one-dimensional lattices, which nonetheless allow for a straightforward generalization to simple cubic ones.

We will use a tight-binding approximation which has the double advantage of showing clearly the underlaying dynamics and allowing for a simple parametrization of the surface/interface phenomena through the boundary conditions. It is a simple matter anyway to translate our picture in terms of e.g. Krönig-Penney model.

In order to keep the model as simple as possible we assume that states belonging to different sites are orthogonal, so that any overlapping effects between neighbouring orbitals are assumed to be integrally due to the interaction and are thus embodied into the diagonal and hopping terms. We also assume, consistently with this approximation, that hopping is only possible between nearest-neighbours. Finally, since we are not concerned with band overlapping, we can start from just one orbital per site.

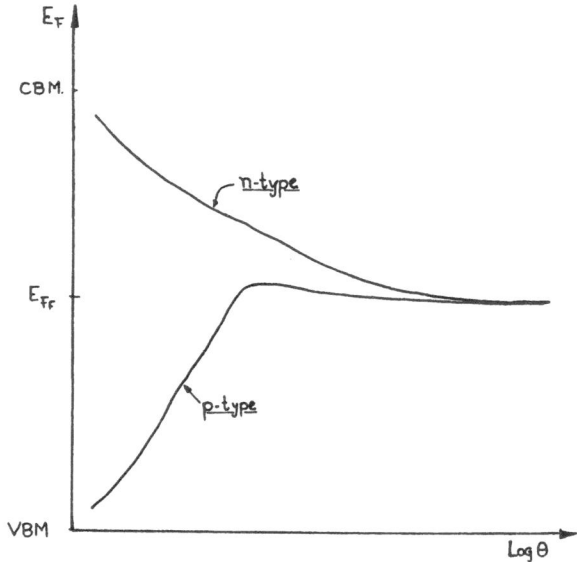

Fig. 1 Position of the Fermi states within the band gap for n- and p-type doped semiconductors, as a function of nominal coverage at room temperature (drawn schematically).

SURFACE STATES

The eigenvalue problem for one electron in a chain of N equal atoms can be cast, in the tight-binding approximation, in the form of a coupled set of equations

$$\lambda \, c_j = c_{j+1} + c_{j-1} \qquad\qquad j = 1,...,N \qquad\qquad (1)$$

which for a variety of boundary conditions leads to the solution $\lambda = 2 \cos\theta$, i.e. the known band spectrum

$$E(k) = \alpha + 2 \gamma \cos \theta \qquad\qquad (2)$$

where $\alpha = <\psi_j|H_0+V|\psi_j>$ and $\gamma = <\psi_j|V|\psi_{j\pm1}>$ are the Coulomb and hopping integrals, and the lattice parameter has been taken to be unity.

In order to account for the effect of a discontinuity in the crystal structure, it suffices to assume that the value of the Coulomb integral for the atom in the 0th position (call it α') differs from that of the rest. An appropriate set of boundary conditions is now the following:
- at the Nth position we take $c_N = 0$ (which allows us to take the limit $N \to \infty$);
- at the surface $(E-\alpha')/\gamma \, c_0 = c_1$ which, calling $\Delta = (\alpha - \alpha')/\gamma$, can be written as

$$(\lambda+\Delta) \, c_0 = c_1 \tag{3a}$$

By taking $c_j \propto \sin(N-j)\theta$ (which incorporates the condition $c_N = 0$) we may rewrite (3a) in the form

$$\lambda = -\Delta + \cos\theta - \cotg N\theta \, \sin\theta \tag{3b}$$

The eigenvalue equation (1) still gives (2) as solution, as can be explicitly verified for this choice of c_j (it is the set of allowed θ-values what depends upon the boundary conditions). By equating (2) and (3b) we rewrite the boundary condition as

$$\Delta + \sin(N+1)\theta \, / \sin N\theta = 0 \tag{3c}$$

Let us first analyze the meaning of Δ: it compares the difference between Coulomb integrals α and α' with the bandwidth γ (2). Since α, α' and (assuming an s-state) γ are usually negative, a value of $\Delta > 0$ means that the level at site 0 is higher that the center of the band, and viceversa. In what respects to the size of Δ we see that as far as $|\Delta|<1$ will α' lie inside the band (2) whereas for $|\Delta|>1$ that level is not degenerate with any one corresponding to the bulk. In this situation, and since nothing forces us to take $|\Delta|\leq1$, there will also appear localized states at $\theta=i\varepsilon$ and $\theta=\pi+i\varepsilon$ (bonding and antibonding states), whose energies will respectively be

$$E(\varepsilon)_{B,A} = \alpha \pm 2\gamma \cosh\varepsilon \tag{4}$$

Since from (1) and (3) it follows that $|c_j|^2 = |c_0|^2 \, e^{-2j\varepsilon}$, we see that $e^{-1}=1/\ln|\Delta|$ tells us how far inside the crystal is the perturbation $\alpha'\neq\alpha$ at the surface still appreciably felt or, in other words, is a measure of the spatial extension of the corresponding localized states. To reinforce this picture we observe that whereas in this LCAO formulation an extended (band) state is made out of a LC of all atomic states, a LC of only a few AO should suffice to create a state which is somewhat localized at the surface.

Given the fact that $\varepsilon>0$ (since we assume the discontinuity to lie at the left side) we find useful in order to plot (3) and (4) to define the variable $L=e^{-\varepsilon}$, which remains bounded ($0\leq L\leq1$) in the limit $N\to\infty$. In that limit (3) reads $\lambda = \pm L-\Delta$, whereas (4) looks as $\lambda=(E-\alpha)/\gamma = \pm2\cosh\varepsilon = \pm(L+L^{-1})$. The boundary condition defines a straight line (see Fig. 2) whose slope is 1 and which intercepts the λ-axis at Δ. We see again that the condition $|\Delta|\geq1$ must be fulfilled in order to have a localized state.

Now if the atom at the 0th is indeed another element we must also take at that site a new value γ'. Call $\Omega=\gamma'/\gamma$, then (3a) shifts into $(\lambda+\Delta) \, c_0=\Omega \, c_1$. The right value for the slope is now Ω, hence we may have a localized state for $|\Delta|<1$ provided that $\Omega>1$; on the other hand, for $\Omega<1$ we must have at least $|\Delta|>(2-\Omega)$. Obviously, both parameters α' and γ' are not independent.

Note that whereas the value of $\gamma=Kh/m^*a^2$ is obtained by measuring the effective mass $\partial E/\partial v$, a detailed study of the bond between adatom and substrate orbitals is needed in order to find the value of γ'. As a consequence, γ' depends strongly on the geometry of that bond.

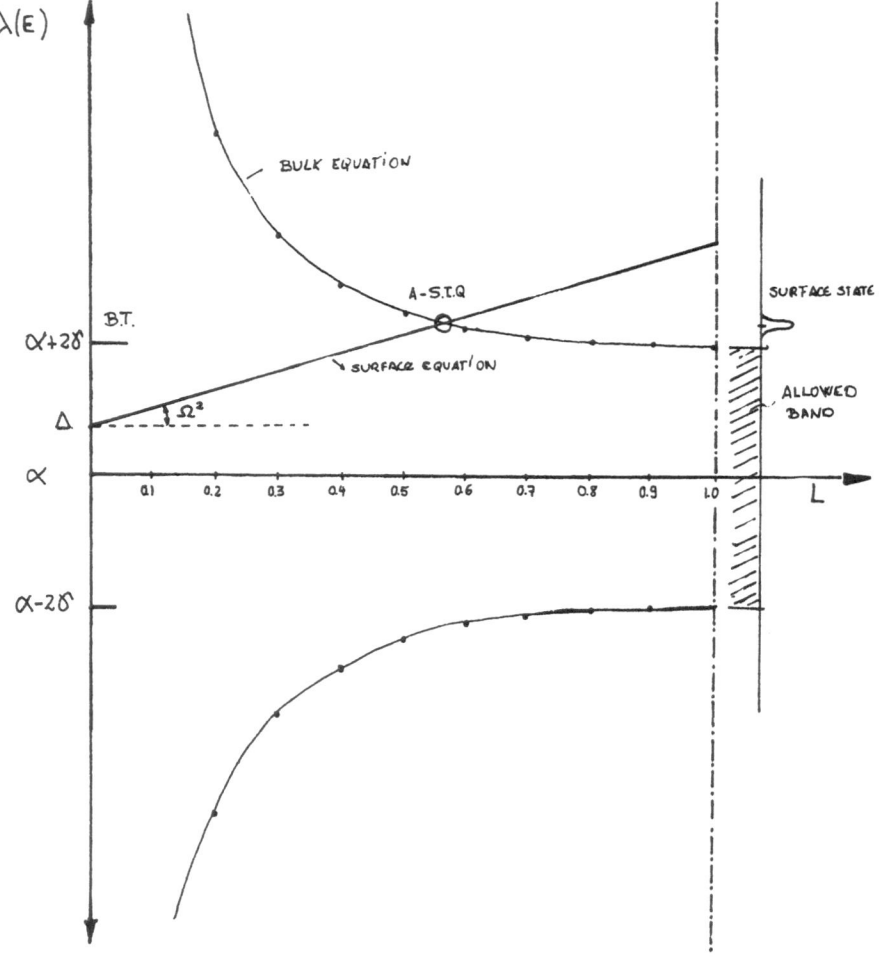

Fig. 2 Graphical solution of a simple 1D surface-state problem. The energy is plotted versus L, where L=exp(-ε) and ε is the damping constant. The surface state appears at the intersection of the bulk- and surface equations.

PINNING OF THE FERMI LEVEL AND SCHOTTKY BARRIER FORMATION

Under the assumptions that led to this model, the final position of the Fermi level has to coincide with that of the localized state. Note that as we have argued, that position is dependent upon the values taken by the parameters, whose meaning is the following:

- Δ_a is a measure of the perturbation introduced by the presence of the adatom;
- Δc parametrizes the effect of the breakdown in traslational symmetry (since the last atom suffers an asymmetric interaction) but can also account for surface relaxation and surface defects generated by excess due to the chemisorption process;
- Ω measures the ratio between the strengths of adatom-substrate and substrate-substrate interaction.

Note in passing that two limiting cases are possible: surface states corresponding to $\Delta_a = 0$ will be totally intrinsic in nature, whereas those appear for $\Delta_c = 0$ are "chemisorption induced states" (CIS). It is convenient to stress that we are identifying the final position of the Fermi level with the one induced by the chemisorbed atom. This we dare to call a "fast pinning mechanism", since it depends exclusively on the interaction of each adatom with the semiconductor, independently of the other adatoms.

There is enough experimental evidence in support of the existence of the before mentioned CIS[12-14], especially when the substrates are kept at about 100k so to reduce surface mobility and thus inhibit cluster formation and make interaction between adatoms negligible. We must note however that much of that evidence is not seen naively, but once it has been corrected for photovoltaic effect. That is particularly true for the available low-dopped n-type samples[10].

The results just drawn are easily translated to the three dimensional case, at least for simple cubic symmetry. The most substantial difference is that localized states now develop into surface bands: their width is $\alpha \pm 8|\gamma|$. Again, it is $\varepsilon(\Delta_c, \Delta_a, \Omega)$ which determines their position relative to the bulk bands.

In order to analyze the remaining differences, let us restrict to two adsorbed atoms. If for some set of parameters is ε small then the introduced surface state will lie near the band and thus will not be very localized. That will give rise to an effective interaction between adatoms, mediated by the crystal, which leads to a splitting of the induced level into two sub-levels (equivalent to the bonding and antibonding states in a diatomic molecule). That splitting increases as the distance between adatoms decreases, forcing the B level to ever lower energy values.

As the covering increases the CIS begin to interact. As a consequence, the induced dipoles by adatoms-substrate charge transfer begin also to interact, provoking a depolarization effect which makes the CIS migrate to the bands according to the nature of both chemisorption and doping. This effect can be expressed as[11]

$$P = P_0 / (1+9\alpha N^{3/2}) \tag{5}$$

where $P_0 = e\delta Q$ is the dipolar moment, α the polarizability and N the number of adatoms. Thus, the final position of the Fermi level results from the competition between:
- the indirect interaction between adatoms, which forces the levels away from the bands, and
- the dipolar interaction which has the opposite action on the levels.

The second interaction overcomes the first one for a critical value of the covering. That gives rise to the overshoot that can be seen in some experiments, e.g. by following the evolution of Fermi level in GaAs with covering by Cs at low temperatures. [14]

Let us agree in classifying chemisorption process into "cationic" and "anionic", according to whether charge transfer occurs from adatoms to substrate or viceversa. Let us consider anionic chemisorption in a doped crystal of n-type: that means that its electronic affinity lies below the Fermi level of the substrate.

The negative charge transferred to the surface generates a space charge region or depletion layer in its vicinity (the width of this depletion layer will be smaller, the larger the concentration of impurities is). That charge distribution generates in turn a potential which gives rise to band bending.

Because of this potential barrier, anionic chemisorption in an n-type semiconductor gives rise to a decrease in surface conductivity. Contrarily, for cationic chemisorption in an n-type material, since the donor level of the adatom lies above the Fermi level there will be no band bending.

CONCLUSIONS

We stress the following conclusions:

1-   We have shown that the Schottky barrier height is almost independent on the chemical nature of the adatom for covalent semiconductors.

2-   In the Defect Model the role of the adatom consists in creating an intrinsic defect in the substrate which is responsible for pinning. In contrast to that, in our model this independence arises after taking into account the adatom-substrate interaction.

3-   We propose two stages during formation: we see that states induced by chemisorption are responsible for a fast Fermi level movement toward midgap, whereas it is dipole interaction which establishes the final barrier height at higher coverages (the overshoot found in p-samples is an evidence for the triggering of this mechanism).

ACKNOWLEDGEMENTS

This work is partially supported by the National Council for Scientific and Technical Research (CONICET) of Argentina.

REFERENCES

+   Physics Department, Faculty of Exact and Natural Sciences, Funes 3350
*   Institute of Materials Science and Technology (INTEMA-CONICET) and Physics Department, Faculty of Engineering, J.B. Justo 4302

1.   W. Schottky, Z. Phys. **113**, 367 (1939).
2.   J. Bardeen, Phys. Rev. **71**, 717 (1947).
3.   W.E. Spicer, P.W. Chye, P.R. Skeath, C.Y. Su and I. Lindau, J. Vac. Sci. Technol. **16**, 1019 (1979).
4.   L.J. Brillson, Phys. Rev. Lett. **40**, 260 (1978).
5.   J. Tersoff, Phys. Rev. Lett. **52**, 465 (1984).
6.   J.L. Freeouf and J.M. Woodall, J. Vac. Sci. Technol. **21**, 570 (1984).
7.   J.E. Klepeis and W.A. Harrison, J. Vac. Sci. Technol. **B7**, 4 (1989).
8.   E.H. Rhoderick and R.H. Williams, *Metal-Semiconductor Contacts*, 2nd Edition (Clarendon, Press, Oxford, 1988).
9.   J.H. Weaver, "Synchrotron Radiation Photoemission Studies of Interfaces", Chapter 2 in *Analysis and Characterization of thin Films*, edited by K.N. Tu and R. Rosenberg (Academic Press, New York, 1987).
10.   C.M. Aldao, G.O. Waddill, P.J. Benning, C.Capasso and J.H. Weaver, Phys. Rev. **B41**, 6092 (1990).
11.   J. Levine and G. Itopoulos, Surf. Sci. **1**, 174 (1964).
12.   K. Stiles, A. Kahn, D. Kilday and G. Margaritondo, J. Vac. Sci. Technol. **B5**, 967 (1987).
13.   F. Barteis, L. Sirkamp, H. Clemens and W. Mönch, J. Vac. Sci. Technol. **B6**, (1988).
14.   R. Cao, K. Miyano, T. Kendelewicz, K. Chin, I. Landau and W. Spicer, J. Vac. Sci. Technol. **B5**, 988 (1985).

SCATTERING OF ATOMIC BEAMS IN SEMIQUANTUM GASES

Eugene P. Bashkin and Sergey B. Stepanyantz

*Institut für Theoretische Physik, Universität zu Köln
D–5000 Köln 41, Germany

INTRODUCTION

A low density system at rather high temperature whose particles obey the classical Boltzmann–Maxwell statistics and which however exhibits fundamentally quantum-mechanical macroscopic properties is usually called a semiquantum gas. The most typical semiquantum gases which are actively being studied in experiment are spin-polarized atomic hydrogen H↓ and deuterium D↓, gaseous ^{3}He↑, the system of ^{3}He-atoms dissolved in superfluid ^{4}He etc. Quantitatively one can expect a rarefied Boltzmann gas to display a quantum–mechanical behavior within the temperature range [1]:

$$\epsilon_d << T << \hbar^2/mr_0^2 , \quad Nr_0^3 << 1, \tag{1}$$

where ϵ_d is the quantum degeneracy temperature, m the mass of a particle, r_0 the interaction range (of the order of an atomic size) and N the atomic density. Under this condition a gas may possess a highly nontrivial spectrum of collective excitation (like spin waves[1-5]). Moreover a very important mechanism of inelastic interaction between a particle in the gas and collective fluctuations of thermodynamical variables comes into effect if the criterion (1) is fulfilled[6]. Microscopically such an interaction may be interpreted as the inelastic scattering of paramagnetic particles with thermally excited phonons, spin modes and entropy fluctuations in the semiquantum gas. E.g. under certain conditions the exchange interaction between a particle of a gas and fluctuations of macroscopic magnetization strongly affects the thermal conductivity and viscosity of a spin–polarized semiquantum gas[7].

An extremely important feature of the phenomenon in question is that a new scattering mechanism is created even at distances less than the mean free path. This gives a possibility to experimentally observe the effect even in the case where direct elastic collisions between particles do not occur, e.g., a gas in the Knudsen regime (e.g. in the case of atomic H↓ in a magnetic trap[8] or in direct experiments with H-beams scattered in H↓[9]).

An extra channel of inelastic scattering provides us with the idea to use a cooled atomic beam as a tool to study the spectrum of collective excitations and correlation properties in a semiquantum gas. In fact by letting a testing beam pass through a semiquantum gas as a target and then measuring the intensity of scattered particles at different scattering angles one can obtain exhausting information about correlation place in neutron optics.

The physical pattern of the effect is very simple. Everyone knows from optics that once a light wave enters a medium the speed of light changes due to the phenomenon of refraction. The refraction index of a media may fluctuate and then a light wave experiences the scattering with these fluctuations. For the case in question, the quantum–mechanical refraction of a wave function (a plane wave) of an incident particle propagating through a semiquantum gas, is considered. If the criterion (1) is valid the real part of the effective refraction index for a ψ–function turns out to be much larger than the imaginary part. Actually it means that the real correction to the self–energy of an incident particle, associated with the effect of quantum–mechanical refraction, considerably exeeds the attenuation due to the finite mean free path. Fluctuations of the quantum–mechanical refraction index can be expressed in terms of fluctuations of the parameters characterizing the thermodynamic state of a gas. It is the effect that causes an extra inelastic scattering of particles of a testing beam with collective modes in a semiquantum target.

SCATTERING PROBABILITIES

Let us consider a semiquantum gas of particles with spin 1/2 as a target. Spin of the testing particle in an incident beam is also assumed to be equal to 1/2. In this case the hamiltonian of the inelastic interaction of a particle with a macroscopic fluctuation field in the target takes the form:

$$H(\vec{r},t) = g_1 N(\vec{r},t) + \frac{1}{\beta} g_2 \vec{\sigma}\vec{M}(\vec{r},t) - \beta_0 \vec{\sigma}\vec{B}(\vec{r},t), \tag{2}$$

where $N(\vec{r},t)$, $\vec{M}(\vec{r},t)$ and $\vec{B}(\vec{r},t)$ are the fluctuating macroscopic variables, namely the atomic density, magnetization and magnetic field; $\vec{\sigma}$ are the Pauli matrices, β and β_0 are the magnetic moments of a particle of the target and of a testing particle respectively. The two first terms in Eq.(2) are of purely exchange origin. The latter Zeemann term, of course, contributes essentially less than the exchange corrections but it results in a very typical anisotropy of the differential cross–section which can be observed in experiment. The coupling constants g_1 and g_2 can be expressed in terms of the scattering amplitudes. If both a beam and a target consist of identical particles the quantities g_1 and g_2 have the simple form:

$$g_1 = -g_2 = \frac{2\pi\hbar^2}{m} a, \tag{3}$$

where a is the s–wave scattering length. If particles of a beam and of a target are distinquishable the coupling constants may be calculated as

$$g_1 = \frac{\pi\hbar^2}{M}(3a_+ + a_-) \, , \; g_2 = \frac{\pi\hbar^2}{M}(a_+ - a_-). \tag{4}$$

Here M is the reduced mass and a_+ and a_- are the s–wave scattering lengths for the triplet and singlet scattering respectively.

The fluctuating magnetic moment $\vec{M}(\vec{r},t)$ induces the fluctuations of a magnetic field $\vec{B}(\vec{r},t)$ which are determined by the Maxwell equations in the magnetostatic limit:

$$curl\vec{h} = 0 \, , \; div\vec{B} = 0 \, , \; \vec{B} = \vec{h} + 4\pi\vec{M}. \tag{5}$$

The solution of Eq.(5) can be easily found in the form of the Fourier transformants:

$$B_i(\vec{k},\omega) = 4\pi\left[M_i(\vec{k},\omega) - \frac{k_i k_l}{k^2} M_l(\vec{k},\omega)\right] \tag{6}$$

along the direction of the equilibrium (or quasi– equilibrium) vector of the spin polarization of the target. The scattering probability for the transition between an initial state $|\vec{p}, \gamma >$ and a final state $< \vec{p}', \delta|$ is given by the well known formula of quantum mechanics:

$$dw = \frac{1}{\hbar^2} | \int_{-\infty}^{\infty} < \vec{p}', \delta|H|\vec{p}, \gamma > e^{-i\omega t} dt|^2 \frac{d^3 p'}{(2\pi\hbar)^3}. \tag{7}$$

Here the Greek indices numerate spin states of a particle (the z–projection of spin) and the transition frequency is given by the usual relationship:

$$\hbar\omega = \frac{p^2 - p'^2}{2m_0} - \beta_0 H(\sigma_{\gamma\gamma}^z - \sigma_{\delta\delta}^z), \tag{8}$$

where m_0 is the mass of a scattered particle of a beam. Inasmuch as we are interested in the inelastic scattering of particles with the thermal fluctuations, the expression (7) should also be averaged over fluctuations. In order to obtain the differential cross–section one has to normalize the initial wave function of an incident particle to a unit flux density. Following this procedure one can easily find that

$$d\sigma_{\nu\nu} = \frac{1}{2\pi^2\hbar^5} \frac{m_0^2}{N} \frac{p'}{p} \left\{ g_1^2 G_{00}(\vec{q}, \omega) + (4\pi\beta_0)^2 \frac{sin^2 2\phi}{4} S_{xx}(\vec{q}, \omega) \right.$$
$$+ (\frac{g_2}{\beta} - 4\pi\beta_0 sin^2\phi)^2 S_{zz}(\vec{q}, \omega) \tag{9}$$
$$\left. + g_1 \sigma_{\nu\nu}^z (\frac{g_2}{\beta} - 4\pi\beta_0 sin^2\phi) \left[G_{0z}(\vec{q}, \omega) + G_{z0}(\vec{q}, \omega) \right] \right\} d\epsilon' \frac{d o'}{4\pi}$$

where $d\epsilon' = p'dp'/m_0$, ϕ is the angle between the transfered momentum $\hbar\vec{q} = \vec{p} - \vec{p}'$ and the vector of spin polarization so that $q_z = qcos\phi$. The following structure factors are also introduced in Eq.(9):

$$< \delta M_i \delta M_k >_{\vec{q}, \omega} = S_{ik}(\vec{q}, \omega), \quad < \delta N \delta N >_{\vec{q}, \omega} = G_{00}(\vec{q}, \omega),$$
$$< \delta M_i \delta N >_{\vec{q}, \omega} = G_{i0}(\vec{q}, \omega), \quad < \delta N \delta M_i >_{\vec{q}, \omega} = G_{0i}(\vec{q}, \omega). \tag{10}$$

These are just Fourier transformants of the corresponding correlation functions in the $(\vec{r}, t)$–space. The expressions (9–10) yield the cross–section of the process whereby spin of an incident particle is conserved. Similar calculations for the spin–flip transition give rise to the result:

$$d\sigma_{\mu\nu} = \frac{1}{2\pi^2\hbar^5} \frac{m_0^2}{N} \frac{p'}{p} \left\{ \left[(\frac{g_2}{\beta} - 4\pi\beta_0)^2 + (\frac{g_2}{\beta} - 4\pi\beta_0 cos^2\phi)^2 \right] S_{xx}(\vec{q}, \omega) \right.$$
$$- 2\sigma_{\mu\nu}^y (\frac{g_2}{\beta} - 4\pi\beta_0)(\frac{g_2}{\beta} - 4\pi\beta_0 cos^2\phi) S_{yx}(\vec{q}, \omega) \tag{11}$$
$$\left. + (4\pi\beta_0)^2 \frac{sin^2 2\phi}{4} S_{zz}(\vec{q}, \omega) \right\} d\epsilon' \frac{d o'}{4\pi}, \quad \mu \neq \nu.$$

Thus the problem of quantitatively calculating the differential cross–section reduces to the problem of finding the appropriate dynamic structure factors.

CORRELATION FUNCTIONS AND STRUCTURE FACTORS

The hydrodynamic limit of a small momentum transfer $ql << 1$ where l is the mean free path, is indeed the most interesting case. In this situation one deals with correlation functions can be obtained by means of the hydrodynamic equations only. The linearized equations of hydrodynamics can be presented in the form:

$$\frac{\partial \alpha}{\partial t} = D\Delta\alpha + \frac{2k_T}{T}\Delta T + \frac{2k_P}{P}\Delta P \,,$$

$$\frac{\partial T}{\partial t} + \frac{T}{c_P}\left(\frac{\partial S}{\partial P}\right)_{T,\alpha}\frac{\partial P}{\partial t} - \frac{k_T}{c_P}\left(\frac{\partial \mu}{\partial \alpha}\right)_{P,T}\frac{\partial \alpha}{\partial t} = \lambda\Delta T \,,$$

$$\frac{\partial \rho}{\partial t} + \rho div\vec{v} = 0 \,,$$

$$\rho\frac{\partial \vec{v}}{\partial t} = -\nabla P + (\zeta + \frac{4\eta}{3})\nabla div\vec{v}$$

(12)

where the pressure P, entropy S and density ρ (all quantities are normalized to the unit mass) are introduced; $\vec{v}$ is the macroscopic velosity, η and ζ are the transport coefficients of the first and the second viscosity respectively, λ is the thermometric conductivity and D is the spin diffusion coefficient. The quantities k_T and k_P play the role of the thermomagnetic and baromagnetic diffusion coefficients respectively. It is also convenient to introduce the degree of polarization, α, defined as

$$N_+ - N_- = N\alpha \,, \quad N_+ + N_- = N$$

(13)

where N_+ and N_- are the numbers of particle per unit volume in a target with spin $(1/2)$ and $(-1/2)$ respectively. The macroscopic magnetization M_z can obviously be calculated as $M_z = \beta N\alpha$. The notation "c_P" in Eqs.(12) is used for the specific heat. The chemical potential, μ, the physical meaning of which is given by the thermodynamic identity

$$dE = TdS + \frac{P}{\rho^2}d\rho + \frac{1}{2}\mu d\alpha$$

(14)

can be easily expressed in terms of the chemical potentials of different spin components

$$\mu = \frac{1}{m}(\mu_+ - \mu_-).$$

(15)

In order to calculate the structure factors one should carry out the following procedure[10]. First, one has to multiply the linearized hydrodynamic equations (12) by a perturbation of a thermodynamic variable δP, δT, $\delta\alpha$ or $\vec{v}$. After this one obtains the system of equations similar to Eqs.(12), which contain all possible pair products, $\delta A(\vec{r},t)\delta B(\vec{r}',t')$, where $A, B = P, T, \alpha, \vec{v}$. Averaging this system yields the equations determining all $(\vec{r},t)$–dependent pair correlation functions.

Second, a special type of the Fourier transformation

$$< AB >^+_{\vec{k},\omega} = \int_0^\infty dt \int_{-\infty}^\infty e^{i\omega t - i\vec{k}\vec{r}} < A(\vec{r},t)B(0,0) >$$

(16)

should be performed. This results in the algebraic system of linear equations for the quantities $< AB >^+_{\vec{k},\omega}$. This system of equations also contains explicitly the initial conditions

$$< A(\vec{r},t)B(0,0) >_{t\to+0} = < A(\vec{r})B(0) >$$

(17)

for which the hydrodynamic spatial correlation functions must be used, e.g.

$$< \delta T(\vec{r})\delta T(0) > = \frac{T^2}{\rho c_V}\delta(\vec{r}) \,,$$

$$< \delta\alpha(\vec{r})\delta\alpha(0) > = \frac{(1-\alpha)T}{N(\partial\mu_+/\partial\alpha)_{P,T}}\delta(\vec{r}) \equiv \Gamma_\alpha\delta(\vec{r}).$$

(18)

factors by means of the obvious relationship:

$$< AB >_{\vec{k},\omega} = < AB >^+_{\vec{k},\omega} + < AB >^+_{\vec{k},-\omega} . \tag{19}$$

There exists one more important circumstance which significantly facilitates the calculation of dynamic form–factors. Since fluctuations of pressure. P, and velocity, $\vec{v}$, propagate through a fluid with the speed of sound, u, and perturbations of magnetization, α, and temperature, T, spread in accordance with the equations of thermal conductivity and spin diffusion, one can assume with a very good accuracy that at frequencies $\omega \sim Dq^2 \sim \lambda q^2 << qu$ only isobaric fluctuations of α and T occur. In this limiting case, we obtain after some algebra:

$$< \delta\alpha\delta\alpha >_{\vec{q},\omega} = \frac{2\Gamma_\alpha}{\Delta} Dq^2 [\omega^2 + \lambda q^4(\lambda + \gamma)] ,$$

$$< \delta T\delta T >_{\vec{q},\omega} = \frac{2T^2}{\rho c_V \Delta} [\lambda q^2(\omega^2 + D^2 q^4) + \gamma\omega^2 q^2] ,$$

$$< \delta\alpha\delta T >_{\vec{q},\omega} + < \delta T\delta\alpha >_{\vec{q},\omega} = \frac{2k_T q^2}{\Delta}(\omega^2 - D\lambda q^4)\left(\frac{2k_T}{\rho c_V} + \frac{D}{c_P}\frac{\partial\mu}{\partial\alpha}\Gamma_\alpha\right) , \tag{20}$$

$$\Delta = (\omega^2 - D\lambda q^4)^2 + \omega^2 q^4(D + \lambda + \gamma)^2 , \gamma = \frac{2k_T^2}{c_P T}\left(\frac{\partial\mu}{\partial\alpha}\right)_{P,T}.$$

The expressions (20) provide us with all the necessary information to find the form–factors G_{00}, S_{zz}, G_{0z} and G_{z0} determining the cross–section $d\sigma_{\nu\nu}$:

$$G_{00}(\vec{q},\omega) = \left(\frac{\partial N}{\partial\alpha}\right)^2_{P,T} < \delta\alpha\delta\alpha >_{\vec{q},\omega} + \left(\frac{\partial N}{\partial T}\right)^2_{P,\alpha} < \delta T\delta T >_{\vec{q},\omega}$$

$$+ \left(\frac{\partial N}{\partial\alpha}\right)_{P,T}\left(\frac{\partial N}{\partial T}\right)_{P,\alpha}\left[< \delta\alpha\delta T >_{\vec{q},\omega} + < \delta T\delta\alpha >_{\vec{q},\omega}\right] ,$$

$$G_{0z}(\vec{q},\omega) + G_{z0}(\vec{q},\omega) = 2\beta N\left(\frac{\partial N}{\partial\alpha}\right)_{P,T} < \delta\alpha\delta\alpha >_{\vec{q},\omega} \tag{21}$$

$$+ \beta N\left(\frac{\partial N}{\partial T}\right)_{P,\alpha}\left[< \delta T\delta\alpha >_{\vec{q},\omega} + < \delta\alpha\delta T >_{\vec{q},\omega}\right] ,$$

$$S_{zz}(\vec{q},\omega) = \beta^2 N^2 < \delta\alpha\delta\alpha >_{\vec{q},\omega} .$$

These formulae for structure factors are valid for any paramagnetic fluid. In the case of a rarefied gas Eqs.(20-21) are drastically simplified because all thermodynamic derivatives can be calculated in an explicit form and many of them vanish (at least in the main approximation of a perfect gas). These computations are very simple and they will not be considered here. All final results also become simpler in the limiting cases $\alpha \to 0$ and $\alpha \to 1$. For instance, if $\alpha \to 1$ all correlators reduce to the expressions typical for one–component system (all spins "up"):

$$< \delta\alpha\delta\alpha >_{\vec{q},\omega} \approx (1 - \alpha)\frac{T}{N}\left(\frac{\partial\alpha}{\partial\mu_+}\right)_{P,T} \frac{2Dq^2}{\omega^2 + D^2 q^4} ,$$

$$< \delta T\delta T >_{\vec{q},\omega} \approx \frac{2T^2}{\rho c_V}\frac{\lambda q^2}{\omega^2 + \lambda^2 q^4} , \tag{22}$$

$$< \delta\alpha\delta T >_{\vec{q},\omega} \approx < \delta T\delta\alpha >_{\vec{q},\omega} \approx 0.$$

At high frequencies $\omega \sim qu >> \lambda q^2 \sim Dq^2$ there appear two more peaks of scattering due to fluctuations of pressure (and velocity). Following the similar method

211

form:

$$< \delta P \delta P >_{\vec{q},\omega} = \frac{\rho T u^3 \delta}{(\omega \mp qu)^2 + u^2 \delta^2} \, , |\omega \mp qu| \sim u\delta \tag{23}$$

where the renormalized sound absorption is given by the expressions:

$$\delta = \frac{q^2}{2qu} \left[\zeta + \frac{4\eta}{3} + \frac{\kappa \rho^2 u^2}{T} \left(\frac{\partial T}{\partial P} \right)_{S,\alpha} - \rho u^2 C_\alpha \right] ,$$

$$C_\alpha = 2 \left[\frac{k_T}{T} \left(\frac{\partial T}{\partial P} \right)_{S,\alpha} + \frac{k_P}{P} \right] \left[\left(\frac{\partial \rho}{\partial \alpha} \right)_{P,S} - \rho^2 \left(\frac{\partial T}{\partial P} \right)_{S,\alpha} \frac{F_\alpha}{2T} \right] , \tag{24}$$

$$F_\alpha = 2k_T \left(\frac{\partial \mu}{\partial \alpha} \right)_{S,\alpha} - T \left(\frac{\partial \mu}{\partial T} \right)_{P,\alpha} \, , \kappa = \rho c_P \lambda.$$

The main contribution to the cross–section $d\sigma_{\nu\nu}$ at frequencies $\omega \sim qu$ is determined with a good accuracy by the density–density structure function:

$$G_{00}\vec{q},\omega = \frac{1}{m^2 u^4} < \delta P \delta P >_{\vec{q},\omega} \, , u^2 = \left(\frac{\partial P}{\partial \rho} \right)_{S,\alpha}. \tag{25}$$

The inelastic scattering of paramagnetic particles of a beam with fluctuations of transverse magnetization is described in terms of the magnetic form–factors $S_{ik}(\vec{q},\omega), i, k = x, y$ in Eqs.(9,11). These structure factors can be calculated with the aid of the Boltzmann transport equation or using some sort of a quasihydrodynamic approach[1,6]:

$$S_{xx}(\vec{q},\omega) = S_{yy}(\vec{q},\omega) = \frac{2\beta^2 N\alpha}{1 - e^{-\hbar\omega/T}} \left[L(\omega - \omega'_q, \omega''_q) - L(\omega + \omega'_{-q}, \omega''_{-q}) \right] ,$$

$$S_{yx}(\vec{q},\omega) = -S_{xy}(\vec{q},\omega) = -i \frac{2\beta^2 N\alpha}{1 - e^{\hbar\omega/T}} \left[L(\omega - \omega'_q, \omega''_q) + L(\omega + \omega'_{-q}, \omega''_{-q}) \right] , \tag{26}$$

$$L(x,y) = \frac{y}{x^2 + y^2} .$$

where ω'_q and ω''_q are the spectrum and the attenuation of transverse spin fluctuations respectively[1-2]:

$$\omega'_q = bq^2 \, , \omega''_q = \frac{b}{\gamma_{int}} q^2 \, ,$$

$$b = D_0 \frac{\gamma_{int}}{1 + \gamma_{int}^2} \, , \gamma_{int} = 4\pi\hbar \frac{aN\alpha}{T} D_0. \tag{27}$$

Here D_0 is the spin diffusion coefficient at $\alpha = 0$. The formulae obtained above completely determine the inelastic scattering of slow particles with collective modes in a semiquantum system.

All statements of this paper can be applied to describe the propagation of slow neutrons through a polarized paramagnetic fluid. Some of these results have already been published[11]. A more detailed consideration of the problem in question and the applications to neutron optics will be published elsewhere.

ACKNOWLEDGEMENTS

This work has been supported in part by the Deutsche Forschungsgemeinschaft under Grant No.Ri 267/14-1.

212

*The permanent address: Kapitza Institute for Physical Problems, 117334 Moscow, U.S.S.R.

REFERENCES

1. E. P. Bashkin, Soviet Phys. USPEKHI, **29(3)**, 238 (1986); Scviet Phys. JETP Lett., **33(1)**, 8 (1981).
2. C. Lhuillier and F. Laloe. J. Phys.(Paris), **43**, 197, 225, 833 (1982).
3. B. R. Johnson et al, Phys. Rev. **52**, 1508 (1984); 53, 302 (198=).
4. L. R. Levi and A. E. Ruckenstein. Phys. Rev. **52**, 1512 (1984`;53,302 (1984).
5. P. J. Nacher et al, J. Phys. Lett.(Paris), **45**, L–441 (1984).
6. E. P. Bashkin. Soviet Phys. JETP Lett. **49(6)**, 363, (1989).
7. E. P. Bashkin. Phys. Rev. B (1991), to be published.
8. N. Masuhara et al, Phys. Rev. Lett. **61**, 935 (1988).
9. I. F. Silvera. Private communication.
10. E. M. Lifshitz and L. P. Pitaevskii, Statistical Physics, part 2, Pergamon (1980).
11. E. P. Bashkin. Soviet Phys. JETP. **69(6)**, 1139 (1989).

ATTEMPTS TO CALCULATE THE STRUCTURE AND DYNAMICS OF MACROMOLECULES

Clas Blomberg

Department of Theoretical Physics
Royal Institute of Technology
100 44 Stockholm, Sweden

INTRODUCTION

Large molecules, macromolecules, with maybe 1000 atoms or more provide a class of important problems in condensed matter physics which can be regarded as being between proper molecular physics and the physics of truly macroscopic objects. In our group the main emphasis has been the study of biological macromolecules such as proteins or nucleic acids (DNA) with the ultimate aim to understand their biologically relevant properties from a physical point of view. However, many features are common to general macromolecules and include artificial polymers such as polyethylene (long pure hydrocarbon chains), polyvinyls, polyesters etcetera. It may be wise to work with a broad scope on these problems, and sometimes choose problems that are the most suitable ones for illuminating a certain property.

The basic physical feature, which distinguishes these systems, is that they are build up by atoms along long chains with a high degree of flexibility. The length of the chain should involve very many atoms (usually at least 100-1000, but typical macromolecular features may be present in chains of the lengths of some tens of atoms.). The chain usually has branches with atomic groups of molecular dimensions (with up to some tens of atoms). In the problems we are interested in, these branches should not be coupled together with molecular, covalent bonds (although couplings by weaker bonds are allowed), which would reduce the flexibility. The macromolecules can then be regarded as true one-dimensional systems.

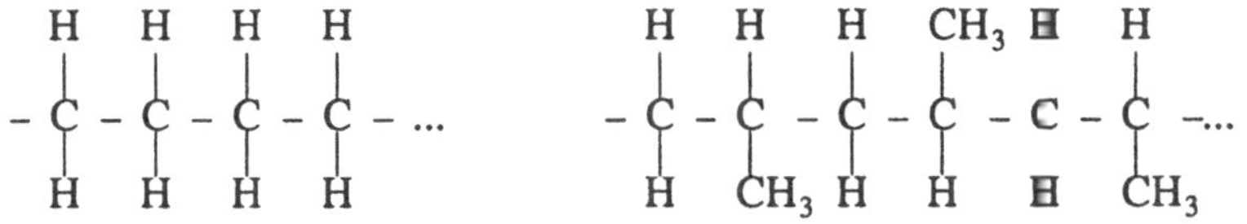

Fig. 1. Typical (simple) chain macromolecules: polyethylene (left) and polypropylene (right), the latter with side group branches. The molecules can rotate around all bonds of the main chain.

Condensed Matter Theories, Vol. 7, Edited by A.N. Proto and J.L. Aliaga, Plenum Press, New York, 1992

The single bonds along the main molecule chain allow free rotations, and the main restriction on the rotation is given by the interaction between different molecule groups. When interaction energies between atoms that are close to each other along the chain are taken into account, one gets a rotation energy as function of rotation angle with usually a low number (2 or 3) of energy minima with not too large energy barriers in between. It is then easy to get transitions between the minima, but one can to a large extent limit the rotation angles to the values at the minima. The attainable configurations of the macromolecule chain with such minimum angles are referred to as *conformations*.

Thus, the main aim of the macromolecular physics is to study conformations and conformation changes, i.e. the dynamics of the molecules with rotations along the bonds of the main chain.

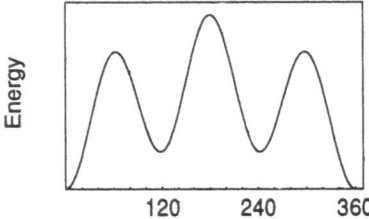

Fig. 2. Typical energy as function of angle for rotations around single bonds

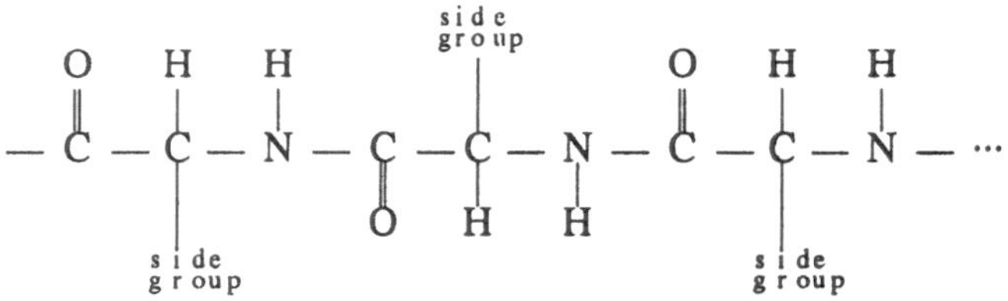

$$-C-C-N-C-C-N-C-C-N-\cdots$$

Fig. 3. The basic structure of a protein chain. The side groups vary along the chain in a manner that is specific for a certain protein. The chain can be rotated around the bonds of the carbon atoms that binds the side groups.

The biologically interesting macromolecules are the proteins and nucleic acids. The protein chains are built up of amino acid units with one part that is common for all amino acids and repeated along the protein chains, and side group branches that are specific for the amino acids and which vary in a definite manner along the chain.

Important for the properties of the proteins is the fact that two atomic groups of the common unit are electrically polar; the C-O and the N-H groups. These form dipolar interaction, and in particular, they can form hydrogen bonds with each other or with other, similar polar groups. The main chain can be rotated around the two bonds to the carbon atom that is connected to the specific side group. The side groups are chosen among 20 different ones for the main proteins of a living cell, and these groups have widely different properties. Some are hydrocarbon chains without strong dielectric moments, some have strong dielectric moments and other are normally charged in water. We can refer to these as non-polar and polar side groups.

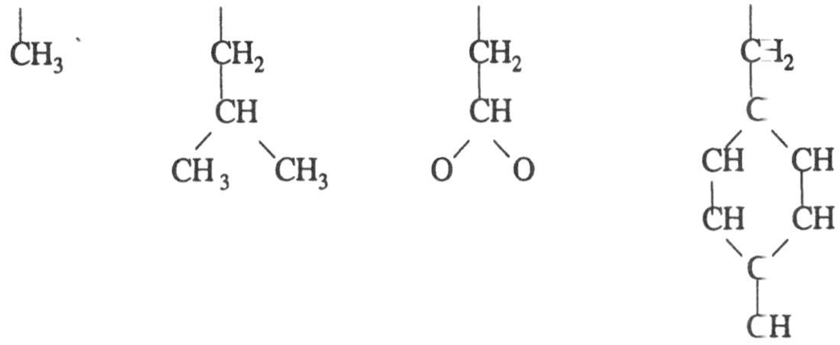

Fig. 4. Examples of amino acid side groups: a. alanine, a small, non-polar group, b. leucine, a large, non-polar group, c. aspartate, a charge (acidic) group, d. tyrosine, a large, polar group.

(Nucleic acids involve somewhat larger atomic groups, but because of lack of space, we will not discuss them here.)

Proteins constitute the main machinery of a living cell. As enzymes, they mediate and regulate all important processes of the cell. They act as receptors of external influences, chemicals or physical influences as light, they control the transport of substances in or out of the cells and they govern the energy-providing processes of the cells as well as the synthesis of all biological substances including themselves.

These very specific tasks rely on the very specific structure of the proteins. The structure is stabilized by relatively strong bonds involving the atoms of the common, main chain and the side groups. The bonds (which will be discussed later) are mainly due to electrostatic forces and involve energies of the order of 0.1eV, large enough to be stable but also (and that is important) of an order that can be broken and reformed by thermal fluctuations during reasonable time scales (less than a second). The specific sequence of side groups (specific amino acids) along the protein determines the structure. As there are some hundred units along a normal protein chain and 20 possibilities for each unit, the number of possible amino acid sequences and possible structures is virtually infinite.

The dynamics of the proteins and the possibility to change structure are very important for their biological roles. An enzyme can ordinarily be found in an inactive structure, but as some triggering substance becomes bound to it, it may change into an active structure, and open a certain pathway to make a certain chemical process possible. Receptor proteins change structures when receiving a certain influence (e.g. light) and may then produce some substance that is used in further reactions of the cell.

Thus, the problems of interest are to understand the structure from the knowledge of the amino acid sequence, and from that to get information about the dynamics and possibilities to structure changes. The ultimate aim is to understand the biological significance. To achieve this, we want to be able to confidently calculate the structure and the dynamics. The main method for this purpose which will be discussed here is the 'Molecular dynamics'. In this, one studies the dynamics of the system by numerical integration of the classical equations of motion for the entire system, for which purpose powerful methods have been developed. We will here discuss these methods with a specific example.

(For a general review of the basic methods, see McCammon and Karplus, 1983, while a more detailed description of the methods is found in van Gunsteren and Karplus, 1982. See also the review by Jaenicke, 1987.)

Here, we will describe calculations for a specific protein, bacterie-rhodopsin, which governs light-driven processes in certain photosynthetic bacteria. These calculations are made by our group in collaboration with F. Jähnig, Tübingen. This is a case where there seems to be good possibilities to calculate a reliable structure. (Jähnig, 1990, Jähnig and Edholm, 1990, 1991).

The use of molecular dynamics is limited as the calculations cannot provide dynamics for time scales longer than about 10^{-10}s. To describe processes for longer times, in particular transitions between relatively stable states, it is important to look for other methods, and in particular stochastic methods are being developed. We will finish this work with a brief discussion of some of the possibilities for that. (See Northrup et al, 1982, Blomberg 1989.)

Some of the emphasis will be put upon difficulties and weaknesses of the present methods. This shall not be taken as a criticism of the methods but rather to point out where physically important problems remain, and where further efforts shall be put.

THE RULES OF THE GAME

Thus, our problem is to make calculations to get information about the entire structure and the dynamic features from the knowledge of the atomic composition. For this, we need knowledge of the total energy with all possible interactions included.

The aim is to use as complicated description that possibly can be handled and calculate the most favorable structure as well as its dynamics. This means that we have to tackle a problem with maybe 1000-10000 degrees of freedom. This is possible only within a classical description. A complete quantum mechanical description is impossible, and quantum mechanical calculations are only used for refining details of the classical description or for processes such as electron transfer which are purely quantum phenomena. The classical description appears to be satisfactory for the main calculations as the molecules and its atomic groups are sufficiently large to essentially be classical objects.

Thus, the primary problem for the macromolecule is to make a calculation of energies of its structures and to describe the complete motion of all its constituents under their mutual interaction forces.

The connectivity of the molecule chain can be treated as constraints or, maybe better, represented by potentials, preferably harmonic ones of the form $c(|x_i-x_{i-1}|-x_0)^2$ which have strong minima at the equilibrium distance (x_0). Also, valence angles and some rotation angles are relatively stiff and can likewise either be treated by fixed constraints or with harmonic energies. In practice, there is no essential difference in the calculations between a constrained description or one with harmonic potentials. We will show some more explicit expressions in the example later. (For a more detailed discussion and presentation of the potentials, see van Gunsteren and Karplus, 1982.)

The potentials used are common ones: Lennard-Jones type that represents general, unspecific interactions for all types of atoms, and electrostatic ones for polar groups. The latter ones may require some comments. Many atomic groups are electrical dipoles (as are the surrounding water molecules). This is usually represented by putting fractional charges at specific locations. Then, direct

Coulomb potentials are used. Further polarizability effects are usually not taken into account but are represented by dielectric constants. The use of dielectric constants can be somewhat arbitrary and is a clear weakness in these calculations. It is desirable to get to a more realistic description, and attempts to do this are at present under way. (In calculations one has for instance to assign a value of the 'dielectricity constant of the protein.)

A particular type of electrostatic interaction is provided by the hydrogen bonds which are known to be highly relevant for the structure of biological molecules. In principle, a hydrogen bond is a quantum mechanical effect. (The electron orbital of a hydrogen bound to an oxygen or nitrogen atom is pushed from the hydrogen by the repulsion from another polar group, which strengthens the dipole moment.) Usually, in the classical description one uses a phenomenological form for the hydrogen bond. Many different forms are used, and it appears· as if parameters are chosen in a suitable way, all these provide satisfactory results, at least in certain situations. The chosen forms are quite different: There are those which completely ignore the hydrogen bond but represent everything by fractional charges and electrostatic interactions. Usually, an expression with a relatively sharp energy minimum as function of distance is used. Some (but far from all) of the expressions include a direction dependence in the potential. (This is probably realistic, but evidently not necessary for providing satisfactory results). Clearly, these expressions are not completely satisfactory. One way to improve the description, and which is being developed at present is to introduce some kind of dynamic polarizability (i.e. the charge distribution will depend on the surrounding charges).

For the correct description of a single macromolecule, it is also important to consider its interaction with the surrounding medium. In a living cell, this is essentially water which, as well known, has very strongly polar molecules. Their electrostatic interactions with the polar groups of the macromolecule are very relevant. Further, for any atomic group of the macromolecule in contact with water, hydrogen bonds can be formed to water molecules with about the same strengths as within the macromolecule. This leads to the conclusion, which now is clear from the molecular calculations, but not always pointed out in general descriptions: *The hydrogen bonds may not always be of importance for the molecule structure as they are alternatively formed between molecule groups or with water*. This may mean that polar interactions at the molecule surface (or any parts with contact to water) may not be important for the structure. On the other hand, *interactions with non-polar groups appear to be crucial for establishing a firm structure*. Non-polar groups have unfavorable interactions to water (they are 'hydrophobic') and an important part of making the macromolecule compact is due to the tendency to minimize the surface of non-polar groups towards the water.

The most ambitious calculations include surrounding water molecules. This can provide suitable descriptions of electrostatic features but it is difficult also in this way to account for the hydrophobic effect. The latter is rather of an entropic nature than energetic: the macromolecule becomes surrounded by a relatively stiff water shell of low entropy and higher free energy than the bulk water. This introduces an unfavorable free energy which is difficult to account for in an appropriate manner in the molecular calculations. Usually, it is treated by a phenomenological expression.

For a molecule in a non-polar surrounding, the polar interactions within the macromolecule become more important, and hydrogen bonds will in a higher degree determine the structure. In a living cell, an important non-polar surrounding is found in the interior of membranes. These are relatively thin structures with polar heads towards the water and an interior, about 20 Ångström thick, constituted by non-polar hydrocarbon chains in a double layer (each with about 20 carbon atoms). Such membranes surround the cells and certain cell organelles and are crucial as they delimit these structures in a sharp way. The transport through this thin but highly non-polar medium is highly restricted and to a large part regulated by proteins that stretch through the membrane and the non-polar medium. In the example given later, we will discuss the features of a protein in the membrane layer.

CALCULATION METHODS

The first goal of the calculations for biological macromolecules is to establish the 'best structure', i.e. the one with lowest free energy. The *free energy* aspect is essential for the interaction with the surrounding medium (e.g. the hydrophobic effect discussed above is an entropic effect) but is usually represented by some simplified expression. The problem is then to minimize a free energy expression under the variation of all degrees of freedom of the molecule. If there were a simple straightforward minimum, this would not necessarily be a too formidable task. However, the main problem is that in the multidimensional space of all the molecule degrees of freedom, there is *an immensely large number of local free energy minima*. An energy minimization method starting from some initial structure only leads to the nearest local minimum. In principle, *all free energy minima should be investigated,* which in reality is not possible.

Methods that to some extent can explore different energy minima are frequently used. *Monte Carlo methods* are methods that strictly look for the structure of lowest free energy by some algorithm that will provide correct statistical mechanical ensemble results when used for sufficiently long time (which may be unattainable for these problems).

Another type of method, which we will emphasize here is *Molecular Dynamics,* where the complete dynamics of the system for some time is considered. This can be made under the constraints of some statistical mechanical ensemble, and then the time average represents the statistical mechanical average. One expects the structure to stabilize in the most favorable form. One here uses energy expressions as discussed in the previous section and considers the classical motion by a numerical solution of Newtons equations of motion.

Now, this method is limited by the complexity of the problem which in a sense is too complex, too huge and with too many possibilities to fold into locally stable structures. In actual calculations with a molecule for 1000-10000 atoms, one needs a numerical time step of about 10^{-15} - 10^{-14}s determined by the fast oscillations of the atoms. Because of the numerical stiffness of the equations, it has not been possible to extend this to a sufficiently large extent. With the largest supercomputers available today, it is then possible to make a numerical integration to time scales of the order of 10^{-10}s. This makes it possible to get some relevant information about the dynamics but it is far to small to provide any firm conclusions about a complete folding and stabilization. In fact, important dynamics of real macromolecules including folding processes take much longer times. Relevant time scales for real molecule can be of the order 10^{-6} - 10^{-3}s, many orders of magnitude higher than what can be accomplished by molecular dynamics calculations. During such time scales, even relatively firm structures can be opened and the molecule chain can be refolded into various structures many times before a final stabilization.

An unprejudiced calculation of the 'best structure' is essentially hopeless. However, there are possibilities to improve the calculations and to simplify the search. One can recognize certain common structure elements, usually referred to as 'secondary structures', which are stabilized by hydrogen bonds between groups of the common molecule chain. If these can be recognized, and there are methods for that, one may get a suitable starting-point for the search of the complete structure. The most important secondary structures are the α-helix and the β-sheets, where the latter consist of extended regions of the chains that are coupled together with hydrogen bonds. What also is relevant for the example of the next section is that the calculations may be more reliable for a protein in a non-polar surrounding (such as the membranes), see (Jähnig, 1990).

Before ending this section, we will mention some further problems that occur here.

As said in the last section, the potential expressions are not quite satisfactory, and their uncertainties may well disrupt the actual differences between the most favorable structures. This can mean that it may not be worthwhile to extend the calculations to extreme times if not at the same times the basic energy expressions are refined.

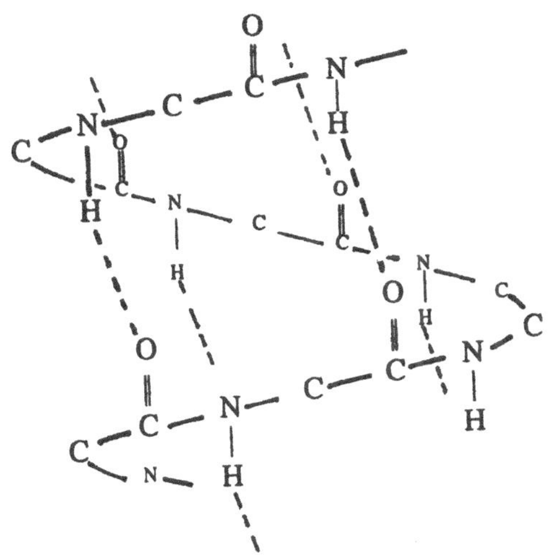

Fig. 5. Schematic picture of the stabilizing bonds along the protein main chain in an α-helix. Side groups are not shown in this figure. They are attached to the unmarked carbons. The hydrogen bonds are the dashed lines.

Other difficulties are connected to limitations due the the finite size, and statistical mechanical requirements. In particular, this makes it difficult to get a completely satisfactory picture of the surroundings. Statistical mechanical ensembles for finite size systems are not equivalent and this may further complicate the problem. For instance, the pressure can be calculated in alternative ways which in a correct ensemble are equivalent, but for the finite size systems these provide different results.

Molecular dynamics calculations provide in a straightforward way energy averages. On the other hand, free energies are more complicated to calculate. Still, there is a great demand to be able to calculate these. One important problem is to calculate the binding free energy for a ligand (which can be some drug) to a macromolecule (a protein or DNA). Such calculations are much desired for instance by the farmaceutical industry.

A common way to make such calculations is the following. One compares the binding free energies of two drugs that differ by some atomic groups. The primary calculation then considers the difference when the drugs are bound to the macromolecule by a varying potential of the form:

$$U(\lambda) = (1-\lambda)V_1 + \lambda V_2$$

where V_1 is the full interaction energy between drug 1 and the macromolecule and V_2 the corresponding between drug 2 and the macromolecule. If then λ goes from 0 to 1 the interaction changes from that of drug 1 to drug 2. The difference in free energy in the binding of drugs 1 and 2 are then given by an integral

$$\Delta F = \int_0^1 U(\lambda)\ d\lambda$$

This therefore means that many values of the interaction energy U shall be calculated to provide a reasonable estimate of the integral. Although such calculations are being performed in some extent it must be said that they at present are not particularly accurate and provide results with an uncertainty which may be equal to the actual number or even larger.

EXAMPLE: THE STRUCTURE OF BACTERIERHODOPSIN

In our group in Stockholm, in collaboration with F. Jähnig, Tübingen, these methods are used for investigating the structure of a relatively large protein, bacterierhodopsin that plays an important role in light absorption and proton transport for certain photosynthetic bacteria. (Jähnig and Edholm, 1990, 1991). The protein extends through the surrounding membrane of these bacteria, and will facilitate the transport of protons through the membrane when illuminated. While we have performed these calculations, detailed information about the structure has been obtained from electron microscopy measurements (Henderson et al, 1990)

This protein has two different surroundings: part of the protein is immersed in the non-polar membrane, and another part is stretching out into water (at both sides of the membrane).

These proteins in a cell membrane appear to be advantageous objects for the structure calculations. It seems to be more straightforward to predict the

inside cell

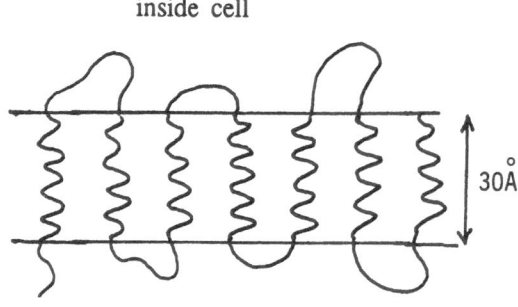

30Å

outside cell

Fig. 6. Schematic, general overview of the bacterierhodopsin molecule, showing the main chain with the helical parts that are immersed in the membrane, and the connecting parts.

structure in a non-polar surrounding, and thus it is possible to get a more confident structure for these proteins with the use of some experimental information. (See the discussion by Jähnig, 1989). Moreover, these proteins are difficult to cristallize, and a detailed structure as obtained from x-ray crystallography can be difficult to get, which of course increases the demand of reliable calculations.

It was early recognized that seven different pieces of the protein are in the membrane with smaller or larger pieces connecting them through the water surrounding. The interaction between the protein and the non-polar membrane is quite different from the interaction to water. Two different important features determine this: There is a general, non-specific interaction of van der Waals type that in general is attractive and thus favors a large surface. The polar interactions with the membrane are small, which makes the electrostatic interactions within the protein more important. It is important that the hydrogen bonds between protein groups are as efficient as possible, and these play a larger role in stabilizing the structure than in the water surrounding.

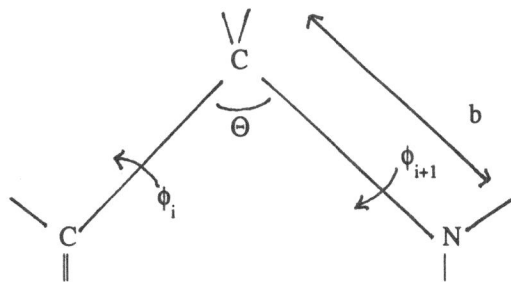

Fig. 7. The relevant angular variables of the protein chain.

There is a general belief that these facts strongly favor the α-helix structure for proteins in non-polar surroundings. The α-helix (see figure 5) means that the protein chain is stabilized in a helical structure by hydrogen bonds between groups of the main chain that are separated by 4 amino acid units. This is an energetically favorable conformation with a large number of hydrogen bonds. (It is a common structure element also in other proteins in the cell fluids, but then usually mixed with extended conformations). It is suggested that most proteins in membranes have a basic α-helical structure, and most information about membrane proteins confirms this. The helix structure can be modified (we will shortly show an example of this). Probably, it is also possible to have extended protein segments through a membrane which are coupled to each other by hydrogen bonds.

So, the basis for the calculations is to start with an initial structure of seven straight α-helixes of the membrane with extended segments in the water between these. Then, the time development of the structure is studied by molecular dynamics as described above. Explicit expression for the energy interactions that was used in these calculations are given below with the most important variables shown in figure 7. (Except for the hydrophobic contribution, these are commonly used expressions, and are suggested by van Gunsteren and Karplus, 1982.):

$$E = E_B + E_\Theta + E_\phi + E_{vdW} + E_{ES} + E_{HB} + E_{Hph}$$

where

$E_B = \frac{1}{2}K_b(b - b_0)^2$ is the bond potential; b is bond length

$E_\Theta = \frac{1}{2}K_\Theta(\Theta - \Theta_0)^2$ is the potential for the valence angle Θ

$E_\phi = \frac{1}{2}K_\phi[1 - \cos(3\phi - \delta)]$ is the potential for the rotation angle ϕ

$E_{vdW} = \dfrac{A}{r^{12}} - \dfrac{B}{r^6}$ is the van der Waals interaction which is cut off at a certain distance

$E_{ES} = \dfrac{q_i q_j}{4\pi\varepsilon\varepsilon_0 r}$ is the electrostatic Coulomb interaction, also with a cut off at a certain distance.

$E_{HB} = \dfrac{A'}{r^{12}} - \dfrac{B'}{r^{10}}$ is the hydrogen bond interaction for which a potential with a sharp minimum without direction dependence is used.

E_{Hph} is the hydrophobic interaction which takes the interaction with the solvent into account.

The complete form that is used for the hydrophobic interaction is rather complex, and we will not show the complete form here. It has two contributions: each amino acid along the chain is represented by a certain 'hydrophobicity' which refers to the gain in free energy to remove the amino acid with its side group from water to the non-polar membrane. Further, it contains a term which depends on location inside the membrane.

The dynamics is then calculated by the numerical integration of Newtons equations for all atoms of this long chain. The canonical ensemble is maintained as the average kinetic energy is calculated at regular time steps and then re-stored by a certain energy transfer to a given value. A calculation for the dynamics up to 25ps took about 7 hours CPU time on a Cray XMP.

Results of the calculations are shown in figure 8. The original straight helix structures are noticeably deformed (although the main features of the con-formations and the hydrogen bonds remain). In particular, some of the helixes are significantly bent. At a time of about 25 ps (corresponding to a real time scale), one sees a clear stabilization of this structure.

In the first calculations of the structure, one made use of some experimental information that was available at that time. Since then, a more de-tailed resolution have been obtained by electron microscopy methods (Henderson et al, 1990). The large features of the calculations, in particular the bending and deformation of the helixes agree well with the observed structure. As for the detailed structure (the actual position of the single atoms) there are deviations which are about the same as the experimental uncertainties, about 3 Å. It seems that the protein structure really attains the equilibrium structure from the α-helix starting point during the time scales of the calculations. It can be said here that relatively small errors of certain variables, in

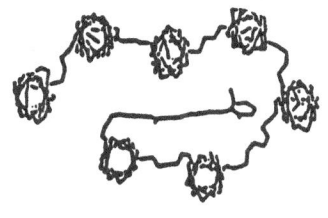
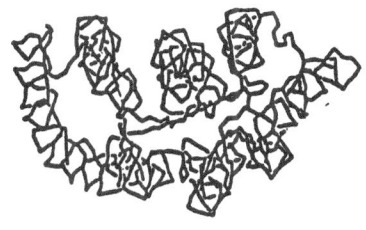

a.   b.

Fig. 8. Results for the structure of bacterierhodopsin as seen perpendicular to the membrane.
a. The initial structure with seven straight α-helixes.
b. The structure that becomes stabilized after about 25ps. In this structure, there is a clear deformation of the helixes.

particular the angles, can give rise to large deviations further along the chains without changing the important features of the structure. In that way large uncertainties in the detailed positions can be introduced and this applies both to the calculated as to the experimentally determined structures.

WHERE TO GO THEN: THE STOCHASTIC APPROACH

As already pointed out, these calculations are severely limited by the time scales. Many processes of these macromolecules take much longer times than are possible to reach with the detailed calculations. For describing these, other methods must be used. The primary method proposed for this aim is to pick out a limited number of degrees of freedom that are relevant for the problem of study and to treat the rest of the system as a less specific random influence.

Important problems that take long times usually involve a transition between two relatively stable states and can be considered as a structural redistribution. One then has to find a path in the configuration space that connects the two states and passes an unfavorable state, which should be a minimum energy barrier. The transition is essentially determined by this barrier and the features of the path that transverses it.

The mathematical description of the transition involves two different problems: first to account for the transition path and to calculate relevant parameters, in particular the height of the (minimum) energy barrier and the features of the path close to it. The second problem is to make a stochastic description of the transition where most of the degrees of freedom are treated as a random influence.

With given energy functions, the first problem is straightforwardly formulated, but some intuitive reasoning is necessary to find the proper path. It is certainly not straightforward to do this. Usually one picks out a small number of relevant variables which are varied to provide the transitions, while other variables at each step are determined by an energy minimization, and thus strictly follow the relevant ones. With a proper description this should give the important features of the transition.

The second stochastic problem is mathematically more complex, but has been treated extensively in the last decade. For a review, see (Blomberg, 1989). In many cases, the transition can be regarded as a quasi-one-dimensional transition with one relevant degree of freedom, the one that determines the actual transition over the minimum energy barrier. Then, well-known expressions for a one-dimensional transition can be used. If the transition is considered to be described as a Brownian motion, the expressions of Kramers (1940) can be used, see (Blomberg 1979, and Landauer and Swanson, 1961).

As a relatively simple example, consider the problem in (Blomberg, 1979). There, the rotation of a segment of three successive bonds in a polyethylene chain (a simple, unbranched hydrocarbon chain) was treated. It was argued that this rotation requires relatively little space and thus could occur also in a relatively dense polymer material where most large movements are not possible.

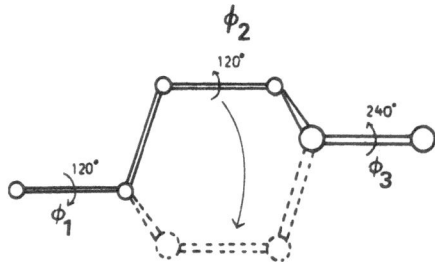

Fig. 9. The segment rotation of the model in the text. The three angles ϕ_1, ϕ_2, ϕ_3 change $120°$ during the rotation. Other variables are not changed after the complete transition, but vary during the rotation. In particular, valence angles have to be deformed.

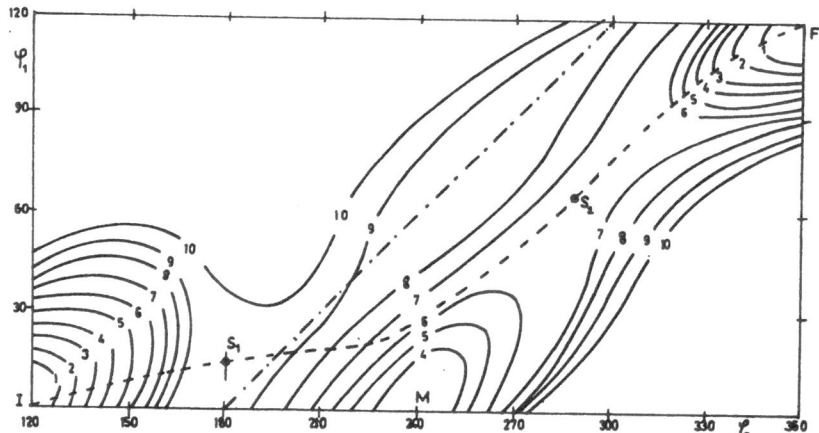

Fig. 10. A map of the energy as function of two of the main angles during the rotation where other variables at each point are supposed to attain the energetically most favorable values. The transition path is the dashed curve. The curves show points with equal energy with values in kcal/mol.

For the rotation, we use a constraint that atoms at some distance from the segment should be fixed during the rotation (this represents the influence from other molecules). A number of angles and also valence angles vary during the rotation, although they will attain the same values at the initial and final positions. In the calculations of the transition path, we select two main angles for describing the movement and choose the others by an energy minimization at each step.

There are some features that may complicate this picture. One is that the simplest Brownian motion picture with a small number of degrees of freedom may not be adequate. It is then necessary to extend the description and to treat it as a non-Markoffian stochastic process. Some methods for this are available (Grote, Hynes, 1982), but still, there remains severe problems. One possibility is to extend the problem in a suitable way with more variables to provide a true Markoffian process.

What is important here is that the molecular dynamics calculations can provide important information about the stochastic description, and how it and its parameters should be interpreted in the most satisfactory way. This problem is discussed by Cartling (1989).

Another problem that one can encounter here is that it may not be sufficient to use a quasi-one-dimensional picture. As an example, consider a problem that we studied some years ago (Blomberg and Edholm, 1984). It does involve a transition in a long hydrocarbon chain between an extended state and a bent, kinked state (see figure 11). Such a transition may occur in a medium of hydrocarbon chains with limited possibilities for large movements of a chain. In particular we had in mind motions in a cell membrane which is a packed agglomeration of hydrocarbon chains. There is a considerable motion of these chains in the membrane, although due to other chains, motions that take a large space are not possible.

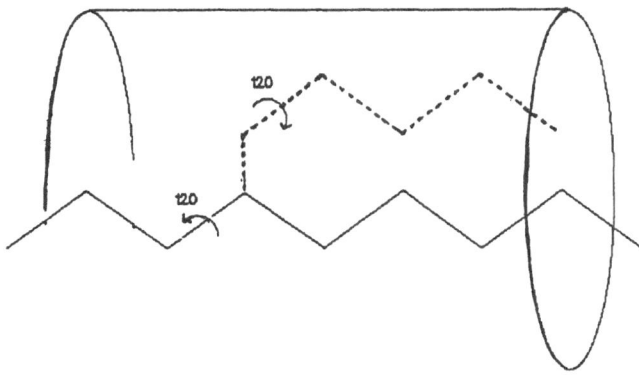

Fig. 11. The kink transition of a simple polymer chain. The constraint that the chain is not allowed to extend outside a suitably chosen cylinder represents the influence of other molecule chains.

The hydrocarbon chain can be rotated around the single bonds. The interactions between neighbouring groups along the chains provide three energy minima for each bond, one lowest and two symmetric ones of somewhat higher energy with energy barriers in between. (Cf. figure 2). For the kink motion, the molecule could rotate around two bonds simultaneously, which gives a motion through relatively little space with a transition barrier twice that of a single bond. A transition that involves first the rotation of one bond, then that of the other bond gets a low rotation barrier, but the rotation will take much space, and thus be hindered by the interaction with other molecule chains.

For this problem several rotations are involved. It is possible to for the transition energy to be distributed in a favorable suitable way among the bonds, which provides a transition which does not take much space into account and which can have a barrier not much higher than that of a single rotation. However, for that a number of rotation angles are confined to very specific values, and very little phase space is available for this transition. The proper way to treat the problem is to investigate several paths, their energy barriers and phase space to get a satisfactory description. This can be expressed by a rate distribution of barriers, f(E), proportional to the phase space that is available to get a barrier energy E. This then provides a formula:

$$rate = \int e^{-E/kT} f(E) \, dE$$

In the mentioned work, we made such calculations. The interaction between molecule chains was represented by a suitably chosen cylinder around the molecule, so that conformations within the cylinder would be possible, but those outside it were excluded. No further energy was introduced. In that way we could calculate a distribution of energy barriers and an apparent barrier if the rate (as is usually done by the experimentalists) is analyzed through an Arrhenius plot. This provided a value that was between the barrier of a single bond and the double value. In fact that was what was seen and interpreted as this kind of transition in NMR-experiments of membrane molecules.

REFERENCES

Blomberg, C., and Edholm, O., 1984, Analysis of constrained motions in hydrocarbon chains with applications to kinks in lipid bilayers, Mol. Phys., 51:485.

Blomberg, C., 1979, Kinetics of small segment motion in macromolecules with applications to hindered motion in polyethylene. Chem. Phys., 37:219.

Blomberg, C., 1989, Beyond the fluctuating enzyme:; The Brownian motion picture of internal molecular motion, J. Mol. Liq., 42: 1.

Cartling, B., 1989, From short-time molecular dynamics to long-time stochastic dynamics of proteins, J. Chem. Phys., 91:427.

Grote, R.F., and Hynes, J.T., 1980, Stable states picture of chemical reactions II Rate constants for condensed and gas phase reaction models, J. Chem. Phys., 73:2715.

van Gunsteren, W.F., and Karplus, M., 1982, Effect of constraints on the dynamics of macromolecules. Macromolecules, 15:1528.

Henderson, R., Baldwin, J.M., Ceska, T.W., Zemlin, F., Beckman, E., and Downing, K.H., 1990, J. Mol. Biol., 213:899.

Jähnig, F., 1990, Structure predictions of membrane proteins are not that bad, Trends Biochem. Sci., 15:93.

Jähnig, F. and Edholm, O., 1990, Can the structure of proteins be calculated?, A. Phys. B.-Condensed Matter, 78:137.

Jähnig, F. and Edholm, O., 1991, Attempts to calculate the structure of a membrane protein, submitted for publication.

Jaenicke, R., 1987, Folding and association of proteins, Prog. Biophys. Molec. Biol., 49:117.

Kramers, H.A., 1940, Brownian motion in a field of force and the diffusion model of chemical reactions, Physica, 7:284.

Landauer, R., and Swanson, J.A., 1961, Frequency factors in the thermally activated process, Phys. Rev., 121:1668.

McCammon, J.A., and Karplus, M., 1983, The dynamic picture of protein structure, Acc. Chem. Res., 16:187.

Northrup, S.H., Pear, M.R., Lee, C.Y., McCammon. J.A., and Karplus, M., 1982, Dynamical theory of activated processes in globular proteins, Proc. Natl. Acad. Sci. USA, 79:4035.

THE N-REPRESENTABILITY PROBLEM AND THE LOCAL-SCALING VERSION
OF DENSITY FUNCTIONAL THEORY

Eugene S. Kryachko

Institute for Theoretical Physics
Kiev, 25130, USSR

Eduardo V. Ludeña

Chemistry Center, Venezuelan Institute
for Scientific Research, IVIC
Apartado 21827, Caracas 1020-A, Venezuela

I. INTRODUCTION

The appeal of density functional theory for dealing with quantum mechanical many-body problems, arises from its simplicity. The usual methods for attaining approximate solutions to the Schrödinger equation require extremely complicated procedures which grow as n^7 or even at higher rates with respect to the number n of basis functions (which, of course, is larger than the number N of electrons). In density functional theory, this dependence is of merely n^3 or even less.

The calculation of numerical results using density functional methods, relies on particular forms of the energy density functionals and on adequately tailored exchange-correlation potentials. In addition, several approximations are introduced for the numerical handling of these problems as well as for the interpretation of results. The shortcomings with respect to exact results are interpreted in terms of deficient modelling of the exchange-correlation or kinetic energy terms.

A commonly held belief is that density functional theory is solidly grounded on the Hohenberg-Kohn theorems or their extensions and that hence there is no need to worry about its foundations. It is the purpose of this paper to discuss the importance of functional N-representability in order to construct a version of density functional theory which stands in a one to one correspondence with the usual quantum mechanical variational principle from which the Schrödinger equation is obtained. In Section II, we briefly review the importance of functional N-representability in the quantum mechanical problem formulated in terms of the reduced 2-matrix. In Section III, we analyze this problem with respect to the Hohenberg-Kohn formulation of density functional theory. In Section IV, we succinctly discuss some relevant aspects of the local-scaling version of density functional theory, as it includes explicitly functional N-representability.

II. THE FUNCTIONAL N-REPRESENTABILITY PROBLEM IN DENSITY FUNCTIONAL THEORY

In conventional quantum mechanics, the lowest eigenvalue E_o of the N-particle Schrödinger equation $\widehat{H}\Psi_o = E_o\Psi_o$ is reached at the extremum of the functional

$$E[\Phi] \equiv \langle \Phi | \widehat{H} | \Phi \rangle \tag{1}$$

where Φ is an arbitrary (normalized) N-particle wavefunction belonging to the antisymmetrized Hilbert space $\mathcal{L}_N$. Thus, we have

$$E_o \equiv \inf_{\Phi \in \mathcal{L}_N} \{E[\Phi]\} = E[\Phi]|_{\Phi=\Psi_o \in \mathcal{L}_N} \tag{2}$$

Therefore, for any trial N-particle wavefunction $\Phi_{trial} \in \mathcal{L}_N$, where $\Phi_{trial} \neq \Psi_o$, the following strict inequality $E[\Phi_{trial}] > E_o$ holds and guarantees that the expectation value of the Hamiltonian operator with respect to any approximate N-particle wavefunction $\Phi_{trial} \in \mathcal{L}_N$ yields an upper bound to the exact ground state energy E_o.[1,2]

A. N-representability and the 2-matrix

For Coulomb systems such as atoms, molecules, and condensed matter in general, the Hamiltonian operator $\widehat{H}$ contains only one- and two-particle terms

$$\widehat{H} = \sum_{i=1}^{N}\left[-\frac{1}{2}\nabla_{\vec{r}_i}^2 + \hat{v}(\vec{r}_i)\right] + \sum_{i=1}^{N-1}\sum_{j=i+1}^{N}\frac{1}{|\vec{r}_i - \vec{r}_j|} \tag{3}$$

and hence it can be rewritten as a sum of the two-body operators $\widehat{K}$:

$$\widehat{H} = \sum_{i=1}^{N-1}\sum_{j=i+1}^{N}\widehat{K}(\vec{r}_i,\vec{r}_j) \tag{4}$$

where

$$\widehat{K}(\vec{r}_i,\vec{r}_j) = \frac{1}{N-1}(\hat{h}(\vec{r}_i) + \hat{h}(\vec{r}_j)) + \frac{1}{|\vec{r}_i - \vec{r}_j|} \tag{5}$$

with $\hat{h}(\vec{r}) = -\frac{1}{2}\nabla_{\vec{r}_i}^2 + \hat{v}(\vec{r}_i)$. This, in turn allows us to establish an exact equivalence between the energy as a functional of $\Phi \in \mathcal{L}_N$ and the energy as a functional of the reduced 2-matrix:

$$E[\Phi] = \mathcal{E}[D_\Phi^2] \equiv \mathrm{Tr}_2[\widehat{K}\, D_\Phi^2]$$
$$= \int d^4 x_3 \cdots \int d^4 x_N \widehat{K}(\vec{r}_1,\vec{r}_2) D_\Phi^2(x_1,x_2;x_1',x_2')|_{x_1'=x_1,x_2'=x_2} \tag{6}$$

In the above expression, x stands for the spatial and spin coordinates, i.e., $x \equiv \vec{r}, s$ and the reduced 2-matrix is defined by

$$D_\Phi^2(x_1,x_2;x_1',x_2') = \frac{N(N-1)}{2}\int d^4 x_3 \cdots \int d^4 x_N \Phi^*(x_1,x_2,\ldots,x_N)\Phi(x_1',x_2',\ldots,x_N) \tag{7}$$

As compared to a wavefunction $\Phi(x_1,\ldots,x_N)$ which depends upon $3N$ spatial coordinates and N spin coordinates, the reduced 2-matrix $D_\Phi^2(x_1,x_2;x_1',x_2')$ depends only upon 12 spatial and 4 spin coordinates. If we consider that for systems of chemical or biological interest, the number of electrons N can get into the thousands, indeed the possibility of expressing the energy as an exact functional of D^2 becomes an alluring one, as it gets rid of the N-particle dependence.

But since the exact ground-state energy E_o corresponds to the extremum of a variational process, let us consider the equivalent variational problem for $\mathrm{Tr}_2[\widehat{K}\, D_\Phi^2]$:

$$E_o \equiv \inf_{D^2 \in \mathcal{P}_N^2} \{\mathrm{Tr}_2[\widehat{K}\, D_\Phi^2]\} \tag{8}$$

Notice that since $\widehat{K}$ is explicitly defined by Eq.(5), the variation only requires a proper characterization of the set $\mathcal{P}_N^2$ as the subdomain of all 2-matrices whose preimages are in $\mathcal{L}_N$ under the mapping (7). This requirement guarantees, of course, that at all steps of the variation $D^2 = D^2_\Phi$, namely, that D^2 is N-representable in the sense that it comes from an N-particle wavefunction $\Phi \in \mathcal{L}_N$. It is important to notice that in this case the 2-matrix N-representability is equivalent to the energy functional N-representability which is defined as the one to one correspondence between the energy as a functional of the N-particle wavefunction and the energy as a functional of the 2-matrix:

$$E[\Phi] \Longleftrightarrow \mathcal{E}[D^2] \equiv \mathrm{Tr}_2[\widehat{K}\, D^2_\Phi], \qquad D^2 \in \mathcal{P}^2_N \tag{9}$$

Hence, for the 2-matrix, functional N-representability[3-5] is attained by stipulating the necessary and sufficient conditions for characterizing $\mathcal{P}^2_N$. However, the solution to the N-representability problem for the 2-matrix is not an easy matter; in fact this is still an open problem.[6]

In general the difficulty lies in that not enough conditions are known and thus only a less restricted domain $\tilde{\mathcal{P}}^2_N$ can be defined. Clearly, the following inclusion relationship is satisfied $\mathcal{P}^2_N \subset \tilde{\mathcal{P}}^2_N$ and hence variation of D^2 over $\tilde{\mathcal{P}}^2_N$ leads to an energy lower than E_0. This shows of course that the optimal 2-matrix for such a problem cannot come from a wavefunction $\Phi \in \mathcal{L}_N$, namely, that neither the 2-matrix nor the energy as a functional of the 2-matrix are N-representable. It would be feasible, nevertheless to transform the above procedure into a truly variational one if we had at our disposal the means to make the conditions on $\tilde{\mathcal{P}}^2_N$ tighter such that the following relation were obeyed: $\mathcal{P}^2_N \subseteq \tilde{\mathcal{P}}^2_N$. In view of the fact that the 2-matrix N-representability problem remains unsolved, no such variational development is as yet possible.

But what would happen if instead of searching for the optimal D^2 over $\tilde{\mathcal{P}}^2_N$ - a procedure that we know beforehand yields a D^2 different from $D^2_{\Psi_*} \in \mathcal{P}^2_N$ - we introduce, based on physical considerations, some particular form for D^2 which depends upon certain parameters, or better even, upon certain functions which could be optimized variationally? One could justify this procedure by asserting that the choice of a particular approximate form for D^2 is in itself an additional constraint. Unfortunately, unless one can show that such a restriction transforms $\tilde{\mathcal{P}}^2_N$ into $\mathcal{P}^2_N$, one loses the variational character of the whole procedure and hence, although D^2 - or the functions upon which it depends - may be optimized variationally, there is no guarantee that the energy should be either a lower bound or an upper bound to the exact energy. This conclusion, as we shall see later, sets up important restrictions on any theory which attempts to represent the energy as a functional of some particular function, such as for example, the one-particle density.

B. N-representability and the one-particle density

Taking into account that the Hamiltonian operator given in Eq.(3) contains only one- and two-particle operators, we can rewite the energy explicitly as a functional of the reduced 1- and 2-matrices:[7-10]

$$E[\Phi] = \frac{1}{2}\int d^4 x_1 \nabla_{\vec{r}_1} \nabla_{\vec{r}'_1} D^1_\Phi(x_1, x'_1)|_{x_1 = x'_1} + \int d^4 x_1 v(\vec{r}_1) \rho_\Phi(x_1)$$
$$+ \frac{1}{2}\int d^4 x_1 \int d^4 x_2 \frac{D^2_\Phi(x_1, x_2; x_1, x_2)}{|\vec{r}_1 - \vec{r}_2|} \tag{10}$$

The 1-matrix $D^1_\Phi(x_1, x'_1)$, is defined by

$$D^1_\Phi(x_1; x'_1) \equiv N \int d^4 x_2 \cdots \int d^4 x_N \Phi^*(x_1, x_2, \ldots, x_N) \Phi(x'_1, x_2, \ldots, x_N) \tag{11}$$

The one-particle density $\rho_\Phi(x)$, is the diagonal part of the 1-matrix, $(\rho_\Phi(x) = D^1_\Phi(x, x))$. Only the diagonal part of the 2-matrix $D^2_\Phi(x_1, x_2; x_1, x_2)$ appears in this expression. Also let us notice that except for the term denoting the interaction with the external potential $v(\vec{r})$, which depends upon the one-particle density, all the other terms depend upon the 1- and the 2-matrices. These objects are non-local operators. It is possible, however, to rewrite them as a product of a local part (expressed in terms of the one-particle density) times a non-local contribution proper. In this vein, let us introduce the following factorizations:

$$D^1_\Phi(x_1;x'_1) \equiv \rho_\Phi^{1/2}(x_1)\rho_\Phi^{1/2}(x_2)\widetilde{D}^1_\Phi(x_1;x'_1) \tag{12}$$

where $\widetilde{D}^1_\Phi$ is the non-local part of the 1-matrix[11,12] and

$$D^2_\Phi(x_1,x_2;x_1,x_2) \equiv \frac{1}{2}\rho_\Phi(x_1)\rho_\Phi(x_2)[1+f^{XC}(x_1,x_2;x_1,x_2)] \tag{13}$$

where f^{XC} is the non-local exchange-correlation factor.[13] Substituting these expressions into Eq.(10), we obtain

$$
\begin{aligned}
E[\Phi] = {}& \frac{1}{8}\int d^4x_1 \frac{[\nabla_{\vec{r}_1}\rho_\Phi(x_1)]^2}{\rho_\Phi(x_1)} \\
& + \frac{1}{2}\int d^4x_1\, \rho_\Phi(x_1)\nabla_{\vec{r}_1}\nabla_{\vec{r}'_1}\widetilde{D}^1_\Phi(x_1,x'_1)|_{x_1=x'_1} + \int d^4x_1 v(\vec{r}_1)\rho_\Phi(x_1) \\
& + \frac{1}{2}\int d^4x_1 \int d^4x_2 \frac{\rho_\Phi(x_1)\rho_\Phi(x_2)}{|\vec{r}_1-\vec{r}_2|} \\
& + \frac{1}{2}\int d^4x_1 \int d^4x_2 \frac{\rho_\Phi(x_1)\rho_\Phi(x_2)f^{XC}_\Phi(x_1,x_2)}{|\vec{r}_1-\vec{r}_2|}
\end{aligned}
\tag{14}
$$

The first term in the right-hand side of this equation is of course the well known von Weizsäcker contribution to the kinetic energy.[14] Notice that except for the non-local contribution to the kinetic energy and for the term containing the exchange-correlation factor, the remaining terms of the energy functional depend exclusively upon the one-particle density $\rho(x)$.

Equation (14) is still an exact expression for the energy. It goes, however, far beyond Eq.(10) in the sense that it shows that a large portion of the total energy can be written as a functional of the one-particle density alone. The question then becomes whether it is possible to express the non-local contributions also as functionals of $\rho(x)$. A partial answer to this question, has been known ever since the early days of quantum mechanics: one may indeed construct approximate expressions - within the framework of the homogeneous electron gas - that are functionals of the one-particle density. Thus, for example, within the context of the homogeneous electron gas model, the non-local contribution to the kinetic energy is approximated by the Thomas-Fermi term:[15-19]

$$\frac{1}{2}\int d^4x_1\, \rho_\Phi(x_1)\nabla_{\vec{r}_1}\nabla_{\vec{r}'_1}\widetilde{D}^1_\Phi(x_1,x'_1)|_{x_1=x'_1} \sim \frac{3}{10}(3\pi)^{2/3}\int d^3\vec{r}\rho(\vec{r})^{5/3} \tag{15}$$

Similarly, since in the homogeneous electron gas model the non-local contribution arising from Coulomb electronic correlation is not included, f^{XC} goes into the simpler f^X. The exchange energy is, therefore, approximated by

$$\int d^4x_1 \int d^4x_2 \rho(x_1)\rho(x_2)\frac{f^X(x_1,x_2)}{|\vec{r}_1-\vec{r}_2|} \sim C\int d^4x_1[\rho(x_1)]^{4/3} \tag{16}$$

where the value of the constant C depends upon the averaging procedure used.[20-23]

It is evident that many different expressions for the energy as an approximate functional of the one-particle density can be advanced.[24] For instance when the Thomas - Fermi kinetic energy term is included but both the von Weizsäcker and the exchange-correlation terms are neglected, we obtain the Thomas-Fermi energy functional:

$$
\begin{aligned}
E[\Phi] \sim {}& \mathcal{E}_{TF}[\rho(\vec{r})] \\
= {}& \frac{3}{10}(3\pi)^{2/3}\int d^3\vec{r}\rho(\vec{r})^{5/3} + \int d^3\vec{r} v(\vec{r})\rho(\vec{r}) \\
& + \frac{1}{2}\int d^3\vec{r}_1 \int d^3\vec{r}_2 \frac{\rho(\vec{r}_1)\rho(\vec{r}_2)}{|\vec{r}_1-\vec{r}_2|}
\end{aligned}
\tag{17}
$$

The von Weizsäcker term is strictly zero for a homogeneous electron gas. Hence, its presence in the approximate energy expression reflects the incorporation of inhomogeneous corrections. Other choices of approximate terms lead to the Thomas-Fermi-Dirac and the Thomas-Fermi-Dirac-von Weizsäcker energy density functional models.[24]

Let us consider now whether any one of these approximate models satisfies the functional N-representability condition. An easy test that shows that they do not is to find some density for which $\mathcal{E}[\rho(\vec{r})]$ yields a value lower than the exact ground state energy E_0. In fact, such situations have been found even for energy functionals of the Thomas-Fermi-Dirac-λ-von Weizsäcker type, where λ is a constant determining the weight of the von Weizsäcker term.[24]

The fact that the Thomas-Fermi energy functional is not N representable simply means that there does not exist any wavefunction $\Phi \in \mathcal{L}_N$ such that $E[\Phi]$ can be made to yield $\mathcal{E}_{TF}[\rho(\vec{r})]$. For this reason we have written in Eq.(17) just ρ and not ρ_Φ.

Variation of $\mathcal{E}_{TF}[\rho(\vec{r})]$ with respect to the one-particle density, subject to the normalization constraint $\int d^3\vec{r}\rho(\vec{r}) = N$ leads to the Euler-Lagrange equation of motion for $\rho(\vec{r})$. At the extremum point of variation we obtain the optimum value of the energy $\mathcal{E}_{TF}[\rho_{opt}(\vec{r})]$. The fact that we use a variational procedure does not mean, however, that the Thomas-Fermi theory (or any other theory based on an approximate functional) satisfies the quantum mechanical variational principle. In fact, the lack of functional N-representability gives rise to variational energies which lie above or below the exact one.

A source of some confusion in density functional theory has come from the fact that it is possible to state the necessary and sufficient conditions guaranteeing that an approximate density comes from a wavefunction $\Phi \in \mathcal{L}_N$, namely that the one-particle density be N-representable (see Eq.(32)).[25-27] Even though the variation of an approximate energy density functional may be carried out in the domain of N-representable one-particle densities, this does not make the functional N-representable. This can be verified by the fact that when a one-particle density $\rho_\Phi(\vec{r})$ coming a wavefunction $\Phi \in \mathcal{L}_N$ (such as the Hartree-Fock density) is introduced into an approximate energy density functional, there are cases where the energy goes below E_0.

Although for the 2-matrix, N-representability of D^2 is equivalent to N - representability of the functional $\mathcal{E}[D^2]$, the same is not true for the one-particle density $\rho(x)$. There exists a many to one correspondence between N-particle wavefunctions $\Phi_{\rho_i} \in \mathcal{L}_N$ and one-particle densities $\rho_i(x) \in \mathcal{N}$, where $\mathcal{N}$ is the set of densities whose preimages are in Hilbert space. Let us distinguish these wavefunctions by means of the superindex $[j]$. This many to one correspondence can be expressed as follows:

$$\Phi_{\rho_i}^{[1]}, \ldots, \Phi_{\rho_i}^{[j]}, \ldots \Longrightarrow \rho_i(x) \tag{18}$$

Assuming that these N-particle wavefunctions are not degenerate, we have

$$E[\Phi_{\rho_i}^{[1]}] \neq \ldots \neq E[\Phi_{\rho_i}^{[j]}] \neq \ldots \tag{19}$$

where the energy functional $E[\Phi]$ is defined as in Eq.(1). It is clearly seen from Eq.(19), that there cannot exist an approximate energy functional $\mathcal{E}[\rho_i(x)]$ which is N-representable in view of the many to one correspondence between N-particle wavefunctions yielding the same density and $\mathcal{E}[\rho_i(x)]$. In order that the energy - expressed as a functional of the one-particle density - be N-representable, that is, that it stand in a one to one correspondence with a particular N-particle wavefunction in Hilbert space, it is necessary to build-in into the functional the wavefunction provenance of the density. In Section IV, we show how such N-representable functionals can be constructed by resorting to local-scaling transformations of the density.

III. FUNCTIONAL AND DENSITY v AND N-REPRESENTABILITY IN THE HOHENBERG-KOHN APPROACH

A. The v-representability problem

There are several basic assumptions, which we shall make explicit here, underlying the proof of the Hohenberg-Kohn theorems.[28–30] Consider an $\widehat{H}_o$-family of Hamiltonians

$$\widehat{H}^{v_k} = \widehat{H}_o + \sum_{i=1}^{N} v_k(\vec{r}_i) \qquad (20-a)$$

where the zeroth-order Hamiltonian is the sum of the total kinetic energy and electron-electron interaction operators:

$$\widehat{H}_o = \widehat{T} + \widehat{U}_{e-e} \qquad (20-b)$$

and where $v_k(\vec{r}_i)$ is an "external" potential. In its usual interpretation, $v_k(\vec{r}_i)$ is taken to be $v_{Coulomb}(\vec{r}_i)$, namely, the Coulomb interaction potential between an electron at position $\vec{r}_i$ and M nuclei (each one at position $\vec{R}_I$):

$$v_{Coulomb}(\vec{r}_i) = \sum_{I=1}^{M} \frac{Z_I}{|\vec{r}_i - \vec{R}_I|} \qquad (20-c)$$

However, in the context of the Hohenberg-Kohns theorems, one must consider a set of external potentials $\{v_k(\vec{r}_i)\}$ such that they differ from each other by more than a constant: $v_k(\vec{r}) - v_l(\vec{r}) \neq const$ for $k \neq l$. Of course, one of these potentials will be the usual Coulomb one. The first assumption is that each one of the Hamiltonians $\widehat{H}^{v_k}$ has a ground state. This means that for each one of the external potentials the following Schrödinger equation is satisfied:

$$\widehat{H}^{v_k} \Psi_o^{v_k} = E_o^{v_k} \Psi_o^{v_k} \qquad (20-d)$$

Since through Eq.(11) we can extract from each one of these ground-state wavefunctions the corresponding one-particle density , we obtain the set A_v^N of one-particle densities defined as

$$A_v^N \equiv \left\{ \rho_o^{v_k} \mid \Psi_o^{v_k} \Longrightarrow \rho_o^{v_k} ; \, \widehat{H}^{v_k} \Psi_o^{v_k} = E_o^{v_k} \Psi_o^{v_k} \right\} \qquad (21)$$

Any density belonging to A_v^N is said to be pure-state "v-representable".

The second assumption is that a continuous path going from an arbitrary reference external potential $v_{ref}(\vec{r})$ to the final Coulomb potential $v_{Coulomb}(\vec{r})$ can be set up such that along this path, a continuous set of Hamiltonian operators $\widehat{H}^{v_k}$ which have ground-state wavefunctions and hence yield v-representable one-particle densities can be constructed.[31] An external potential $v_k(\vec{r})$ belonging to this continuous path may be defined as

$$v_k(\vec{r}) = k \, v_{in}(\vec{r}) + (1-k) v_{Coulomb}(\vec{r}), \qquad 0 \leq k \leq 1 \qquad (22)$$

where we interpret k as a continuous index. Notice that A_v^N contains all densities $v_k(\vec{r})$ corresponding to all possible paths generated by the arbitrary choice of $v_{ref}(\vec{r})$.

The first Hohenberg-Kohn theorem[28] establishes a one to one correspondence between an external potential $v_k(\vec{r})$ and the one-particle density $\rho_o^{v_k}$. This means, clearly, that from the sole knowledge of a particular $\rho_o^{v_k}(\vec{r})$, one can obtain uniquely the external potential $v_k(\vec{r})$. The second theorem[32] states that, in view of theorem 1, one can write the Hamiltonian corresponding to this particular external potential and hence, by solving the Schrödinger equation, obtain the ground-state energy. In this sense, the one-particle density implies the energy and thus, we can consider the exact ground-state energy to be a functional of the exact ground-state one-particle density

$$E_o^{v_k} = \langle \Psi_o^{v_k} | \widehat{H}^{v_k} | \Psi_o^{v_k} \rangle$$
$$\equiv \mathcal{E}_{HK}^{v_k}[\rho_o^{v_k}(\vec{r})] \qquad (23)$$

Using this equivalence, a variational principle can be established by introducing the inequality

$$\langle \Psi_o^{v_k} | \widehat{H}^{v_{Coulomb}} | \Psi_o^{v_k} \rangle \geq \langle \Psi_o^{v_{Coulomb}} | \widehat{H}^{v_{Coulomb}} | \Psi_o^{v_{Coulomb}} \rangle \qquad (24)$$

Noticing that $\widehat{H}^{vCoulomb} = \widehat{H}^{v_k} + \sum_i (v_k(\vec{r}_i) - v_{Coulomb}(\vec{r}_i))$ and using Eq.(23), we can rewrite Eq.(24) as follows:

$$\mathcal{E}_{HK}^{v_k}[\rho_o^{v_k}(\vec{r})] + \int d^3\vec{r}(v_{Coulomb}(\vec{r}) - v_k(\vec{r}))\rho_o^{v_k}(\vec{r}) \geq E_o^{vCoulomb} \tag{25}$$

Breaking up the Hamiltonian $\widehat{H}^{v_k}$ appearing in Eq.(23) according to Eq.(20-a) we get

$$F_{HK}^{v_k}[\rho_o^{v_k}(\vec{r})] + \int d^3\vec{r}(v_{Coulomb}(\vec{r}))\rho_o^{v_k}(\vec{r}) \geq E_o^{vCoulomb} \tag{26}$$

where

$$F_{HK}^{v_k}[\rho_o^{v_k}(\vec{r})] \equiv \mathcal{E}_{HK}^{v_k}[\rho_o^{v_k}(\vec{r})] - \int d^3\vec{r}(v_k(\vec{r}))\rho_o^{v_k}(\vec{r}) \tag{27}$$

Clearly, $F_{HK}^{v_k}$ may also be written as

$$F_{HK}^{v_k}[\rho_o^{v_k}(\vec{r})] = \langle \Psi_o^{v_k}|\widehat{H}_o|\Psi_o^{v_k}\rangle \tag{28}$$

Since according to Eq.(20-b) $\widehat{H}_o$ does not contain the external potential, it is the same for all N-particle Coulomb systems. For this reason it has been customary to regard $F_{HK}^{v_k}$ as the universal functional $F_{HK}[\rho(\vec{r})]$. Defining

$$E_{HK}[\rho(\vec{r})] \equiv F_{HK}[\rho(\vec{r})] + \int d^3\vec{r}v_{Coulomb}\rho(\vec{r}) \tag{29}$$

the following variational principle ensues[28-30]

$$\frac{\delta}{\delta\rho(\vec{r})}\left\{E_{HK}[\rho(\vec{r})] + \mu\left(\int d^3\vec{r}\rho(\vec{r}) - N\right)\right\}_{\rho(\vec{r})\in A_v^N} = 0 \tag{30}$$

where μ is the Lagrange multiplier introducing the normalization condition on the density. Notice that the density variation must be performed within the set A_v^N of v-representable one-particle densities.

In order to apply the Hohenberg-Kohn variational procedure embodied in Eq.(30), one needs to define the necessary and sufficient conditions for characterizing the set A_v^N. These, as yet are not known. Furthermore, it is necessary to construct the "universal" functional $F_{HK}[\rho(\vec{r})]$. Notice that this functional is only defined along the continuous paths satisfying Eq.(22) and that it is undefined otherwise. But the existence of these continuous paths is in itself a supposition. Hence, the rigorous construction of this "universal" functional lies on very feeble grounds.[32]

B. The functional N-representability problem

Apparently, the N-representability problem does not occur within the Hohenberg-Kohn version of density functional theory.[33] But let us take a closer look at the basic relationships upon which this theory is erected. In Eq.(23), the definition of the energy functional $\mathcal{E}_{HK}^{v_k}[\rho_o^{v_k}(\vec{r})]$ involves the exact ground-state N-particle wavefunction $\Psi_o^{v_k}$. According to Eq.(18), we find a many to one correspondence between N-particle wavefunctions in Hilbert space and one-particle densities. Furthermore, (Eq.(19), non-degenerate wavefunctions yield different values for the expectation value of the Hamiltonian. Hence, in a strict sense, we must include in the definition of the functional in Eq.(23), the fact that this functional refers to the exact ground-state wavefunction, as otherwise, the functional is ill-defined:

$$E_o^{v_k} \equiv \mathcal{E}_{HK}^{v_k}[\rho_o^{v_k}(\vec{r}); \Psi_o^{v_k}]. \tag{31-a}$$

Now, *mutatis mutandis*, the "universal functional" becomes

$$F_{HK}^{v_k}[\rho_o^{v_k}(\vec{r})] \equiv F_{HK}^{v_k}[\rho_o^{v_k}(\vec{r}); \Psi_o^{v_k}] \tag{31-b}$$

235

Thus, although believed to be the same for all N-particle systems, - hence its appellative of "universal," - in reality this object differs for each physical system under consideration because its functional form depends upon the particular ground-state wavefunction corresponding to a particular external potential. Furthermore, according to Eq.(20-d), $\Psi_o^{v_k}$ is the solution to a Schrödinger equation for a particular $v_k(\vec{r})$. In general, each one of these potentials implies some particular boundary conditions which in turn define a particular Hilbert space $\mathcal{L}_N^{v_k}$. The variational principle for the Hohenberg-Kohn functional (Eq.(30)), involves a continuous path over a collection of Hilbert spaces. Thus, not every arbitrary function $\Phi^{v_k} \in \mathcal{L}_N^{v_k}$ need belong to this continuous path. As a consequence, a one-particle density corresponding to this wavefunction through the reduction given by Eq.(11), is N-representable but is not v-representable. Thus, not all densities belonging to the Hohenberg-Kohn-Lieb set

$$\mathcal{N}_{HKL}^N = \left\{ \rho(\vec{r}) | \rho(\vec{r}) \geq 0, \quad \int d^3\vec{r} = N, \quad \nabla\rho(\vec{r}) \in L^2(R^3), \quad \int d^3\vec{r}[\nabla\rho(\vec{r})^{1/2}]^2 < \infty \right\}$$
(32)

may be regarded as suitable candidates for the energy variation with respect to $\rho(\vec{r})$ in the Hohenberg-Kohn approach.

An important reformulation of the Hohenberg-Kohn procedure which bypasses the v-representability problem has been advanced by Levy[34] and Lieb.[35] It succinctly states that the exact energy E_o can be attained by means of a two-step constrained variation. In the first step, for a *fixed* density $\rho^{v_{Coulomb}}(\vec{r})$ belonging to the Hohenberg-Kohn-Lieb set of one-particle densities, the "universal" functional is obtained by carrying out the following minimization leading to the Levy-Lieb functional

$$F_{LL}^{v_{Coulomb}}[\rho^{v_{Coulomb}}(\vec{r})] = \min_{\substack{\Phi_\rho^{v_{Coulomb}} \in \mathcal{L}_N^{v_{Coulomb}} \\ \Phi_\rho^{v_{Coulomb}} \Longrightarrow \rho^{v_{Coulomb}}(\vec{r})}} \left\{ \langle \Phi_\rho^{v_{Coulomb}} | \widehat{H}_o | \Phi_\rho^{v_{Coulomb}} \rangle \right\}$$
(33)

In writing this equation, we have stressed the role of the particular external potential $v_{Coulomb}$, as the problem is defined in this case in the Hilbert space corresponding to this potential, namely, the Hilbert space whose bases satisfy the particular boundary conditions stipulated by this potential. In what follows, however, for simplicity in the notation, we shall drop this index. In the second step, the infimum of the total energy functional is sought with respect to variations of the one-particle density:

$$E_o = \inf_{\rho(\vec{r}) \in \mathcal{N}_{HKL}^N} \left\{ F_{LL}[\rho(\vec{r})] + \int d^3\vec{r}v(\vec{r})\rho(\vec{r}) \right\}$$
(34)

The wavefunction dependence of the functional $F_{LL}[\rho(\vec{r})]$ is more evident in this case. Clearly, from Eq.(33) we have

$$F_{LL}[\rho(\vec{r})] \equiv F_{LL}[\rho(\vec{r}); \Phi_\rho^{min}]$$
(35)

In this way, a one to one correspondence is established between the energy as a functional of the N-particle wavefunction Φ_ρ^{min} and the energy as a functional of the one-particle density:

$$E[\Phi_\rho^{min}] = \mathcal{E}_{LL}[\rho(\vec{r}); \Phi_\rho^{min}]$$
$$\equiv F_{LL}[\rho(\vec{r}); \Phi_\rho^{min}] + \int d^3\vec{r}v(\vec{r})\rho(\vec{r})$$
(36)

Let us notice that a very important characteristic of the Levy-Lieb energy functional $\mathcal{E}_{LL}[\rho(\vec{r}); \Phi_\rho^{min}]$ is its dynamical nature. In fact, when the density is varied from $\rho(\vec{r})$ to $\rho'(\vec{r}) = \rho(\vec{r}) + \delta\rho(\vec{r})$ the functional must change to $\mathcal{E}_{LL}[\rho(\vec{r}) + \delta\rho(\vec{r}); \Phi_{\rho+\delta\rho}^{min}]$ in order to maintain functional N-representability. When this condition is not met, as in the case of a "fixed" functional which upon density variation yields $\mathcal{E}_{LL}[\rho(\vec{r}) + \delta\rho(\vec{r}); \Phi_\rho^{min}]$, the quantum mechanical variational character of the Levy-Lieb procedure is lost and the energy thus calculated can lie either above or below the exact value E_0. For this reason, the use of non-N-representable energy density functionals cannot be justified by resorting to the Levy-Lieb constrained variational method. This fact has been clearly understood by Cioslowski[36] who has advanced an implementation of the Levy-Lieb two-step variational procedure which is based on "dynamical" energy functionals. This point is further discussed in Section IV.

C. Non-N-representable approximate energy functionals

Improvements on approximate energy density functional theories have been chiefly motivated by physical considerations; the N-representability problem has not been dealt with adequately mostly because of the confusion mentioned in Section II.B. regarding functional and one-particle density N-representabilities, and also because of the prevalent belief that the variational character of approximate density functional theories is solidly based on the Hohenberg-Kohn theorems. In order to understand the kinds of improvements brought upon density functional theory, let us remember that it was born within the homogeneous gas model and that for this reason it was only natural to try to remove the drawbacks of the local density approximation by systematically incorporating inhomogeneous terms. This led to what is known as the "gradient expansion" of the energy functional.[37-41] Although early attempts were not successful [for critical appraisals, see Vosko and Macdonald[42], and Perdewsp43] quite recently adequate functionals have been constructed in this fashion by including cut-off terms[44-48] or by carrying out wavevector analysis[49,50] which include higher order correlation terms.[51-54] In this manner, accurate functionals for the exchange potential have been constructed and important computational strategies have been implemented.[55-61]

The origin of Coulomb correlation functionals can be traced back to Wigner[62-64] and to the refinements brought about by Gombás.[65] These works form the basis for the developments of Lie and Clementi[66] and of Colle and Salvetti[67-70] [see also Lee et al.[71]] But just as in the case of exchange, Coulomb correlation functionals have also been highly influenced by the homogeneous electron model. In the absence, however, of an exact analytical expression for the correlation energy, several approximations such as those of von Barth and Hedin,[72] Gunnarsson and Lundqvist[73] and Vosko, Wilk and Nusair,[74] among others, have been advanced. The latter are based on the accurate Monte Carlo results of Ceperley and Alder.[75] These approximate functionals have been modified[76,77] to include the self-interaction correction.[78]

Still another way to describe the exchange-correlation potential is related to the adiabatic parameter λ which turns on the electron-electron interaction; within this approach, several methods have been developed for modelling the λ-averaged pair-correlation function.[54,79-81]

IV. FUNCTIONAL N-REPRESENTABILITY IN THE LOCAL-SCALING TRANSFORMATION APPROACH

Because of their similarity with scaling transformations

$$\vec{r} \equiv (x, y, z) \quad \xrightarrow{\hat{f}_\lambda} \quad \vec{f}(\vec{r}) \equiv (\lambda x, \lambda y, \lambda z) \tag{37}$$

the more general transformation of the type

$$\vec{r} \equiv (x, y, z) \quad \xrightarrow{\hat{f}} \quad \vec{f}(\vec{r}) \equiv (\frac{f(\vec{r})}{r} x, \frac{f(\vec{r})}{r} y, \frac{f(\vec{r})}{r} z) \tag{38}$$

are called local-scaling transformations. Their importance in density functional theory can be appreciated when one notices that - as it has been shown by Bader,[82,83] - an important topological feature of the scalar field $\rho(\vec{r})$ is the presence of closed equidensity surfaces. Since local-scaling transformations are the simplest ones which carry closed curves in $\mathcal{R}^3$ into other closed curves in $\mathcal{R}^3$, they preserve the topological features of the one-particle density.[84-86]

Let us consider consider explicitly the effect of a local-scaling transformation on a density $\rho(\vec{r})$. This is given by

$$\rho(\vec{r}) \xrightarrow{\hat{f}} \hat{f}(\rho(\vec{r})) \equiv \rho_f(\vec{r}) = J\{\vec{f}(\vec{r}); \vec{r}\}\rho(\vec{f}(\vec{r})) \tag{39}$$

where $\rho_f(\vec{r})$ is the transformed density and $J\{\vec{f}(\vec{r}); \vec{r}\}$ is the Jacobian of the transformation. When the latter is explicitly written, Eq.(39) becomes a first-order differential equation In the case of an atom, along a chosen path $\Omega_o \equiv (\theta_o, \phi_o)$, this equation becomes

$$\frac{df^3(r, \theta_o, \phi_o)}{dr} = \frac{3r^2 \rho_f(r, \theta_o, \phi_o)}{\rho(f(r, \theta_o, \phi_o), \theta_o, \phi_o)} \tag{40}$$

Scanning over all paths Ω, one obtains the total local-scaling transformation $f(\vec{r}) \equiv f(r, \theta, \phi)$.[84-86]

Let us consider now the effect of these transformations on the set $\{\Phi_\rho^{[k]}(1, \ldots, N); \quad k = 1, 2, \ldots\}$ formed by all N-particle wavefunctions $\Phi_\rho^{[k]}$ that yield the same density $\rho(\vec{r})$ (see Eq.(18)). The N-particle transformation operator $\hat{F}^N \equiv \underbrace{\hat{f} \ldots \hat{f}}_{N-\text{times}}$ acting on a wavefunction $\Phi_\rho^{[k]}$ yields the transformed wavefunction $\Phi_f^{[k]} \equiv \hat{F}^N \Phi_\rho^{[k]}$ defined by

$$\Phi_f^{[k]}(\vec{r}_1, \ldots, \vec{r}_N) = \prod_{i=1}^{N} J\left\{\vec{f}(\vec{r}_i); \vec{r}_i\right\}^{1/2} \Phi_\rho^{[k]}(\vec{f}(\vec{r}_1), \ldots, \vec{f}(\vec{r}_N)) \tag{41}$$

which is in a one-to-one correspondence with the one-particle density $\rho_f(\vec{r}) \equiv \rho_{\Phi_f^{[k]}}(\vec{r})$.

As a consequence, these continuous transformations generate a class of N-particle wavefunctions, or an "orbit" $O_\mathcal{L}^{[k]}$ where every wavefunction $\Phi_f^{[k]}(\vec{r}) \in O_\mathcal{L}^{[k]}$ is in a one-to-one correspondence with a one-particle density $\rho_f(\vec{r})$. When the action is applied to other reference wavefunctions, other orbits are generated and as a result, the Hilbert space $\mathcal{L}_N$ is partitioned into disjoint subsets of N-particle wavefunctions or orbits.

In view of the orbit structure of $\mathcal{L}_N$, the variational problem[86,87] given by Eq.(2) is transformed into the double search involving intra-orbit and inter-orbit energy optimizations:

$$E_o \equiv \inf_{\substack{\text{over all orbits} \\ O_\mathcal{L}^{[k]} \subset \mathcal{L}_N}} \left\{ \inf_{\Phi_f^{[k]} \in O_\mathcal{L}^{[k]}} \left\{ E[\Phi_f^{[k]}] \right\} \right\} \tag{42}$$

Furthermore, since within each orbit there is a one-to-one correspondence between $\Phi_f^{[k]} \in O_\mathcal{L}^{[k]}$ and $\rho_f(\vec{r})$, $E[\Phi_f^{[k]}]$ becomes a well-defined functional of $\rho_f(\vec{r})$, namely, $E[\Phi_f^{[k]}] \equiv \mathcal{E}[\rho_f(\vec{r}); \Phi_g^{[k]}]$ where $\Phi_g^{[k]}$ is an arbitrary orbit generating function. Hence, the variational principle given by Eq.(42) can be rewritten as

$$E_o \equiv \inf_{\substack{\text{over all orbits} \\ O_\mathcal{L}^{[k]} \subset \mathcal{L}_N}} \left\{ \inf_{\substack{\Phi_f^{[k]} \in O_\mathcal{L}^{[k]} \\ \Phi_f^{[k]} \Rightarrow \rho_f}} \left\{ \mathcal{E}[\rho_f(\vec{r}); \Phi_g^{[k]}] \right\} \right\} \tag{43}$$

where the energy density functional is given by the counterpart of Eq.(14):

$$\mathcal{E}[\rho_f(\vec{r}); \tilde{\Phi}_g^{[k]}] = \frac{1}{8} \int d^3\vec{r} \frac{[\nabla_{\vec{r}} \rho_f(\vec{r})]^2}{\rho_f(\vec{r})}$$

$$+ \frac{1}{2} \int d^3\vec{r} \rho_f(\vec{r}) \nabla_{\vec{r}} \nabla_{\vec{r}'} \widetilde{D}_g^{1[k]}(\vec{f}_{g,f}^{[k]}(\vec{r}); \vec{f}_{g,f}^{[k]}(\vec{r}'))|_{\vec{r}'=\vec{r}} + \int d^3\vec{r} \rho_f(\vec{r}) v(\vec{r}) \tag{44}$$

$$+ \frac{1}{2} \int d^3\vec{r} \frac{\rho_f(\vec{r}) \mathcal{E}_{XC,g}^{[k]}(([\rho_f(\vec{r})]; \vec{f}(\vec{r}))}{|\vec{r} - \vec{r}'|}$$

In this equation, the exchange correlation energy density is

$$\varepsilon_{XC,g}^{[k]}(([\rho_f(\vec{r})]; \vec{f}(\vec{r})) \equiv \int d^3\vec{r}\,\frac{\rho_f(\vec{r})(1 + f_{XC,g}^{[k]}(\vec{f}_{g,f}^{[k]}(\vec{r}); \vec{f}_{g,f}^{[k]}(\vec{r}')))}{|\vec{r} - \vec{r}'|} \tag{45}$$

The implicit dependence of the above functional on the one-particle density arises from the presence of the transformation function $f_{g,f}^{[k]}(\vec{r})$ in the non-diagonal components of Eq.(44). This characteristic is precisely what lends a "dynamical" character to the local-scaling density functional, as it is precisely through this dynamical transformation that functional N-representability is maintained at all steps of the variation.

Let us show now that Cioslowski's[36] N-representable realization of the two-step Levy-Lieb procedure corresponds to a finite basis representation of the local-scaling density optimization procedure within a given orbit. For this purpose, consider first the action of local-scaling transformations on a set $\{\psi(\vec{r})\}$ of single-particle functions. We have

$$\psi_i(\vec{r}) \xrightarrow{\hat{f}} \hat{f}(\psi_i(\vec{r})) \equiv \psi_{i,f}(\vec{r}) = \left[J\{\vec{f}(\vec{r}); \vec{r}\}\right]^{1/2} \psi_i(\vec{f}(\vec{r})) \tag{46}$$

Extracting from Eq.(39) an explicit expression for the Jacobian in terms of the densities, and substituting this expression in Eq.(46), we obtain

$$\psi_{i,f}(\vec{r}) = \left[\frac{\rho_f(\vec{r})}{\rho(\vec{f}(\vec{r}))}\right]^{1/2} \psi_i(\vec{f}(\vec{r})) \tag{47}$$

Let us now expand the function $\psi_i(\vec{f}(\vec{r}))$ in terms of the one - particle set $\{\psi_p(\vec{r})\}_{p=1}^{m}$:

$$\psi_i(\vec{f}(\vec{r})) = \sum_{p=1}^{m} W_{ip}\psi_p(\vec{r}) \tag{48}$$

The one-electron density, namely, the diagonal part of the 1-matrix is

$$\rho(\vec{r}) = \sum_{i=1}^{m}\sum_{j=1}^{m} D_{i,j}^{1}\psi_i^*(\vec{r})\psi_j(\vec{r}) \tag{49}$$

In view of Eqs.(48) and (49), the one-particle density evaluated at the position vector $\vec{f}(\vec{r})$ becomes

$$\rho(\vec{f}(\vec{r})) = \sum_{i=1}^{m}\sum_{j=1}^{m}\sum_{p=1}^{m}\sum_{q=1}^{m} D_{i,j}^{1}W_{pi}^*W_{jq}\psi_p^*(\vec{r})\psi_q(\vec{r}) \tag{50}$$

where $W_{pq} \equiv (\tilde{S}^{-1/2})_{pq}$ with

$$\tilde{S}_{pq} \equiv \int d^3\vec{r}\,\frac{\rho_f(\vec{r})}{\rho(\vec{f}(\vec{r}))}\psi_p^*(\vec{r})\psi_q(\vec{r}) \tag{51}$$

Equations (48)-(51) are precisely those of Cioslowski's "density-driven" method.[36,88]

REFERENCES

1.- E.S. Kryachko and E.V. Ludeña, Energy Density Functional Theory of Many-Electron Systems, (Kluwer, Dordrecht, 1990).
2.- Ref. 1, Section 2.4.
3.- Ref. 1, Section 4.4.
4.- E.R. Davidson, Reduced density matrices in quantum chemistry. (Academic, New York, 1976, p. 31).
5.- A.J. Coleman, Revs. Mod. Phys. **35**, 668 (1963).

6.- A.J. Coleman, In: Density matrices and density functionals. Erdahl, R., Smith, Jr., V.H. (eds.). Reidel, Dordrecht, p.5, (1987).

7.- Ref. 1, Section 2.2.

8.- K. Husimi, Proc. Phys.-Math. Soc. Japan **22**, 264 (1940).

9.- V.A. Fock, Zh. Eksp. Teor. Fiz. **10**, 961 (1940).

10.- P.-O. Löwdin, Phys. Rev. **97**, 1474 (1955).

11.- J.L. Gázquez and E.V. Ludeña, Chem. Phys. Lett. **83**, 145 (1981).

12.- E.V. Ludeña, J. Chem. Phys. **76**, 3157 (1982).

13.- R. McWeeny, Revs. Mod. Phys. **32**, 335 (1960).

14.- C.F.von Weizsäcker, Z. Phys. **96**, 431 (1935).

15.- Ref. 1, Section 5.1.

16.- L.H. Thomas, Proc. Cambridge Phil. Soc. **23**, 542 (1927).

17.- E. Fermi, Atti Accad. Nazl. Lincei **6**, 602 (1927).

18.- W. Macke, Phys. Rev. 100, 992 (1955); Ann. Phys. (Leipzig) **17**, 1 (1955).

19.- E.V. Ludeña, Int. J. Quantum Chem. **23**, 127 (1983).

20.- Ref. 1, Section 4.2.e.

21.- J.C. Slater, Phys. Rev. **81**, 385 (1951); Adv. Quantum Chem. **6**, 1 (1972).

22.- R. Gáspár, Acta. Phys. Hung. **3**, 263 (1954).

23.- W. Kohn, and L.J. Sham, Phys. Rev. **140A**, 1133 (1965).

24.- Ref. 1, Section 5.3.d.

25.- T.L. Gilbert, Phys. Rev. **B12**, 2111 (1975).

26.- J.E. Harriman, Phys. Rev. **A 24**, 680 (1981).

27.- Ref. 1, Section 6.3.

28.- Ref. 1, Sections 6.1. and 6.2.

29.-P. Hohenberg and W. Kohn, Phys. Rev. **136B**, 864 (1964).

30.-P. Hohenberg, W. Kohn and L.J. Sham, Adv. Quantum Chem. **21**, 7 (1990).

31.- H. Nakatsuji and R.G. Parr, J. Chem. Phys. **63**, 1112 (1975).

32.- Ref. 1, Section 6.4.

33.- P.-O. Löwdin, In: Density matrices and density functionals. Erdahl, R., Smith, Jr., V.H. (eds.). Reidel, Dordrecht, p.21, (1987).

34.- M. Levy, Proc. Natl. Acad. Sci. USA **76**, 6062 (1979).

35.- E.H. Lieb, Int. J. Quantum Chem. **24**, 243 (1983).

36.- J. Cioslowski, Adv. Quantum Chem. **21**, 303 (1990).

37.- D.A. Kirzhnitz, Zh. Eksp. Teor. Fiz. **32**,115 (1957).

38.- C.H. Hodges, Can. J. Phys. **51**, 1428 (1973).

39.- D.R. Murphy, Phys. Rev. **A24**, 1682 (1981).

40.- W. Yang, Phys. Rev. **A34**, 4575 (1986).

41.- B. Grammaticos and A. Voros, Ann. Phys. **129**, 153 (1979).

42.- S.H. Vosko and L.D. MacDonald, In: Condensed Matter Theories, P. Vashishta, R.K. Kalia and R.F. Bishop, Eds., Plenum, New York, Vol. 2, p. 101 (1987).

43.- J.P. Perdew, In: Condensed Matter Theories, P. Vashishta, R.K. Kalia and R.F. Bishop, Eds., Plenum, New York, Vol. 2, p. 89 (1987).

44.- J.P. Perdew, Phys. Rev. Lett. **55**, 1665 (1985).

45.- A.D. Becke, J. Chem. Phys. **84**, 7184 (1986).

46.- A.D. Becke, Phys. Rev. **A 38**, 3098 (1988).

47.- A.D. Becke, J. Chem. Phys.. **88**, 1053 (1988).

48.- A.D. Becke, J. Chem. Phys. **88**, 2547 (1988).

49.- D.C. Langreth and J.P. Perdew, Phys. Rev. **B 15**, 2884 (1977).

50.- D.C. Langreth and J.P. Perdew, Phys. Rev. **B 21**, 5469 (1980).

51.- D.C. Langreth and M.J. Mehl, Phys. Rev. Lett. **47**, 446 (1981).

52.- D.C. Langreth and M.J. Mehl, Phys. Rev. **B 28**, 1809 (1983); erratum ibid. **29**,2310(1984).

53.- J.P. Perdew, Phys. Rev. **B 33**, 8822 (1986).

54.- A.C. Pedroza, Phys. Rev. **A 33**, 804 (1986).

55.- J.W.D. Connolly, In: Semiempirical methods of electronic structure calculations. Part A: Techniques. Segal, G.A. (ed.). Plenum, New York, p. 105 (1977).

56.- E.J. Baerends, D.E. Ellis and P. Ros, Chem. Phys. **2**, 41 (1973).

57.- H. Sambe and R.H. Felton, J. Chem. Phys. **62**, 1122 (1975).

58.- B.I. Dunlap, J.W.D. Connolly and J.R. Sabin, J.R. J. Chem. Phys. **71**, 3396 (1979).

59.- B.I. Dunlap and N. Rösch, J. chim. phys. **86**, 671 (1989).

60.- P.M Boerrigter, G. Te Velde and E.J. Baerends, Int. J. Quantum Chem. **33**, 87 (1988).

61.- A.D. Becke, J. Chem. Phys. **88**, 2547 (1988).

62.- E.P. Wigner, Phys. Rev. **46**, 1002 (1934).

63.- E.P. Wigner, Trans. Faraday Soc. **34**, 678 (1938).

64.- E.P Wigner and F. Seitz, Phys. Rev. **46**, 509 (1934).

65.- P. Gombás, Theret. Chim. Acta **5**, 112 (1966).

66.- G.C. Lie and E. Clementi, J. Chem. Phys. **60**, 1275 (1974).

67.- R. Colle and O. Salvetti, Theoret. Chim. Acta **37**, 329 (1975).

68.- R. Colle and O. Salvetti, Theoret. Chim. Acta **53**, 55 (1979).

69.- R. Colle and O. Salvetti, J. Chem. Phys. **79**, 1404 (1983).

70.- R. Colle and O. Salvetti, In: Density matrices and density functionals. Erdahl, R., Smith, Jr., V.H. (eds.). Reidel, Dordrecht, p.545 (1987).

71.- R. Colle and O. Salvetti, J. Chem. Phys. **93**, 534 (1990).

72.- U. von Barth and L. Hedin, J. Phys. C: Solid State Phys. **5**, 1629 (1972).

73.- O. Gunnarsson and B.I. Lundqvist, Phys. Rev. **B 13**, 4274 (1976).

74.- S.H. Vosko, L. Wilk and M. Nusair, Can. J. Phys. **58**, 1200 (1980).

75.- D.M. Ceperley and B.J. Alder, Phys. Rev. Lett **45**, 566 (1980).

76.- H. Stoll, C.M.E. Pavlidou and H. Preuss, Theoret. Chim. Acta **49**, 143 (1978).

77.- H. Stoll, E. Golka and H. Preuss, Theoret. Chim. Acta **55**, 29 (1980).

78.- J.P. Perdew and A. Zunger, Phys. Rev. **B 23**, 5048 (1981).

79.- J. Alonso and L. Girifalco, Solid State Commun. **24**, 135 (1977).

80.- O. Gunnarsson and R.O. Jones, Phys. Scr. **21**, 394 (1980).

81.- O. Gunnarsson and R.O. Jones, Phys. Rev. **B 31**, 7588 (1985).

82.- R.F.W. Bader, Y. Tal, S.E. Anderson and T.T. Nguyen-Dang, Isr. J. Chem. **19**, 8 (1980).

83.- R.F.W. Bader, T.T. Nguyen-Dang, and Y. Tal, Rep. Prog. Phys. **44**, 893 (1981).

84.- Ref. 1, Sections 7.1.-7.3.

85.- E.S. Kryachko and E.V. Ludeña, Phys. Rev. **A 35**, 957 (1987).

86.- E.S. Kryachko and E.V. Ludeña, Phys. Rev. **A 43**, 2179 (1991).

87.- T. Koga, Y. Yamamoto and E.V. Ludeña, Phys. Rev. **A 43**, 5814 (1991).

88.- E.S. Kryachko and E.V. Ludeña, to be published.

ATOMIC AND IONIC INFORMATION ENTROPIES

M.C. Donnamaría

Instituto de Física de Líquidos y Sistemas Biológicos, IFLYSJB
C.C.: 565 (1900) La Plata, Argentina

A.N. Proto

Grupo de Sistemas Dinámicos, UBA - Centro Universitario Regional
Norte. C.C.: 2 (1638) V. Lopez, Argentina

INTRODUCTION

The Density Functional Theory (DFT) has some advantages over the quantum theory of atoms, molecules and solid state[1-6], resulting from the one-particle picture of many body systems, and it simplifies computational problems. Usually this approximate statistical theory is a tempting alternative to the exact quantum theory either when the accuracy of the latter is not necessary, or when the labor involved is not completely justified. On the other hand theoretical information approaches to different problems have been attempted since the maximum entropy principle (MEP)[7-9] was posed by Jaynes[7] in 1957, based on Shanon's accomplishments[8]. Applications in chemical kinetics[9], quantum mechanics[10-12] and density functional theory (analysis of electron densities of atoms and molecules)[13-14] clearly suggest that Jaynes MEP is a powerful tool.

In a previous paper, Gadré[13] combines the capacities of both formalisms and calculated information entropies within the context of Thomas-Fermi[1-4] (TF) density functional for neutral atoms using the numerical solution of the TF equation. He shows that the sum of the information entropy in coordinate and momentum space follows the Bialynicki-Byrula and Mycielski (BBM) universal bound and also justifies its use as a potential measure of wavefunction quality. However, as it has been already pointed out, the "exact" TF leads to non-realistic results. This failure is due, on the one hand, to the "infinite" electron distribution of the TF equation and on the other, to the electronic self-interaction. The first problem can be avoided by choosing a trial electron parametric function for the density with an adequate dependence on the distance[15,16] and the other via the Amaldi correction as it was proved previously by M.C. Donnamaría et al.[4,5,16,17]. Besides, using Thomas-Fermi-Amaldi procedure (TFA) negative and positive ions can be easily included, retaining the simplicity of the TF density functionals, as it was shown in Refs. 16,17.

In this paper we will apply the information entropy concept to analyze the behavior of some electron densities, in coordinate and momentum space, into the density functional formalism[1-6], avoiding the limitations imposed by the simple TF theory by using both the TFA procedure, and trial electronic functions. In doing so we not only reobtain the universal bound of BBM for positive and negative ions, but also check the quality of different trial electronic density functions.

INFORMATION ENTROPIES FOR POSITIVE AND NEGATIVE IONS

In the Amaldi correction to the TF theory spurious electron self-interaction is removed introducing the term $(N - 1/N)$ which appears as a simple factor in the energy density functional

$$E_{TFA} = 2.8712 \int \rho(r)^{5/3} dv + \int \rho(r) V_N \, dv$$

$$+ \frac{N-1}{2N} \int \frac{\rho(r)\rho(r')dvdv'}{|r - r'|} \tag{1}$$

ρ being the electron charge density in the atom, r the distance from the nucleus, $dv = 4\pi r^2$, N is the number of electrons and V_N the electron-nucleus interaction, which in an atom with nuclear charge Z is $(- Z / r)$. Instead of using, as Gadré, the "exact" universal solution[3] for the TF equation which is solvable numerically, only for neutral atoms, our ansatz was to use trial electronic density functions with adequate distance dependence, particularly those of Jensen and Wu, exhaustively analized in previous papers[4-5,18].

Jensen function

For the Jensen's function there is

$$\rho = \frac{N}{A} \frac{e^{-x}}{x^3} (1 + cx)^3 \tag{2}$$

with

$$x = Z^{1/6} \left(\frac{1}{a_B}\right)^{1/2} r^{1/2} \quad \text{and} \quad A = \frac{16\pi L_0}{Z\lambda^3} \tag{3}$$

c and λ are variational parameters obtainable from the minimization of the energy density functionals while L_0 is a polynomial in c. In order to obtain the information entropies, we write in the coordinate space:

$$S_\rho = - \int \rho(r) \ln \rho(r) \, d\bar{r} \tag{4}$$

By replacing ρ, Eq. (2), into the latter expression of S, with x from Eq. (3):

$$S_{\rho J} (N) = -\frac{N}{2L_0}\left[\int E_1\left(\ln N + \ln Z + 3\ln \lambda - \ln B \right) dx + \right.$$

$$\left. + \int E_1\left(-x - 3\ln x + 3\ln (1 + cx) \right) dx \right] \tag{5}$$

$$E_1 = x^2 e^{-x} (1 + cx)^3, \qquad B = 16\pi L_0 \tag{6a-b}$$

$$L_0 = 1 + 9c + 36c^2 + 60c^3 \tag{6c}$$

For neutral atoms, $N = Z$, Eq. (5) takes on the particular expression:

$$S_{\rho J} (N) = -\frac{N}{2L_0}\left[\int E_1\left(2\ln N + 3\ln \lambda - \ln B \right) dx + \right.$$

$$\left. + \int E_1\left(-x - 3\ln x + 3\ln (1 + cx) \right) dx \right] \tag{7}$$

Through the E_{TFA} minimization a particular set of (c, λ) is obtained for each element. One special case is that of the limit energy[4] with $c = 0.265$ and $l = 10.91$. For this universal values the previous expression becomes:

$$S_{\rho J} (N) = N (5.087 - 2 \ln N) \tag{8}$$

Analogously, the entropy relation in the momentum space is:

$$S_{\gamma J} (N) = N (1.671 + \ln N) \tag{9}$$

and the total entropy

$$S_{TJ} = S_{\rho J} (N) + S_{\gamma J} (N) = N (6.758 - \ln N) \tag{10}$$

Wu function

The other suggested trial density: the Wu's[17] function has the following expression:

$$\phi_w = (1 + mx^{1/2} + nx)^2 e^{-2mx^{1/2}} \tag{11}$$

From previous paper[17] we have obtained a modified Wu's function in the limit energy case with the following appropriate parameters, $m = 1.0305$, $n = -0.11503$, which determine the optimum density. For this case we use the equivalent Gadré expression for the entropy[13]

$$S_{\rho W}(N) = -N \int \phi_W^{3/2} x^{1/2} \left(\ln k + \frac{3}{2} \ln \phi_W - \frac{3}{2} \ln x \right) dx \tag{12}$$

and

$$S_{\rho W}(N) = -N \left[\int E_2(2\ln N - 2.166)dx + \right.$$

$$\left. + \frac{3}{2} \int E_2 \left(2\ln(1 + mx^{1/2} + nx) - 2mx^{1/2} - \ln x \right) dx \right] \tag{13}$$

where

$$E_2 = (1 + mx^{1/2} + nx)^3 \, e^{-3mx^{1/2}} \, x^{1/2} \tag{14}$$

and for the universal parameters above mentioned

$$S_{\rho W}(N) = N (4.704 - 1.998 \ln N) \tag{15}$$

For the entropy relation in the momentum space:

$$S_{\gamma W}(N) = N (1.863 + \ln N) \text{ is obtained} \tag{16}$$

and the total entropy from Eqs. (15) and (16) is:

$$S_{TW} = S_{\rho W}(N) + S_{\gamma W}(N) = N (6.758 - \ln N) \tag{17}$$

From Eqs. (7), (10) and (17) we can conclude: a) the TF-Gadré's relation $S_r(N) = a N + b N \ln N$ does not remain valid any more for Jensen's trial density function, as the charge number appears explicitly, unless for the case of $N = Z$. b) in the case of E_{TF} unique values for the optimum parameters, (c, λ) or (m, n), are obtained. Instead, in the case of E_{TFA} the minimization gives a particular set of them for each element (neutral atom or ion) also for those belonging to the same isoelectronic series, and c) the total entropy in all the cases, even those concerning non-neutral atoms, obeys the universal bond of BBM, which states that $S_\rho + S_\gamma \geq N (6.43 \, 2 \ln N)$.

RESULTS AND CONCLUSIONS

We have chosen a set of 40 atoms and 16 singly and doubly charged ions with a noble gas electronic structure to test the quality of the TFA-trial function procedure presented above. In previous papers[4,5,16,17] we minimized the TFA energy density functionals, associated with the Jensen and Wu functions, Eqs. (2) and (11). Thus, there is a set of (c, λ)$_{Jensen}$ and (m, n)$_{Wu}$ for each atomic species[16-18]. With that choice of the parameters we have used the optimum ρ to compute the TFA information entropies, for neutral atoms, with the Jensen function, Eqs. (7)-(10) and with the Wu function, Eqs. (13)-(17). Regarding ions the treatment is more restrictive: we can only use the Jensen function to calculate the TFA-entropy, Eq. (7) as it is not possible to calculate the information entropies associated to the Wu ions, since expressions depending on the logarithm of negative arguments appear.

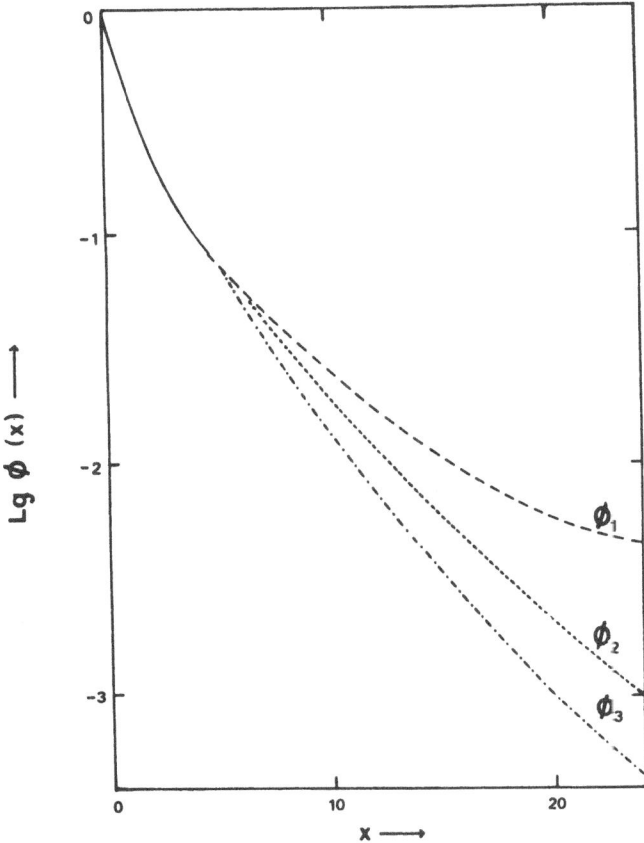

Fig. 1. Screening functions versus N.

In Fig. 1 the behavior of the different screening functions in the limit energy is analyzed regarding x. With ϕ_1 we denote the "exact" TF solution, as given numerically by Kobayashi et al[4], with ϕ_2 we represent the ϕ_{Jensen} function, obtained from Eq. (2), while ϕ_3, Eq. (8), is correlated with ϕ_{Wu}. In Fig. 2, it is possible to see the trend of the information entropies, with N, in the coordinate space, $S_\rho(N)$, Eqs. (8) and (15).

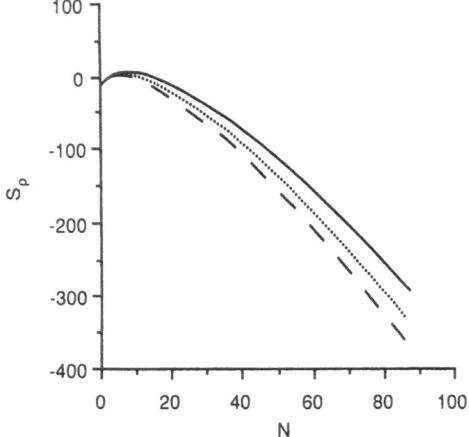

Fig. 2. Information entropy in the coordinate space for the density functions. Solid line: $S_{\rho Gadré}$, dotted line: $S_{\rho Jensen}$, dashed line: $S_{\rho Wu}$.

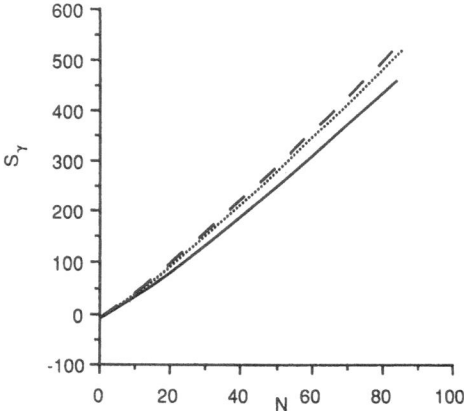

Fig. 3. Information entropy in the momentum space for the density functions. Solid line: $S_{\gamma Gadré}$, dotted line: $S_{\gamma Jensen}$, dashed line: $S_{\gamma Wu}$.

In Fig. 3 we investigate the entropies in the momentum space, $S\gamma(N)$, as given by Eqs. (9) and (16), while in Fig. 4 the behavior of the total entropies is described, Eqs. (10) and (17). In all the cases the solid curve is S_G,(Gadré) and it is related to the "exact" numerical TF solution. The dotted curve represents S_{Jensen} and the dashed one is S_{Wu}.

There is strict correlation between S_ρ and $S\gamma$, and the physical meaning is transparent: the more concentrated the density in the coordinate space the lower its entropy S_ρ and the uncertainty in localizing a particule in the momentum space is high ($S\gamma$). Regarding ions from Fig. 5 we can observe a particular feature of S_{TFA} with N. In our formalism S_{TF}, Eq. (8), is obtained as a particular case of S_{TFA}, Eq. (7), when $N = Z$, (neutral atoms) and (c, λ) are the abovementioned universal parameters. Furthermore, using the general expression for the

TFA-entropy, Eq. (7), we obtain a family of curves[18]: S_i, with i = $1^+, 2^+, 1^-, 2^-$. S_o is correlated with neutral atoms, S_1^+ and S_2^+ represent the single and double charged positive ions, while S_1^- and S_2^- show the behaviour of single and double charged negative ions. For the same value of N it is possible to obtain different entropies within each isoelectronic series.

As far as isoelectronic series are concerned Gadré et. al.[13] have presented values for the respective information entropies, but using in Eq. (7) near-Hartree-Fock atomic densities (NHF)[19]. In Ref. 14 it is observed for Be(1s) and Ne(1s) isoelectronic series, that S_ρ diminishes with increasing Z, but it is not possible to infer any other general trend for all the negative and positive ions. We have obtained the same characteristic behaviour but without taking recourse on a particular NHF density for each neutral atom or ion. Instead, in our case this result responds to the natural trend shown in Fig. 5. and it is the consequence of using trial-functions.

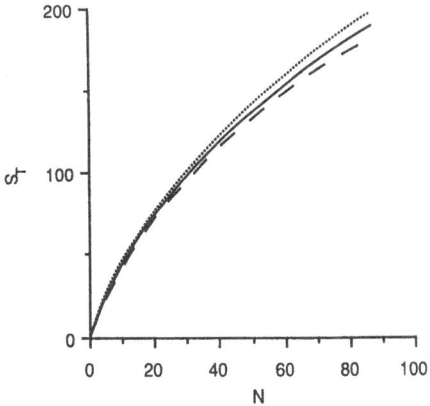

Fig. 4.   Total information entropy for the density functions. Solid line: $S_{TGadré}$, dotted line: $S_{TJensen}$, dashed line: S_{TWu}.

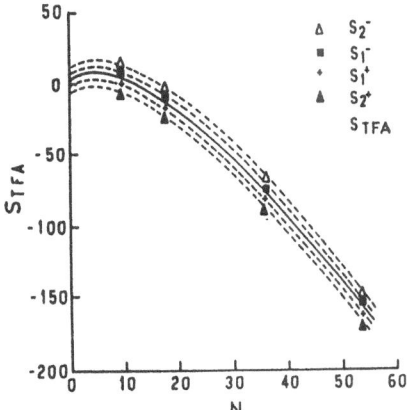

Fig. 5.   TFA-information entropies for neutral atoms and their isoelectronic series.

To sum up on the whole, the information entropy concept is applied in combination with the Amaldi correction to the Thomas-Fermi theory and trial screening functions for electron densities thus allowing the measure of information content to be extended also to positive and negative ions. The Bialynicki-Birula and Mycielsky (BBM) lower bound remains valid even when the information entropy is N ln Z dependent.

The authors gratefully acknowledge the Comisión de Investigaciones Científicas de la Provincia de Buenos Aires for its support. Also one of the authors (M.C.D.) is indebted to the Workshop Trust Foundation for its finacial support which has enabled her to attend the meeting.

REFERENCES

1. R.M. Dreizler and J. da Providencia Eds., Density Functionals Methods on Physics, NATO, ASI Series, Serie B, Physics (1984).

2. N.H. March,Theory of the Inhomogeneous Electron Gas, Edited by S. Lundqvist and N. H. March (Plenum, New York 1983).

3. P. Gombás, Die Statistische Theorie des Atoms und Ihre Anwendungen, (Springer, Verlag, Vienna, 1949).

4. M.C. Donnamaría, E.A. Castro and F.M. Fernández, Int. J. Quantum Chem, **20** 1005 (1982).

5. M.C. Donnamaría, E.A. Castro and F.M. Fernández, J. Chem. Phys., **78**, 5013 (1983).

6. M.C. Donnamaría, R.E. Cachau and E.A. Castro, Teochem, **210**, 121 (1990).

7. E.T. Jaynes, Phys. Rev.**106**, 620 (1957); Phys. Rev. **108** (1957).

8. C.E. Shannon, Bell. System Tech. **27**, 623 (1948).

9. R.D. Levine and R.B. Bersntein, in Dynamics of mollecular collisions, (ed.y) W.H. Miller (Plenun, New York, 1976).

10. A.N. Proto, Maximum Entropy Principle and Quantum Mechanics, Proc. XIII International Workshop on Condensed Matter Theories, edited by Valdir Aguilera Navarro (Plenum Press, 1989).

11. J. Aliaga and A.N. Proto. Proc. XIV International Workshop on Condensed Matter Theories, edited by S. Fantoni and S. Rosati (Plenum Press, 1990).

12. J. Aliaga, G. Crespo and A.N. Proto, Phys. Rev **A42**, 618 (1990) and Phys. Rev. **A42**, 4325 (1990).

13. S.R. Gadré, Phys. Rev. **A30** (1), 620 (1984).

14. S.R. Gadré, S.B. Sears, S.J. Chakravorty and R.D. Bendale, Phys. Rev. **A32**, 2602 (1985).

15. P. Csavinszky, Phys. Rev. **166**, 53 (1968).

16. M.D. Glossman, M.C. Donnamaría, E.A. Castro and F.M. Fernández, Bol. Soc. Chil. Quim. **24** (4), (1983): Match **14**, 247 (1983).

17. M.D. Glossman, M.C. Donnamaría, and F.M. Fernández, Acta Physica Slovaca **37**(5), 298 (1987); J. Physique **46**, 173 (1985).

18. M.C. Donnamaría and A.N,.Proto, Match **26** (1991).

19. Clementi and C. Roetti, At. Data Nucl. Data Tables **14**, 177 (1974).

SUPERCONDUCTIVITY IN $Ba_{1-x}K_xBiO_3$ CUBIC OXIDES

Wei Jin, M. H. Degani[1], Rajiv K. Kalia, and Priya Vashishta

Concurrent Computing Laboratory for Materials Simulations
and Department of Physics and Astronomy
Louisiana State University, Baton Rouge, Louisiana 70803-4001, USA

C.-K. Loong

Intense Pulsed Neutron Source
Argonne National Laboratory, Argonne, Illinois 60439-4814, USA

ABSTRACT

Superconductivity in $Ba_{1-x}K_xBiO_3$ is investigated within the framework of Eliashberg theory using a model of the Eliashberg function, $\alpha^2F(\omega)$. The phonon density of states (DOS) in $Ba_{1-x}K_xBiO_3$ is studied using inelastic neutron scattering and molecular-dynamics (MD) simulations. The model of $\alpha^2F(\omega)$ is based upon the MD phonon density of states. The function $\alpha(\omega)$ is constructed using information from electron tunneling experiments, with the premise that the electron-phonon coupling constant $\lambda \approx 1$ and that strong electron-phonon coupling exists for high energy (30-60 meV) phonon modes. From a study of the reference oxygen isotope-effect exponent in the phonon DOS, the oxygen isotope-effect exponent in T_c, $2\Delta/k_BT_c$, and electron tunneling spectra, we conclude that $Ba_{1-x}K_xBiO_3$ is a weak to moderate coupling BCS superconductor. The coupling of electrons to high energy oxygen phonons provide a reasonable description of superconductivity in this material within the framework of the Eliashberg theory.

1. INTRODUCTION

Superconductivity in potassium doped $BaBiO_3$ was first discovered by Mattheiss et al.[1] in 1988. Subsequently the structure of the superconducting material was determined by Cava et al.[2] Hinks et al.[3] gave a detailed account of synthesis, structure, and transition temperature as a function of x. Unlike other high-temperature copper oxide superconductors,[4] $Ba_{1-x}K_xBiO_3$ contains no copper, and in the superconducting phase it has a cubic perovskite structure,[5] as shown in Fig. 1. In addition, local magnetic moments do not exist in these materials.[6, 7] They are diamagnetic.[6, 7] According to neutron diffraction measurements,[5] potassium atoms are randomly distributed over the barium sites. The Bi-O-Bi bonds in the potassium doped system form an orthorhombic or a simple-cubic perovskite structure (depending on x) with each bismuth surrounded by six neighboring oxygen atoms.[5] The undoped compound, $BaBiO_3$, is also diamagnetic and has a body centered monoclinic structure.

---

[1] Permanent Address: Instituto de Física e Química de São Carlos-USP, São Carlos, Brazil.

Condensed Matter Theories, Vol. 7, Edited by A.N. Proto
and J.L. Aliaga, Plenum Press, New York, 1992

A number of investigations, experimental[5-17] as well as theoretical,[18-25] have been carried out on $Ba_{1-x}K_xBiO_3$. Structural properties, electric, magnetic, thermal and optical responses of this system have been studied.[5-17] Infrared reflectivity,[9] Raman scattering,[10] inelastic neutron scattering,[11, 12] and electron tunneling[13, 14] experiments have been carried out. Pei *et al.*[5] have investigated the crystalline structures for the entire composition range of this material by neutron powder diffraction and the phase diagram as a function of temperature and composition has been determined. The highest critical temperature ($T_c \sim 30$ K) is observed near x = 0.4.

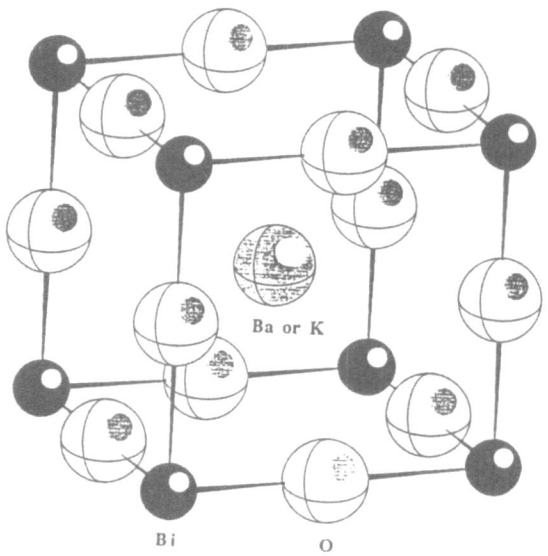

Fig. 1 Unit cell of cubic superconductor $Ba_{0.6}K_{0.4}BiO_3$.

2. MOTIVATION

In this paper our motivation is to show that $Ba_{1-x}K_xBiO_3$ is a weak to moderate coupling BCS superconductor and the coupling of electrons to high energy oxygen phonons provide a reasonable description of superconductivity in this material. This claim is supported by molecular-dynamics simulations, inelastic neutron scattering (INS) experiments, oxygen isotope-effect in T_c, electron tunneling spectra, and calculations of superconducting properties within the framework of the Eliashberg theory.

Phonon density of states in the insulating $BaBiO_3$ and in the doped $Ba_{0.6}K_{0.4}BiO_3$ systems have been studied by inelastic neutron scattering experiments and MD simulations.[11] Neutron scattering experiments[11] reveal three broad bands centered around 15, 35 and 61 meV. Molecular-dynamics simulations[11, 12, 22] show that phonons below 20 meV are due to Ba, K, and Bi whereas phonons above 20 meV are entirely due to oxygen. Softening by ~ 5 meV of oxygen phonons in the potassium doped system relative to the undoped $BaBiO_3$ has also been observed.[12] Using the ^{16}O and ^{18}O phonon DOS from MD simulations and INS measurements, the reference isotope-effect exponent - which characterizes the isotope shift in the phonon DOS - is found to be $\alpha_{or} = 0.42 \pm 0.05$ ($<\omega> \sim M_0^{-\alpha}$or where $<\omega>$ is the first frequency moment of the phonon DOS and M_0 is the mass of the oxygen isotope).

Oxygen isotope-effect on the superconducting transition temperature has been observed in the $Ba_{1-x}K_xBiO_3$ system. Batlogg *et al.*[8] measured the oxygen isotope-effect and found the exponent

$\alpha_o = 0.22 \pm 0.03$ ($T_c \sim M_o^{-\alpha_o}$). Hinks et al.[15] have determined a considerably larger value of the isotope-effect exponent, $\alpha_o = 0.41 \pm 0.03$, for the $Ba_{0.625}K_{0.375}BiO_3$ system. Kondoh et al.[6] also find a larger value of the exponent, $\alpha_o = 0.35 \pm 0.05$, than Batlogg et al.[8] The value of $\alpha_o \approx 0.35 - 0.41$ from T_c is close to the reference isotope-effect exponent from phonon DOS, α_{or} $= 0.42 \pm 0.05$. Since the strength of electron-phonon coupling effects, indicated by $\Delta\alpha_o = \alpha_o - \alpha_{or}$, is small, it implies that $Ba_{1-x}K_xBiO_3$ is a weak to moderate coupling superconductor.

The electron tunneling experiments on polycrystalline $Ba_{1-x}K_xBiO_3$ by Zasadzinski et al.[13] revealed well resolved structures in the high energy range, 30 - 60 meV. There is good agreement between the observed structures in the second derivative of the tunneling current and the positions of peaks in the phonon DOS from MD simulations[11, 22] and neutron scattering experiments.[11] Zasadzinski et al. have estimated from the tunneling data that $\lambda \approx 1$. In a recent tunneling experiment on thin films of $Ba_{0.6}K_{0.4}BiO_3$, Sato et al.[17] find the ratio $2\Delta/k_BT_c = 3.7 \pm 0.5$ where Δ is the superconducting energy gap. Using point contact junctions, Huang et al.[14] have recently carried out electron tunneling experiments on $Ba_{0.625}K_{0.375}BiO_3$ and inverted their data to obtain $\alpha^2F(\omega)$ for this material. They have shown that high energy phonons are involved in superconductivity, that the electron-phonon coupling constant $\lambda = 1$, and $2\Delta/k_BT_c = 3.8 \pm 0.1$. Schlesinger et al.[9] have measured the superconducting energy gap of $Ba_{0.6}K_{0.4}BiO_3$ using infrared reflectivity. They obtain $\Delta = 4.35$ meV for $T_c = 29$ K and $\Delta = 3.85$ meV for $T_c = 26$ K. An energy gap ratio of $2\Delta/k_BT_c = 3.5 \pm 0.5$ is obtained.

These observations of a substantial oxygen isotope-effect, high energy phonon modes in inelastic neutron scattering experiments, and images of phonons up to 60 meV in electron tunneling spectra suggest that $Ba_{1-x}K_xBiO_3$ is a BCS superconductor[26] in which electron-phonon interaction plays an important role.[20] In the following sections, we shall discuss each of these aspects of superconductivity in $Ba_{1-x}K_xBiO_3$ in some detail.

3. MOLECULAR-DYNAMICS SIMULATION

Molecular-dynamics simulations were performed for $BaBiO_3$ in the orthorhombic phase (a=6.2000Å, b=6.1561Å, c=8.6948Å) for a system of 540 particles at the experimental density of 7.88 g/cm^3. Effective interparticle interactions used in the simulations include steric repulsion between ions, Coulomb interactions due to charge transfer effects, and charge-dipole interactions due to large electronic polarizability of O^{--} ions. For $Ba_{0.6}K_{0.4}BiO_3$ the simulations were performed for a system of 625 particles at the experimental density of 7.33 g/ cm^3 in the cubic phase (a = 4.3160Å). The $Ba_{0.6}K_{0.4}BiO_3$ system was obtained by randomly replacing 40% of the Ba sites with K atoms. Before calculating the phonon DOS, it was ensured that the systems were dynamically stable in the appropriate symmetries at the correct experimental densities. Phonon DOS was calculated using three methods: (1) the Fourier transform of the velocity autocorrelation function, (2) the equation of motion method, and (3) direct diagonalization of the dynamical matrix. The results of all these three calculations are in agreement with one another.

Molecular-dynamics simulation results for partial and total phonon DOS of $Ba_{0.6}K_{0.4}Bi^{16}O_3$ and $Ba_{0.6}K_{0.4}Bi^{18}O_3$ are shown in Fig. 2. There are four significant features in the MD spectra: (1) the DOS below 20 meV is mainly due to Ba, K, and Bi vibrations, whereas the region above 20 meV is entirely due to oxygen, (2) the peak at 11 meV in the total DOS is mainly due to Ba and Bi, whereas K and Bi contribute to the peak at 15 mev, (3) in the DOS of the ^{16}O material, beyond 20 meV where the oxygen contribution is dominant, there is a broad band from 25-43 meV, a peak around 51 meV, a band between 54 and 65 meV, and small peaks at 67 and 73 meV. (4) The overall shape of the oxygen DOS and the total DOS for the ^{18}O system is similar to that for the ^{16}O system, except that the phonon spectrum above 20 meV is shifted to lower energies by 3-4 meV in the ^{18}O system.

4. NEUTRON SCATTERING EXPERIMENT

Inelastic neutron scattering experiments were performed using the HRMECS chopper spectrometer at the Intense Pulsed Neutron Source (IPNS) of Argonne National Laboratory. Pulsed spallation neutron sources have large fluxes of epithermal neutrons and are particularly well suited for investigation of the high-energy (20-80 meV) oxygen phonons. Polycrystalline samples of $Ba_{0.6}K_{0.4}Bi^{16}O_3$ were initially prepared by a melt-process technique.[3] A two-step

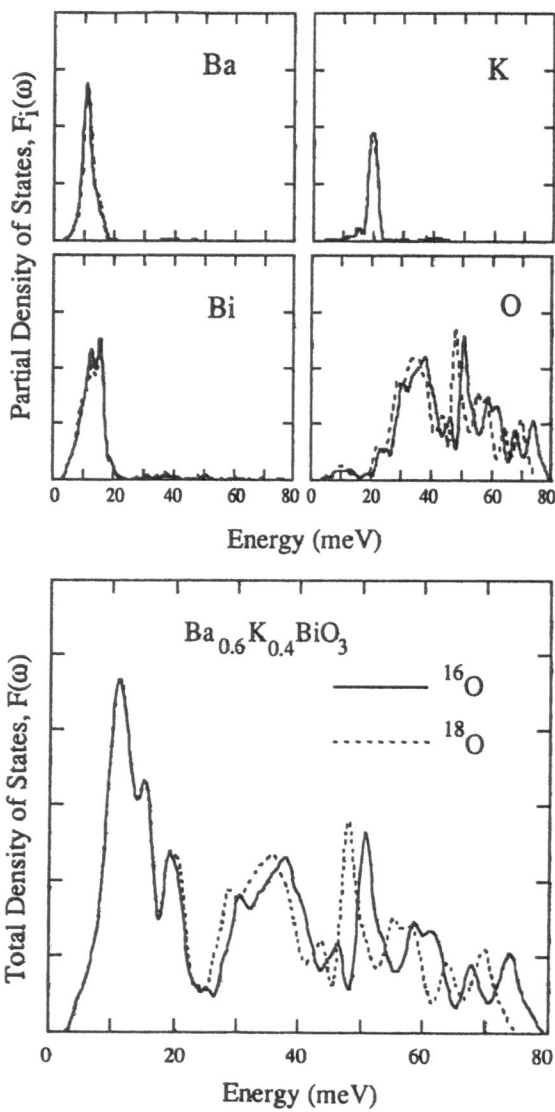

Fig. 2 MD partial and total phonon density of states for $Ba_{0.6}K_{0.4}Bi^{16}O_3$ and $Ba_{0.6}K_{0.4}Bi^{18}O_3$.

process[15] was used to produce the $Ba_{0.6}K_{0.4}Bi^{18}O_3$ sample through isotope exchange. The phonon density of states of $Ba_{0.6}K_{0.4}BiO_3$ upon oxygen isotope substitution has been determined by inelastic neutron scattering experiments.[12] Fig. 3(a) shows the measured generalized phonon DOS, $G(\omega)$, of ^{16}O and ^{18}O samples of $Ba_{0.6}K_{0.4}BiO_3$ at 15 K. In the ^{16}O sample there are broad bands centered around 15, 30, and 60 meV and overlapping features around 26, 30, 42 and 55 meV. The overall shape of $G(\omega)$ for the ^{18}O sample is similar to that for the ^{16}O sample, except that the spectrum above 20 meV is shifted to lower energies by 3-4 meV. This indicates the spectrum above 20 meV arises mainly from the vibrations of oxygen atoms, in agreement with the predictions of MD simulations.

To compare with the neutron data, we calculate the neutron-weighted MD phonon DOS, $G(\omega)$, using the relation $G(\omega) = \sum_i c_i\sigma_i F_i(\omega)/M_i$, where c_i, M_i, σ_i, and $F_i(\omega)$ are the concentration, mass, neutron scattering cross section, and partial phonon DOS for the ith atomic species. This relation, obtained under the incoherent approximation, applies very well to the present neutron scattering experiments. In Fig. 3 we show that all of the four features observed in the MD simulations (Sec. 3) agree with the INS experimental phonon DOS to within about 10%. In addition, there is a semiquantitative agreement between the positions of the peaks in the MD phonon DOS and the peaks observed in the second derivative of the tunneling current in electron tunneling experiments.[13, 14]

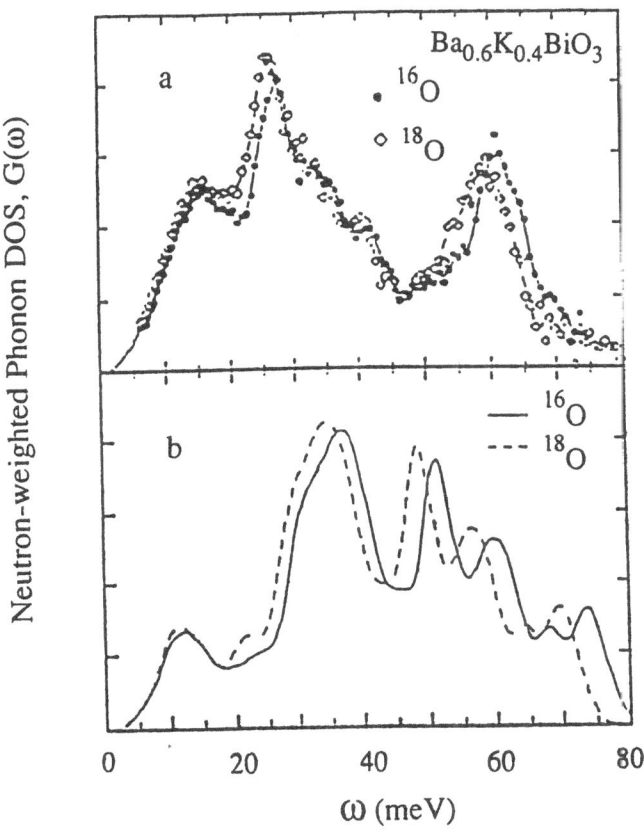

Fig. 3 Neutron-weighted phonon DOS, $G(\omega)$, for ^{18}O and ^{16}O samples: (a) inelastic neutron scattering experiments, and (b) calculated values based on partial DOS from molecular-dynamics simulations. The MD DOS, $G(\omega)$, has been convolved with the experimental resolution function.

5. ELIASHBERG GAP EQUATIONS

The microscopic relation between the electron-phonon interaction and properties of a phonon mediated superconductor can be described by the Eliashberg theory.[27-33] For isotropic superconductors, the Eliashberg function, $\alpha^2F(\omega)$, which is an electron-phonon weighted phonon density of states, and the Coulomb pseudopotential μ^* describing the repulsive Coulomb interactions between electrons, are central to the theory.[28] The Eliashberg function can be viewed as a product of the phonon density of states, $F(\omega)$, and an effective electron-phonon matrix element square $\alpha^2(\omega) = \alpha^2F(\omega)/F(\omega)$ which represents the strength of the electron-phonon coupling. The dimensionless electron-phonon coupling constant, λ, is defined by,

$$\lambda = 2 \int_0^\infty \frac{\alpha^2F(\omega)}{\omega} d\omega . \tag{1}$$

(a) Eliashberg Gap Equation Near T_c

Near the transition temperature T_c, the Eliashberg gap equations can be linearized because the gap approaches zero. The linearized equations for the critical temperature T_c in the imaginary-frequency formalism can be written as[31, 32]

$$\Delta(i\omega_n) = \sum_{m = 0, \pm1,\cdots}^{|\omega_m| \le \omega_c} \frac{S(n, m)}{|2m+1|} \Delta(i\omega_m) , \tag{2}$$

where

$$S(n, m) = \lambda(n - m) - \mu^*(\omega_c) - \delta_{n, m} \sum_{l = 0, \pm1, \cdots}^{|\omega_l| \le \omega_c} \lambda(n - l)\operatorname{sgn}(\omega_n)\operatorname{sgn}(\omega_l) . \tag{3}$$

In Eqs. (2) and (3), $\Delta(i\omega_n)$ is the energy-dependent gap function and ω_n are the Matsubara frequencies, $\omega_n = \pi k_B T_c (2n+1)$, $n = 0, \pm1, \pm2,....$ The frequency summations in Eqs. (2) and (3) are cutoff at a maximum value, ω_c. The Coulomb pseudopotential μ^* depends upon the cutoff ω_c. The function $\lambda(n-m)$ is related to $\alpha^2F(\omega)$ through the relation,

$$\lambda(n - m) = 2 \int_0^\infty \frac{\omega\, \alpha^2F(\omega)\, d\omega}{\omega^2 + (\omega_n - \omega_m)^2} . \tag{4}$$

The numerical method for solving Eq. (2) is discussed by Bergman and Rainer,[30] and by Allen and Mitrovic.[32] Using this scheme the transition temperature, T_c, can be calculated for a given $\alpha^2F(\omega)$ and μ^*. For given $\alpha^2F(\omega)$ and T_c from experiments, one can determine the value of μ^*.

(b) Eliashberg Gap Equations at Finite Temperature

The Eliashberg gap equations at a temperature, T, can be written as[28, 32]

$$\omega[1- Z(\omega, T)] = \int_0^\infty d\omega' \, \text{Re}\left[\frac{\omega'}{\sqrt{\omega'^2 - \Delta^2(\omega', T)}}\right] \int_0^\infty d\Omega \, \alpha^2 F(\Omega) \, K_+(T, \omega, \omega', \Omega) \,, \qquad (5)$$

and

$$\Delta(\omega, T)Z(\omega, T) = \int_0^\infty d\omega' \, \text{Re}\left[\frac{\Delta(\omega', T)}{\sqrt{\omega'^2 - \Delta^2(\omega', T)}}\right]$$

$$\times \left\{ \int_0^\infty d\Omega \, \alpha^2 F(\Omega) \, K_-(T, \omega, \omega', \Omega) - \mu^* \, \theta(\omega_c - \omega') \tanh\left(\tfrac{1}{2}\beta\omega'\right) \right\} \,, \qquad (6)$$

where $\beta = 1/k_B T$. The kernels $K_\pm(T, \omega, \omega', \Omega)$ are defined as

$$K_\pm(T, \omega, \omega', \Omega) = \left[f(\omega') + n(\Omega)\right]\left[\frac{1}{\omega + \omega' - \Omega + i0^+} \pm \frac{1}{\omega - \omega' + \Omega + i0^+}\right]$$

$$+ \left[f(-\omega') + n(\Omega)\right]\left[\frac{1}{\omega + \omega' + \Omega + i0^+} \pm \frac{1}{\omega - \omega' - \Omega + i0^+}\right] \,, \qquad (7)$$

where $f(\omega)$ and $n(\Omega)$ are the Fermi and Bose distributions, respectively. In Eqs. (5) and (6), the functions $\Delta(\omega, T)$ and $Z(\omega, T)$ are the complex temperature- and energy-dependent gap and renormalization functions, respectively. The integrals over K_+ have both imaginary and principal parts. These non-linear coupled integral equations can be solved numerically by iterative methods.[34] At a temperature T, the gap edge $\Delta_0(T)$ is defined by the equation $\text{Re}[\Delta(\Delta_0(T), T)] = \Delta_0(T)$. In the weak-coupling BCS limit, $\Delta(\omega,T)$ is taken as a constant, $\Delta(T)$, and given by the BCS gap equation.[26, 33]

6. MODEL OF THE ELIASHBERG FUNCTION $\alpha^2 F(\omega)$

We have constructed[25] a model for $\alpha^2 F(\omega)$ using the MD phonon density of states, $F(\omega)$, and information from electron tunneling experiments. In general, different phonon modes may have different contributions to the weighting factor $\alpha(\omega)$.[33] However, structures in $F(\omega)$ will be manifested in $\alpha^2 F(\omega)$. We divide the phonon energy from 0 to ω_{max} (= 80 meV) into several intervals (bands). This division is chosen such that each band contains either one or several peaks of $F(\omega)$. The detailed manner of this division is not important as long as it is qualitatively consistent with the intensities of peaks in the derivative of the experimental tunneling conductance. We divide the phonon energy into five bands: 0-23 meV, 27-43 meV, 47-52 meV, 56-63 meV, and

67-80 meV. The electron-phonon weighting factor, $\alpha(\omega)$, is modeled as a simple function of ω. In the model for $\alpha(\omega)$, these bands are given weights, A_1, A_2, A_3, A_4, and A_5, respectively. Linear interpolation is used to connect the region between the bands.

In electron tunneling experiments, the dips in the second derivative of the tunneling current with respect to the applied voltage correspond to peaks in the Eliashberg function. The amplitudes of these dips should vary roughly as $\alpha^2F(\omega)/\omega^2$. This allows us to estimate the constants A_i's, apart from an overall scaling factor, in accordance with the relative amplitude of the dips in the second derivative of the tunneling current in the experimental tunneling spectrum. The most interesting feature of Zasadzinski *et al.*'s tunneling experiment[13] is that there is clear evidence of phonon images at high energies in the range of 30-60 meV. Even though the quality of the tunneling data[13] is not good enough for inversion,[35] and the accuracy of the experiment is poorer at higher energies, it is clear that the high energy phonon structure is unambiguously present. This observation, combined with the fact that the intensity in the second derivative of the tunneling

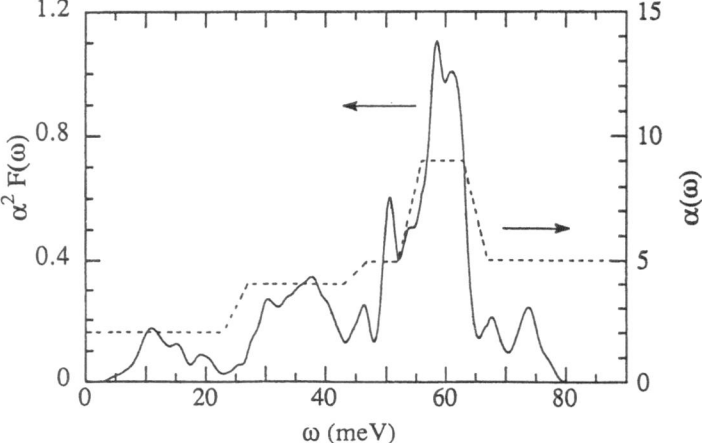

Fig. 4 Electron-phonon coupling function, $\alpha(\omega)$ and Eliashberg function $\alpha^2F(\omega)$ for $Ba_{0.6}K_{0.4}Bi^{16}O_3$.

current is roughly proportional to $1/\omega^2$ and the low energy region does not totally overwhelm the second derivative, establishes the fact that strong electron-phonon coupling is present for high energy phonons. We take the following ratios in our model: A_1: A_2: A_3: A_4: A_5 = 1.0:2.0:2.5:4.5:2.5. The overall scaling factor for these A_i's is determined by fixing the value λ to the experimental value, i.e., $\lambda = 1$, through Eq. (1).

The electron-phonon coupling constant λ can also be estimated[16] from measurements of upper and lower critical magnetic fields H_{c2} and H_{c1}. Recent magnetic field measurements[16] give $\lambda \approx 0.9$-1.1. Tunneling experiments[13, 14] also give $\lambda \approx 1$. Furthermore, the ratio $2\Delta/k_BT_c$ is another measure of the degree of the strong electron-phonon coupling. The observed[9, 14, 17] value of this ratio (3.5~3.8), which is close to the weak-coupling BCS value of 3.52, suggests that λ cannot be too large. Therefore $\lambda = 1$ is a reasonable experimental value. The function $\alpha(\omega)$ and $\alpha^2F(\omega)$ corresponding to $\lambda = 1$ is shown in Fig. 4.

Following Allen and Dynes,[31] the nth moment of the phonon frequency is defined by

$$\langle \omega_n \rangle = \frac{2}{\lambda} \int_0^\infty \frac{\omega^n \alpha^2 F(\omega) d\omega}{\omega} , \quad (n = 0, 1, 2, \cdots),$$ (8)

and the logarithmic mean-phonon-frequency is defined by[31]

$$\omega_{\ln} = \exp\left\{ \frac{2}{\lambda} \int_0^\infty \frac{\alpha^2 F(\omega)}{\omega} \ln(\omega) d\omega \right\}.$$ (9)

Using our model of $\alpha^2 F(\omega)$, we obtain $\langle \omega_1 \rangle = 39.2$ meV, $\langle \omega_2 \rangle^{1/2} = 44.2$ meV, and $\omega_{\ln} = 31.1$ meV. We note that both the first moment $\langle \omega_1 \rangle$ and $\omega_{\ln}$ are large (30 - 40 meV). Roughly speaking,[32] this is the origin of the high critical temperature T_c of this superconductor. The area of the $\alpha^2 F(\omega)$ curve is given by $0.5\lambda\langle \omega_1 \rangle = 19.6$ meV. This area is also large compared to that of any conventional low temperature superconductor.[30, 32] The total electron-phonon coupling constant is moderate to weak, $\lambda = 1$. This is due to large electron-phonon matrix elements corresponding to high energy (30 - 60 meV) phonons.

7. RESULTS AND DISCUSSION

(a) Relationship between λ and μ^*

We shall now explore the interrelationship between λ, μ^*, and T_c, within the context of our model. Given $\alpha^2 F(\omega)$ and $\lambda = 1$, we determine μ^* by requiring that the calculated T_c from the linearized Eliashberg equations be equal to the experimental value ($T_c^{EXPT} = 29.5$K for $x = 0.4$). First, for a fixed value of $\lambda = 1$, we calculate the values of μ^* for T_c in the range of 20 - 45 K with a cutoff $\omega_c = 201\pi k_B T_c = 1,605$ meV. The results are shown in Fig. 5. Clearly, T_c depends strongly on μ^*. The dependence of μ^* on λ for a fixed value of $T_c = 29.5$ K is shown in Fig. 6. We note that for λ between 0.95 and 1.10, the corresponding value of μ^* is between 0.09 and 0.16. These values of λ and μ^* are quite reasonable, considering the fact that the experimental estimates of λ is ~1 for $Ba_{1-x}K_x BiO_3$, and $\mu^* = 0.1$-0.13 is reasonable for almost all of the known low-temperature superconductors.[32]

(b) Oxygen Isotope Effect in Phonon Density of States

The isotope-effect is manifested in the phonon DOS when the mass of oxygen is changed by isotopic substitution. Let us define the first frequency moments, $\langle \omega \rangle$ and $\langle \tilde{\omega} \rangle$, as,

$$\langle \omega \rangle = \int_0^\infty \omega F(\omega) d\omega \bigg/ \int_0^\infty F(\omega) d\omega ,$$ (10)

$$\langle \tilde{\omega} \rangle = \int_0^\infty \omega G(\omega) d\omega \bigg/ \int_0^\infty G(\omega) d\omega .$$ (11)

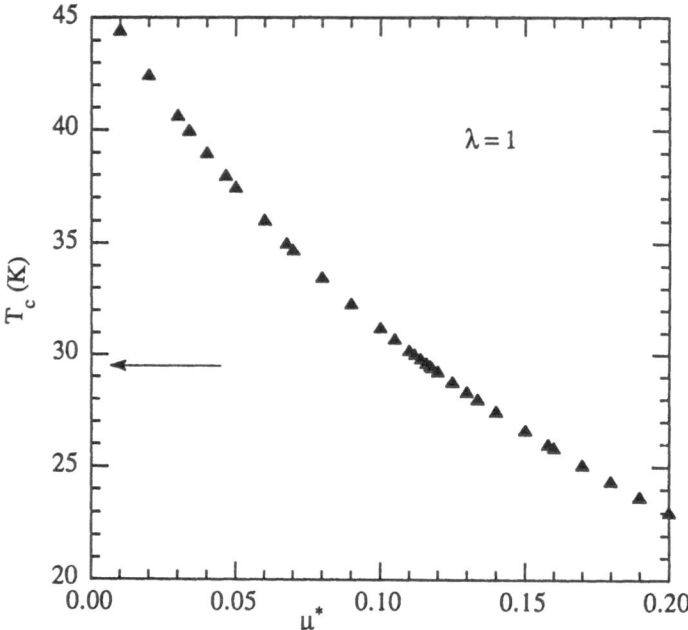

Fig. 5 T_c versus μ^* for a fixed value of $\lambda = 1$. The filled triangles are solutions of the Eliashberg gap equations. The arrow indicates the experimental T_c for $Ba_{0.6}K_{0.4}Bi^{16}O_3$.

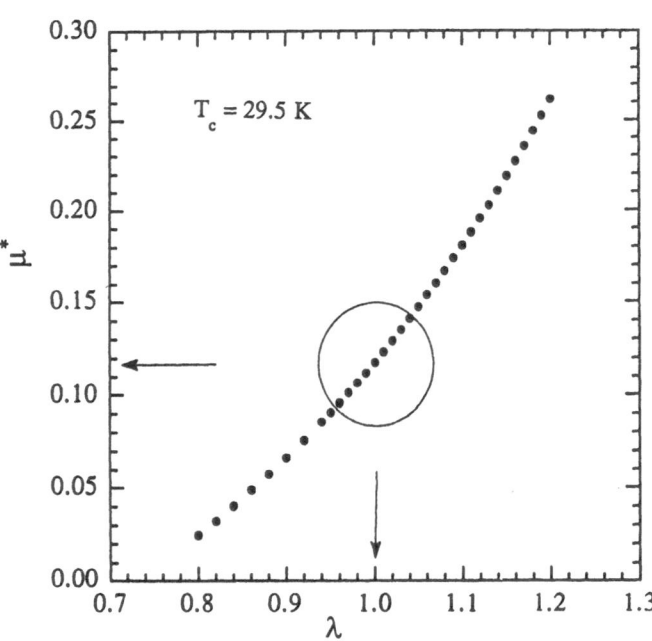

Fig. 6 λ versus μ^* for fixed $T_c = 29.5$ K. The values of μ^* for $\lambda = 0.93$ - 1.07 are enclosed in the circle. The arrows indicate the values of μ^* and of $\lambda = 1$.

For a monatomic system in which M is the mass of each atom, $\langle\omega\rangle$ and $\langle\tilde{\omega}\rangle$ behave as $M^{-1/2}$. For a multicomponent system such as $Ba_{1-x}K_xBiO_3$, we define the reference oxygen isotope-effect exponents as,

$$\alpha_{or} = -\frac{\partial\ln\langle\omega\rangle}{\partial\ln M_o}, \tag{12}$$

$$\tilde{\alpha}_{or} = -\frac{\partial\ln\langle\tilde{\omega}\rangle}{\partial\ln M_o}, \tag{13}$$

where M_o is the oxygen mass. α_{or} is generally smaller than 0.5 and it depends on the masses of other atoms in the unit cell.[12]

Inelastic neutron scattering experiments measure the neutron cross section weighted phonon DOS, $G(\omega)$. From the measured $G(\omega)$ for the ^{16}O and ^{18}O samples, as shown in the Fig. 3 (a), we can calculate the first frequency moments, $^{16}\langle\tilde{\omega}\rangle$ and $^{18}\langle\tilde{\omega}\rangle$, of $^{16}G(\omega)$ and $^{18}G(\omega)$. The mass variation of $\langle\tilde{\omega}\rangle$ gives the neutron weighted reference isotope-effect exponent $\tilde{\alpha}_{or} = 0.49 \pm 0.05$. Using the MD partial phonon DOS and neutron cross sections we construct the MD $G(\omega)$, as shown in Fig. 3 (b) and determine the MD value of $\tilde{\alpha}_{or} = 0.48$. This excellent agreement between the neutron scattering result and the MD result for the reference isotope-effect exponent confirms the reliability of our MD simulations.

For $Ba_{0.6}K_{0.4}BiO_3$, the mass variation of the phonon DOS, $F(\omega)$, gives a value of the reference isotope-effect exponent, $\alpha_{or} = 0.42$.

(c) Oxygen Isotope Effect in T_c

Once μ^* is determined from $^{16}T_c$ for the ^{16}O material, the same μ^* can be used to calculate $^{18}T_c$ from Eq. (2) for the system containing ^{18}O. The model $\alpha^2F(\omega)$ for the ^{18}O system is obtained by using the MD phonon DOS for the system in which only the mass of ^{16}O is replaced by the mass of ^{18}O. Using the calculated $^{18}T_c$ for the ^{18}O system, we determine the oxygen isotope-effect exponent $\alpha_o = 0.36 \pm 0.04$. This is in reasonably good agreement with the experimental values of $\alpha_o = 0.41 \pm 0.03$ of Hinks et al.[15] and $\alpha_o = 0.35 \pm 0.05$ of Kondoh et al.[6] from T_c measurements.

(d) Solutions of the Gap Equations at Finite Temperature

We have solved the coupled integral equations (5) and (6) with a cutoff $\omega_c \approx 200$ meV and μ^* = 0.15. The solutions of the renormalization function $Z(\omega, T)$ and the gap function $\Delta(\omega, T)$ at T = 2 K are shown in Fig. 7. The imaginary parts of the gap $\Delta(\omega, T)$ and the renormalization function $Z(\omega, T)$ are zero below the gap edge $\Delta_0(T)$. The structure in the Eliashberg function, Fig. 4, and of the phonon DOS, $F(\omega)$, are reflected in the real and imaginary parts of $\Delta(\omega, T)$ and $Z(\omega, T)$. We obtain a zero temperature gap edge $\Delta_0(0) = 4.749$ meV. The ratio $2\Delta_0(0)/k_BT_c$ calculated from the Eliashberg equations is 3.75 ± 0.1, in agreement with the values 3.5 ± 0.5 and 3.8 ± 0.1 obtained from infrared[9] and tunneling[14, 17] experiments.

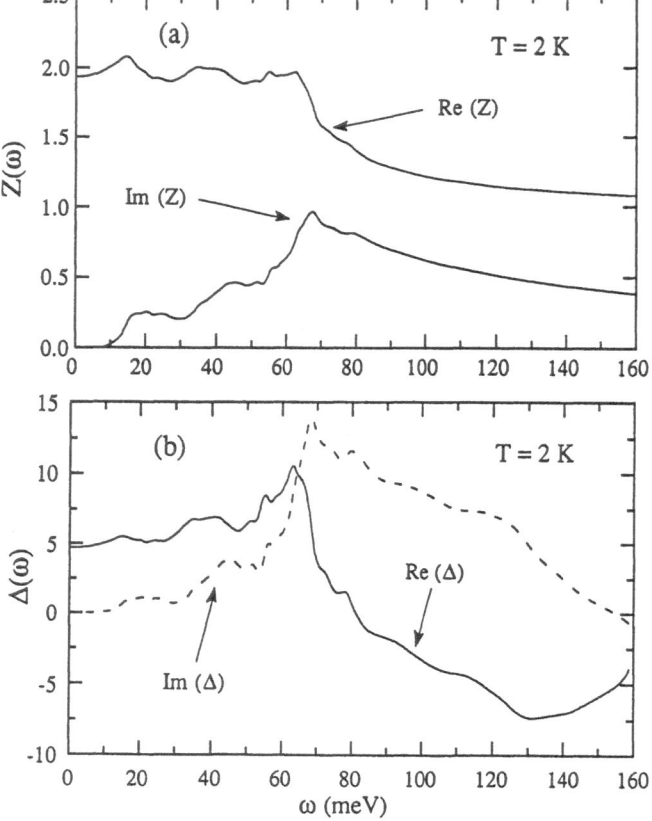

Fig. 7 Real and imaginary parts of $Z(\omega, T)$ and $\Delta(\omega, T)$ at $T = 2$ K.

(e) Electron Tunneling

The tunneling current in a superconductor-insulator-normal metal junction can be computed from[28]

$$
I(V) \propto \int_{-\infty}^{\infty} d\omega \, \text{Re} \left[\frac{\omega}{\sqrt{\omega^2 - \Delta^2(\omega, T)}} \right] [f(\omega) - f(\omega + V)] , \tag{14}
$$

where V is the applied voltage. In Fig. 8 we show the calculated normalized tunneling conductance, $\sigma(V)$, and the tunneling second derivative, $d\sigma(V)/dV$, at $T = 2$ K. Strong features in the energy range 30-60 meV can be seen in Fig. 8 (b). The overall features in the calculated results are similar to the experimental tunneling spectra[13] which are also measured near 2 ~ 5 K. However, the amplitude of the calculated results around 60 meV is considerably larger than in the experimental spectra.[13] This is due, in part, to experimental difficulties in observing structures at high voltages.[13, 14] It is also possible that the relative weight given to the frequency band around 60 meV in our model is larger than needed. Reducing the weight around 60 meV in our model of $\alpha(\omega)$, keeping $\lambda = 1$ fixed, has the effect of increasing relative weights of other parts of the spectrum in $\alpha^2 F(\omega)$. Such a modified model does not change the results in any significant manner. All the conclusions discussed in this paper remain unchanged.[36] This issue can only be settled by tunneling experiments with better resolution at higher energies.

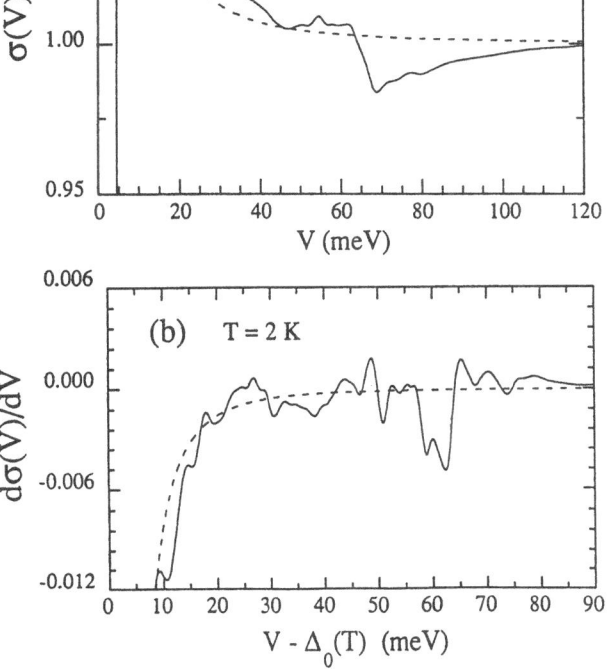

Fig. 8 Calculated tunneling conductance and tunneling second derivative at T = 2 K. The dashed curves are the BCS results.

8. CONCLUDING REMARKS

In this section we would like to remark on two points, (1) how reliable are the various formulas for calculating T_c of this material, and (2) what T_c is in the strong coupling limit $\lambda = 2 - 3$, given the fact that the proposed model [$\alpha^2 F(\omega)$ with $\lambda = 1$ and $\mu^* = 0.1173$] provides a reasonable description of the superconducting properties of $Ba_{1-x}K_xBiO_3$.

To address these questions we have calculated T_c as a function of λ for a fixed value of $\mu^* = 0.1173$. Nothing is changed except the value of λ by scaling the function $\alpha^2 F(\omega)$. Linearized Eliashberg gap equations are solved to determine T_c. Results are shown in Fig. 9. For comparison, we also show in Fig. 9, the approximate results obtained from the McMillan-Allen-Dynes [29, 31] formula for T_c,

$$T_c = \frac{\omega_{ln}}{1.20} \exp\left\{ - \frac{1.04\,(1 + \lambda)}{\lambda - \mu^*(1 + 0.62\lambda)} \right\}, \tag{15}$$

where ω_{ln} is defined in Eq. (9). In Fig. 9, for $\lambda \geq 2$, there are large deviations between the exact solutions and the values of T_c from Eq. (15). In general, Eq. (15) underestimates T_c. For example, $T_c \approx 23$ K from Eq. (15) with $\lambda = 1.0$ and $\mu^* = 0.1173$, which is close but lower than the exact value (29.5 K) from the gap equations.

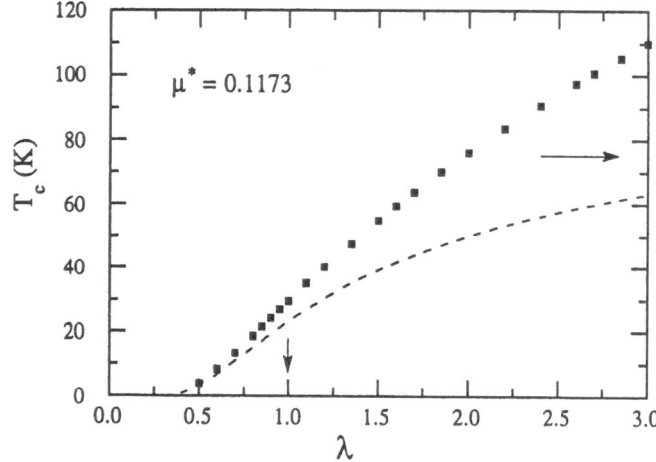

Fig. 9 T_c as a function of λ for a fixed value of $\mu^* = 0.1173$. The filled squares are from the solution of Eliashberg gap equations. The dashed curve is based on the McMillan-Allen-Dynes equation, Eq. (15). The arrow at $\lambda = 1$ corresponds to $T_c = 29.5$ K from the Eliashberg gap equations and 23.05 K from Eq. (15). For $\lambda = 3$ and $\mu^* = 0.1173$, Eq. (15) gives $T_c = 63.11$K, and with $\lambda = 3$ and $\mu^* = 0$ a value of $T_c = 75.16$ K (indicated by an arrow).

The point of the above calculation is to demonstrate from the solution of the Eliashberg gap equations that when there is substantial coupling of the carriers to the high energy phonons, there is no inherent problem in having a $T_c \approx 100$ K in the strong coupling limit, $\lambda = 3$. Approximate formulas to estimate T_c lead to the opposite conclusion.[33] We wish to emphasize that no claim is made whether a material with $\lambda = 3$ exists, or that layered copper oxides are the strong coupling limit of this material.

To summarize, we have studied the nature of phonons in the cubic $Ba_{1-x}K_xBiO_3$ using MD simulations and INS. We have constructed a model of the Eliashberg function, $\alpha^2F(\omega)$, for $Ba_{0.6}K_{0.4}BiO_3$. Superconducting properties calculated using the Eliashberg equations are consistent with experiments. We conclude that coupling of electrons to high energy oxygen phonons provides a reasonable description of superconductivity in this material within the framework of the Eliashberg theory.

ACKNOWLEDGMENTS

This work was supported in part by a grant from Louisiana Education Quality Support Fund under grant number LEQSF-(1991-92)-RD-A-05. Travel to the workshop for Priya Vashishta was supported by the U.S. Army Research Office. W.J. would like to acknowledge the support, during his stay at the Argonne National Laboratory, from the National Science Foundation (DMR 88-09854) through the Science and Technology Center for Superconductivity. M.H.D. would like to thank Fundacão de Amparo a Pesquisa do Estado de São Paulo, Brazil for a research fellowship. C.-K.L. was supported by the U.S. DOE, BES-Materials Sciences, under Contract No. W-31-109-ENG-38.

REFERENCES

1   L. F. Mattheiss, E. M. Gyorgy, and D. W. Johnson, Jr., Phys. Rev. B 37, 3745 (1988).

2   R. J. Cava, B. Batlogg, J. J. Krajewski, R. Farrow, L. W. Rupp Jr., A. E. White, K. Short, W. F. Peck, and T. Kometani, Natural 332, 814 (1988);   R. J. Cava and B. Batlogg, MRS Bulletin, 14, No. 1, 49 (1989).

3   D. G. Hinks, B. Dabrowski, J. D. Jorgensen, A. W. Mitchell, D. R. Richards, S. Pei, and D. Shi, Nature 333, 836 (1988);   D. G. Hinks, D. R. Richards, B. Dabrowski, A. W. Mitchell, J. D. Jorgensen, and D. T. Marx, Physica C 156, 477 (1988);   D. G. Hinks, A. W. Mitchell, Y. Zheng, D. R. Richards, and B. Dabrowski, Appl. Phys. Lett. 54, 1585 (1989).

4   *Physical Properties of High Temperature Superconductors*, Vols. I and II, edited by D. M. Ginsberg (World Scientific, Singapore, 1989, 1990).

5   S. Pei, J. D. Jorgensen, B. Dabrowski, D. G. Hinks, D. R. Richards, A. W. Mitchell, J. M. Newsam, S. K. Sinha, D. Vaknin, and A. J. Jacobson, Phys. Rev. B 41, 4126 (1990).

6   S. Kondoh, M. Sera, Y. Ando, and M. Sato, Physica C 157, 469 (1989).

7   Y. J. Uemura, B. J. Sternlieb, D. E. Cox, J. H. Brewer, R. Kadono, J. R. Kempton, R. F. Kiefl, S. R. Kreitzman, G. M. Luke, P. Mulhern, T. Riseman, D. L. Williams, W. J. Kossler, X. H. Yu, C. E. Stronach, M. A. Subramanian, J. Gopalakrishnan, and A.W. Sleight, Nature 335, 151 (1988).

8   B. Batlogg, R. J. Cava, L. W. Rupp, Jr., A. M. Mujsce, J. J. Krajewski, J. P. Remeika, W. F. Peck, Jr., A. S. Cooper, and G. P. Espinosa, Phys. Rev. Lett. 61, 1670 (1988).

9   Z. Schlesinger, R. T. Collins, J. A. Calise, D. G. Hinks, A. W. Mitchell, Y. Zheng, B. Dabrowski, N. E. Bickers, and D. J. Scalapino, Phys. Rev. B 40, 6862 (1989).

10   K. F. McCarty, H. B. Radousky, D. G. Hinks, Y. Zheng, A. W. Mitchell, T. J. Folkerts, and R. N. Shelton, Phys. Rev. B 40, 2662 (1989).

11   C. K. Loong, P. Vashishta, R. K. Kalia, M. H. Degani, D. L. Price, J. D. Jorgensen, D. G. Hinks, B. Dabrowski, A. W. Mitchell, D. R. Richards, and Y. Zheng, Phys. Rev. Lett. 62, 2628 (1989).

12   C. K. Loong, D. G. Hinks, P. Vashishta, W. Jin, R. K. Kalia, M. H. Degani, D. L. Price, J. D. Jorgensen, B. Dabrowski, A. W. Mitchell, D. R. Richards, and Y. Zheng, Phys. Rev. Lett. 66, 3217 (1991).

13   J. F. Zasadzinski, N. Tralshawala, D. G. Hinks, B. Dabrowski, A. W. Mitchell, and D. R. Richards, Physica C 158, 519 (1989);   J. F. Zasadzinski, N. Tralshawala, J. Timpf, D. G. Hinks, B. Dabrowski, A.W. Mitchell, and D. R. Richards, Physics C 162-164, 1053 (1989).

14   Q. Huang, J. F. Zasadzinski, N. Tralshawala, K. E. Gray, D. G. Hinks, J. L. Peng, and R. L. Greene, Nature 347, 369 (1990).

15   D. G. Hinks, D. R. Richards, B. Dabrowski, D. T. Marx, and A. W. Mitchell, Nature 335, 419 (1988).

16   W. K. Kwok, U. Welp, G. W. Grabtree, K. G. Vandervoot, R. Hulscher, Y. Zheng, B. Dabrowski, and D. G. Hinks, Phys. Rev. B 40, 9400 (1989).

17   H. Sato, H. Takagi, and S. Uchida, Physica C 169, 391 (1990).

18   L. M. Mattheiss and D. R. Hamann, Phys. Rev. Lett. 60, 2681 (1988).

19   C. M. Varma, Phys. Rev. Lett. 61, 2713 (1988);   in *High Temperature Superconductors*, edited by T. Akachi, J. A. Cogordan, and A. A. Valladares (World Scientific, Singapore, 1989), p. 35.

20   P. B. Allen, Nature 339, 428 (1989).

21   A. A. Aligia, M. D. Nuñez Regueiro, and E. R. Gagliano, Phys. Rev. B 40, 4405 (1989).

22   P. Vashishta, M. H. Degani, R. K. Kalia, in *Correlations in Electronic and Atomic Fluids*, edited by P. Jena, R. K. Kalia, P. Vashishta, and M. P. Tosi (World Scientific, Singapore, 1990), p. 223;   M. H. Degani, R. K. Kalia and P. Vashishta, in *Strongly Coupled Plasma Physics*, edited by S. Ichirmaru, (Elsevier Science Publishers, 1990), p. 385; in *Condensed Matter Theories*, Vol. 5, edited by V. C. Aguilera-Navarro (Plenum Press, New York, 1990), p. 151.

23   M. Shirai, N. Suzuki, and K. Motizuki, J. Phys. Condens. Matter $\underline{1}$, 2939 (1989); *ibid.*, $\underline{2}$, 3553 (1990); Solid State Commun. $\underline{73}$, 633 (1990); O. Navarro and R. Escudero, Physica C $\underline{170}$, 405 (1990).

24   W. Jin, C. K. Loong, R. K. Kalia, and P. Vashishta, Bull. Am. Phys. Soc. $\underline{36}$, 525 (1991); W. Jin, C. K. Loong, D. G. Hinks, P. Vashishta, R. K. Kalia, M. H. Degani, D. L. Price, J. D. Jorgensen, and B. Dabrowski, Mater. Res. Soc. Sym. Proc. $\underline{209}$, 895 (1991).

25   W. Jin, M. H. Degani, R. K. Kalia, and P. Vashishta (to be published).

26   J. R. Schrieffer, *Theory of Superconductivity* (Benjamin, New York, 1964).

27   G. M. Eliashberg, Soviet Phys. JETP $\underline{11}$, 696 (1960); *ibid.*, $\underline{12}$, 1000 (1961).

28   D. J. Scalapino, in *Superconductivity*, edited by R. D. Parks (Dekker, New York, 1969), Vol. 1, p. 449.

29   W. L. McMillan, Phys. Rev. $\underline{167}$, 331 (1968).

30   G. Bergmann and D. Rainer, Z. Physik $\underline{263}$, 59 (1973).

31   P. B. Allen and R. C. Dynes, Phys. Rev. B $\underline{12}$, 905 (1975).

32   P. B. Allen and B. Mitrovic, in *Solid State Physics*, Vol. 37, edited by H. Ehrenreich, F. Seitz and D. Turnbull (Academic, New York, 1982), p. 1.

33   J. P. Carbotte, Rev. Mod. Phys. $\underline{62}$, 1027 (1990).

34   P. Vashishta and J. P. Carbotte, J. Low Temp. Phys. $\underline{10}$, 551 (1973); $\underline{18}$, 457 (1975).

35   W. L. McMillan and J. M. Rowell, in *Superconductivity*, edited by R. D. Parks (Dekker, New York, 1969), Vol. 1, p. 561.

36   W. Jin (unpublished).

NON CONVENTIONAL SUPERCONDUCTIVITY IN $BaPb_{1-x}Bi_xO_3$ AND $Ba_{1-x}K_xBiO_3$

A.A. Aligia and M. Baliña

Centro Atómico Bariloche
8400 Bariloche, Argentina

ABSTRACT

We show that the usual scheme for conventional superconductivity, which starts from one particle Bloch states derived fron band-structure calculations neglecting correlations, and the electron-phonon interaction, is not consistent with the experimental evidence in BiO based superconductors. If electron-electron and electron-electron-phonon correlations are included, a non-conventional pairing mechanism is possible in which, both, phonons and excitons play an essential role. We study the conditions for the existence of pairing and its competition against segregation using a combination of exact calculations in a small cluster and perturbation theory in the hopping. We review a simplified model for the electronic structure of $Ba(Pb,Bi)O_3$ and show that the whole picture is consistent with most of the available experimental information.

1- INTRODUCTION

In 1975, Sleight et al.[1] discovered what they called "high-temperature superconductivity in the $BaPb_{1-x}Bi_xO_3$ system". The superconducting critical temperature (T_c) of up to 13K was considered exceptionally high for an oxide. In 1986, Bednorz and Müller[2] obtained $T_c \sim 30K$ in another perovskite, $La_{2-x}Ba_xCuO_4$, opening an era of considerable research effort in these materials. The high value of T_c led many researches to suspect that the mechanism responsible for superconductivity should be different from the usual electron-phonon one, particularly after for different CuO_2 based perovskites, T_c was increased successively, reaching values as high as 110K [3]. Several of the exotic pairing mechanisms proposed, were based on the magnetism of the Cu^{+2} ions[4-6]. When superconductivity near 30K was discovered[7,8] in another non-magnetic[9], Cu-free perovskite, $Ba_{1-x}K_xBiO_3$, it became clear that if the pairing mechanism is of similar nature in all superconducting perovskites, the magnetism should not be an essential part of it. One possibility is an excitonic-like pairing mechanism[10-13]. Other point of view is that the electron-phonon mechanism leads to T_c of the order of 30K or larger in some of these materials[14]. Using the Eliashberg equations to calculate several properties of the cuprate superconductors, Carbotte[15] concludes that it is necessary to postulate another pairing mechanism, particularly for those materials with higher T_c.

In any case, unless several cuprate superconductors are conventional ones, it is clear that the magnitude of T_c alone is not a good classification criterion. The T_c of $BaPb_{1-x}Bi_xO_3$ (13K)[1] and that of the electron-doped cuprate $Nd_{2-x}Ce_xCuO_4$ (24K)[16] are less than or of the same order of magnitude as the T_c of several systems considered as conventional superconductors, like A15 and Chevrel phases. Also the T_c of $La_{2-x}(Sr,Ba)_xCuO_4$ and $Ba_{1-x}K_xBiO_3$ are very similar. Physical properties other than T_c in conventional superconductors are extremely varied, but those of highest T_c are characterized by a large density of states at the Fermi level $\rho(\varepsilon_F)$, what leads to a comparatively large value of the electron-phonon coupling parameter[15] λ and thus, to a high T_c. Instead, all perovskite superconductors have a low value of $\rho(\varepsilon_F)$ (in particular for $BaPb_{1-x}Bi_xO_3$ and $Ba_{1-x}K_xBiO_3$, it is more than an order of magnitude smaller than in A15 or Chevrel phases[9,17]), the parent compounds are semiconductors, the number of carriers is very low[18,19], and T_c scales with the number of carriers[20]. Thus, it is not unreasonable to expect a similar pairing mechanism acting in all superconducting perovskites, and that it is related to Coulomb correlations, since the latter are expected to be important in systems with a small number of carriers because of the large screening length. In fact, there is evidence that the Coulomb interactions are crucial in determining the structure of several perovskites, in particular $YBa_2Cu_3O_{6+x}$[21-23]. They allow also for a strong-coupling explanation of the plateaus in T_c and other properties of this compound[21,24].

In this work we address the question of the mechanism responsible for superconductivity in relation with normal properties of BiO_3 based perovskites. In the next section we describe the model, which is treated neglecting correlations in section 3 and in the strong-coupling limit in section 4. In both treatments, the phonons are an essential part of the problem, in agreement with experiment[14,25-31]. The first treatment corresponds to the conventional one for superconductivity and leads to a charge density wave and a composition dependence of the breathing mode that disagrees with several experiments. The approach of section 4 leads to a pairing mechanism with an electronic energy scale in which charge fluctuations accompanied by displacements of O atoms provide the mechanism, as proposed before[12,13]. In section 5, we use a simplification of this model introduced by Sofo et al.[32,33] to describe the electronic structure of $BaPb_{1-x}Bi_xO_3$. Section 6 contains a comparison with experiments (structural studies, reflectivity, X-ray absorption and photoemission, resistivity, Hall and Seebeck effects, Raman and inelastic neutron spectra), and a short discussion.

2- MODEL

The band structure calculations[34-36] show that the band that crosses the Fermi level is composed of 6s orbitals of the metallic atoms (Bi or Pb) and O 2p orbitals. Except for small distortions,[37-40] the Bi and Pb atoms form a simple cubic lattice and each O atom bisects a segment joining two nearest-neighbor (NN) Bi atoms. The effect of distortions is included in the electron-phonon interaction. Thus, including only Bi-O NN hopping, the Hamiltonian takes the form:

$$H = H_{os} + H_h + H_{e-e} + H_{ph} + H_{e-ph} \tag{2.1}$$

$$H_{os} = \sum_{i\sigma} \varepsilon_{iM} c_{i\sigma}^+ c_{i\sigma} + \varepsilon_0 \sum_{j\sigma} p_{j\sigma}^+ p_{j\sigma} \tag{2.2}$$

$$H_h = \sum_{\delta\sigma} t_\delta p_{i+\delta\sigma}^+ c_{i\sigma} + h.c. \tag{2.3}$$

$$H_{e-e} = U_M \sum_i n_{i\uparrow} n_{i\downarrow} + U_0 \sum_j n_{j\uparrow} n_{j\downarrow} + U_{MO} \sum_{i\delta} n_i n_{i+\delta} + \frac{U}{2} \sum_{i\delta\neq\delta'} n_{i+\delta} n_{i+\delta'} \tag{2.4}$$

$$H_{ph} = \sum_{rq} \hbar\omega_{rq} (b_{rq}^+ b_{rq} + 1/2) \tag{2.5}$$

where:

$$n_{i\sigma} = c_{i\sigma}^+ c_{i\sigma} \quad , \quad n_{j\sigma} = p_{j\sigma}^+ p_{j\sigma} \quad , \quad n_i = n_{i\uparrow} + n_{i\downarrow} \tag{2.6}$$

i (j=i+δ) label metal (O) sites. δ labels the six vectors which connect a metal (Bi or Pb) atom with its NN O atoms. $c_{i\sigma}^+$ creates a hole in the 6s orbital at site i with spin σ. The $p_{j\sigma}^+$ have a similar meaning for the O 2p orbitals. H_{os} contains the on-site energies, H_h is the hopping term, H_{e-e} contains the interaction terms and H_{ph} describes the vibrational energy. Here b_{rq}^+ creates one of the phonon modes (labeled by r) with wave vector q.

The dominant part of the electron-phonon interaction in the weakly correlated limit is derived from the distance dependence of the hopping integral, replacing t_δ in Eq. (2.3) by:

$$t_\delta \rightarrow t_\delta + \nabla_\delta t_\delta \cdot (u_{i+\delta} - u_i) \tag{2.7}$$

where u_i is the displacement operator of the atom located at site l and is expressed in terms of phonon operators by[41]:

$$u_1 = \left(\frac{\hbar}{2NM_\kappa}\right)^{1/2} \sum_{rq} \omega_{rq}^{-1/2} e^{iq.R_1} e_{rq\kappa}(b_{rq}+b_{r-q}^+) \qquad (2.8)$$

where κ is the label of the metal ion or one of the three O atoms per unit cell corresponding to site 1, M_κ is the mass of this atom and $e_{rq\kappa}$ is the polarization vector.

When interatomic Coulomb repulsions are not well screened, the main term of H_{e-ph} has the form of a Coulomb force on each O atom times its displacement:

$$H_{e-ph} = -f\sum_j (n_{j+\delta}-n_{j-\delta})(2-n_j)u_j \qquad (2.9)$$

where f is a constant force and u_j is the displacement of the O atom at site j towards the metal atom at site $j+\delta$.

The Hamiltonian given by Eqs (2.1) to (2.6) and (2.9) is studied in Section 4. In the next section we take H_{e-ph} given by the modifications that Eqs (2.7) and (2.8) introduce in Eq. (2.3) and neglect the correlations in H_{e-e}.

3- A CONVENTIONAL SUPERCONDUCTOR

By this title, we understand a system in which the superconductivity is due to the electron-phonon interaction *and* the correlations play a secondary role. These can be taken into account through the correction parameter μ^* in the Eliashberg equations[15], Weber[42] took this point of view to explain the superconductivity with $T_c \sim 35K$ in $La_{2-x}(Sr,Ba)_x CuO_4$. Aligia et al.[43,44] obtained similar results using only one orbital per site. While the magnitude of T_c could be explained, these theories predicted a breathing mode instability which has not been observed. The case of BiO_3 superconductors is similar. However, a breathing mode distortion in $BaBiO_3$ has been observed[37-40]. This distortion is such that if the simple cubic lattice of metal atoms is divided in two equal interpenetrating f.c.c. sublattices, the regular octahedra composed of the 6 NN O atoms of the metal atoms of one sublattice contract, while the other expand. In addition, the quantity $\alpha^2 F(\omega)$, which is a measure of the effective attractive electron-electron interaction[15], has been determined from tunneling measurements in $Ba_{1-x}K_x BiO_3$[26], and reflects the phonon density of states $F(\omega)$, which in turn has been determined by inelastic-neutron measurements.[14] Using these data, the isotope effect could be predicted[25-27] and a consistent explanation of these three experiments was given[25]. The main conclusion is that $Ba_{1-x}K_x BiO_3$ is a weak-coupling superconductor (this means that λ, which is the integral of $2\alpha^2 F(\omega)/\omega$ is near 1 or less) with strong interaction between electrons and high-energy phonons involving oxygen vibrations[25,27]. The average energy of these phonons is $<\omega>\sim 30-40$ meV. Taking $\lambda \sim 1$, the Eliashberg equations lead to the observed T_c and $2\Delta/kT_c \sim 4$ [25,45]. The last value is in agreement with tunneling measurements[26] and inside the error bars of infrared experiments[46]. Thus, these results are consistent with an explanation of superconductivity based only on electron-phonon interaction. However, the value $2\Delta/kT_c = 3.5\pm0.1$ found in $BaPb_{1-x}Bi_xO_3$ as well as other experiments carried out in BiO_3 superconductors[9] suggest a value of $\lambda \sim 3/4$ and the presence of another pairing mechanism in both systems. The tunneling[26] and isotope effect[9,25,27] experiments are consistent not only with a pure electron-phonon mechanism, but also with an electronic mechanism which involves vibrations of O atoms, as discussed in section 4, or with the coexistence of exotic and electron-phonon mechanisms[15,44]. A study of the electron-phonon interaction should help in narrowing the range of different possibilities. Here we show that if correlations are neglected and the electron-phonon interaction is so strong that it alone leads to $\lambda \sim 1$, it modifies the phonon spectrum in a way that is not consistent with experiments.

We consider the Hamiltonian given by Eqs (2.1) to (2.8) with $\varepsilon_{iM} = \varepsilon_M$ for all sites, and H_{e-e} in the Hartree-Fock approximation. This is equivalent as taking $H-H_{e-e}$ with renormalized parameters. The result for the eigenenergies of the bonding band (the one which crosses the Fermi energy) is:

$$E_k = (\varepsilon_M+\varepsilon_0)/2-r(\varepsilon_k) \tag{3.1}$$

$$r(\varepsilon) = +\left[(\varepsilon_M-\varepsilon_0)^2/4+6t^2-t\varepsilon\right]^{1/2} \tag{3.2}$$

$$\varepsilon_k = 2t\left[\cos(k_x x)+\cos(k_y y)+\cos(k_z z)\right] \tag{3.3}$$

Here $t = |t_\delta|$. The form of the electron-phonon interaction in terms of this eigenstates is a trivial generalization of the two-dimensional result given in Ref. 44. Since the expressions are rather long, we do not reproduce them here. For no added holes ($n=0$, corresponding to $BaBiO_3$), the Fermi surface is given by the equation $\varepsilon_k=0$ and is a cube. This implies that the system is unstable under charge density waves of wave vector $Q=(\pi/a,\pi/a,\pi/a)$ and in particular to the observed breathing mode[37-40]. For simplicity we take into account only this mode in the adiabatic approximation. Calling u the magnitude of the displacement of any O atom, we can thus write:

$$H_{ph}+H_{e-ph} = \frac{1}{2}Ku^2+u(\Sigma_{k\sigma}F_k a_{k\sigma}^+ a_{k+Q\sigma}+h.c.) \tag{3.4}$$

where $a_{k\sigma}^+$ creates a hole in the bonding state with energy given by Eqs (3.1) to (3.3) and F_k can be deduced from H_{e-ph}[44]. To be able to get a simple analytical solution we approximate F_k by its value on the Fermi surface for $n=0$. Using the three-dimensional extension of H_{e-ph} given in Ref. 44 we obtain:

$$|F_k| = 2b\frac{\partial t}{\partial d}\Big|_{MO} = F \quad , \quad k/\varepsilon_k = 0 \tag{3.5}$$

where d_{MO} is the metal-O distance and

$$b = \left[1+(\varepsilon_M-\varepsilon_0)^2/24t^2\right]^{-1/2} \tag{3.6}$$

Diagonalizing Eq. (3.4) and adding all eigenenergies up to the Fermi level for n added holes per original unit cell, we obtain for the ground state energy:

$$E(u)-E(0) = \int_{\mu_0}^{\infty} d\varepsilon \, \rho_0(\varepsilon)\Big\{|r(\varepsilon)-r(-\varepsilon)|/2$$

$$-\left[(r(\varepsilon)-r(-\varepsilon))^2/4+F^2u^2\right]^{1/2}\Big\}+\frac{1}{2} Ku^2 \tag{3.7}$$

where $\rho_0(\varepsilon)$ is the density of states of the dispersion relation given by Eq. (3.3) and μ_0 is given by:

$$n = \int_0^{\mu_0} d\varepsilon \, \rho_0(\varepsilon) \tag{3.8}$$

The largest contribution to the integral of Eq. (3.7) comes from values of $\varepsilon \lesssim Fu$. Expanding $r(\varepsilon)$ up to second order in ε and replacing $\rho_0(\varepsilon)$ by a constant $1/W$ extending from $-W$ to W, the integral can be done analytically. The optimum distortion u is obtained minimizing Eq. (3.7):

$$\frac{\partial E(u)}{\partial u} = (K-IF^2)u = 0 \tag{3.9}$$

With the above assumption I takes the form:

$$I = \frac{1}{Wb} \ln\left[\frac{Wb+(W^2b^2+F^2u^2)^{1/2}}{Wbn+(W^2b^2n^2+F^2u^2)^{1/2}}\right] \tag{3.10}$$

$u = 0$ for $n \geq n_c$ where:

$$n_c = e^{-1/\lambda'} \quad , \quad \lambda' = F^2/KWb \tag{3.11}$$

For $n \leq n_c$:

$$\left(\frac{u}{u_0}\right)^2 = \frac{(n_c-n)(1-nn_c)}{n_c} \quad , \quad u_0 = \frac{2Wbn_c}{F(1-n_c^2)} \tag{3.12}$$

u_0 is the distortion for $n=0$. For the frequency we get if $n \geq n_c$:

$$\omega^2 = \frac{1}{M_0} \frac{\partial^2 E(u)}{\partial u^2} = \omega_0^2 \left[1 + \lambda' \ln(n) \right] \quad , \quad \omega_0^2 = K/M \tag{3.13}$$

and if $n \leq n_c$:

$$\omega^2 = 2\omega_0^2 \lambda' q \left\{ \left[n + (n^2 + 4q)^{1/2} \right]^{-1} \left[n^2 + 4q \right]^{-1/2} \right.$$

$$\left. - 1 + (1 + 4q)^{1/2} \right]^{-1} \left[1 + 4q \right]^{-1/2} \right\} , \qquad q = (2Wbu/F)^2 \tag{3.14}$$

λ' is of the order of the pairing parameter λ. In fact if one calculates the pairing potential V as in Ref. 47 for example, assuming that the electron-phonon coupling for all relevant phonons has the same strength as for the breathing mode (an overestimation), and multiplies the result by the density of states per spin $\rho(\varepsilon_F)/2$ at the Fermi level for $n=0$ and $u=0$, one obtains $\lambda = \lambda'/2$. $\lambda \sim 1$ implies also $\lambda' \sim 1$. The results for u and ω taking $\lambda'=1$ are shown in Fig. 1. As has been shown in the two-dimensional case[48], taking realistic band structures or interactions that modify the perfect nesting (0-0 hopping for example), the results are not modified qualitatively.

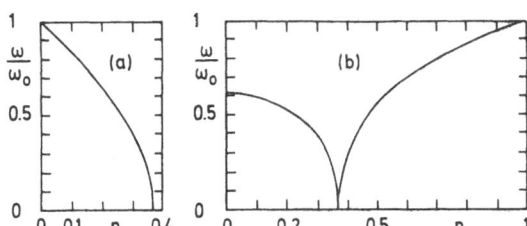

FIG. 1. (a) Breathing mode displacement and (b) Frequency of this mode as a function of the number of added holes per metal atom.

For $Ba_{1-x}K_x BiO_3$, $n=x$. Inelastic neutron experiments and molecular dynamics simulations[14] show that the frequency of this breathing mode is around 35 meV, and that the oxygen phonon modes have only ~5 meV less energy for x=0.4 than for x=0. In addition, X-ray absorption experiments[49], show that for x=0.2, the local displacements remain $\pm u_0$ (as for x=0), but with a certain degree of disorder, such that neutron or X-ray diffraction experiments see a small or zero value of the long-range staggered displacement[37-40]. Evidence of local distortions around Bi atoms, instead of an homogeneous charge-density wave, also exists for $BaPb_{1-x}Bi_x O_3$ from reflectivity[31], X-ray absorption[50] and Raman experiments[29]. Thus, it is clear that the conventional picture should be modified.

The above mentioned evidence of the existence of local modes points towards a strong interaction of the form of Eq. (2.9). In Fact, Yu et al.[51] considered this interaction neglecting the O states (n_j=0) in the Hartree-Fock approximation, and obtained that the solution with local breathing displacements has less energy than the homogeneous solution. In addition, two added holes tend to go to the metal atom at the center of a new contracted O octahedron, forming a bipolaron. In fact, local breathing distortions give rise to an effective on-site attraction[52], characteristic of bipolaronic theories for superconductivity[53,54]. The fact that the O states should be considered, as indicated by the band structure calculations[34-36] and experimental evidence[49,55,56], points towards two-band models with on-site attraction in one of them[57], which are able to explain the "marginal Fermi liquid" behavior of high T_c superconductors[58]. However, if O holes are allowed, a simple mapping of the original model into another with effective on-site attractions is not possible.

Dolores Núñez Regueiro and Aligia[12,13] proposed an extension of the excitonic pairing mechanism as the origin of high-T_c superconductivity in BiO_3 based superconductors. This mechanism, originally proposed for superconducting cuprates[10], has the interatomic Coulomb repulsion as an essential ingredient. It has also been extended to electron-doped systems[59,60]. The studies for BiO_3 systems carried out in Refs 12 and 13, have the shortcoming that the possibility of segregation (tendency of added holes to form bound states of three or more holes instead of two), has not been studied. Although this possibility is unlikely if O-O repulsion U_2 or hopping t is large enough, these energies work also against pairing. For cuprates, the competition between pairing and segregation has been studied recently[11]. In addition, terms of order t^4, not included in Ref. 12 become important when for example the energy of a state with 2 added O holes at 2 nearest neighbors (NN) of a given metal atom with two holes (Fig. 3(c) of Ref. 12) becomes nearly degenerate to that of the "biexciton" (Fig. 3(d) of Ref. 12) in which the 2 holes at the metal ion of the previous state move to 2 NN O atoms, so that there are 4 holes in 4 of the 6 O atoms of the octahedra that surrounds a metal atom without holes. However, the onset of the pairing mechanism was given precisely by the equality of these two energies. These difficulties are avoided in the present treatment.

In this section we consider the Hamiltonian given by Eqs (2.1) to (2.6) and (2.9). For simplicity we treat the phonons in the antiadiabatic approximation: for each eigenstate of the electronic part of H_{e-ph} (see Eq. (2.9)), we minimize $H_{e-ph} + Ku_j^2/2$. This is consistent with the experimental evidence in favor of local distortions[29,31,49,50]. The result for the displacement and energy gain is:

$$u_j = f(n_{j+\delta} - n_{j-\delta})(2-n_j)/K \quad , \quad \Delta E_j = -Ku_j^2/2 \qquad (4.1)$$

Depending on the values of the charges of the O atom and its two NN metal atoms, there are 7 possible displacements and 4 corresponding energy gains. We take $\varepsilon_{iM} = \varepsilon_M$ for all i and define:

$$A = 8f^2/K \quad , \quad \Delta = \varepsilon_O - \varepsilon_M \quad , \quad E_{BP} = U_M - 2\Delta - 6A - U_2 \qquad (4.2)$$

A is the largest possible energy gain due to an O displacement at site j $n_j=0$, $|n_{j+\delta} - n_{j-\delta}| = 2$). According to band structure calculations[34-36], $\Delta<0$. Thus, in absence of O distortions it is more favorable to add holes in O atoms. However, when $E_{BP}<0$, and assuming for the moment t=0, for one hole per metal atom (corresponding to $BaBiO_3$), the ground state is the "disproportionated" state ($Ba_2Bi^{+3}Bi^{+5}O_6$) in which the simple cubic lattice of metal atoms is divided into two equal f.c.c. sublattices, the NN metal atoms for one sublattice are in the other, all holes are in one sublattice, and all O atoms are distorted towards the atoms of this sublattice as experimentally observed[37-40]. Both, the charge disproportionation and the O distortions together stabilize the ground state. Neither of them can be considered as cause or effect. The magnitude of the charge disproportionation is reduced for t>0 as holes hop to the NN O atoms. As Ba is replaced by K, holes are added into the system and they go mainly to O atoms. The question that we want to address is if these holes have a tendency to group in pairs or to segregate.

We have calculated the energy gain when starting from the parent compound, n holes are added to the 7-atom cluster formed by a contracted octahedra of O atoms and the central $Bi^{+4+\nu}$ ion ($0<\nu<1$), solving exactly this cluster and using perturbations in the hopping t for the rest of the system. Calling E(n) the resulting energy, and defining:

$$e(n) = \frac{E(n)-E(0)}{n} \qquad (4.3)$$

the value of n for which e(n) is minimum determines the number of holes which like to come together. This generalizes the concept of binding energy used in Refs 12,13. The binding energy defined there is $E_b = 2(e(2)-e(1))$. We have calculated e(n) for $1\le n\le 4$. For n>4 e(n) increases sharply. The Hamiltonian restricted to the 7-atom cluster is invariant under any permutation of its 6 O atoms. Thus, the space symmetry group is S_6, which is much larger than O_h. The use of this symmetry reduces considerably the computation time. States that transform according to the irreducible representations of S_6 were obtained diagonalizing once the full matrix. Double hole occupation at the oxygen site was disregarded ($U_O \to \infty$). Calling $G(n)$ the ground state energy of the 7-atom cluster with n+2 holes, $|c_i(n)|^2$, i=0,1,2 the probability of finding exactly i holes in the metal atom of this cluster, and $D_i(n)$, the diagonal energies (energies for $|t_\delta|=t=0$) of the cluster with i holes in the central atom, we have obtained:

$$E(n) = G(n)-t^2 \Big(n|c_2(n)|^2(15A/4-\Delta-2U_{MO}-(n-1)U_2+D_2(n)-G(n))^{-1}+$$
$$+(n+1)|c_1(n)|^2(61A/16-\Delta-U_{MO}-nU_2+D_1(n)-G(n))^{-1}+ \qquad (4.4)$$
$$+(n+2)|c_0(n)|^2(7A/2-\Delta-(n+1)U_2+D_0(n)-G(n))^{-1}\Big)$$

The diagonal energies are together with the number of particles n+2 and the strength of the hopping t the input of the exact diagonalization of the cluster. They are given by:

$$D_0(n) = 6A+(n+2)\Delta+U_2(n+1)(n+2)/2 \qquad (4.5)$$

$$D_1(n) = 3(25+n)A/16+(n+1)\Delta+(n+1)U_{MO}+U_2 n(n+1)/2 \qquad (4.6)$$

$$D_2(n) = 3nA/4+n\Delta+2nU_{MO}+U_2 n(n-1)/2 \qquad (4.7)$$

The small variations of t due to its distance dependence[48,12] were neglected.

In general $D_1(n)>min[D_0(n),D_2(n)]$. If this is true, the result for t=0 takes a simple form:

$$e(n)=3A/4+\Delta+2U_{MO}+U_2(n-1)/2+min\ [0, |E_{BP}|/n-2(U_{MO}-U_2+3A/8)] \qquad (4.8)$$

Note that if $U_2=0$ and e(2)<e(1), then also e(4)<e(3)<e(2). Thus the region of parameters for which pairing was obtained for small t in Ref. 12 corresponds actually to segregation. When U_2 is included, pairing is favored if the last member of Eq. (4.8) is negative for n=2 and $|E_{BP}|/3<U_2<|E_{BP}|$. This conditions are fulfilled by the parameters used in Fig. 2(a).The physical mechanism of this process is quite simple: if a hole is added at an O atom of $Ba_2Bi^{+3}Bi^{+5}O_3$ without modifying the rest of the charge distribution it cost energy $3A/4+\Delta+2U_{MO}$. If the Bi-O charge repulsion U_{MO} is large enough, it is more convenient to break a bipolaron and promote the two holes of the NN Bi^{+5} of the O^{-1} to two others O atoms, NN of this Bi atom. In this way, the energy $|E_{BP}|$ is lost but $2(U_{MO}-U_2+3A/8)$ is gained. When further O holes are added, they prefer to go to other NN of this particular, now Bi^{+3} ion, because neither U_{MO} nor $|E_{BP}|$ should be paid. This would lead to segregation. However, as more holes are added the O-O repulsion increases quadratically, so that for certain parameters, it is not convenient to add more than two holes to one particular O octahedra that surrounds an originally Bi^{+5} ion. As seen in Fig. 2(a), when hopping t is turned on, it favors $e(1)$, because the diagonal energies $D_1(n)$ are more similar among them for n=1. For large enough t $e(1)<e(2)$. In the region of parameters in which pairing is favored, the wave function of the ground state has total spin S=0, is 9-fold degenerate and is dominated by states of the form of the "biexciton", (4 holes moving around an empty Bi site); $|c_0(2)|^2 \geq 0.8$ for the parameters used in Fig. 2(a).

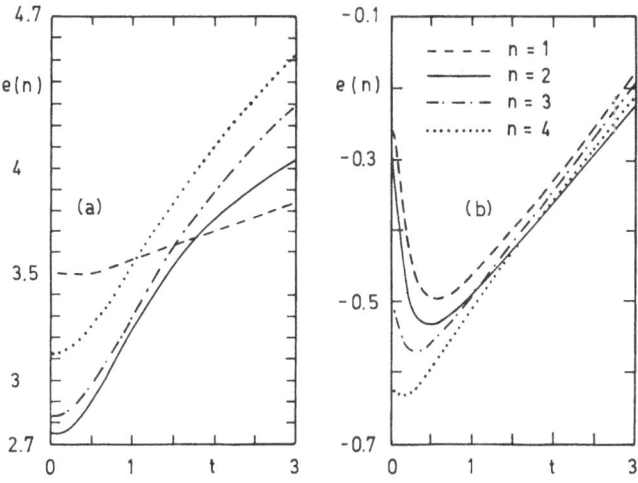

FIG. 2. Energy per particle necessary to add a large number of particles in groups of n. Parameters in an arbitrary unit of energy (of the order of 0.5 eV) are: (a) $A=1$, $\Delta=-1$, $U_M=2.5$, $U_{MO}=3$, $U_2=1$, (b) the same but $U_{MO}=U_2=0$.

A different pairing regime, which has not been described or proposed before, is obtained when $D_2(2) \approx D_0(2)$ and $D_1(2)-D_0(2)$ is small, so that large hopping favors the states with two added holes in the 7-atom clusters due to delocalization effects, although for t=0 no pairing is obtained. This possibility is shown in Fig. 2(b). The effect of the hopping on pairing is the opposite as that of the previous case. Also no interatomic repulsions are required. However, even when $U_{MO}=U_2=0$, interatomic correlations are present in the electron-electron-phonon interaction contained in the Eq. (2.9). The ground state for two added particles is also in this case a singlet that belongs to a 9-dimensional representation of S_8. However in this case $|c_0(2)|^2 \approx |c_2(2)|^2 \approx |c_1(2)|^2/2 \approx 0.25$; the proportion of biexitons is smaller and that of two O holes around a doubly occupied Bi atom larger. Since 4 hopping processes are needed to displace the latter state to an equivalent place, while the biexciton requires 8, a larger mobility and therefore higher T_c is expected in this case.

In both cases ((a) and (b) in Fig. 2), pairing is related to the possibility of breaking a bipolaron (double occupied Bi site and distortions around it) and promote the holes to NN O atoms (because of interatomic repulsion or hopping). Thus $|E_{BP}|$ cannot be very large. We believe that the effect of substitution of Bi by Pb or Ba by K, reduces the effective value of $|E_{BP}|$, making superconductivity possible until for positive E_{BP} or very low $|E_{BP}|$ the effective attraction disappears again. This view is supported by detailed calculations in $BaPb_{1-x}Bi_xO_3$[33].

5- ELECTRONIC STRUCTURE OF $BaPb_{1-x}Bi_xO_3$

Contrary to $Ba_{1-x}K_xBiO_3$, which was discovered recently as a high T_c superconductor[7,8], $BaPb_{1-x}Bi_xO_3$ has been studied during 16 years[1] and there is a good deal of experimental information available. In particular, a detailed picture of the electronic structure has emerged from optical studies[31]. However, a calculation of the electronic excitations of the system including correlations is extremely difficult. If these are neglected and the system treated as in section 3, the result is not in agreement with the experiments. Also the disorder of the metal ions (Bi or Pb) introduces an additional complication. However, Sofo et al[32,33] were able to explain the experimental results making certain simplifying assumptions, but keeping the essential ingredient: the effect of local O displacements. The most serious of the assumptions made is that only virtual O hole occupations are allowed. However, the effects of O holes are discussed in Ref. 33. The second assumption is that O atoms between any two Bi atoms are always displaced by the same amount, while they are not displaced if one of its nearest neighbor (NN) metal atoms is Pb. This assumption exaggerates the displacements u in the Pb reach limit , since according to Eq. (4.1) u is proportional to the difference between the charges of the NN metal ions, and this difference is expected to be small in the Pb reach limit. The resulting Hamiltonian in terms of electrons takes the form (it is simpler to talk about electrons than holes in this section):

$$H=\Sigma \ \varepsilon_{iM} c^+_{i\sigma} c_{i\sigma} - t'\Sigma \ c^+_{i+2\delta,\sigma} c_{i,\sigma} \tag{5.1}$$

where the meaning of the operators and energies is the same as in section 2 interchanging electrons and holes. t' is an effective hopping (of order t^2)

280

between two metal NN ions, mediated by the O ion in between. $\varepsilon_{1M}=\varepsilon_{Bi}+mA$, $-6\leq m\leq 6$, where m depends on the number and sign of the O distortions around the Bi atom. The sign was assumed the same as for $BaBiO_3$ for any Pb content. the problem was reduced to that of calculating the electronic structure of an alloy of 8 components in a lattice with 2 inequivalent sites. In turn, this problem was solved using the coherent potential approximation (CPA) replacing the density of states of the simple cubic system by the semielliptical form with half-width W, as usual in CPA treatments. The total density of states for parameters slightly different than those used in Refs 32 and 33 is represented in Fig. 3. For those parameters an important part of the band (of the order of 5% of the total weight) was splited for x=0.05. This fact was criticized[62]. The only correct criticism of Ref. 62 is that the charge disproportionation is exaggerated for small x.

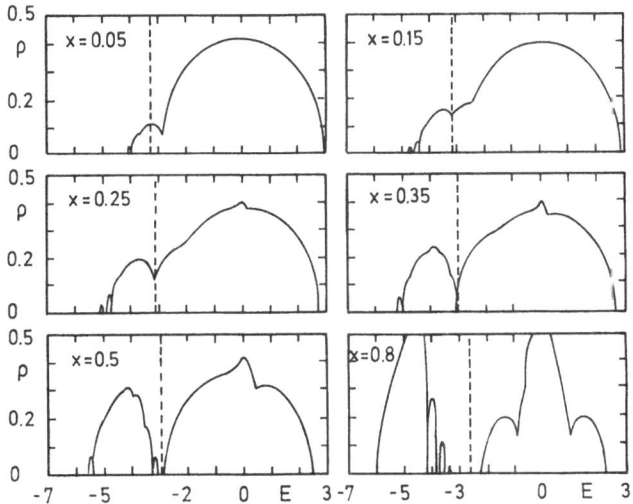

FIG. 3. Total density of metal states of Ba(Pb,Bi)O for different Bi content. Parameters in an arbitrary unit of energy (of the order of 1eV) are: $\varepsilon_{Pb}=0$, $\varepsilon_{Bi}=-2.4$, A=0.4 and W=3. The dashed line indicates the Fermi energy.

As a consequence of the O distortions and related charge disproportionation, a pseudogap opens in the spectral density of states for $x\geq0.15$, which becomes a real gap for $x\approx0.3$. For larger values of x, there are states of small weight inside the gap due to statistically less probable configurations of Bi atoms surrounded by more Pb atoms than the average. For each Bi content x, there are precisely x electrons in the metal bands, since Bi^{+4} brings one 6s electron and Pb^{+4} has none.

Superconductivity in $BaPb_{1-x}Bi_xO_3$ would be due to self doping: at the metal insulator transition, lowering x, $|E_{BP}|$ becomes small enough so that biexcitons turned to be energetically favored, and states of mainly Pb character provide the holes to the biexcitons.[33]

In this section we compare the implications of the model for the pairing mechanism (section 4) and the simplified model of the previous section, as well as the assumptions made, with the available experimental data. The discussion is divided according to the different physical ingredients.

a) *Electron-phonon interaction*: the strength of this interaction is clearly high. This is consistent with the Fano effect for high energy phonons seen in Raman experiments[28-30], and the evidence in favor of the existence of bipolarons deduced from Raman experiments[29,30], and from the two gap structure observed by "pump and probe" optical spectroscopy in $BaBiO_3$. Further evidence comes from reflectivity measurements[31,56] in $BaPb_{1-x}Bi_xO_3$. The tunneling measurements [26] in $Ba_{1-x}K_xBiO_3$ suggest a direct participation of phonons in the pairing mechanism. Pairing based on an on-site attraction in Bi ions due to an exotic non-linear screening[64] seems unlikely. However the electron-phonon coupling parameter λ is small[9,25-27] due to the low density of states at the Fermi level[9], and superconductivity due to an effective attraction of one particle Bloch states in an homogeneous system is unlikely[9], particularly due to the implications for phonon softening and lattice instabilities (section 3).

b) *Local O displacements*: The evidence in favor of local distortions, rather than an homogeneous charge-density wave, was enumerated in the final paragraph of section 3. In addition, the electronic structure of $BaPb_{1-x}Bi_xO_3$ explained in section 5, and derived on the basis of local O displacements is in very good agreement with optical experiments and its interpretation[31].

c) *Charge disproportionation*: The pairing mechanisms of section 4 require an important difference between the charge of inequivalent Bi atoms in $BaBiO_3$. This difference is not known. Using Zachariasen's formula Chaillout et al.[38] proposed charges 3.4 and 4.4. Evidence of inequivalent Bi sites is also provided by EXAFS[49,65] and photoemission measurements[66] but not by XPS[67].

d) *Holes in O atoms*: This makes a difference between the model for pairing studied here and others based exclusively on bipolarons[53,54]. EXAFS experiments[49,55] provide evidence that K doping creates holes in O 2p orbitals in $Ba_{1-x}K_xBiO_3$. Reflectivity studies[56] in the semiconducting phase of $BaPbB_{1-x}Bi_xO_3$ show that there are more O holes than in other perovskites and that its amount increases towards the insulator-superconductor transition.

e) *Doping and carriers*: While as mentioned above, replacement of Ba by K creates holes in O atoms, the situation is more complicated in $BaPb_{1-x}Bi_xO_3$. starting from $BaPbO_3$, replacement of Pb by Bi introduces electrons in the metal band. This is consistent with Hall measurements[19] which show a linear increase with x in the number of negative carriers n_e, for x<0.10 in $BaPb_{1-x}Bi_xO_3$. However n_e increases more than linearly for x>0.12 and decreases again for x>0.25 until the material becomes insulating for x≥0.35. This is consistent with the presence for x>C.12 of positive heavy carriers, like the biexcitons mentioned in section 4. Charge neutrality requires a larger n_e and the number of heavy holes n_h does not alter directly the Hall effect because of its lower mobility. This picture is confirmed by thermopower measurements[68] and by the x dependence of T_c[19], which seems to correlate directly with n_h as in other high T_c perovskites[20]. Also the intensity of a 100 cm^{-1} phonon mode observed in Raman experiments shows a similar x dependence as T_c; the promotion of holes from Bi to O atoms that occurs when the pairing conditions are satisfied, should alter the force constants between the atoms involved.

While the Hall effect in $Ba_{1-x}K_xBiO_3$ shows the presence of light negative carriers[70,71], the thermopower experiments are contradictory[70-73]. The measurements of resistivity in these systems are affected by the presence of inhomogeneities[73] and grain boundaries. The most recent thermopower measurements[73] obtain a result very similar to that seen in the in-plane direction of $YBa_2Cu_3O_{6+x}$, reinforcing the similarities between CuO_2- and BiO_3- based superconducting perovskites (see section 1).

f) *Optical and resistivity gaps*: While reflectivity experiments weight all states similarly, in activation energy experiments, the states are weighted exponentially according to their distance to the Fermi energy. Thus, it is clear from Fig. 3 (in particular for x=0.8) that a much smaller gap is expected in resistivity measurements in agreement with experiment[17,31]. Also as noted in Ref. 31, as a consequence of the presence of a pseudogap, the optical experiments see a gap for x>0.15 also in the metallic phase. In presence of a pronounced pseudogap (depression of the density of states at the Fermi level), a decrease of the resistivity for increasing temperature is expected, as experimentally observed[31].

Discussion: In summary, a model in which both, localized O vibrations and displacements, and interatomic correlaticns are essential, provides a pairing mechanism with an electronic energy scale and is consistent with the experimental information available for $Ba_{1-x}K_xBiO_3$ and $BaPb_{1-x}Bi_xO_3$. Interatomic repulsion seem also to be essential to describe the electronic[21,24] and atomic[21-23] structure of $YBa_2Cu_3O_{6+x}$.

An important point, which was not addressed here is the semiconductor-superconductor transition for increasing x in $Ba_{1-x}K_xBiO_3$. An explanation of this transition based also on local phonon modes was made by D. Nguyen Manh et al.[74]. However, these authors started with an artificial separation of the Bi atoms in two species with concentrations such that the system is insulating for small x. The insulating behavior is probably due to the fact that the biexcitons (or bipolarons[51]) remain pinned near defects or K atoms for small x, as suggested by the results of Yu et al.[51] A deeper study of the superconductivity is difficult but also necessary.

ACKNOWLEDGEMENTS

We are indebted to J.O. Sofo for useful discussions and the computation program with the help of which Fig.3 was made. A.A.A. is indebted to M.D. Núñez Regueiro and M. Núñez Regueiro for the interchange of ideas concerning the pairing mechanism and the neutron measurements of Ref.14. M.B. is a fellow of Consejo Nacional de Investigaciones Científicas y Técnicas (CONICET).

REFERENCES

1. A.W. Sleight, J.L. Gillson and P.E. Bierstadt, Solid State Commun. 17:27 (1975)
2. J.G. Bednorz and K.A. Müller, Z.Phys.B 64:189 (1986)
3. See for example G. Leyva et al., Solid State Commun. 78:887 (1991); A.Y. Khan et al. Mod.Phys.Lett. 5:771 (1991)
4. J.R. Schrieffer, X.G. Wen and S.C. Zhang, Physica C 153-155:21 (1988)
5. P. Lederer, D. Poiblanc and T.M. Rice, Phys.Rev.Lett. 63:1519 (1989)
6. A.G. Rojo and G.S. Canright, Phys.Rev.Lett. 66:949 (1991)
7. L.F. Mattheiss, E.M. Gyorgy and D.W. Johnson, Jr., Phys.Rev.B 37:3745 (1988)
8. R.J. Cava et al., Nature 332:814 (1988)
9. B. Batlogg et al., Phys.Rev.Lett. 61:1670 (1988)
10. P.B. Littlewood, C.M. Varma and E. Abrahams, Phys.Rev.Lett. 63:2602 (1989) and references therein.
11. M. Grilli et al., Phys.Rev.Lett. 67:259 (1991)
12. M.D. Núñez Regueiro and A.A. Aligia, Phys.Rev.Lett. 61:1889 (1988)
13. A.A. Aligia, M.D. Núñez Regueiro and E.R. Gagliano, Phys.Rev.B 40:4405 (1989)
14. C.K. Loong et al., Phys.Rev.Lett. 62:2628 (1989)
15. J.P. Carbotte, Rev.Mod.Phys. 62:1027 (1990)
16. H. Takagi, S. Uchida and Y. Tokura, Phys.Rev.Lett. 62:1197 (1989)
17. S. Tajima et al., Phys.Rev.B 32:6302 (1985)
18. N.P. Ong et al., Phys.Rev.B 35:8807 (1987)
19. T.D. Thanh, A. Koma and S. Tanaka, Appl.Phys. 22:205 (1980)
20. Y.J. Vemura et al., Phys. Rev. Lett. 62:2317 (1989)
21. A.A. Aligia, J. Garcés and H. Bonadeo, Phys. Rev. B 42:10266 (1990)
22. A.A. Aligia, H. Bonadeo and J. Garcés, Phys.Rev.B 43:542 (1991) and references therein.
23. A.A. Aligia, Phys.Rev.Lett. 65:2475 (1990)
24. A.A. Aligia, Solid State Commun. 78:739 (1991)

25. P. Vashishta et al., this conference.
26. Q. Huang et al., Nature 347:369 (1990)
27. C.K. Loong et al., Phys.Rev.Lett. 66:3217 (1991)
28. K.F. Mc Carty et al., Phys.Rev.B 40:2662 (1989)
29. S. Sugai, Solid State Commun. 72:1187 (1989)
30. S. Sugai, Y. Enomoto and T. Murakami, Solid State Commun. 72:1193 (1989)
31. S. Tajima et al., Phys.Rev.B 35:696 (1987)
32. J.O. Sofo, A.A. Aligia and M.D. Núñez Regueiro, Phys.Rev.B 39:9701 (1989)
33. J.O. Sofo, A.A. Aligia and M.D. Núñez Regueiro, Phys.Rev.B 40:6955 (1989)
34. L.F. Mattheiss and D.R. Hamann, Phys.Rev.B 28:4227 (1983)
35. L.F. Mattheiss and D.R. Hamann, Phys.Rev.Lett. 60:2681 (1988)
36. D.A. Papaconstantopoulos et al., Phys.Rev.B 40:8844 (1989)
37. D.E. Cox and A.W. Sleight, Solid State Couumn. 19:969 (1976)
38. C. Chaillout et al., Solid State Commun. 65:1363 (1988)
39. L.F. Schneemeyer et al., Nature 335:421 (1988)
40. S. Pei et al., Phys.Rev.B 41:4126 (1990)
41. G.K. Horton and A.A. Maradudin, "Dynamical Properties of Solids", North-Holland Publishing Company, Amsterdam, Vol.1 (1974)
42. W. Weber, Phys.Rev.Lett. 58:1371 (1987)
43. A.A. Aligia et al., Int.J.Mod.Phys.B 1:951 (1987)
44. A.A. Aligia et al., Solid State Commun. 65:501 (1988)
45. F. Marsiglio and J.P. Carbotte, Solid State Commun. 63:419 (1987)
46. Z. Schlesinger et al., Phys.Rev.B 40:6862 (1989)
47. C. Kittel, "Quantum Theory of Solids", John Wiley and Sons, New York, London, Sydney, Chapter 8 (1963)
48. A.A. Aligia, Phys.Rev.B 39:6700 (1989)
49. S. Salem-Sugui et al., Phys.Rev.B 43:5511 (1991)
50. J.B. Boyce et al., Phys.Rev.B 41:6306 (1990)
51. J. Yu, X.Y. Chen and P. Su, Phys.Rev.B 41:344 (1990)
52. T.M. Rice and L. Sneddon, Phys.Rev.Lett. 47:689 (1981)
53. B.K. Chakraverty and J. Ranninger, Philosophical Magazine B 52:669 (1985)
54. R. Micnas, J. Ranninger and S. Robaszkiewicz, Rev.Mod.Phys. 62:113 (1990)
55. S.M. Heald et al., Phys.Rev.B 40:8828 (1989)
56. S. Uchida et al., J.Phys.Soc.Jpn 54:4395 (1985)
57. S. Robaszkiewicz, R. Micnas and J. Ranninger, Phys.Rev.B 36:180 (1987)
58. B.R. Alascio, R. Allub, C.R. Proetto and C.I. Ventura, Solid State Commun. 77:949 (1991) and references therein.
59. J. Lorenzana and C.A. Balseiro, Phys.Rev.B 42:936 (1990)
60. G.A. Medina and M.D. Núñez Regueiro, Phys.Rev.B 42:8073 (1990)
61. B. Velický, S. Kirkpatrick and H. Ehrenreich, Phys.Rev. 175:747 (1968)
62. Y. Inada and C. Ishii, J.Phys.Soc.Jpn. 59:2124 (1990)
63. J.F. Federici et al., Phys.Rev.B 42:923 (1990)
64. C.M. Varma, Phys.Rev.Lett. 61:2713 (1988)
65. A. Balzarotti et al., Solid State Commun. 49:887 (1984)
66. T.J. Wagener et al., Phys.Rev.B 40:4532 (1989)
67. G.K. Wertheim, J.P. Remeika and D.N.E. Buchman, Phys.Rev.B 26:2120 (1982)
68. T. Tani, T. Itoh and S. Tanaka, J.Phys.Soc.Jpn. Suppl. A 49:309 (1980)
69. S. Sugai et al., Phys.Rev.Lett. 55:426 (1985)
70. M. Sera, S. Kondoh and M. Sato, Solid State Commun. 68:647 (1988)
71. H. Sato et al., Nature 338:241 (1989)
72. M. Pekala et al., Phys.Status Solidi B 155 K123 (1989)
73. C. Uher, S.D. Peacor and A.B. Kaiser, Phys.Rev.B 43:7955 (1991)
74. D. Nguyen Manh, D. Mayou and F. Cyrot-Lackmann, Eurcphys.Lett. 13:167 (1990)

THE ESSENTIAL SINGULARITY IN BCS SUPERCONDUCTIVITY THEORY AND THE GAP-TO-T_c RATIO

V.C. Aguilera-Navarro* and M. de Llano**

Physics Department, North Dakota State University
Fargo, North Dakota 58105 USA

ABSTRACT

The Cooper pair equation can be solved exactly for the free-electron density-of-states in 1 and 3 dimensions, just as in the original Cooper treatment which effectively reduces the problem to 2 dimensions. The essential singularity in coupling, often attributed to the two-dimensional character of the BCS model interaction which is nonzero only very near the Fermi surface, appears in *all three cases*. Weak-coupling BCS theory is shown to admit of a more general formula where T_c approaches zero somewhat faster than with the familiar BCS T_c-formula. This allows recent empirical values for both organic and ceramic superconductors of the gap-to-T_c ratio in excess of the universal BCS value of 3.53 not to be inconsistent with weak electron-boson coupling.

INTRODUCTION

The phenomenon of Cooper electron pairing[1] in a metal is the central ingredient in the Bardeen-Cooper-Schrieffer (BCS) theory[2] of super-conductivity which has been so successful in describing *low-temperature* superconductors. Although the BCS formalism---at least via the electron-phonon mechanism---has thus far not proven as satisfactory for

---

* On leave from Instituto de Física Teórica-UNESP, 01405 São Paulo, Brazil with a FAPESP grant.

** U.S. Army Research Office travel grant and NATO research grant are thankfully acknowledged.

the new high-temperature ceramic superconductors, experimental evidence steadily accumulates[3] in support of the presence of current-carrier pairing very similar, if not identical, to Cooper pairing.

Cooper pairing in D dimensions emerges from a constrained two-electron Schrödinger equation with a net attractive pair potential V(r) due to attractive electron-phonon effects dominating the electron-electron Coulomb repulsion. The constraint precludes two-electron wavefunction components corresponding to states occupied by the N-2 inert, background electrons. In the momentum representation one defines

$$
V_{k'k} \equiv L^{-D} \int d^D r \, e^{-i k' \cdot r} V(r) e^{i k \cdot r} \tag{1}
$$

The matrix element (1) in the celebrated BCS model interaction is taken to be

$$
V_{k',k} = \begin{cases} -V & \text{if } E_F < \epsilon_k, \ \epsilon_{k'} < E_F + \hbar \omega_D \\ 0 & \text{otherwise,} \end{cases} \tag{2}
$$

where $V > 0$ is a coupling strength, $\epsilon_k \equiv \hbar^2 k^2 / 2m$, $E_F \equiv \hbar^2 k_F^2 / 2m$ is the Fermi energy separating occupied from vacant states, and $\hbar \omega_D$ is the Debye energy, the maximum energy an electron can absorb from a lattice vibrational phonon in scattering to a vacant state. This leads to the Cooper pair equation for the eigenvalue E,

$$
1 = V \sum_k{}' (2\epsilon_k - E)^{-1} , \tag{3}
$$

with the prime on the summation sign signifying the constraints in (2). To solve (3) one can convert the sum to an integral over k, and then to an integral over energy $\mathcal{E}$ by the usual prescription

$$
\sum_k \longrightarrow \left(\frac{L}{2\pi}\right)^D \int d^D k \equiv \int d\mathcal{E} g(\mathcal{E}), \tag{4}
$$

where the density-of-states (DOS) $g(\mathcal{E})$ for $\mathcal{E} = \hbar^2 k^2 / 2m$ is[4]

$$
g(\mathcal{E}) = C_D \mathcal{E}^{\frac{D-2}{2}} \tag{5}
$$

with C_D constant. The rhs of (3) thus becomes

$$V\int_{E_F}^{E_F+\hbar\omega_D} d\mathcal{E}\, \frac{g(\mathcal{E})}{2\mathcal{E}-E} \quad \alpha \quad Vg(E_F) \int_{E_F}^{E_F+\hbar\omega_D} d\mathcal{E}\, \frac{d\mathcal{E}}{2\mathcal{E}-E} \cdot \tag{6}$$

The last member follows from the fact that in a metal $\hbar\omega_D/E_F \simeq 10^{-3} \ll 1$, which means that the BCS model interaction (2) is nonzero only in a spherical shell about the Fermi sphere which is *less than* 10^{-3} thick (in units of k_F). This would *apparently* convert the Cooper pair problem into a *two-dimensional* one, a surmise additionally supported by the fact that from (5) $g(\mathcal{E})$ is a constant independent of $\mathcal{E}$ for $D = 2$. The last member of (6) is then just a logarithm, which upon exponentiation allows one to solve for E, namely

$$E = 2E_F - \frac{2\hbar\omega_D}{e^{2/\lambda}-1} \equiv 2E_F - \Delta \tag{7}$$

where $\lambda \equiv g(E_F)V$, and the (positive) binding pair-energy Δ for weak coupling reduces to

$$\Delta \xrightarrow[\lambda\to 0]{} 2\hbar\omega_D e^{-2/\lambda} \cdot \tag{8}$$

This has an *essential singularity* about $\lambda = 0$ and *cannot* be expanded perturbatively as a power series in λ.

In the many-electron BCS theory[2] of superconductivity there appears a temperature-dependent "gap (order) parameter" $\Delta(T)$, which is $\neq 0$ for $T < T_c$ (superconducting phase) and is $= 0$ for $T > T_c$ (normal phase). The gap parameter $\Delta(T)$ satisfies[5] the implicit equation

$$1 = \lambda \int_0^{\hbar\omega_D} \frac{d\mathcal{E}}{\sqrt{\mathcal{E}^2 + \Delta^2(T)}} \tanh\frac{\sqrt{\mathcal{E}^2 + \Delta^2(T)}}{2k_B T} , \tag{9}$$

where $\mathcal{E} \equiv \epsilon_k - \mu$, with ϵ_k the single-electron spectrum and μ the chemical potential. For $T = 0$ (9) reduces to the exact integral

$$1 = \lambda \int_0^{\hbar\omega_D} \frac{d\mathcal{E}}{\sqrt{\mathcal{E}^2 + \Delta^2(0)}} = \lambda \sinh^{-1}\left[\frac{\hbar\omega_D}{\Delta(0)}\right] , \tag{10}$$

which upon inversion gives

$$\Delta(0) = \frac{\hbar\omega_D}{\sinh(1/\lambda)} \xrightarrow[\lambda\to 0]{} 2\hbar\omega_D e^{-1/\lambda} , \qquad (11)$$

and which is similar to, though exponentially larger than, the weak-coupling Cooper binding energy (8). Since $\Delta(T_c) = 0$, (9) also supplies us with an implicit equation for T_c, namely,

$$1/\lambda = \int_0^{\hbar\omega_D} \frac{d\mathcal{E}}{\mathcal{E}} \tanh \frac{\mathcal{E}}{2k_B T_c} \equiv \int_0^Z dx \frac{\tanh x}{x} \qquad (12)$$

where $Z \equiv \hbar\omega_D/2k_B T_c \equiv \Theta_D/2T_c$, with Θ_D the Debye temperature. Integrating (12) by parts leaves

$$1/\lambda = [\tanh x \, \ln x]_0^Z - \int_0^Z dx \, \text{sech}^2 x \, \ln x. \qquad (13)$$

For elemental superconductors $T_c \lesssim 10K$ while $\Theta_D \approx 300$ K, empirically, so that $Z \equiv \Theta_D/2T_c \gg 1$ is a moderately justified assumption. In this case (13) simplifies to

$$1/\lambda \xrightarrow[Z\gg 1]{} \ln Z - \int_0^\infty dx \, \text{sech}^2 x \, \ln x \; = \; \ln(\Theta_D/2T_c) - (\ln \tfrac{\pi}{4} - \gamma), \qquad (14)$$

where $\gamma \approx 0.5772$ is the Euler constant. Exponentiating (14) and solving for T_c gives the famous BCS T_c-formula

$$T_c = \frac{2e^\gamma}{\pi} \Theta_D e^{-1/\lambda} \approx 1.13 \, \Theta_D e^{-1/\lambda}, \qquad (15)$$

which is valid for weak-coupling and *vanishingly small* T_c/Θ_D. In the next section we lift this latter restriction to arrive at a more general (high T_c) BCS T_c-formula. The form of (15) is qualitatively correct in that it explains: a) the existence of immeasurably small T_c's, (which would be associated with very small values of λ), and b) why good superconductors are generally *bad conductors*, since larger λ also means larger Joule (resistivity) effects. On the other hand, crudely speaking, the electron-phonon coupling parameter λ cannot[6] surpass a value of about 1/2 without the ionic lattice itself becoming unstable, i.e., melting. This means that $e^{-1/\lambda} < 0.135$. Taking $\Theta_D \approx 300$ K we see that T_c according to (15) *cannot* predict critical (transition) temperatures *greater than about 40 K*, a limit which has been called the "phonon barrier" of standard superconductivity theory. This is well below the highest reproducible value $T_c = 125$ K found in the TlBaCaCuO ceramic compounds since early 1988, and in turn well below the ideal 300 K of room temperatures.

If high-temperature superconductivity is to be understood---and perhaps, more significantly, if room-temperature T_c's are ever to be achieved---it is necessary to better understand the *origin* of the peculiar factors $e^{-2/\lambda}$ in (8) and $e^{-1/\lambda}$ in (11) at best, and/or the origin of the *nature* of the underlying dynamical mechanisms at worst. A very similar (essential) singularity emerges in the ordinary binding of a *single* particle in a quantum well in 2D to which we now turn.

QUANTUM BINDING IN 1, 2, AND 3 DIMENSIONS
It is well-known that in 3D a critical threshold value of an attractive well depth and/or range are needed to support a bound state at all, whereas a 1D or 2D attractive well, no matter how shallow and/or short-ranged, will *always* have a bound state. Furthermore, by expanding the familiar tangent implicit equation for the ground-state energy E of a 1D rectangular well of depth V_0 and range a, one finds that

$$E \xrightarrow[V_0 a^2 \to 0]{} -2ma^2 V_0^2/\hbar^2 + O(V_0^3). \qquad (1D) \qquad (16)$$

Similarly, since in 3D the reduced radial wavefunction differential equation is identical with the 1D one for the *odd* solutions, in 3D one has a cotangent implicit equation for a spherical well of depth V_0 and radius a which on expansion about small $V_0 a^2 - \hbar^2\pi^2/8m \equiv \eta > 0$, gives

$$E \xrightarrow[\eta \to 0]{} -m\eta^2/2\hbar^2\pi^2 + O(\eta^3). \qquad (3D) \qquad (17)$$

Both (16) and (17) are *perturbative* expressions in the appropriate "smallness" parameter. The 2D case, on the other hand, is entirely different in this regard since it gives[7]

$$E \xrightarrow[v_a^2 \to 0]{} - \frac{\hbar^2}{2ma^2} e^{-2\hbar^2/mv_o^2 a^2}, \qquad (2D) \qquad (18)$$

i.e., with the same singularity structure as (8) for the Cooper pair binding energy or of (11) for the BCS gap energy parameter.

Given the result (18) and the "2D-ness" of the Cooper pair problem with the BCS model interaction (which is non-zero only very near the Fermi surface), it is tempting to attribute the presence of the singular factor $e^{-2/\lambda}$ in Cooper's result (8)---and subsequently in the BCS T_c-formula (15)---to a dimensionality effect. This idea was recently traced by Rau[8] to Weisskopf[9], who discussed it in a colloquium[10]. It has since found its way into the textbook literature[11]. We next discuss three counterexamples[12] whereby this is seen *not* to be the case.

COOPER PAIRING IN 1D AND 3D: EXACT TREATMENT

By exact here we mean *not* assuming constancy for the DOS $g(\mathscr{E})$ in the first integral of (6). We analyze first the 1D case, with the BCS model interaction (2). Using (3) to (6) for $D = 1$ we have

$$\frac{1}{C_1 V} = \int_{E_F}^{E_F + \hbar\omega_D} \frac{d\mathscr{E}}{\mathscr{E}^{1/2}(2\mathscr{E} - E)} . \qquad (19)$$

Introducing dimensionless quantities, let $\lambda \equiv g(E_F)V$ again, $\epsilon \equiv E/2E_F \equiv 1 - \Delta/2E_F \equiv 1 - \delta$ and $\nu \equiv \hbar\omega_D/2E_F$. As in (7), the quantity Δ stands for the (positive) binding energy of the pair. The integral (19) is tabulated[13], and gives

$$\frac{2}{\lambda} = \frac{1}{\sqrt{\epsilon}} \ln \left\{ \left[\frac{\sqrt{1 + 2\nu} - \sqrt{\epsilon}}{\sqrt{1 + 2\nu} + \sqrt{\epsilon}} \right] \left[\frac{1 + \sqrt{\epsilon}}{1 - \sqrt{\epsilon}} \right] \right\} . \qquad (20)$$

The crucial point in both 1D and 3D is now this: *the essential singularity* $e^{-2/\lambda}$ *arises* **before** *the "2D-ness" limit of* $\nu \equiv \hbar\omega_D/2E_F \ll 1$ *is taken.* To see this in 1D, note that for λ small the rhs of (20) diverges for $\epsilon \equiv 1 - \delta \longrightarrow 1^-$, or equivalently $\delta \ll 1$. Writing $\epsilon \equiv 1 - \delta$

in (20) and expanding for $\delta \ll 1$, exponentiating, and solving for δ we get

$$\delta \xrightarrow[\lambda \to 0]{} 4\left[\frac{\sqrt{1 + 2\nu} - 1}{\sqrt{1 + 2\nu} + 1}\right] e^{-2/\lambda} \tag{21}$$

in accordance with the stated assertion. If we *further* assume that $\nu \equiv \hbar\omega_D/2E_F \ll 1$ and expand (21) about this parameter value, one finally obtains

$$\delta \xrightarrow[\substack{\lambda \to 0 \\ \nu \to 0}]{} 2\nu e^{-2/\lambda}, \tag{22}$$

which is *precisely* of the Cooper form (8), except that $g(E_F)$ now refers to the 1D DOS.

Finally, going back to 3D and to the BCS model interaction (2), the first integral in (6) using (5) for $D = 3$ can in fact also be performed *exactly*,[13] and gives

$$\frac{2}{\lambda} = 2\left[\sqrt{1 + 2\nu} - 1\right] + \sqrt{\epsilon}\ln\left\{\left[\frac{\epsilon + 1 + 2\nu - 2\sqrt{\epsilon(1 + 2\nu)}}{\epsilon - 1 - 2\nu}\right]\left[\frac{\epsilon - 1}{\epsilon + 1 - 2\sqrt{\epsilon}}\right]\right\}. \tag{23}$$

Again, for $\lambda \ll 1$, $\epsilon \equiv 1 - \delta \to 1^-$, or $\delta \ll 1$ so that expanding and exponentiating one gets

$$\delta \xrightarrow[\lambda \to 0]{} 8\left[\frac{1 + \nu - \sqrt{1 + 2\nu}}{2\nu}\right] e^{-2/\lambda} \xrightarrow[\nu \to 0]{} 2\nu e^{-2/\lambda}, \tag{24}$$

the last term being of the Cooper form (8) and where the singularity $e^{-2/\lambda}$ emerged *before* the 2D-ness limit $\nu \to 0$ is taken.

ATTRACTIVE DELTA-POTENTIAL COOPER PAIRS

The essential singularity (24) also appears in the one-dimensional Cooper problem for an attractive delta function interaction $-v_0\delta(x_1 - x_2)$ between spin-1/2 fermions, and sheds light on the large and small observed coherence lengths (pair sizes) in low- and high-temperature, respectively, superconductivity.[14] In this case, instead of the integral in (6) one has for the eigenvalue E the equation

$$1 = \frac{V_o}{L} C_1 \int_{E_F}^{\infty} d\mathcal{E} \; \frac{\mathcal{E}^{-1/2}}{2\mathcal{E} - E} , \tag{25}$$

where $g(\mathcal{E}) = C_1 \mathcal{E}^{-1/2}$ is the 1D DOS, with $C_1 \equiv \sqrt{m/2}\; L/\pi\hbar$. Introducing the dimensionless variable $\epsilon \equiv E/2E_F$, the integral in (25) gives

$$\frac{1}{2\sqrt{E_F}} \; \frac{1}{\sqrt{\epsilon}} \; \ln\left[\frac{1 + \sqrt{\epsilon}}{1 - \sqrt{\epsilon}}\right] \qquad \text{if} \quad \epsilon > 0, \tag{26}$$

$$\frac{1}{\sqrt{E_F}} \; \frac{1}{\sqrt{\epsilon}} \; \left[\frac{\pi}{2} - \tan^{-1}\frac{1}{\sqrt{\epsilon}}\right] \qquad \text{if} \quad \epsilon < 0. \tag{27}$$

Note that (26) is just (20) for $v \equiv \hbar\omega_D/2E_F \to \infty$, in which case the integral in (19) becomes that in (25). The Fermi energy $E_F \equiv \hbar^2 k_F^2/2m$ is simply related to the number density $\rho \equiv N/L = 2k_F/\pi$. Defining the dimensionless ratio $\tau \equiv \hbar^2\rho/mv_o$, the prefactor in (25) becomes $2\sqrt{E_F}/\tau\pi^2$ so that (25) to (27) lead to the two implicit equations for ϵ given by

$$\pi^2\tau\sqrt{\epsilon} = \ln\left[\frac{1 + \sqrt{\epsilon}}{1 - \sqrt{\epsilon}}\right] \qquad \text{if} \quad \epsilon > 0 \tag{28}$$

$$\pi^2\tau\sqrt{|\epsilon|} = \pi - 2 \tan^{-1}\frac{1}{\sqrt{\epsilon}} \qquad \text{if} \quad \epsilon < 0. \tag{29}$$

Graphically it is easy to see[14] that continuous nontrivial solutions, $\epsilon(\tau) \neq 0$, exist when $\tau > 2/\pi^2$ with $\epsilon > 0$, and when $0 \leq \tau < 2/\pi^2$ with $\epsilon < 0$. Consider the two extremes given by $\tau \to 0$ and $\tau \to \infty$. The first extreme $\tau \equiv \hbar^2\rho/mv_o \to 0$ is equivalent to finite v_o and ρ (or equivalently E_F) $\to 0$; it should correspond to treating the two fermions *in a vacuum*. In this case (29) becomes, since $|\epsilon| \equiv E/2E_F \to \infty$,

$$\pi\tau\sqrt{|\epsilon|} \simeq 1 \qquad (\tau \to 0), \tag{30}$$

which on squaring leaves $E \simeq -2m^2v_o^2 E_F/\pi^2\hbar^4\rho^2 = -mv_o^2/4\hbar^2$. This is the well-known elementary 1D quantum-mechanical result for the sole ground-state energy of two particles of mass m (or *one* of reduced mass m/2) interacting via an attractive delta-function potential of strength v_o. On the other hand for $\tau \to \infty$ it is again convenient to use the definition $\delta \equiv \Delta/2E_F$, where $\Delta \equiv 2E_F - E$ is the (positive) binding energy of the weakly-bound Cooper pair. Then (28) can be expanded for $1 - \epsilon \equiv \delta \to 0^+$ and gives $\pi^2 \tau \simeq -\ln(\delta/4)$, or (exponentiating) we again see the essential singularity in v_0

$$2E_F - E \equiv \Delta \simeq 8E_F e^{-\pi^2 \hbar^2 \rho/m v_0} \qquad (\tau \to \infty). \qquad (31)$$

The two extremes $\tau = 0$ and $\tau = \infty$ just discussed for the 1D Cooper *two*-fermion problem with attractive delta-function potential are labeled "strong" and "weak" coupling in the bottom panel of Figure 1 (where the symbol v stands for the number of distinct fermion species). This panel schematically displays the ground-state, zero-temperature energy-per-particle (thick curve) of the exactly-soluble[15] *many-fermion* assembly interacting via attractive pair delta-function potentials. The dashed curve is the ideal Fermi gas (IFG) result, which in 1D is quadratic in ρ, as opposed to the $\rho^{2/3}$ behavior of 3D depicted in the top panel for a typical 3D fermion liquid such as liquid ^{3}He or nuclear matter, where weak (strong) coupling accompanies low (high) density. Precisely the *opposite* correlation rules in the 1D many-fermion model shown in the bottom panel. This renders it useful to study the clustering of weakly-interacting fermions (quarks, nucleons or electrons) into mesons[16], baryons[17], alphas[18]---and now "point Cooper pairs" or, more appropriately, "bipolaron" dimers. These dimers can be viewed as a limit for high-temperature superconductive Cooper pairs, in a regime where weak coupling but low carrier densities prevail.

GENERALIZED BCS T_c-FORMULA

It is straightforward[19] to lift the restriction $Z \equiv \Theta_D/2T_c \gg 1$ imposed on (13) in order to arrive at the BCS T_c-formula (15). For non-infinite Z (13) can still be exponentiated, and after some rearrangement leads to the somewhat more general BCS T_c-formula

$$T_c = C(\Theta_D/2T_c)\ \Theta_D\ \exp\left[-\frac{1}{\lambda}\coth\ (\Theta_D/2T_c)\right], \qquad (32)$$

which is really an implicit equation in T_c instead of the explicit one (15). The dimensionless coefficient $C(\Theta_D/2T_c)$ is defined as

$$C(Z) \equiv \frac{1}{2}\ \exp\left[-\coth Z \int_0^Z dx\ \mathrm{sech}^2 x\ \ln x\right], \qquad (33)$$

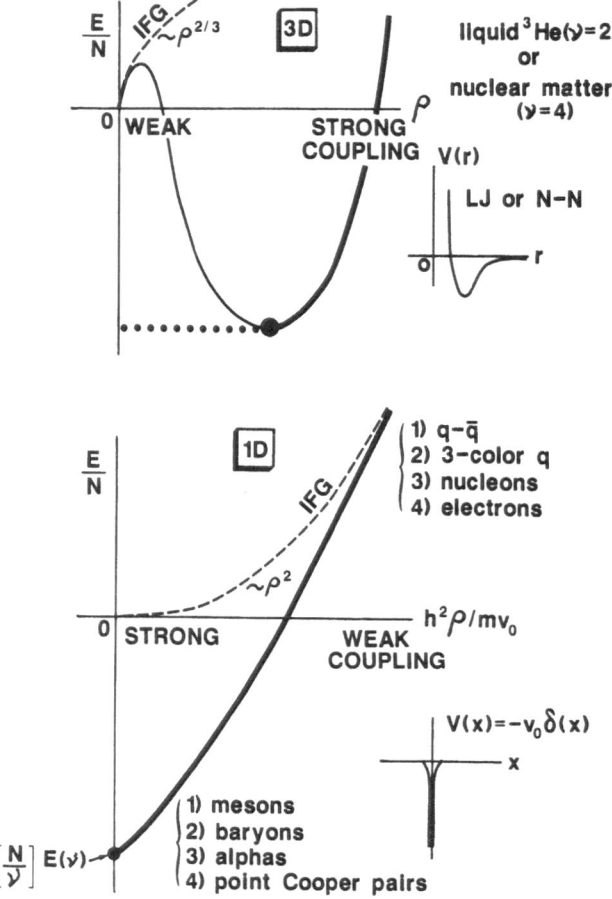

Figure 1. Comparison of ground-state energy-per-particle *vs* density curves for a typical 3D Fermi liquid (like [3]He with 2 species or nuclear matter with 4) [top panel] and for the 1D many-fermion fluid interacting pairwise via attractive delta-function potentials of strength v_o [bottom panel]. IFG refers to the ideal Fermi gas.

and can be computed numerically.   But first let us check the limit $Z \rightarrow \infty$, namely

$$C(Z) \xrightarrow[Z \rightarrow \infty]{} \frac{1}{2} \exp\left[-\int_0^\infty dx \; \mathrm{sech}^2 \; x \; \ln x \right] = \frac{2e^\gamma}{\pi} \propto 1.13, \qquad (34)$$

where the value of the integral in (14) was used.   Figure 2 displays the behavior of $C(Z)$ for $Z < \infty$; in the figure the function $\coth Z$ is also given, as a frame of reference.   Thus (32) reduces to (15) when $Z \rightarrow \infty$, as it should.   However, (32) is more accurate, and vanishes *faster*, than (15) as $\lambda \rightarrow 0$.   This fact has crucial consequences for the BCS universal

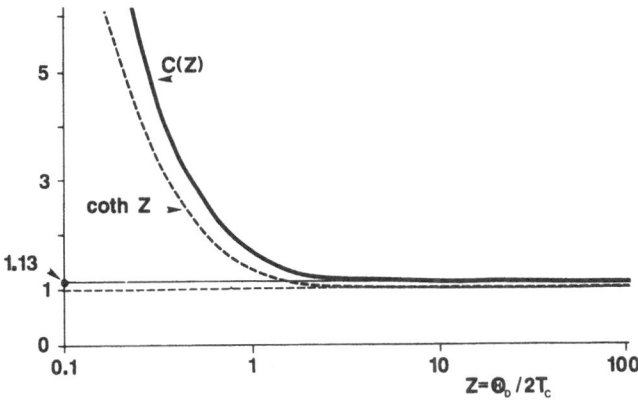

Figure 2.   Illustration of how rapidly the two functions $C(Z)$, eq. (35), and $\coth Z$, approach their asymptotic values of 1.13 and 1, respectively, as $Z \equiv \theta_D/2T_c$ increases.   Note that the abscissa is on a log scale.

constants, beginning with the gap-to-T_c ratio.   This has physical implications related to the presence of weak or strong electron-boson coupling that are currently being surmised in the literature from discrepancies between empirical results (for both ceramic and organic superconductors) and the BCS universal constants.

BCS UNIVERSAL CONSTANTS

Using the assumption $Z \equiv \theta_D/2T_c \gg 1$ which leads to the BCS T_c-formula for weak coupling (15), and the expression (11) for the gap energy $\Delta(0)$, also valid for weak coupling, one can write the familiar gap-to-T_c ratio

$$\frac{2\Delta(0)}{k_B T_c} = \frac{4\hbar\omega_D e^{-1/\lambda}}{(2e^\gamma/\pi)k_B\theta_D e^{-1/\lambda}} = \frac{2\pi}{e^\gamma} \approx 3.53 , \tag{35}$$

which is a universal (dimensionless) constant since the λ-dependence cancels out *exactly*. Similarly, if $C_n(T)$ and $C_s(T)$ are the normal and superconducting specific heats, and $H_c(T)$ the critical magnetic field, one can deduce[5,20] at least *four* other BCS universal constants, namely:

$$\frac{T_c C_n(T_c)}{H_c^2(0)} = \frac{e^{2\gamma}}{6\pi} \approx 0.168 \tag{36}$$

$$\left[\frac{C_s(T) - C_n(T)}{C_n(T)}\right]_{T_c} = \frac{12}{7\zeta(3)} \approx 1.43 \tag{37}$$

$$\left[\frac{T}{C_n(T)}\frac{d}{dT}\left\{C_s(T) - C_n(T)\right\}\right]_{T_c} \approx 3.77 \tag{38}$$

$$\min_{0\leq T/T_c \leq 1}\left\{\frac{H_c(T)}{H_c(0)} - \left[1 - \left(T/T_c\right)^2\right]\right\} \approx -0.037. \tag{39}$$

Here $\zeta(3) \approx 1.202$ is the Riemann zeta function.

Using the last member of (11) for $\Delta(0)$ (weak coupling) and the more general T_c formula (32) we now have

$$\frac{2\Delta(0)}{k_B T_c} = \frac{4}{C(\theta_D/2T_c)} \exp\left[\frac{1}{\lambda}\left[\coth\left(\theta_D/2T_c\right) - 1\right]\right] , \tag{40}$$

with a λ-dependence surviving for *all* values $\theta_D/2T_c < \infty$. The gap-to-T_c ratio, even for weak coupling, is *not* a universal constant. [A non-universal gap-to-T_c ratio, and indeed also for all quantities (36) to (39), also emerges from *strong*-coupling Eliashberg theory[20] where the coupling constant is eliminated in favor of a smallness parameter

298

analogous to our Z^{-1}.] Table 1 lists T_c, θ_D and $Z \equiv \theta_D/2T_c$ for several low- and high-temperature superconductors, with RTSC referring to an imaginary substance with $\theta_D = T_c = 30K$. For the cuprates, the lowest reported θ_D is given. Figure 3 displays $2\Delta(0)/k_B T_c$ according to (40) as a function of $1/\lambda$ for the materials of Table 1. Note that a small value of λ (less than 1/2) can always be found that will give a gap-to-T_c ratio *larger* than the BCS weak-coupling *and* zero-T_c value of 3.53. (The value $\lambda = 1/2$ has been marked in the figure as a vertical dashed line because simple arguments[6] give electron-phonon-induced lattice instability for $\lambda > 1/2$.) Open circles mark intersections with the value of 8 for the gap-to-T_c ratio reported[24] by various experimental groups using different techniques to measure $2\Delta(0)$, the energy gap in the excitation spectrum.

Table 1. Critical temperatures T_c, Debye temperatures θ_D, both in kelvins [K], and the ratio $Z \equiv \theta_D/2T_c$, for several super-conductors. "RTSC" refers to an imaginary substance. For the cuprate compounds, the smallest reported θ_D value is listed.

superconductor	T_c [K]	θ_D [K]	Z
Al	1.17	385	164.5
Pb	7.2	85	5.9
YBaCuO	90	$\geq 288^{21}$	1.6
TlCaBaCuO	125	$\geq 363^{22}$	1.45
BiSrCaCuO	100	$\geq 230^{23}$	1.15
"RTSC"	300	300	0.5

Thus, empirical values of $2\Delta(0)/k_B T_c$ in excess of the weak-coupling (*and* zero T_c/θ_D) BCS value of 3.53 *do not imply strong-coupling* of a single (or joint) boson-exchange interaction in metallic, ceramic or organic superconductors. This conclusion is consistent with various recent experiments[25] suggesting the absence of strong-coupling in the cuprate superconductors. Equations (36) to (39), like (35), are strictly valid only if $T_c/\theta_D = 0$; their generalization to higher T_c values is in progress and will be reported later.

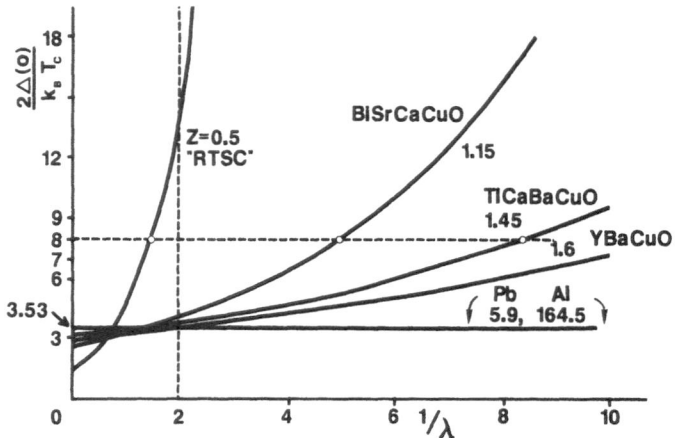

Figure 3. Variation with inverse coupling $1/\lambda$ of the gap-to-T_c ratio $2\Delta(0)/k_B T_c$ as given by eq. (40), for several typical-superconductor $Z \equiv \theta_D/2T_c$ values taken from Table 1. The case "RTSC" refers to an imaginary substance. Open circles mark intersections at a ratio value of 8.

CONCLUSIONS

The essential singularity $e^{-2/\lambda}$ in coupling λ appearing in the weak-coupling limit of the Cooper pair binding energy---and consequently in the BCS formula for T_c---is also familiar from 2D quantum binding where λ is proportional to the well depth times its range squared. Since the BCS model interaction reduces a 3D pairing problem to a 2D one, Cooper pairing and 2D quantum binding would *appear* to be related. But this is seen *not* to be the case through exact solution of the Cooper pair problem in both 1D and 3D (for free-electron density-of-states) where the essential singularity persists. A further counterexample is provided by the 1D fermion fluid of spin-1/2 particles interacting via attractive delta-function potentials, which illustrates the smooth crossover from "large" (low-T_c) to "small" (high-T_c, "bipolaronic") Cooper pairs.

A T_c formula more general than the well-known BCS T_c formula for weak coupling *and* zero T_c values allows one to conclude that empirical values of the gap-to-T_c ratio in excess of the BCS value of 3.53 are *not* inconsistent with weak electron-boson coupling.

300

REFERENCES

1. L.N. Cooper, Phys. Rev. **104**, 1189 (1956).

2. J. Bardeen, L.N. Cooper and J.R. Schrieffer, Phys. Rev. **108**, 1175 (1957).

3. W.A. Little, Science **24**, 1390 (1988) and references therein.

4. G. Burns, *Solid State Physics* (Academic, N.Y., 1985) p. 238.

5. A.L. Fetter and J.D. Walecka, *Quantum Theory of Many-Particle Systems* (McGraw-Hill, N.Y., 1971) p. 447.

6. J.M. Blatt, *Theory of Superconductivity* (Academic, N.Y., 1964) p. 206.

7. L.D. Landau and E.M. Lifshitz, *Quantum Mechanics* (Pergamon, N.Y., 1977) p. 163.

8. A.R.P. Rau, Am. J. Phys. **58**, 904 (1990).

9. V.F. Weisskopf, CERN Rept. # 79-12 (1979).

10. A.R.P. Rau, priv. comm.

11. D.L. Goodstein, *States of Matter* (Dover, N.Y., 1985) p. 391.

12. C. Esebbag, J.M. Getino, M. de Llano, S.A. Moszkowski, U. Oseguera, A. Plastino and H. Rubio, *The essential singularity in Cooper pairing in 1-, 2-, and 3-D* (to be published).

13. I.S. Gradshteyn and I.M. Ryzhik, *Table of Integrals, Series and Products* (Academic, N.Y., 1980) p. 71.

14. M. Casas, C. Esebbag, A. Extremera, J. M. Getino, M. de Llano, A. Plastino, and H. Rubio, Phys. Rev. **A** (in press).

15. J.B. McGuire, J. Math. Phys. **5**, 622 (1964); M. Gaudin, Phys. Lett. **A 24**, 55 (1967); C.N. Yang, Phys. Rev. **168**, 1920 (1968).

16. S. Tosa, Phys. Rev. **C 34**, 2302 (1986).

17. D.S. Koltun, Phys. Rev. **C 36**, 2047 (1987).

18. T. Kebukawa, Phys. Rev. **C 35**, 794 (1986).

19. V.C. Aguilera-Navarro and M. de Llano, *High-T_c BCS Gap-to-T_c Ratio* (to be published).

20. J.P. Carbotte, Rev. Mod. Phys. **62**, 1027 (1990).

21. A. Junod, in *Physical Properties of High-Temperature Superconductors*, vol. II, ed. by D.M. Ginsburg (World Scientific, Singapore, 1990).

22. A.D. Kulkarni, F.W. de Wette, J. Prade, U. Schroeder and W. Kress, Phys. Rev. **B 43**, 5451 (1991).

23. R. Micnas, J. Ranninger and S. Robaszkiewicz, Rev. Mod. Phys. **62**, 113 (1990).

24. B. Goss Levi, Phys. Today, March 1990, p. 20.

25. Z. Schlesinger *et al*, Nature **343**, 242 (1990); S.D. Brorson *et al*, Sol. State Comm. **74**, 1305 (199); C. Thomsen *et al*, Sol. State Comm. **75**, 219 (1990).

SEARCH OF SUPERCONDUCTIVITY IN METAL CLUSTERS

M. Barranco[†], E. S. Hernández[§], R. J. Lombard[‡] and Ll. Serra[¶]

[†] Departament ECM, Facultat de Física, Universitat de Barcelona
E-08028 Barcelona, Spain
[§] Departamento de Física, Facultad de Ciencias Exactas y Naturales, Universidad de Buenos Aires, 1428 Buenos Aires, Argentina
[‡] Division de Physique Théorique, Institut de Physique Nucléaire
F-91406 Orsay Cedex, France[1]
[¶] Departament de Física, Universitat de les Illes Balears, E-07071 Palma de Mallorca, Spain

ABSTRACT

We have investigated the possibility of interpreting the even-odd alternations in the mass spectra and ionization potentials of metal clusters as evidence of pairing correlations among the electrons. In particular, we have applied a spherical BCS model to some Al clusters.

1. INTRODUCTION

In metal clusters, the disappearance of some sub-shell effects as well as the presence of even-odd alternations in the mass spectra and ionization potentials (IP) [1-6] qualitatively resemble the mass and binding energy systematics in nuclei [7]. In the case of Al_N, a clear even-odd difference shows up in the IP for small clusters, $N \leq 30$ [1]. The same happens for Ag clusters [2], for Na particles in several N-regions [3], and also for some small K clusters [5, 6].

These experimental features cannot be explained by the spherical jellium model. According to several prior investigations [8, 9, 10], the deformation of the one-body mean-field seems to account for most of the experimental findings. However, it is

---

[1]Unité de Recherche des Universités Paris 11 et Paris 6 Associée au CNRS

worthwhile noticing that in nuclei, the nucleon separation energies and odd-even mass differences demand the consideration of additional correlations in angular momentum space in order to appropriately interpret the data, which are better described in terms of the competition between the single-particle (sp) energy gap in a mean field approximation, and the pairing correlations that destabilize the particle-hole vacuum. If the pairing field is strong enough, the ground state (gs) of the nuclear droplets becomes stable again due to the formation of a non-negligible amount of Cooper pairs [7,11-13] and in medium and heavy nuclei, say $A \geq 150$, both static quadrupole deformations and pairing correlations coexist [7, 13, 14]. While the even-odd differences in the nucleon separation energy are a valuable aid to fix the strength of the pairing interaction, other data, i.e. , the energy gap in the spectrum of intrinsic excitations, moments of inertia of deformed nuclei, two-particle intensities in quasi-elastic nucleus-nucleus collisions and pairing vibrations [7] provide complementary constraints. The paired nucleon system is thus well described by the BCS theory [12] adapted to finite Fermi systems[7, 13, 23].

Since the Cooper instability [11, 12] seems to be a basic structural characteristic of both macroscopic Fermi systems and nuclear droplets, it is very tempting to explore the possibility of a pairing origin for the regularities in the mass spectra and IP's of metallic clusters –recalling that monovalent metals (e.g. Na, Ag) do not exhibit superconductivity [15]–. Some experimental evidence of the coexistence of superconducting and normal Al_N clusters has been reported [16] together with the suggestion that the critical temperature for small particles should be larger than for the bulk. Consequently, as first step towards a more fundamental approach, we select the spherical BCS model commonly used in nuclear physics [7, 13] to estimate the size of the superconducting gap and the critical temperature induced by electron-electron pairing correlations in Al clusters, as functions of the force strength.

In section 2 we briefly review the BCS formalism. The results obtained for several Al_N clusters are presented in section 3. Finally, an outlook is given in section 4.

2. THE SPHERICAL BCS FORMALISM

We shall recall here the BCS equations essential to the purpose discussed in the introduction [13], considering the simplest case, namely that of a constant pairing force of strength G acting on a few sp levels around the Fermi surface. With the above hypothesis, the hamiltonian of paired electrons reads,

$$H = \sum_{k>0} \varepsilon_k(a_k^+ a_k + a_{-k}^+ a_{-k}) - G \sum_{k,k'>0} a_k^+ a_{-k}^+ a_{k'} a_{-k'} , \qquad (1)$$

$(G > 0)$, with a_k^+ denoting the creation operator of a single electron in the state $\mid k >$ characterized by quantum numbers (n, ℓ, j, m) and ε_k is the single-electron energy. The state $\mid -k >$ is the time-reversed of $\mid k >$. The formulae will be written in a jj-coupling scheme, although if as usual, one neglects the electron spin-orbit interaction, the states with $j = \ell + 1/2$ and $j = \ell - 1/2$ are degenerate. In this scheme we have

$| -k >=| n, \ell, j, -m >$. Furthermore, both states $| k >$ and $| -k >$ are supposed to be simultaneously occupied or empty and, as can be inferred from (1), to have the same Kohn-Sham (KS) sp energy.

Particle number conservation in enforced on the average by resort to a Lagrange multiplier λ representing the chemical potential. The new hamiltonian $H' = H - \lambda N$ is approximately diagonalized by the Bogoliubov-Valatin transformation,

$$\left. \begin{array}{l} \alpha_k^+ = u_k a_k^+ - v_k a_{-k} \\ \alpha_{-k}^+ = u_k a_{-k}^+ + v_k a_k \end{array} \right\} \quad k > 0 , \tag{2}$$

where α_k^+ is the qp creation operator, and the u_k and v_k satisfy $u_k^2 + v_k^2 = 1$.

The minimum energy in the spectrum of elementary excitations of the paired electron system is roughly twice the energy gap Δ given by [13],

$$\Delta = G \sum_{k>0} u_k v_k . \tag{3}$$

The expectation value of the electron number operator in the qp vacuum $| O >_\alpha$ is

$$N_e = 2 \sum_{k>0} v_k^2 , \tag{4}$$

and its variance,

$$(\Delta N_e)^2 = 4 \sum_{k>0} u_k^2 v_k^2 . \tag{5}$$

The occupation probability v_k^2 of pairs in orbital $| k >$ is

$$v_k^2 = \frac{1}{2} \left(1 - \frac{\varepsilon_k - \lambda}{E_k} \right) , \tag{6}$$

with the qp energy

$$E_k = \sqrt{(\varepsilon_k - \lambda)^2 + \Delta^2} . \tag{7}$$

Equation (6) represents a Fermi-like distribution with spreading proportional to the gap Δ around the chemical potential λ. This broadening of the Fermi surface of the correlated system, relative to the chemical potential itself, measures the ratio of paired electrons with respect to the total number N_e. The BCS model holds in the weak-coupling limit, thus ensuring a low proportion of Cooper pairs; in turn, the coherence length (actually the quantum size of a correlated pair),

$$\xi = \frac{\hbar v_F}{\Delta}, \tag{8}$$

with the velocity v_F associated to the last occupied orbit in the mean field, should be not smaller than the "diameter of the system. In infinite metals, the coherence length corresponding to the zero temperature gap is of the order of 10^{-4} cm, a number to be kept in mind for future considerations. Finally, the chemical potential and gap are given by the equations,

$$
\begin{cases}
N_e = \sum_{k>0} \left(1 - \dfrac{\varepsilon_k - \lambda}{E_k} \right), \\
\dfrac{2}{G} = \sum_{k>0} \dfrac{1}{\sqrt{(\varepsilon_k - \lambda)^2 + \Delta^2}} = \sum_{k>0} \dfrac{1}{E_k}.
\end{cases}
\tag{9}
$$

Equations (9) always possess a trivial solution $u_k v_k = 0$ for all k, which corresponds to the KS case. Whenever a non trivial solution of these equations exists, we have a superconducting gs associated to the qp vacuum $\mid 0 >_\alpha$ whose energy is given by

$$
E_{gs} = 2 \sum_{k>0} \varepsilon_k v_k^2 - \frac{\Delta^2}{G}.
\tag{10}
$$

The last term is the pairing energy E_p. It does not correspond exactly to the energy gain when the pairing interaction is turned on: some energy is lost in the mean field, because the v_k^2 are no longer zero or one. Equation (10) corresponds to the extreme independent particle model, since the model Hamiltonian (1) does not explicitly contain the interaction energy $-U/2$ stored in the Fermi sea. Should one take this contribution into account from the start, the gs energy acquires the form,

$$
E_{gs} = E_{KS} + 2 \left\{ \sum_{k>F} \varepsilon_k v_k^2 - \sum_{0<k\leq F} \varepsilon_k u_k^2 \right\} - \frac{\Delta^2}{G},
\tag{11}
$$

where F denotes the unperturbed Fermi level and the KS energy is

$$
E_{KS} = 2 \sum_{0<k\leq F} \varepsilon_k - \frac{U}{2}.
\tag{12}
$$

It should be mentioned that the only solution to the BCS equations is the trivial one if the strength G is smaller than a critical value G_c. On the other hand, a very intense pairing field may yield a gap Δ comparable to the chemical potential $\mid \lambda \mid$ itself, thus the whole BCS picture becomes meaningless. A predictive calculation must rely on the formation of a relatively small amount of Cooper pairs, a condition that may pose an upper limit to the permitted intensity of the pairing strength.

In order to size the magnitude of thermal effects, we have used a crude model in which the most important perturbation is experienced by the valence electrons rather than by the ions. In this case, the only modification to the zero temperature (T) BCS formalism involves the gap equation, which acquires the form [17, 18]:

$$
\frac{2}{G} = \sum_{k>0} \frac{1}{E_k} \tanh(\beta E_k / 2),
\tag{13}
$$

where we have introduced $\beta = 1/T$. In this context, for a given pairing strength G, eq. (13) possesses a non-vanishing solution $\Delta(T)$ only if the temperature is lower than a critical one, $T_c(G)$ [21]. When T goes to zero, eq. (13) reduces to the second one in (9). All the other zero temperature formulae keep their meaning at finite T.

3. RESULTS

As in light nuclei, the constant pairing force approximation is expected to be un-appropriate for small clusters, for which departure from spherical symmetry should be more relevant than any pairing effect. This is substanciated by the deformed calcula-tions of refs. [8-10].The situation could change if one takes into account deformation and pairing on the same footing, adopting for example a Nilsson-plus-BCS reference scheme. On the other hand, some experimental indications have been reported [16] that for Al_N clusters with N larger than 100, the critical temperature for the normal-to-superfluid phase transition may decrease as a function of cluster size. We will show in the present section that this experimental trend can be justified in the frame of a simple one-shell, spherical BCS model. For this sake, it is sufficient to work out the Al_N clusters either neutral or ionized with $N \sim 30$ atoms, since the general behaviour for higher masses can be straightforwardly anticipated from these results.

In practice, given a cluster, we have first solved the KS equations, determining the sp energies ε_k. For partially filled orbitals (n, ℓ), the standard spherical average over the $(2\ell + 1)$ m-substates [19] has been made. We have used the same local density functional as in ref. [20]. We have chosen the sp levels entering the gap and chemical potential equations as those within a narrow band of width 2δ chosen as the energy difference between the first empty KS orbit and the last completely occupied one below the Fermi level.

In the absence of the experimental information needed to fix the pairing constant in an unambiguous way, we have investigated the highest possible values of G compatible with the experimentally well established major shell closures. It means that for a magic electron number N_e, we have determined the critical $G(N_e)$ as the biggest value of G for which the solution of the zero temperature BCS equations collapse onto the KS solution. Reference figures to be kept in mind are the dimensionless pairing strength for the superconducting bulk,

$$\alpha = N(0)G , \qquad (14)$$

where $N(0) = 3N_e/4\varepsilon_F$ is the density of electron states at the Fermi level. For metallic Al, one has $\alpha = 0.193$ and $N_e = 3N$. One may then temptatively extrapolate (14) to finite systems, setting

$$G(N_e) = \frac{4\alpha\varepsilon_F}{3N_e} , \qquad (15)$$

which gives, for Al_N clusters,

$$G^{Al}(N_e) = \frac{0.11}{N_e} \text{ (au) .} \qquad (16)$$

Table 1. Zero temperature gap, coherence length and critical temperature of Al_{24} and Al_{30} for different pairing strengths G.

| G | Al_{24} | | | Al_{30} | | |
| | $\Delta(T=0)$ | $\xi(T=0)$ | T_c | $\Delta(T=0)$ | $\xi(T=0)$ | T_c |
(au)	(au)	(au)	(K)	(au)	(au)	(K)
10^{-3}	$4.24\ 10^{-3}$	104.3	735.	$3.16\ 10^{-3}$	129.9	516.
10^{-4}	$4.24\ 10^{-4}$	1043.1	73.	$3.27\ 10^{-4}$	1255.6	53.
10^{-5}	$4.24\ 10^{-5}$	10431.	7.3	$3.17\ 10^{-5}$	12952.2	5.1

Numerical explorations of KS closed shell clusters Al_{19}^-, Al_{23}^- and Al_{31}^+ yield critical strengths a few times 10^{-3} au, in agreement with the estimate (16). However, neighbouring non-magic clusters exhibit non-vanishing zero temperature pairing gaps for much smaller values of G. One additionally encounters that the critical temperature for the disappearance of the gap is, for a given cluster, a linear function of G that scales to a few K for $G \sim 10^{-5}$ au. This is illustrated in table 1, where we display the zero T gap, the corresponding coherence length and T_c for several values of G, for the neutral clusters Al_{24} and Al_{30}. The coherence length has been evaluated according to eq. (8) choosing $v_F = w_0 R$, with the characteristic energy $\hbar \omega_0 = 0.98 \varepsilon_F / N_e^{1/3}$.

It is clear that the coherence length is sufficiently large to guarantee that the given cluster is fully correlated; moreover, it scales to the bulk value when the temperature is low enough. It should be however pointed out that the linear trend exhibited by these results is exclusive to the G-range under consideration; for instance, a choice $G \sim 3\ 10^{-3}$ au leads to a T_c higher than 4000 K.

The results of the spherical KS-plus-BCS calculations for Al_N particles can be better interpreted as we realize that for the force intensities under consideration, the pairing field is not capable to couple electrons in different shells. One may then straightforwardly write down the expressions for the sp occupancy v^2 and energy gap Δ (at zero T) of a system with N_v electrons in a partially filled valence orbit with pair degeneracy $\Omega = (2\ell + 1)$, $\Omega > N_v/2$ (see, for example [13]),

$$v^2 = \frac{N_v}{2\Omega}, \tag{17}$$

$$\Delta = \frac{G}{2}\sqrt{N_v(2\Omega - N_v)}. \tag{18}$$

On similar grounds, one can derive an approximate expression for the critical temperature, under the assumption that the qp occupation probability $n(E)$, with E given by eq. (7), takes the form of a Fermi distribution. The standard approach, namely to compute the temperature at which the second derivative of the electron free energy with respect to the gap Δ vanishes [21] easily gives

$$T_c = \frac{G\Omega}{4} . \tag{19}$$

One can check that the critical temperatures in table 1 closely follow this law. Furthermore, one can appreciate that eq. (19) together with assumption (16) for the pairing strength lead to a decrease of T_c with the cluster size, in agreement with experimental findings [16].

4. DISCUSSION and OUTLOOK

We have carried out an exploratory study of the influence that pairing correlations could have on the structure of near-free-electron (NFE) clusters. To this end, we have performed a BCS calculation on top of a KS calculation, following the philosophy of successful mean field+BCS computations in finite nuclei. As an illustrative example, we have applied the method to some Al_N clusters.

We are well aware that the present calculation could raise more questions that it answers. In particular, we do not attempt any justification of the origin of the pairing, but take it for granted and use a sound model, namely the BCS nuclear model, to obtain semiquantitative results. However, as we have indicated in the introduction, the conspicuous even-odd alternation in the IP for some metals in certain N-regions, as well as in the mass spectra of others (see for example ref. [22]) makes it very tempting to try to interpret these experimental facts as a signature of electron pairing correlations. An equivalent nuclear phenomenology inspired the early observations and estimates, by Bohr, Mottelson and Pines [23], and the subsequent development of the BCS model for finite Fermi systems. In such a case, the pairing force is a definite attractive component of the effective nucleon-nucleon interaction acting in spin-singlet, time reversed states, the intensity law $G = 20/A$ being empirical throughout the mass region $150 \leq A \leq 190$. This law is to be contrasted with the rule $G(N_e)$ derived in section 3 for Al_N which is model-dependent and designed for the present study. Futhermore, we have put aside any consideration regarding electrodynamic properties of presumably superconducting clusters, which might become manifest in experimental detections [16].

On the other hand, the self-consistent spheroidal model has been very successful in describing IP and electron affinities [9, 10]. This fact does not invalidate the present study. Indeed, we would like to point out that in the atomic nuclei both deformation and pairing effects do coexist in medium and heavy nuclei [13], whereas in other regions, one aspect prevails over the other (deformation in small-A nuclei, due to the large spacing of the sp energy levels; pairing in even-even nuclei near a magic nucleus). It is also worth to recall that the electronic magic numbers [24] and surface plasmon [25] of NFE metals were successfully predicted by Ekardt before being experimentally confirmed.

Some refinements of the present model can be easily put forth. Still restricting oneself to spherical symmetry, the first step could consist in selfconsistently solving the KS-BCS equations; in this manner, after adding the pairing energy E_p to the KS energy E_{KS}, one obtains better estimates of E^{gs} for even N_e−values.

Going beyond the imposed spherical symmetry, one could apply the BCS method on top of the Nilsson model of ref. [8], as done here for the spherical jellium, or start from the self-consistent spheroidal well, much along the line of the self-consistent KS-BCS approach. It will be very interesting to see the deformation and pairing interplay. As the BCS correlations spread the valence electrons over several sp levels, they tend to make the cluster more spherical.

ACKNOWLEDGEMENTS

We would like to thank J.A. Alonso and L.C. Balbás for useful discussions. This work has been supported by the DGICYT (Spain), grant PB89-0332, and by the IN2P3-CICYT exchange program. One of us (ESH) is also grateful to the Departamento ECM of the Facultad de Fisica, Universidad de Barcelona for warm hospitality and to the Ministerio de Educacion y Ciencia, Spain, for financial support.

REFERENCES

[1] Schriver,K.E., Persson, J.L., Honea, E.C., Whetten,R.L.: Phys. Rev. Lett. **64**, 2539 (1990)

[2] Laiting,K., Cheng,P.Y., Duncan,M.A.: Z. Phys. **13**, 161 (1989)

[3] Kappes,M.M., Schar,M., Rothlisbarer,U., Yeretzian,C., Schemacher,E.: Chem. Phys. Lett. **143**, 251 (1988)

[4] Honea,E.C., Homer,M.L., Persson,J.L., Whetten,R.L.: Chem. Phys. Lett. **171**, 147 (1990)

[5] Saunders,W.A., Clemenger,K., de Heer,W.A., Knight,W.D.: Phys. Rev. B **32**, 1366 (1985)

[6] Bréchignac,C., Cahuzac,Ph.: Z. Phys. D **3**, 121 (1986)

[7] Bohr,A., Mottelson,B.R.: Nuclear Structure. Vols. I (1969) and II (1975), W.A. Benjamin, Reading, Mass.

[8] Clemenger K.: Phys. Rev. B **32**, 1359 (1985)

[9] Ekardt,W., Penzar,Z.: Phys. Rev. B **38**, 4273 (1988)

[10] Ekardt,W., Penzar,Z.: Z. Phys. D **17**, 69 (1990)

[11] Cooper,L.N.: Phys. Rev. **104**,1189 (1956)

[12] Bardeen,J., Cooper,L.N., Schrieffer,J.R.: Phys. Rev. **106**,162 (1957); ibid **108**,1175 (1957)

[13] Ring,P., Schuck,P.: The Nuclear Many-Body Problem, Springer-Verlag, Berlin (1980)

[14] Schrieffer,J.R.: Theory of Superconductivity Benjamin/Cummings Pu. Co., 3rd. ed. (1983)

[15] Pines,D.: Phys. Rev. **109**, 280 (1958)

[16] Komori,F., Goto,T., Wago,K., Kobayashi,S.: J. of Phys. Soc. of Japan **57**, 3868 (1988)

[17] Sano M., Yamasaki,S.: Progr. Theor. Phys. **29**, 397 (1963)

[18] Moretto,L.G.: Nucl Phys **A185**, 145 (1972)

[19] Ishii,Y., Ohnishi,S., Sugano,S.: Phys. Rev. B **33**, 5271 (1986)

[20] Serra,Ll., Garcias,F., Barranco,M., Navarro,J., Balbás, L.C., Mañanes A.: Phys. Rev. B **39**, 8247 (1989)

[21] Leggett,A.J.: Rev. Mod. Phys. **47**, 331 (1975)

[22] Kandler,O., Athanassenas,K., Echt,O., Kreisle,D., Leisner,T., Recknagel E.: Contribution to the 5th Int'l Symposium on small particles and inorganic clusters, Sept 10 - 14 1990 Konstanz, Germany. To be published in Z. Phys. D.

[23] Bohr,A., Mottelson,B.R., Pines,D.: Phys. Rev. **110**, 936 (1958)

[24] Ekardt,W.: Phys. Rev. B **29**, 1558 (1984)

[25] Ekardt,W.: Phys. Rev. Lett. **52**, 1925 (1984)

THE ROLE OF THE SPIN OCCUPANCY IN THE ELECTRON GAS PROBLEM

A. Calles[1], A. Cabrera[1], F. Ramos-Gómez[1], M. L. Marquina[1], E. Yépez[2] and J. J. Castro[3]

[1]Facultad de Ciencias, UNAM, Apdo. Post. 70-646, 0451C México D.F.
[2]Escuela Superior de Física y Matemáticas, IPN, 07738 México D.F.
[3]Departamento de Física, CINVESTAV, IPN, Apdo. Post. 14-740, 07000 México D.F.

Mean field theory is used to define a continuous spin occupancy parameter between the polarized and non-polarized states in the electron gas jellium model. Using an independent particle approximation, the appearance of the charge density wave is discussed as a function of the spin occupancy, temperature and density system. It is found that the introduction of the spin occupancy parameter is equivalent to a scaling in the electronic distance of the system.

INTRODUCTION

The discovery of high Tc superconductivity[1] brought new problems to the already complex phenomenon of superconductivity. Many efforts have since been devoted to finding a theory capable of explaining the properties of these new materials, particularly those based on copper oxides. Special attention has been paid to the understanding of the raising of the transition temperature. This has been done using approaches both within and outside the BCS theory. In particular, and with relevance for the present work, we would like to mention the attempt made by De Llano[2] who assuming an abnormal occupation[3] in the Fermi sphere of the electron gas, he claims to obtain a higher transition temperature using the BCS model. It is not the aim of the present work to discuss such possibility but rather to study, in a general way, the problem of the abnormal occupation in the electron gas. The De Llano's proposal for the abnormal occupation is shown in Figure 1. He solved the problem at zero temperature. We show that a more general abnormal occupation like the one shown in Figure 2 can be dealt for the electron gas system. Furthermore we solved the problem at finite temperature where the complex abnormal occupation is used for the zero temperature energy as a starting point. In order to answer this question we used a mean field approximation to average the occupancy in the momentum space. The following sections include a summary of the jellium model and the mean field approximation used. According to the results it is found that the introduction of the spin occupancy parameter is equivalent to a scaling in the electronic distance implying, in this case, that the system has a set of corresponding states.

Condensed Matter Theories, Vol. 7, Edited by A.N. Proto
and J.L. Aliaga, Plenum Press, New York, 1992

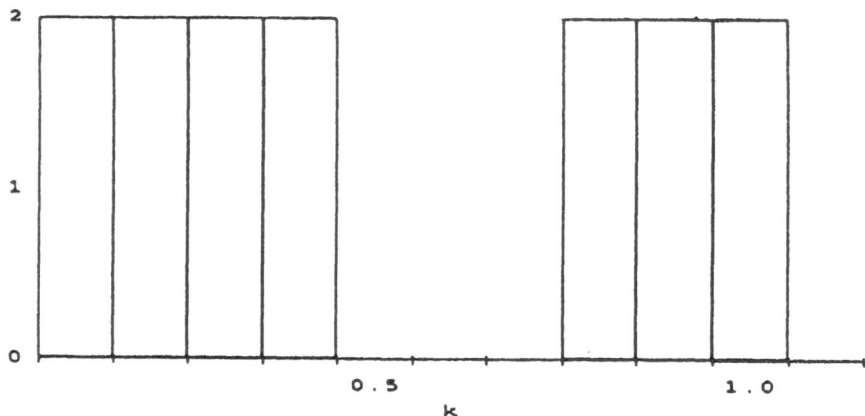

Fig. 1. Abnormal occupation in the $\underline{k}$ space with a gap inside the Fermi sphere.

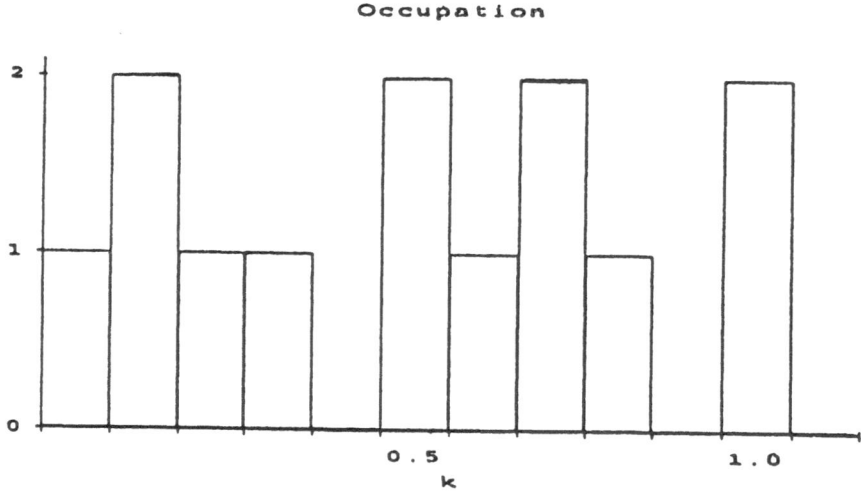

Fig. 2. A complex abnormal occupation in the $\underline{k}$ space.

THE JELLIUM MODEL

The jellium model consists of particles repelling one another with a Coulomb force and immersed in a field of an external charge distribution called the background. The background charge density is not concentrated at a point, but is rather spread with density $\eta(R)$ through the volume V. The model can be regarded as a model of highly condensed matter with the background charge produced by an assemblage of ions and fast moving electrons.

The non-relativistic hamiltonian for the electron gas is:

$$\hat{H} = \hat{H}_{ee} + \hat{H}_{bb} + \hat{V}_{eb} \quad , \tag{1}$$

where the first term represents the electronic hamiltonian, the second one is the background hamiltonian and the third is the interaction between electrons and the background, that is:

$$\hat{H}_{ee} = \sum_k T_{kk} \, a_k^+ a_k^+ + \frac{1}{2} \sum_{kk'} V_{kk'kk'} \, a_k^+ a_{k'}^+ a_{k'} a_k , \tag{2a}$$

$$\hat{H}_{bb} = \frac{1}{2} \int \eta(\underline{R}) \, V(\underline{R}-\underline{R}') \, \eta(\underline{R}') \, d\underline{R} \, d\underline{R}' , \tag{2b}$$

$$\hat{V}_{eb} = - \sum_k \int V_{kk}(\underline{R}) \, \eta(\underline{R}) \, d\underline{R} \, a_k^+ a_k , \tag{2c}$$

where a_k^+ and a_k are the creation and annihilation operators respectively, T_{kk} and $V_{kk}(\underline{R})$ are the matrix elements of kinetic energy and one-particle operators in the basis orbitals $\varphi_k(\underline{r})$ and, finally, $V_{kk'kk'}$ are the matrix elements for the two-electron operator. These matrix elements can be written as:

$$T_{kk} = \int \varphi_k^*(\underline{r}) \, \hat{T} \, \varphi_k(\underline{r}) \, d\underline{r}, \tag{3a}$$

$$V_{kk}(\underline{R}) = \int \varphi_k^*(\underline{r}) \, V(\underline{R}-\underline{r}) \, \varphi_k(\underline{r}) \, d\underline{r}, \tag{3b}$$

$$V_{k_1 k_2 k_3 k_4} = \int \varphi_{k_1}^*(\underline{r}) \, \varphi_{k_2}^*(\underline{r}') \, V(\underline{r}-\underline{r}') \, \varphi_{k_3}(\underline{r}) \, \varphi_{k_4}(\underline{r}') \, d\underline{r} \, d\underline{r}'. \tag{3c}$$

The potential we use is known as the screened coulombic interaction and has the following form

$$V(r) = \frac{e^{-\mu_o \, r}}{r} \tag{4}$$

where μ_o is the so called screening parameter.

The total energy of the system as a function of the temperature can be written as:

$$E = \sum_k T_{kk} \langle \hat{n}_k \rangle_T - \sum_k \int \eta(\underline{R}) \, V_{kk}(\underline{R}) \, d\underline{R} \, \langle \hat{n}_k \rangle_T +$$

$$+ \frac{1}{2} \sum_{kk'} (V_{kk'kk'} - V_{kk'k'k}) \, \langle \hat{n}_k \rangle_T \langle \hat{n}_{k'} \rangle_T +$$

$$+ \frac{1}{2} \int \eta(\underline{R}) \, V(\underline{R}-\underline{R}') \, \eta(\underline{R}') \, d\underline{R} \, d\underline{R}' , \tag{5}$$

where $\langle \hat{n}_k \rangle_T$ is the Fermi distribution function which reduces to the step function when T= 0. To obtain the electron-electron interaction term, we used the independent particle approximation in the form $\langle n_k \, n_{k'} \rangle_T \cong \langle n_k \rangle_T \langle \hat{n}_{k'} \rangle_T$.

The Hartree-Fock equations, which can be obtained from a variation of the energy respect to the basis orbitals, are:

$$T \varphi_k(\underline{r}) + V(\underline{r}) \varphi_k(\underline{r}) + \sum_{k'} V_{kk'}(\underline{r}) \langle \hat{n}_{k'} \rangle_T \varphi_k(\underline{r}) -$$

$$- \sum_{k'} V_{kk'}(\underline{r}) \langle \hat{n}_{k'} \rangle_T \varphi_{k'}(\underline{r}) = \varepsilon_k \varphi_k(\underline{r}) . \tag{6}$$

On the other hand, the best choice[5] for the background density such that the energy, equation (5), is an extremum (in fact a minimum) results:

$$\int \eta(\underline{R}') V(\underline{X}-\underline{R}') d\underline{R}' = \sum_k V_{kk}(\underline{X}) \langle \hat{n}_k \rangle_T ; \tag{7}$$

when substituted in equation (5), the energy reduces to:

$$E = \sum_k T_{kk} \langle \hat{n}_k \rangle - \frac{1}{2} \sum_{kk'} V_{kk'k'k} \langle \hat{n}_k \rangle_T \langle \hat{n}_{k'} \rangle_T . \tag{8}$$

So the only contribution to the electron-electron interaction energy comes from the exchange term. This means that the direct term plus the electron-background interaction plus the background-background energy adds up to zero. If written in terms of the spin occupancy for each k level, the energy is:

$$E = \sum_k T_{kk} O_k \langle \hat{n}_k \rangle - \frac{1}{2} \sum_{kk'} V_{kk'k'k} O_k \langle \hat{n}_k \rangle_T \langle \hat{n}_{k'} \rangle_T, \tag{9}$$

where O_k stands for the occupation of level k after the corresponding sum has been made. So, O_k can only take the values 0, 1, or 2, corresponding to non-occupied, total spin 1/2 or 0, respectively.

The equation (9) is a reduced energy equation with only two terms. To obtain it only required the orbitals to satisfy the theorem in reference [5]. In order to determine the orbitals one should solve the HF equation that comes from the variation of the reduced energy, equation (9), respect to the orbitals with the usual orthonormalization condition. After doing the variation, the reduced Hartree-Fock equations are:

$$T \varphi_k(\underline{r}) - \sum_{k'} V_{kk'}(\underline{r}) \langle \hat{n}_{k'} \rangle_T \varphi_k(\underline{r}) = \varepsilon_k \varphi_k(\underline{r}) . \tag{10}$$

We shall use a Bloch Type function as solution to the equation (7) in the form:

$$\varphi_k(\underline{r}) = \frac{1}{\sqrt{v}} e^{-\underline{k} \cdot \underline{r}} \sum_{n=N_1}^{N_2} C_n e^{i q_o \underline{n} \cdot \underline{r}} , \tag{11}$$

where $\underline{n} = n_x \hat{i} + n_y \hat{j} + n_z \hat{k}$, with n_x, n_y and n_z integers and $q_o \geq 2k_F$, k_F being the Fermi radius defined in the momentum space. A particular case to this equation is the Plane Wave function (PW) known as the trivial solution. Note that depending upon the use of the n_x, n_y and n_z indices one has 1, 2 or 3 components for the CDW part.

The coefficients C_n in equation (11) are obtained through the solution of the algebraic equation that comes from the substitution of equation (10) in the reduced HF equations:

$$\sum_{n_3} \left[\sum_{k_1} T_{k_1+n_1, k_1+n_3} \; O_{k_1} \langle \hat{n}_{k_1} \rangle_T - \right.$$

$$\left. - \sum_{n_2 n_4} \sum_{k_1 k_2} C_{n_2} V_{k_1+n_1, k_2+n_2, k_1+n_4, k_2+n_3} \; O_{k_1} \langle \hat{n}_{k_1} \rangle_T \langle \hat{n}_{k_2} \rangle_T \; C_{n_4} \right].$$

$$\cdot C_{n_3} = \varepsilon \; C_{n_1} . \tag{12}$$

The eigenvalue matrix equation for the coefficients has to be solved self-consistently. The solutions to the algebraic equation for the coulombic and screened interactions, in the one dimensional case for the CDW, were reported in reference [6] at zero temperature and in reference [7] at finite temperature.

MEAN FIELD CONSIDERATIONS

In order to solve the HF equation one must know the specific values for each of the k levels of the system. Once the O_k dependence is given one has to perform again all the k integrals in 3 and 6 dimensions. Since the integrals may not be solvable and in any case the procedure is cumbersome we try another approach. Instead of assuming the details of the O_k function we shall now use a mean field approximation such that the occupation details are substituted by the average occupation of the system so that the eq. (9) takes the form:

$$E = P_0 \sum_k T_{kk} \langle \hat{n}_k \rangle - \frac{1}{2} P_0 \sum_{kk'} V_{kk'k'k} \langle \hat{n}_k \rangle_T \langle \hat{n}_{k'} \rangle_{T'}, \tag{13}$$

So one has a continuous spin occupancy parameter that contains the extreme ferromagnetic ($P_0 = 1$) and antiferromagnetic ($P_0 = 2$) cases.

Another expression that should be modified accordingly, is the Fermi radius which takes the form:

$$k_F = \frac{1.92}{r_s} \; \frac{1}{\left(\dfrac{P_0}{2} \right)^{1/3}} \tag{14}$$

This is substituted in the corresponding energy and HF equations to be solved.

In the following section we show the results for the total energy, specific heat and the separation curve for the PW and CDW solution to the problem.

RESULTS AND DISCUSSION

In this section we present the results for different properties when the localization is considered in 1 dimension for $\mu_0 = 0.815/\sqrt{r_s}$. We are interested mainly in the low temperature regime so that the Sommerfeld approximation can be used for the particle distribution function in the following form:

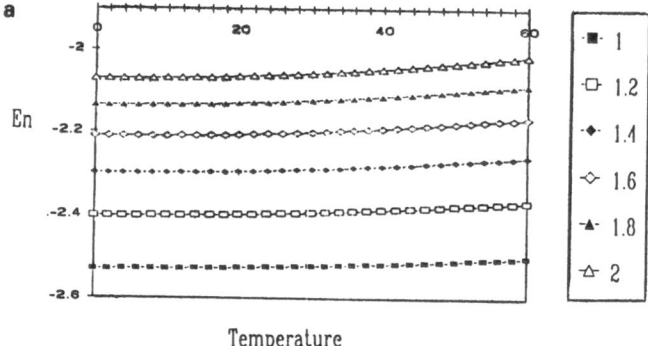

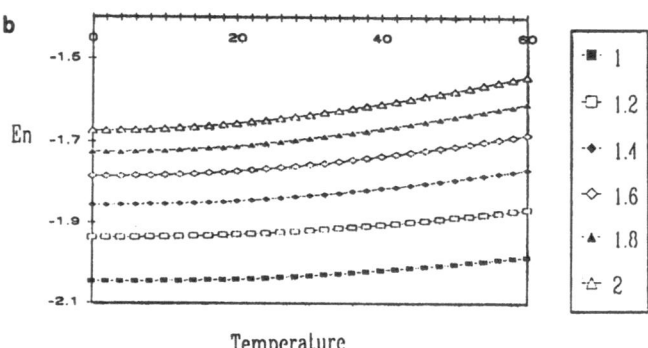

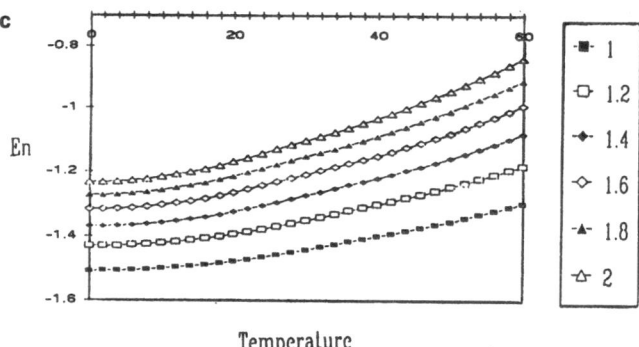

Fig. 3. Energy (10^{-2} meV) as a function of temperature (K) for the different P_o values, (a) $r_s = 30$, (b) $r_s = 40$ and (c) $r_s = 60$.

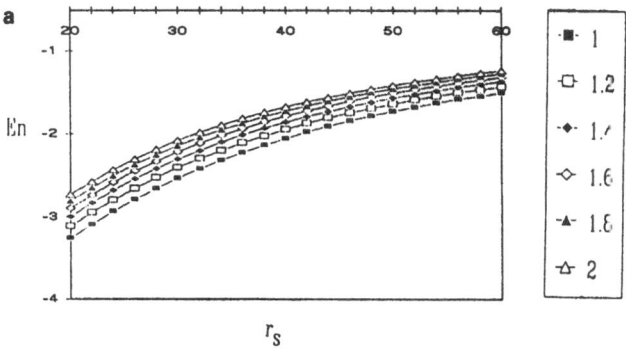

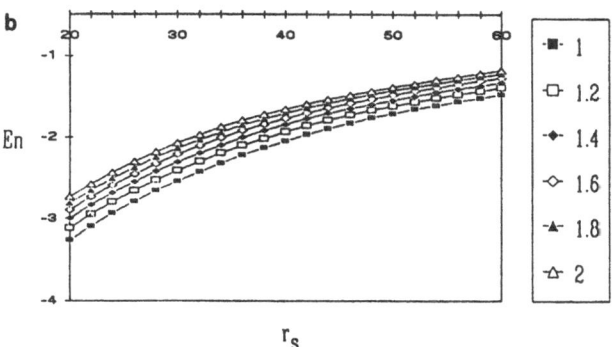

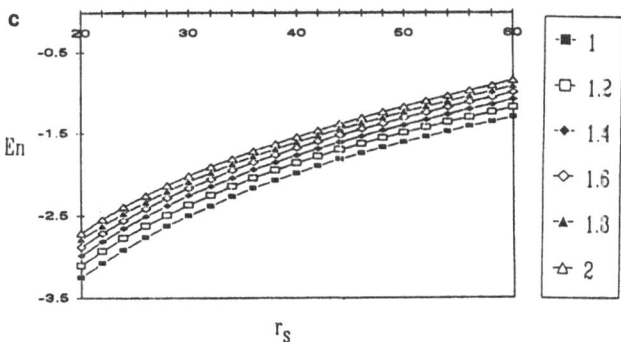

Fig. 4. Energy (10^{-2} meV) as a function of r_s for the different P_o values, (a) T= 0, (b) T= 20 K and (c) T= 60 K.

$$\langle \hat{n} \rangle_{k\,T} \approx \vartheta(\varepsilon_F - \varepsilon_k) - \frac{\pi^2}{6\beta^2} \frac{d}{d\varepsilon_k} \delta(\varepsilon_F - \varepsilon_k), \tag{15}$$

which is introduced in the algebraic HF and energy equations. It should be noticed that the last equation is an expansion of Fermi distribution function in terms of the temperature, not on the energy, and that it represents a good approximation for this system even at room temperature[8]. For the case $T = 0$ eq. (16) reduces to the step function as it should be.

Figures 3 and 4 show the different plots for the internal energy as function of r_s and T, respectively.

Figure 5 shows the separation curves, in a r_s vs T diagram, between the CDW and PW solutions for different occupancy parameters.

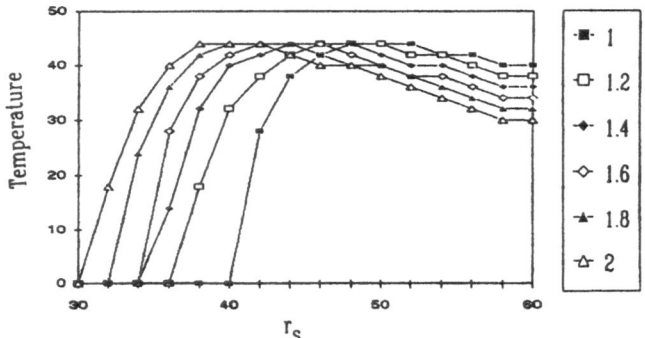

Fig. 5. Separation curve between PW and CDW solutions in a T (K) vs. r_s diagram, for the different P_o values.

In order to calculate thermodynamic properties one can use a free energy such as the Helmholtz function; instead, in this work we use the internal energy $E(T,V,\mu)$ obtained from equation (15) after solving equation (12) for the coefficients. The results for the calculated properties should be equivalent regardless of the thermodynamic potential used.

We performed the calculation for the specific heat of the system through the derivative of the internal energy:

$$C_v = \left(\frac{\partial E}{\partial T} \right). \tag{16}$$

The plots for the specific heats are in figures (6) and (7) as function of r_s and T, respectively. Figure (7) indicates a second order transition. The system changes from the CDW to the PW solutions when the temperature is raised. Above the transition temperature the specific heat has the typical linear metallic behavior.

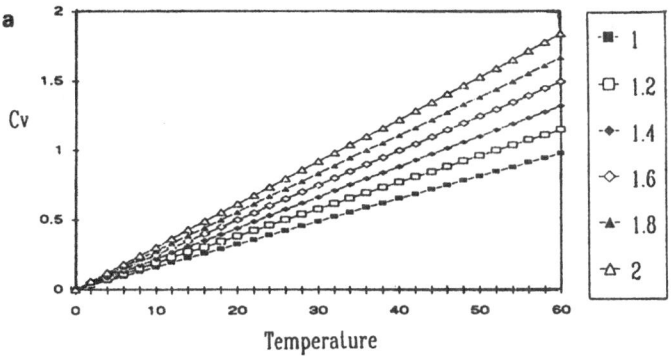

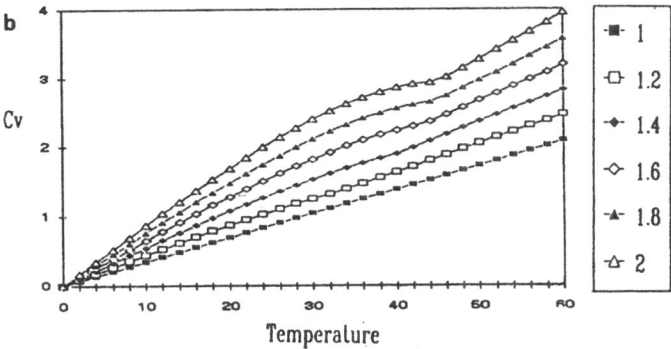

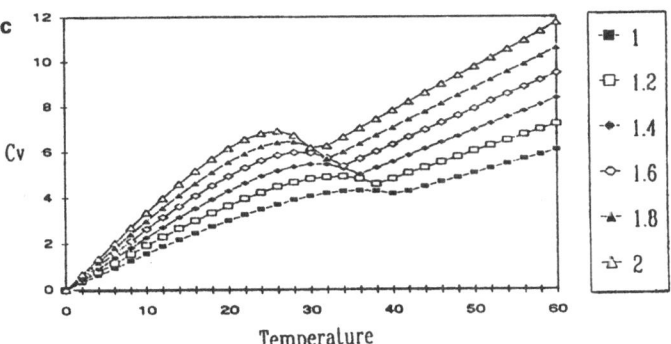

Fig. 6. Specific heat $(10^{-5}$ meV/K) as a function of T (K) for different P_o values, (a) $r_s = 30$, (b) $r_s = 40$ and (c) $r_s = 60$.

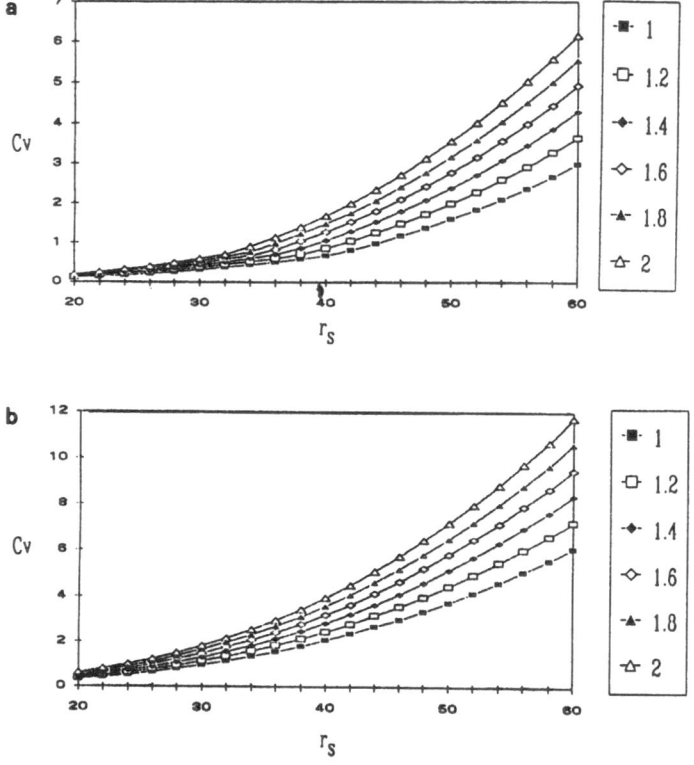

Fig. 7. Specific heat (10^{-5}meV/K) as a function of r_s for the different P_o values, (a) T= 20 K and (b) T= 60 K.

CONCLUSIONS AND COMMENTS

The solutions we obtained have a parametric temperature dependence through the coefficients C_n. In both zero and finite temperature cases, the CDW solution has, from equation (12), a wavelength:

$$\lambda = 1.64 \; r_s \; a_o = 0.8659 \; r_s \; \text{Å}, \tag{17}$$

so, as it can be expected the size of the CDW increases linearly with the distance r_s.

The transition densities r_s are temperature dependent such that the larger the temperature the smaller the r_s transition value. No transition at all was found for temperatures larger than 60K and the largest transition temperature was found at $r_s = 40$.

The mean field theory allows us to solve complex problems in a simpler fashion in order to study macroscopic properties such as specific heats and the transition to the CDW in the electron gas. It was found that although the results obtained using the spin occupancy parameter were different, they are qualitatively similar, suggesting that there was a kind of scaling of the problem. Given a value for the occupancy, this implies that the problem has a set of corresponding states. In the physical system where the electron gas is a reliable model, the occupancy for different situations can be determined observing whether or not there is a shift on the values of the specific heat measurements according to the present model.

ACKNOWLEDGMENTS

Partial support is due the Programa Universitario de Superconductividad de Alta Temperatura de Transición at UNAM and to CONACYT, México. One of us (E.Y.) acknowledge partial support by COFAA-IPN, México.

REFERENCES

1. J.G. Berdnorz and K.A. Müller, Z. Phys. B64:189 (1986).
2. M. de Llano and J.P. Vary, Abnormal occupation, tighter-bound Cooper pairs and high Tc superconductivity, in: "Condensed MATTER THEORIES", V.C. Aguilera-Navarro, ed., Plenum Press, New York (1990).
3. M. de Llano and J.P. Vary, Phys. Rev. C 19:1083 (1979).
4. A.L. Fetter and J.D. Walecka, "Quantum theory of many-particle systems", McGraw-Hill Book Company, New York (1971).
5. A. Cabrera y A. Calles, Rev. Mex. Fís., 33,2: 194 (1987).
6. A. Cabrera, A. Calles, R.M. Méndez-Moreno and M.A. Ortíz, Tech. Rep., 1-85, Fac. de Ciencias, UNAM (1985).
7. A. Cabrera y A. Calles, Rev. Mex. Fís., 32,4: 573 (1986).
8. Neil W. Ashcroft and N. David Marmin, "Solid State Physics", Holt, Rinehart and Winston, New York (1976).

LOCAL PAIR MODEL FOR THE ANALYSIS OF TUNNELING AND PHOTOEMISSION

EXPERIMENTS IN HIGH Tc SUPERCONDUCTORS: SUPERCONDUCTING PHASE

C. I. Ventura[+], B. R. Alascio[++], R. Allub[++]

Centro Atómico Bariloche, 8400-Bariloche
Prov. de Río Negro, Argentina

ABSTRACT

We analyze tunneling and high resolution photoemission experiments in the superconducting phase of high Tc materials. Our model is based on a "mixed Fermion-Boson" Hamiltonian, where a band of paired states overlaps the Fermi level of a wider Fermion band. In a simple mean field approximation, superconductivity is characterized by two coupled order parameters. Related to them two types of fermionic excitations appear, which can respectively be associated to Cooper pair breaking and local pair phase correlation breaking. The former processes produce a gap, and the latter are responsible for the anomalous V-shaped contribution to the tunneling (and photoemission) density of states.

In this paper we proceed with our study of the "mixed Fermion-Boson" model [1,2,3], which already proved to be useful [4] to derive the self-energy proposed by Varma et al. [5] to explain the anomalous normal state properties of high temperature superconductors (HTS), and lead to the observed shape of the tunneling conductance and photoemission spectrum [6]. This was a consequence of anomalous scattering processes inherent to the model and not of the one-particle density of states which shows a cusp at the Fermi level [6].

In the present work we concentrate on predictions for the superconducting phase of a system where a band of localized paired states overlaps the Fermi level of a broader conduction band. In our approach [4,6] disorder is the origin of the band of localized paired states, and local pairs are represented by hard core bosons, as will soon be explained. The superconducting phase for the problem of a localized level was studied by Robaszkiewicz et al. [1] and Kulic [3], among others. Khomskii et al. [7] analized the superconducting phase placing the pair level far above the Fermi level (in which case the superconductivity of the system is BCS like) or placing the conduction band far above the Fermi level (in which case the superconductivity is the Bose condensation of a gas of hard core bosons). Also Sinha et al. [8] studied the former situation and combined the mechanism with conventional electron-phonon interaction to obtain higher critical temperatures. Bar Yam [9] studied a mixed model of fermions and hard core bosons with translation symmetry. The otherwise immobile paired states gain mobility through hybridization with the extended states and the same interaction provides pairing between the extended states, giving rise to superconductivity. Ioffe et al. [10] analized superconductivity in a mixed model in which the hard core boson band arises from considering them mobile. In a series of papers, Friedberg et al.[11] studied a mixed model with translation symmetry, without considering a hard core repulsion for the bosons, which are mobile in their approach.

. Here, we take the simplest mean-field approach to describe the superconducting state. Superconductivity is characterized by two coupled order parameters, one of them corresponding to a BCS-type excitation gap and the other related to the anomalous processes responsible for the V-shaped contribution to the tunneling conductance [12] (and photoemission spectrum [13]). Comparison with results of tunneling experiments [12] shows qualitative agreement , even though we must remark that we do not obtain the "additional structure" observed (on which different experiments do not agree [14]). Recently, Cucolo et al.[15] related this structure with the surface oxygen stoichiometry of $YBa_2Cu_3O_{7-\delta}$ samples through a careful study. Taking into account the experimental resolution in photoemission experiments [13], our results agree with the observations as we will later show.

As in our previous treatment of the normal state [6], we base our description of HTS on the assumption of the existence of extended conduction band states and, at random sites, localized states with strong attractive correlations. In the limit of correlation energy much stronger than the hybridization between both kinds of states we can eliminate single occupation of localized states [6] and represent the system by :

$$H= \sum_{k\sigma} (\varepsilon_k-\mu)c^+_{k\sigma}c_{k\sigma} + \sum_j 2(\Lambda_j-\mu)A^+_jA_j + \frac{I}{N} \sum_{jkk'} \left[e^{-i(\bar{k}+\bar{k}').\bar{R}_j} c^+_{k\downarrow}c^+_{k'\uparrow}A_j + H.c. \right]$$

(1)

Here $c^+_{k\sigma}$ creates an electron with crystal momentum k, spin σ and energy ε_k in the conduction band; A^+_j is the "hard core boson" creation operator for a pair localized at site j with energy $2\Lambda_j$; μ denotes the chemical potential of the system of conduction electrons and localized pairs, and N is the total number of sites. The last term in Eq.(1) represents the effective scattering between pairs of conduction electrons and local pairs.

We now consider the scattering term in (1) in the mean-field approximation (MFA) [1,3], introducing two order parameters which will characterize the superconducting phase. We would like to stress that in the system we are considering neither of the two subsystems, conduction electrons and local pairs, can become superconducting by itself. Superconductivity is mutually induced in the two subsystems via the effective scattering term which couples them, as discussed by Robaszkiewicz et al. [1] and Bar Yam [9].

$$H_{MFA} = \sum_{k\sigma} (\varepsilon_k-\mu)c^+_{k\sigma} c_{k\sigma} - \sum_k \left(\Delta c^+_{k\uparrow} c^+_{-k\downarrow} + \Delta^* c_{-k\downarrow}c_{k\uparrow} \right)$$

$$+ \sum_j 2(\Lambda_j-\mu) A^+_jA_j - \sum_j \left(\square A_j + \square^* A^+_j \right)$$

(2)

where

$$\Delta = \frac{1}{N} \sum_j \langle A_j \rangle \tag{3}$$

$$\square = \frac{1}{N} \sum_k \langle c^+_{k\uparrow} c^+_{-k\downarrow} \rangle \tag{4}$$

The Hamiltonian in MFA (2) consists of two parts: the first has the ordinary BCS form and can be diagonalized by a Bogoliubov transformation. The second is a sum of one-site Hamiltonians, which also admit diagonalization in terms of fermionic quasiparticles by employing the "spin analogy" for the hard core boson operators, and performing a convenient rotation. The diagonal form of Hamiltonian (2) then reads :

$$H_{MFA} = E_{gs} + \sum_k E_k \left(\alpha^+_{k\uparrow} \alpha_{k\uparrow} + \alpha^+_{-k\downarrow} \alpha_{-k\downarrow} \right) + \sum_j 2R_j \gamma^+_j \gamma_j \tag{5}$$

where :

$$E_k = \sqrt{(\varepsilon_k - \mu)^2 + \Delta^2} \tag{6}$$

$$R_j = \sqrt{(\Lambda_j - \mu)^2 + \square^2} \tag{7}$$

$$\gamma^+_j = b_j \left(2A^+_j A_j - 1 \right) + a_j^2 A^+_j - b_j^2 A_j \tag{8}$$

$$a_j^2 = \frac{1}{2} \left(1 + \frac{\Lambda_j - \mu}{R_j} \right) \tag{9}$$

$$b_j^2 = \frac{1}{2} \left(1 - \frac{\Lambda_j - \mu}{R_j} \right) \tag{10}$$

and $\alpha^+_{k\uparrow}$ is the usual BCS-single particle excitation operator, expressed in terms of the conduction state operators through the Bogoliubov transformation. E_{gs} is the energy of the ground state, which in the superconducting phase (non zero Δ and $\square$) can be pictured as a BCS-paired state of conduction electrons and the phase correlated state of local pairs (by this we mean the same combination of zero and double occupation holds for each of the sites with localized orbitals). The order parameters which couple the superconductivity of conduction electrons and local pairs are obtained solving the self-consistent system of equations formed by Eqs. (3), (4) and the total electron number equation for the chemical potential of the system.

In Fig. 1 we show schematically the band configuration employed in this work, for the case of fixed disorder and half-filling.

Although in our model disorder plays an important role to obtain the normal state properties [4,6], it competes with superconductivity. For the second order transition to the superconducting phase we obtain the following expression for the critical temperature (Tc) considering a band configuration like that plotted in Fig. 1:

$$\frac{4WA}{I^2} = \int_0^{W/2T_c} \frac{dx}{x} \tanh(x) \int_0^{A/T_c} \frac{dy}{y} \tanh(y) \tag{11}$$

This expression is valid for half filling, which for fixed disorder gives the highest critical temperature.

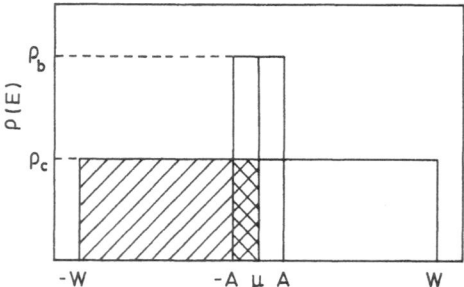

Fig. 1. Schematic plot of the band configuration and filling used in the present work. ρ_c denotes the density of states for conduction electrons, while ρ_b is the density of states for electrons forming local pairs.

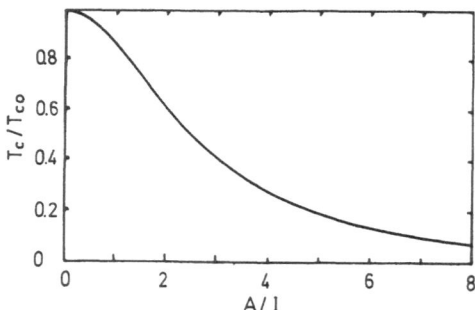

Fig. 2. Critical temperature (Tc) as a function of disorder (A), for I=0.117 eV and W=0.5 eV. Tc_0 is Tc for A=0.

In Fig. 2 we plot the transition temperature as a function of the disorder, as given by Eq. (11). Notice the exponential dependence on disorder obtained for A>>I, which can be derived from Eq. (11) with the usual BCS approximation for the integrals:

$$T_c = \sqrt{2.6\ AW}\ \exp\left[-\left[\frac{4AW}{I^2} + 0.25\ \ln^2\left(\frac{W}{2A}\right)\right]^{1/2}\right] \tag{12}$$

In Fig.3 we plot the temperature dependence of the two coupled order parameters, as used throughout the rest of the present work.

In this context, we describe the tunneling process between a normal metal and the HTS by the following Hamiltonian [3,6]:

$$H = H_{MFA} + \sum_{k\sigma} (\varepsilon'_k - \mu)\, b^+_{k\sigma}\, b_{k\sigma}$$

$$+ T_1 \sum_{kk'\sigma} \left[b^+_{k\sigma}\, c_{k'\sigma} + H.c. \right] + \frac{T_2}{\sqrt{N}} \sum_{\substack{kk'j \\ \sigma}} \left(b^+_{k\sigma}\, c^+_{k'\bar{\sigma}}\, A_j + H.c. \right) \qquad (13)$$

The second term in Eq.(13) refers to the normal metal, with states labeled by the quantum numbers k,σ and energies ε'_k. The third term in (13)

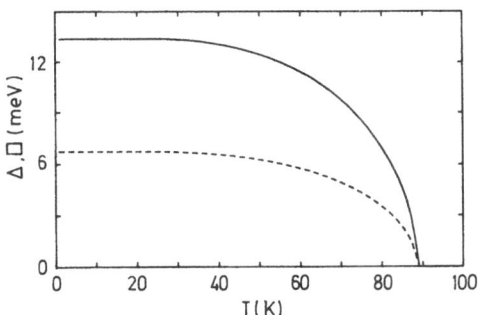

Fig. 3. Superconducting order parameters as a function of temperature, using A=0.1eV, W=0.5eV, I=0.117eV. The critical temperature is Tc=89K. Full line: Δ ,dashed: □

accounts for the usual tunneling processes between a normal metal and a BCS-type superconductor. Basically, at T=0K the tunneling electron enters the superconductor and creates a single-particle excitation thereby breaking a Cooper pair. The last term in (13) describes anomalous tunneling processes involving the scattering between local pairs and conduction electrons. Because of relation (8) between A-operators and γ-operators, two basically different tunneling contributions arise from this term: one analogous to the usual process already mentioned, and truly anomalous contributions involving creation (or destruction at T>0K) of a localized excitation (γ) in addition to the BCS -single particle excitation (α).

As for the normal phase in Ref.[6], we employ the Fermi Golden Rule to calculate the tunneling conductance g(V). We obtain the following expressions for the contributions to the tunneling conductance mentioned above :

i) the usual tunneling contribution between the normal metal and the conduction band states of the HTS,

$$\frac{g_1(V)}{g_0} = \int_{-\infty}^{\infty} dE\ \rho_s(E)\ \left(1-f(E)\right) \left\{ -\frac{\partial f}{\partial(E-eV)} - \frac{\partial f}{\partial(E+eV)} \right\} \tag{14}$$

g_0 denotes the conductance between the two materials in the normal phase, $g_0 = 4\pi e^2 \rho_n \rho_c |T_1|^2/\hbar$ being ρ_n and ρ_c the densities of states of both materials, $\rho_s(E) = |E|/[(E^2-\Delta^2)]^{1/2}$, and $f(E)$ is the usual Fermi distribution. We denote by $\partial f/\partial x$ the derivative of the Fermi function with respect to energy evaluated at energy x.

The next contributions arise from the anomalous tunneling term in Hamiltonian (13), where we assume the local pair energies to be distributed according to the density of states $\rho_b(\Lambda)$.

ii) a contribution analogous to the usual tunneling term,

$$\frac{g_1'(V)}{g_0} = \left|\frac{T_2}{T_1}\right|^2 \left[\frac{g_1(V)}{g_0}\right] \int d\Lambda\ \rho_b(\Lambda) . a^2(\Lambda)\ b^2(\Lambda) \tag{15}$$

iii) an anomalous contribution involving creation of a localized excitation in addition to the BCS-single particle excitation,

$$\frac{g_2(V)}{g_0} = \left|\frac{T_2}{T_1}\right|^2 \int d\Lambda\ \rho_b(\Lambda)\ \left[1-f(2R)\right] \int_{-\infty}^{\infty} dE\ \rho_s(E)\ \left[1-f(E)\right].$$

$$\cdot \left[-a^4 \frac{\partial f}{\partial(E+2R-eV)} - b^4 \frac{\partial f}{\partial(E+2R+eV)}\right] \tag{16}$$

iv) the anomalous contribution involving absorption of a localized excitation (only possible at T>0K),

$$\frac{g_3(V)}{g_0} = \left|\frac{T_2}{T_1}\right|^2 \int d\Lambda\ \rho_b(\Lambda)\ f(2R) \int_{-\infty}^{\infty} dE\ \rho_s(E)\ \left[1-f(E)\right].$$

$$\cdot \left[-b^4 \frac{\partial f}{\partial(E-2R-eV)} - a^4 \frac{\partial f}{\partial(E-2R+eV)}\right] \tag{17}$$

In the normal state we have $a^2=1$ and $b=0$, so that these expressions reduce to our previous results [6].

In Fig.4 we plot the above mentioned contributions and the total conductance for a temperature below Tc. In Fig.5 we plot $g(V)/g(0.1V)$ for different temperatures. We take $|T_2/T_1|^2=1.4$, with these parameters we can reproduce the normal state tunneling results of Ref.[12]. Our results show the effect of gap Δ due to the usual tunneling processes, superimposed to the anomalous V-shaped contribution associated to localized excitations. Comparison with experimental data from Ref.[12] evidences a qualitative agreement with these results, and indicates the absence of both the observed additional structure and zero bias anomaly in our results. However, as discussed in Ref.[14], there is still debate on the presence and eventual intrinsicalness of these features.

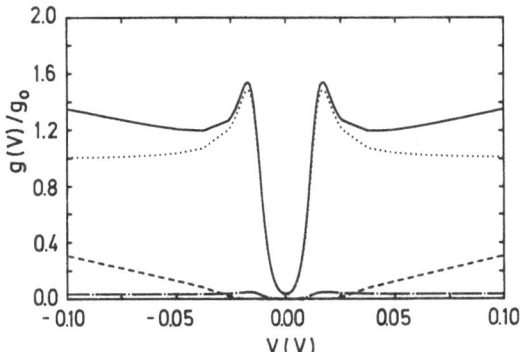

Fig. 4. Voltage dependence of the tunneling conductance contributions at T=30K. Dotted line: $g_1(V)$, dashed line: $g_2(V)$, dot-dashed line: $g_1'(V)$. Full line: total $g(V)$

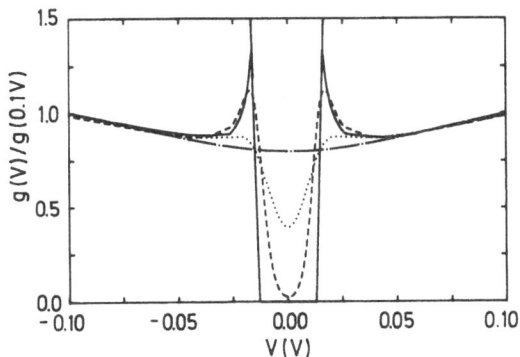

Fig. 5. Voltage dependence of the tunneling conductance at different temperatures. Full line: T=0, dashed line: 30K, dotted line: 65K, dot-dashed line: 100K.

To describe photoemission we consider the following Hamiltonian [3,6]:

$$H = H_{MFA} + \hbar w_q a_q^+ a_q + \sum_{k\sigma} E_k^f f_{k\sigma}^+ f_{k\sigma} +$$

$$+ V_1 \sum_{kk'\sigma} \left(a_q + a_{-q}^+ \right) \left[f_{k\sigma}^+ c_{k'\sigma} + H.c. \right] +$$

$$+ \frac{V_2}{\sqrt{N}} \sum_{\substack{kk'j \\ \sigma}} \left(a_q + a_{-q}^+ \right) \left[f_{k\sigma}^+ c_{k'\bar{\sigma}}^+ A_j + H.c. \right] \qquad (18)$$

where a_q^+ creates a photon of momentum q and energy $\hbar w_q$, and $f_{k\sigma}^+$ creates an emitted photoelectron of energy E_k^f. The last two terms express the interaction of the HTS with the incident radiation and describe two different photoemission processes. One is the usual term describing photoemission of a conduction band electron, therefore involving BCS-type excitations. The last term describes anomalous processes in which the radiation delocalizes a pair, ejecting one of the electrons out of the sample and placing the other in a hole state in the conduction band.

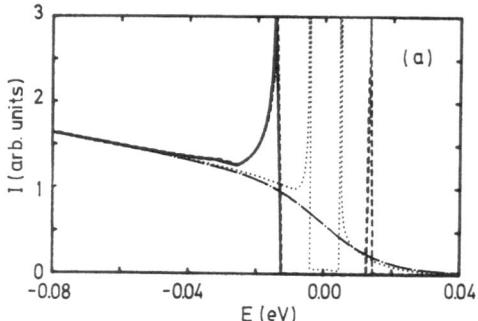

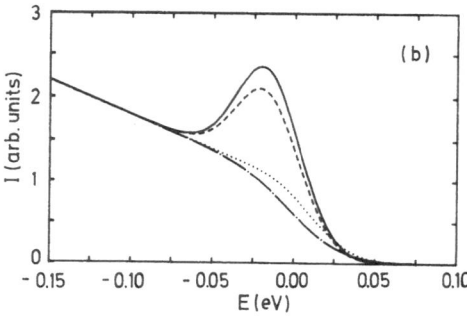

Fig. 6. (a) Photoemission spectrum as a function of initial state energy for different temperatures. Full line: T=0, dashed line: T=30K, dotted line: T=85K, dot-dashed line: T=100K. (b) Convolution of the spectra of (a) with a gaussian of dispersion σ=20meV.

Again, taking account of (8), this process can proceed through three channels : 1) involving only BCS-type excitations, with a contribution analogous to the usual term; 2) involving creation of localized excitations, in addition to BCS-type excitations; or 3) (only at finite temperature) involving destruction of localized excitations.

We employed the Fermi Golden Rule to evaluate the probability per unit time to collect photoelectrons of given energy through the different processes mentioned, which is proportional to the photoemission spectrum :

$$I(\varepsilon) \approx h(\varepsilon) \, \rho_s(\varepsilon) f(\varepsilon) \left[1 + \left|\frac{T_2}{T_1}\right|^2 \int d\Lambda \, \rho_b(\Lambda) \, a^2 b^2 \right] +$$

$$+ \left|\frac{T_2}{T_1}\right|^2 \left\{ \int d\Lambda \, \rho_b(\Lambda) \left[1 - f(2R) \right] b^4 \, \rho_s(\varepsilon + 2R) \, f(\varepsilon + 2R) \, h(\varepsilon + 2R) \right. +$$

$$\left. + \int d\Lambda \, \rho_b(\Lambda) \, f(2R) \, a^4 \, \rho_s(\varepsilon - 2R) \, f(\varepsilon - 2R) \, h(\varepsilon - 2R) \right\} \qquad (19)$$

where ε refers to initial state energies measured from the Fermi level, and:

$$h(\varepsilon) = \theta(\varepsilon - \Delta) + \theta(-\varepsilon - \Delta) \qquad (20)$$

In Fig.6(a) we plot the photoemission spectrum for different temperatures. To enable comparison with the experiment [13], in Fig.6(b) we plot our spectrum convoluted with a gaussian function with dispersion $\sigma = 20$meV, as claimed for the experimental instrumental resolution in Ref.[13]. Our results display the qualitative features shown by experimental photoemission results. We would like to remark that in the convolution of the photoemission spectrum for T=85K, which is near Tc=89K, all gaplike structures of Fig.6(a) have been blurred because of the resolution considered.

In conclusion, in this work we present the results obtained for the tunneling conductance and the photoemission spectrum in the superconducting phase of the "mixed Fermion-Boson" model. This model proved adequate to obtain the self-energy [4] proposed by Varma et al. [5] and the observed V-shape for the normal state tunneling and photoemission densities of states [6], without referring to the one-particle density of states having a cusp at the Fermi level [6,16]. In the superconducting phase of the model two types of fermionic excitations are possible : one involving Cooper pair breaking and the other related to local pair phase correlation breaking. The gap related to the BCS-type one-particle excitations shows up in the tunneling conductance and photoemission spectrum in the usual way, while the localized excitations give rise to the anomalous V-shaped contribution to these two quantities. Differences with experimental results appear in the tunneling conductance, but related to the zero bias anomaly and additional structure observed, facts not yet well established from the experimental point of view [14,15].

References

+ Fellow of the Consejo Nacional de Investigaciones Científicas y Técnicas (CONICET).
++ Member of the Carrera del Investigador Científico (CONICET).
[1] J.Ranninger and S.Robaszkiewicz, Physica 135B, 468 (1985); S.Robaszkiewicz,R.Micnas, and J.Ranninger, Phys.Rev.B36, 180 (1987); R.Micnas, J.Ranninger,and S.Robaszkiewicz, Rev.Mod.Phys.62, 113(1990).
[2] G.M.Eliashberg , Pisma Zh.Eksp.Teor.Fiz. 46, 94 (1987).
[3] I.O.Kulik , Int.J.of Mod.Phys. B2, 851 (1988).

[4] B.R.Alascio and C.R.Proetto, Solid State Comm. 75,217(1990).

[5] C.M.Varma, P.B.Littlewood, S.Schmitt-Rink, E.Abrahams and
A.E.Ruckenstein , Phys.Rev.Lett. 63, 1996 (1989).

[6] B.R.Alascio, R.Allub, C.R.Proetto,and C.I.Ventura, Solid State Comm.
77,949(1991).

[7] D.I.Khomskii and A.K.Zvezdin, Solid State Comm. 66,651(1988).

[8] K.P.Sinha and M.Singh, J.Phys.C : S.S.P.21,L231(1988).

[9] Y.Bar Yam, Phys.Rev.B43,359(1991); Phys.Rev.B43,2601(1991).

[10] L.Ioffe, A.I.Larkin, Yu N.Ovchinnikov, and Yu Lu, in "Strongly
Correlated Electron Systems",ed. by G.Baskaran, A.E.Ruckenstein
E.Tossatti and Yu Lu, World Sci.Publ.Co., Singapore(1989).

[11] R.Friedberg and T.D.Lee, Phys.Lett.A138,423(1989), Phys.Rev.B40,
6745(1989);R.Friedberg,T.D.Lee and H.C.Ren, Phys.Lett.A152,417(1991),
Phys.Lett.A152,423(1991).

[12] M.Gurvitch, J.M.Valles Jr., A.M.Cucolo, R.C.Dynes, J.P.Garno, and
L.F.Schneemeyer , Phys.Rev.Lett. 63, 1008 (1989).

[13] J.M.Imer, F.Patthey, B.Dardel, W.D.Schneider, Y.Baer, Y.Petroff, and
A.Zettl, Phys.Rev.Lett. 62, 336 (1989).

[14] J.Geerk, G.Linker, O.Meyer, Q.Li, R.L.Wang,and X.X.Xi, Phys.C 162-164,
837(1989).

[15] A.M.Cucolo, R.Di Leo, P.Romano, L.F.Schneemeyer,and J.V.Waszczak,
Phys.Rev.B 44,2857(1991).

[16] P.Lee,Proc. of the Symposium on High Temperature Superconductivity,
Los Alamos (1989).

GRAND-CANONICAL DESCRIPTION OF THE HUBBARD HAMILTONIAN

J. Aliaga, A.N. Proto and V. Zunino

Grupo de Sistemas Dinámicos (CITEFA) and Comisión de
Investigaciones Científicas de la Prov. de Buenos Aires
C.C. 2 (1638) V. López, Argentina

I. INTRODUCTION

The electronic properties in a metallic system close to the metal-insulator transition play a crucial role in high T_c superconductor theory. Among others, the Hubbard model has been extensively studied using different methods [1-8]. In particular, the one-dimensional Hubbard Hamiltonian (H-H) is considered as the most simplified model able to treat the main characteristics of the highly correlated electron systems including the metal insulator transition. Numerical studies have shown that at half-filling this model has long-range antiferromagnetic order [3]. Recently [7-8], it has also been studied the behavior of different properties away from half-filling.

The aim of the present effort is to give a thermodynamical description of this model based on the Maximun Entropy Principle (MEP) formalism [9-13]. MEP techniques has been applied to different physical systems and it has proved to be a powerful tool to get dynamical and thermodynamical descriptions. For the present case we have chosen a mean field approximation to the Hubbard model, which allow us to get an exactly solvable density operator. Then, our principal goals in this case are: to obtain a mean-field phase diagram in terms of the band filling, and the specific heat as a function of temperature.

II. MEAN-FIELD APPROACH TO THE HUBBARD HAMILTONIAN

The Hubbard Hamiltonian is given by [1-2]

$$\hat{H} = - t' \sum_{<i;j>;\sigma} \hat{c}_{i\sigma}^{\dagger} \hat{c}_{j\sigma} + U \sum_{i} \hat{n}_{i\uparrow} \hat{n}_{i\downarrow} \tag{1}$$

Condensed Matter Theories, Vol. 7, Edited by A.N. Proto
and J.L. Aliaga, Plenum Press, New York, 1992

where $\hat{c}_{i\sigma}^{\dagger}$, $\hat{c}_{i\sigma}$ creates and destroys an electron of spin σ at site i, respectively, $\hat{n}_{i\sigma} = \hat{c}_{i\sigma}^{\dagger} \hat{c}_{i\sigma}$, t' the hopping parameter between nearest neighbours, and U the on-site Coulomb interaction. In order to obtain an *exactly* solvable model we make the following approximation in the hopping term [4-6]:

$$\hat{c}_{i\sigma}^{\dagger} \hat{c}_{j\sigma} + \hat{c}_{j\sigma}^{\dagger} \hat{c}_{i\sigma} \approx \Delta \, (\hat{c}_{i\sigma} + \hat{c}_{j\sigma}) + (\hat{c}_{i\sigma}^{\dagger} + \hat{c}_{j\sigma}^{\dagger}) \, \Delta^* - |\Delta|^2 - \hat{n}_{i\sigma} - \hat{n}_{j\sigma} \tag{2}$$

where

$$\Delta = < \hat{c}_{i\sigma}^{\dagger} + \hat{c}_{j\sigma}^{\dagger} > . \tag{3}$$

In this approximation, the Hamiltonian can be written in site-diagonal form. The sites are coupled only by the mean-field parameter Δ, to be determined from the minimum of a generalized thermodynamical potential (see below) [5]. Thus, each site can be solved independently in a subspace of four states. The Hamiltonian of the site i reads

$$\hat{H}_i = \alpha \sum_{\sigma} \hat{n}_{i\sigma} - t \sum_{\sigma} (\Delta^* \, \hat{c}_{i\sigma}^{\dagger} + \Delta \, \hat{c}_{i\sigma}) + U \hat{n}_{i\uparrow} \hat{n}_{i\downarrow} + t\Delta^2 \tag{4}$$

where $\alpha = t$, $t = mt'$, and m is the number of nearest neighbours. So, we define the operator $\hat{n}_i$, the number of electrons in the site i; $\hat{r}_i$, the number of pairs in the site i (or, the double occupancy probability); and $\hat{x}_i$ a mean-field hopping interaction between nearest neighbours, as

$$\hat{n}_i = \hat{n}_{i\uparrow} + \hat{n}_{i\downarrow} \tag{5a}$$

$$\hat{r}_i = \hat{n}_{i\uparrow} \hat{n}_{i\downarrow} = \hat{c}_{i\uparrow}^{\dagger} \hat{c}_{i\uparrow} \hat{c}_{i\downarrow}^{\dagger} \hat{c}_{i\downarrow} \tag{5b}$$

$$\hat{x}_i = (\hat{c}_{i\uparrow}^{\dagger} + \hat{c}_{i\downarrow}^{\dagger}) \Delta^* + \Delta \, (\hat{c}_{i\uparrow} + \hat{c}_{i\downarrow}) . \tag{5c}$$

From now onwards we will throw the subindex i away.

In order to evaluate mean values we introduce, at t=0, a Grand Canonical Maximum Entropy Principle density matrix. As can be easily seen, $\hat{n}$ does not commute with the Hamiltonian. So, we introduce a well known quantum method in order to properly evaluate mean values [9-13]. Beginning with the relevant operators $\hat{H}$ and $\hat{n}$, we define the whole set of "relevant operators" (those necessary to describe properly the problem at hand) through the well-known Peletminskii-Zubarev symmetry condition or closure relation

$$[\hat{H}, \hat{O}_i] = i\hbar \sum_{j=0}^{q} \hat{O}_j \, G_{ji} \, . \tag{6}$$

We obtain:

$$[\hat{H}, \hat{n}] = -i\,t\,\hat{p} \tag{7a}$$

$$[\hat{H}, \hat{x}] = -i\,\alpha\,\hat{p} - i\,U\,\hat{L}_- \tag{7b}$$

$$[\hat{H}, \hat{p}] = i\,\alpha\,\hat{x} + i\,U\,\hat{L}_+ - i\,4\,|\Delta|^2\,t\,\hat{w}_1 \tag{7c}$$

$$[\hat{H}, \hat{L}_+] = -i\,(\alpha + U)\,\hat{L}_- \tag{7d}$$

$$[\hat{H}, \hat{L}_-] = i\,(\alpha + U)\,\hat{L}_+ - i\,|\Delta|^2\,t\,\hat{w}_2 \tag{7e}$$

$$[\hat{H}, \hat{w}_1] = i\,2\,t\,\hat{p} \tag{7f}$$

$$[\hat{H}, \hat{w}_2] = i\,8\,t\,\hat{L}_- \tag{7g}$$

where $\hat{p}$ is a mean field electron's current,

$$\hat{p} = i\left((\hat{c}_\uparrow^\dagger + \hat{c}_\downarrow^\dagger)\,\Delta^* - \Delta\,(\hat{c}_\uparrow + \hat{c}_\downarrow)\right) , \tag{8a}$$

and $\hat{L}_+$, $\hat{L}_-$, $\hat{w}_1$, and $\hat{w}_2$ are given by

$$\hat{L}_+ = (\Delta^* \hat{c}_\uparrow^\dagger + \Delta\,\hat{c}_\uparrow)\,\hat{n}_\downarrow + \hat{n}_\uparrow\,(\Delta^* \hat{c}_\downarrow^\dagger + \Delta\,\hat{c}_\downarrow) , \tag{8b}$$

$$\hat{L}_- = i\left((\Delta^* \hat{c}_\uparrow^\dagger - \Delta\,\hat{c}_\uparrow)\,\hat{n}_\downarrow + \hat{n}_\uparrow\,(\Delta^* \hat{c}_\downarrow^\dagger - \Delta\,\hat{c}_\downarrow)\right) , \tag{8c}$$

$$\hat{w}_1 = \hat{1} - \hat{n} - (\hat{c}_\uparrow^\dagger \hat{c}_\downarrow + \hat{c}_\downarrow^\dagger \hat{c}_\uparrow) , \tag{8d}$$

$$\hat{w}_2 = 2\,\hat{n} - 8\,\hat{n}_\uparrow\,\hat{n}_\downarrow - 2\,(\hat{c}_\uparrow^\dagger \hat{c}_\downarrow + \hat{c}_\downarrow^\dagger \hat{c}_\uparrow) . \tag{8e}$$

In order to analyze the meaning of these operators we consider the following basis for the Hilbert space of the site i: $|0\rangle$, $|1\rangle \equiv (1/\sqrt{2})(\hat{c}_\uparrow^\dagger + \hat{c}_\downarrow^\dagger)|0\rangle$, $|2\rangle \equiv (1/\sqrt{2})(\hat{c}_\uparrow^\dagger - \hat{c}_\downarrow^\dagger)|0\rangle$, and $|3\rangle \equiv \hat{c}_\uparrow^\dagger \hat{c}_\downarrow^\dagger|0\rangle$. It can be easily seen that, on the one hand, in this basis, the matrix form of $\hat{n}$ as well as that of $\hat{r}$ are diagonal while on the other hand the matrix form of $\hat{x}$ is nondiagonal.

The operator $\hat{x}$ matrix is broken in two two-dimensional blocks (in the subspaces $\{|0>, |1>\}$ and $\{|2>, |3>\}$, respectively). Each of these blocks is proportional to the real, nondiagonal, Pauli matrix ($\hat{\sigma}_1$). Thus, these two-dimensional subspaces are dynamically decoupled. The matrix form of operator $\hat{p}$ is similar to that of operator $\hat{x}$, but the two-dimensional blocks are imaginary, nondiagonal, Pauli matrices ($\hat{\sigma}_2$). The Operator $\hat{w}_1$ is diagonal and it can be thought of as a $\hat{\sigma}_3$ Pauli matrix in the subspaces $\{|0>, |1>\}$ and $\{|2>, |3>\}$. The operators $\hat{L}_+$, $\hat{L}_-$, and $\hat{w}_2$ are different from zero only in the $\{|2>, |3>\}$ subspace and they are related with the existence of the on-site Coulomb interaction. They are proportional to $\hat{\sigma}_1$, $\hat{\sigma}_2$ and $\hat{\sigma}_3$, respectively. Thus, all the relevant operators are related to the spin dynamics.

Thus, the density matrix results

$$\hat{\rho} = \exp\left(- \lambda_0 \hat{I} - \beta \hat{H} - \lambda_2 \hat{n} - \lambda_3 \hat{x} - \lambda_4 \hat{p} - \lambda_5 \hat{L}_+ - \lambda_6 \hat{L}_- - \lambda_7 \hat{w}_1 - \lambda_8 \hat{w}_2\right), \qquad (9)$$

where the nine Lagrange multipliers, λ_0, β, λ_i, $i=2,...,8$, are determined in order to fulfil

$$< \hat{O}_j / \hat{\rho} > = \mathrm{Tr}\,[\hat{\rho}(t)\,\hat{O}_j] = o_j, \quad j=0,1,...,8\ . \qquad (10)$$

It can be proved that this definition of the density matrix maximizes the entropy $S[\hat{\rho}]$ given (in units of the Boltzmann constant) by

$$S[\hat{\rho}] = - \mathrm{Tr}\,(\hat{\rho}\ln\hat{\rho}\,) = \lambda_0 + \beta <\hat{H}> + \sum_{j=2}^{8} \lambda_j < \hat{O}_j > \qquad (11)$$

subject to the initial constraints determined through Eq. (10). It can also be demonstrated that the operator $\hat{\rho}(t)$ obeys the Liouville equation. Using the Ehrenfest theorem and the closure condition (Eq. (6-7)), we obtain the temporal evolution of the mean values of the operators and the Lagrange multipliers

$$\frac{d < \hat{O}_i >_t}{dt} = - \sum_{j=0}^{8} < \hat{O}_j >_t G_{ji}\,, \qquad (12)$$

$$\frac{d\,\lambda_i}{d\,t} = \sum_{l=0}^{8} G_{il}\,\lambda_l\,, \quad i=1,2,...,8. \qquad (13)$$

These evolutions are oscillatory functions of frequency $\sqrt{\alpha^2+8t^2|\Delta|^2}$ and $\sqrt{(\alpha+U)^2+8t^2|\Delta|^2}$. It can be seen that $\mu \equiv \lambda_2/\beta$, the chemical potential, is an invariant of the motion.

The partition function of the system, λ_0, is evaluated diagonalizing the density matrix over the states $|0>$, $|1>$, $|2>$ and $|3>$. We obtain

$$\lambda_0 = \ln 2 - \beta b - \frac{\alpha\beta+\lambda_2}{2} + \ln (A) \tag{14}$$

where

$$b = t\,\Delta^2 , \tag{15a}$$

$$A = C \cosh k_1 + \cosh k_2 \tag{15b}$$

$$k_1 = \sqrt{Z_1^2 + X_1^2 + Y_1^2} , \tag{15c}$$

$$k_2 = \sqrt{Z_2^2 + X_2^2 + Y_2^2} , \tag{15d}$$

and

$$C = \exp(-(\alpha+U/2)\beta-\lambda_2) \tag{16a}$$

$$Z_1 = \frac{(\alpha+U)\,\beta + \lambda_2 - 2\,\lambda_7 - 8\,\lambda_8}{2} , \tag{16b}$$

$$X_1 = \sqrt{2}\,|\Delta|\,(\beta\,t - \lambda_3 - \lambda_5) , \tag{16c}$$

$$Y_1 = \sqrt{2}\,|\Delta|\,(\lambda_4 + \lambda_6) , \tag{16d}$$

$$Z_2 = \frac{\alpha\,\beta + \lambda_2 - 2\,\lambda_7}{2} , \tag{16e}$$

$$X_2 = \sqrt{2}\,|\Delta|\,(\beta\,t - \lambda_3) , \tag{16f}$$

$$Y_2 = \sqrt{2}\,|\Delta|\,\lambda_4 . \tag{16g}$$

Using Eqs. (13) it can be demonstrated that k_1 and k_2 are invariants of the motion. Now, the entropy of the system (Eq. (11)) can be evaluated

$$S = \ln 2 + \ln(A) - \frac{C \ln(C) \cosh k_1 + C k_1 \sinh k_1 + k_2 \sinh k_2}{A} . \tag{17}$$

We can obtain the mean values of the relevant operators using the equation

$$\frac{\partial \lambda_0}{\partial \lambda_i} = - <\hat{O}_i> . \tag{18}$$

It is also possible to evaluate the specific heat

$$C(\beta,\lambda_2,...,\lambda_8) \equiv -\beta \frac{\partial S}{\partial \beta}$$

$$= -\beta^2 \left(\frac{\partial <\hat{H}>}{\partial \beta} + \sum_{j=2}^{8} \lambda_j' \frac{\partial <\hat{O}_j>}{\partial \beta} \right), \tag{19}$$

where $\lambda_j' \equiv \lambda_j/\beta$ are thermodynamical intensive variables [13].

As it has been stated at the beginning, the mean-field parameter Δ must be determined in a self-consistent way. This can be done introducing the Generalized Thermodynamical Potential, G,

$$G = <\hat{H}> - S/\beta + \sum_{j=2}^{8} \lambda_j' <\hat{O}_j> = -\lambda_0/\beta \tag{20}$$

Thus, the parameter Δ is that which minimizes the thermodynamical potential G. It can be demonstrated that the same value of Δ is obtained if one chooses the values of Δ which maximizes the entropy, subject to the initial constraints determined through Eq. (10); as it should be, due to the complete equivalence between the thermodynamical energetic and entropic descriptions.

In order to make the number of variables smaller we consider Grand-canonical initial conditions, i.e. $\lambda_i = 0$; $i = 3,...,8$. With these elements at hand, we can obtain the phase diagram and the specific heat. This results can be compared with those of Refs. [9-10]. In Fig. 1 we show the variation of the metal-insulator transition temperature T_c as a function of the mean value of the number of particles per site, $<\hat{n}>$. In Fig. 1a we show the results for small Coulomb repulsion ($U = 0$ (A), $U = 3$ t (B), $U = 3.9$ t (C)). Bellow the curve the metallic phase ($\Delta \neq 0$) is obtained. For $U \leq 2$ t we obtain that the maximum transition temperature occurs at half-filling ($<\hat{n}> = 1$). For $U > 2$ t the transition temperature has a maximum located away from half-filling (see the curves labeled (B) and (C) in Fig. 1a). In Fig. 1b we depict the $U > 4$ t cases ($U = 5$ t (D), $U = 6$ t (E), $U = 10$ t (F), $U = 20$ t (G)). For $T < 0.3$ t we obtain a first order transition between a metallic phase and an antiferromagnetic phase at half-filling (dashed line).

The specific heat as a function of the mean value of the number of particles per site is shown in Fig. 2. In Fig. 2a we present the results for $U = 3.9$ t and $T = 1.0$ t (A), $T = 0.5$ t

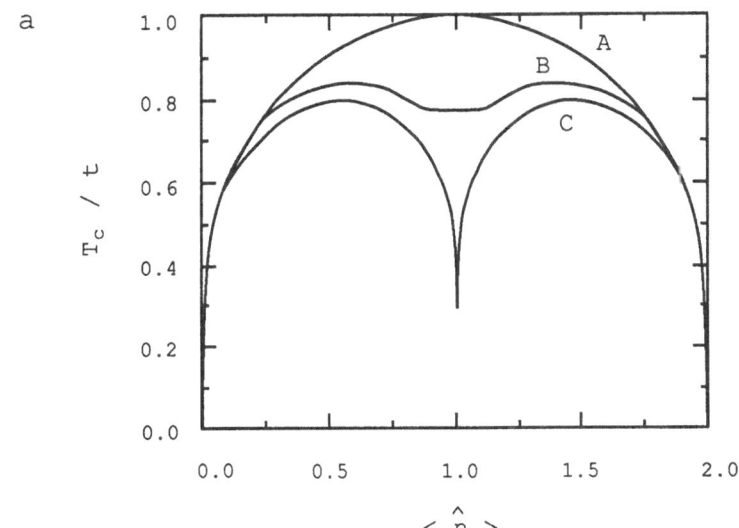

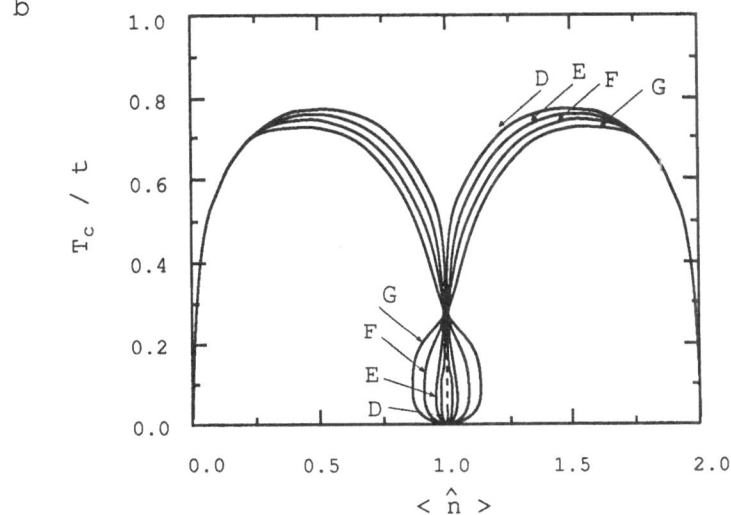

Fig. 1 Metal-insulator transition temperature T_c as a function of the mean value of the number of particles per site, $<\hat{n}>$. In Fig. 1a we show the results for small Coulomb repulsion ($U = 0$ (A), $U = 3\,t$ (B), $U = 3.9\,t$ (C)). In Fig. 1b we depict the $U > 4\,t$ cases ($U = 5\,t$ (D), $U = 6\,t$ (E), $U = 10\,t$ (F), $U = 20\,t$ (G)). For $T < 0.3\,t$ we obtain an antiferromagnetic phase at half-filling (dashed line).

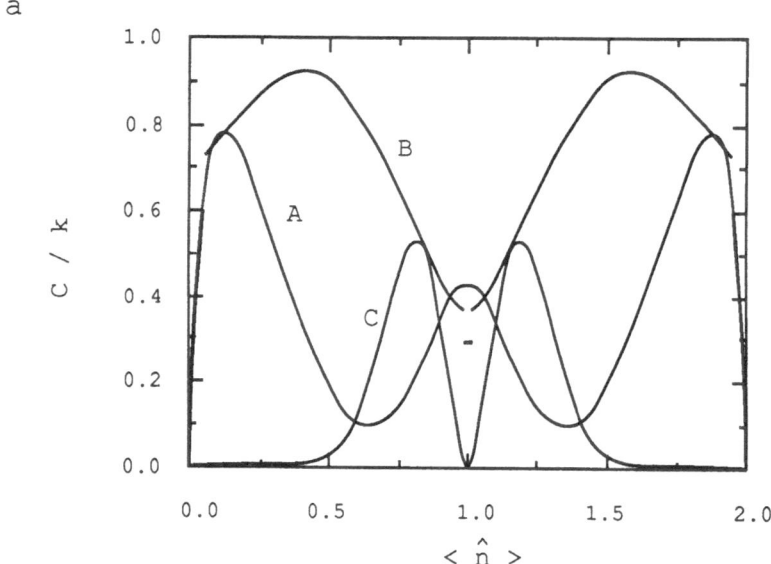

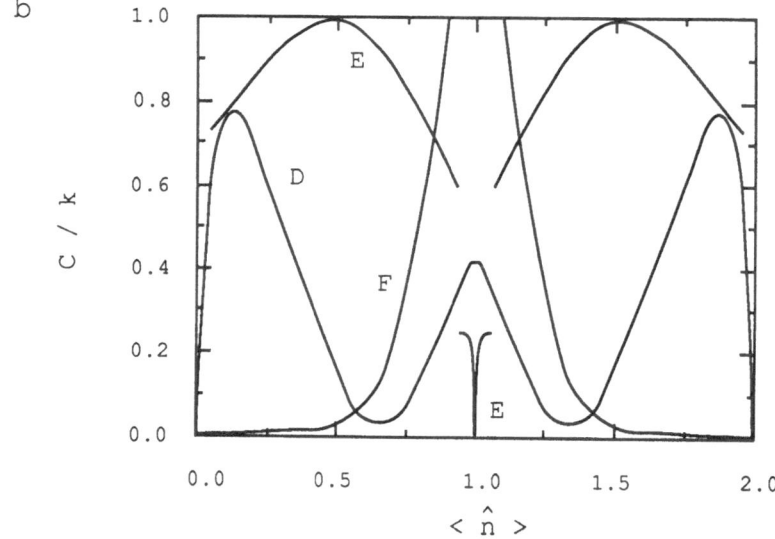

Fig. 2 The specific heat as a function of the mean value of the number of particles per site. In Fig. 2a we present the results for U = 3.9 t and T = 1.0 t (A), T = 0.5 t (B), T = 0.1 t (C). The results for U = 6 t and T = 1.0 t (D), T = 0.5 t (E), T = 0.1 t (F) are shown in Fig. 2b.

(B), T = 0.1 t (C). In the first case, (A), the system is an insulator while in the third case, (C), the system is a conductor, for all values of $<\hat{n}>$. In the second case, (B), three second order transitions occur. At $<\hat{n}> \cong 0$, 1 and 2 we obtain metal-insulator transitions. The results for U = 6 t and T = 1.0 t (D), T = 0.5 t (E), T = 0.1 t (F) are shown in Fig. 2b. In (D) the system is an insulator for all values of $<\hat{n}>$. The (E) case has three second order transitions, at $<\hat{n}> \cong 0$, 1 and 2. In (F) it can be seen a first order transition and the antiferromagnetic gap at $<\hat{n}> \cong 1$.

Thus, we obtain a grand-canonical description of the Hubbard Hamiltonian in the mean-field hopping solution which shows first and second order transitions. This approach allows for the evaluation of the expectation values of the relevant operators, the temporal evolution of the mean values of these operators and the Lagrange multipliers. It is also possible to obtain the mean-square local moment, the energy and kinetic energy per site and the compressibility for different band-fillings.

ACKNOWLEDGEMENTS

J. A. thanks the Argentine National Research Council (CONICET) for its support. A. N. P. and V.Z. acknowledge support from the Comisión de Investigaciones Científicas de la Provincia de Buenos Aires (CIC).

REFERENCES

1. P. W. Anderson, Science **235**, 1196 (1987).

2. J. Hubbard, Proc. Roy. Soc. **A276**, 238 (1963).

3. J. Hirsch and S. Tang, Phys. Rev. Lett. **62**, 591 (1989).

4. N.N. Bogoliubov, N. Cim. 7 (1958) 6; P.D. de Gennes, Superconductivity of Metals and Alloys, Benjamin, N.Y. 1966.

5. Michael Ma and Patrick A. Lee, Phys. Rev. **B32**, 5658 (1985).

6. R. Allub, A. Caro and C. Wiecko, Phys. Rev. **B36**, 8823 (1987); R. Allub, Phys. Rev. **B37**, 7815 (1988).

7. A. Moreo and D. Scalapino, Phys. Rev. B 43, 8211 (1991).

8. A. Moreo, D.J. Scalapino, R.L. Sugar, S. R. White and N.E. Bickers, Phys. Rev. B 41, 2313 (1990).

9. A. N. Proto, "Maximun Entropy Principle and Quantum Mechanics", Procc. XIII International Conference on Condensed Matter Theories, edited by V.C. Aguilera Navarro (Plenum, 1989).

10. A.N. Proto, J. Aliaga, D. R. Napoli, D. Otero, and A. Plastino, Phys. Rev. **A39** (1989) 4223.

11. J. Aliaga and A.N. Proto, Phys. Lett. **A142** (1989) 63.

12. J. Aliaga, G. Crespo and A.N. Proto, Phys. Rev. **A42** (1990) 618.

13. J. Aliaga, D. Otero, A. Plastino and A.N. Proto, Phys. Rev. **A38** (1988) 918.

REACTANT SEGREGATION : THE EFFECT OF STRONG SPACE DISORDER IN

DIFFUSION-LIMITED BIMOLECULAR REACTIONS

H.S. Wio[*]

Centro Atómico Bariloche and Instituto Balseiro[#]
8400 S.C. de Bariloche, Rio Negro, Argentina

M.A. Rodriguez and L. Pesquera

Departamento de Física Moderna, Universidad de Cantabria
E-39005 Santander, Spain

C.B. Briozzo

FAMAF, Universidad Nacional de Córdoba
5000 Córdoba, Argentina

INTRODUCTION

Quite recently, the non-classical kinetics of the diffusion-limited reaction processes for two species annihilation A+B→C (C: inert specie), has attracted considerable attention [1-7]. Such a process has been studied as a model of several different physical and chemical systems (ionic, electron-hole and defect recombination, matter-antimatter annihilation, etc.). An aspect pointed out almost a decade ago by Zeldovich and coworkers was the possibility of macroscopic segregation [8]. Several recent reviews have analyzed this and related aspects of such systems [9-12]. Among other problems, authors have been concerned with the influence of dimensionality, initial conditions, sources, conservation laws, etc. [1,5,6,12]. Also the form of the kinetic or rate equations have received some attention [4,7,11 12,13].

In this contribution we would like to address the effect of spatial correlated disorder on the diffusive properties of the above mentioned systems and, consequently, its effect on the macroscopic segregation characteristic. The analysis of diffusion in random media has been done mainly by means of discrete models [14], but without taking into account the effect of correlated disorder. However, some aspects of correlated disorder in a discrete medium, have been discussed in Ref. 15. Also for discrete systems, and based on operator projection techniques and ordered cumulants, a systematic way of obtaining an effective medium approximation (EMA) and perturbations around it, have been recently introduced [16]. Although the bibliography of continuous models of disorder is meager [17], there is a strong interest in this problem, and the previously refered EMA for discrete systems has been extended to such a case [18]. We will apply here this new procedure, discussing only the one-dimensional problem, the higher dimensional situation will be considered in the future. From the results of Ref. 18, it is clear that the effect of weak disorder will not alter the conclusions of Refs. [6,12], as it comes out that the resulting effective diffusion constant only implies a change in its magnitude but not a time dependence. We then concentrate on the most interesting case of strong disorder [16-18].

THE METHOD

We follow the approach of Refs. [6,12]. This means we assume to be valid the reaction diffusion equations for the sum and difference variables: $\rho(x,t) = 1/2[\rho_A(x,t)+ \rho_B(x,t)]$ and $\gamma(x,t) = 1/2 [\rho_A(x,t) - \rho_B(x,t)]$, where ρ_A and ρ_B are the local densities of the A and B reactants respectively. The equations are:

$$\dot{\gamma}(x,t) = D_o \, \partial_x^2 \, \gamma(x,t) + R_\gamma(x,t) \tag{1a}$$

$$\dot{\rho}(x,t) = D_o \, \partial_x^2 \, \rho(x,t) - k_e \, [\rho^2 - \gamma^2] + R_\rho(x,t) \tag{1b}$$

where R_γ and R_ρ are the associated source terms, and k_e the local (time-independent) rate coefficient. Applying the continuous version of the EMA according to Ref. [18] to Eq. (1a) we arrive at:

$$\dot{\bar{\gamma}}(x,t) = \int^t D(t-t') \, \partial_x^2 \, \bar{\gamma}(x,t') \, dt' + R_\gamma \, (x,t) \tag{2}$$

where the overbar indicates the (EMA) averaged results. The effective, time dependent, diffusion coefficient $D(t)$ is given, in the Laplace transformed space $t \rightarrow s$ (hereafter, the argument will indicate the space in which we are working) by the self consistent condition:

$$\left\langle \frac{\eta - D(s)}{1 + \frac{1}{\bar{D}(s)}\left\{1 - l \, G_1(o,s)\right\}\left[\eta - D(s)\right]} \right\rangle = 0 \tag{3}$$

η being the stochastic variable whose statistical properties are those corresponding to the correlated spatial disorder under consideration, <> indicates the average over all possible randomness, l is the distance among sites in a discrete medium to which the present EMA condition is equivalent, and G_1 the one point Green function of the ordered case [20].

As shown in Refs [6,12] the best way to obtain information regarding the degree of segregation is through the analysis of the correlation function $C(x-x',t) = \langle\gamma(x,t) \, \gamma(x',t)\rangle$. From the knowledge of the initial correlation $\langle\gamma(x,o) \, \gamma(x',o)\rangle$, we can obtain $C(x-x',t)$ as:

$$C(x-x',t) = \int \bar{G}_2(x,x',t \, | \, x_1,x_2,0) \, \langle\gamma(x_1,0) \, \gamma(x_2,0)\rangle \, dx_1 dx_2 \tag{4}$$

where $\bar{G}(x,x',t \, | x_1,x_2 0)$ is the associated two-point Green function. Here we make the following Ansatz for such a Green function:

$$\bar{G}_2(x,x',t| \, x_1,x_2,0) \approx \bar{G}_1(x,t|x_1,0) \, \bar{G}_1(x',t) \, x_2,0) \tag{5}$$

where $\bar{G}_1(x,t|x,0)$ is the one-point Green function, solution of Eq.(2). A thorough discussion of this Ansatz and other aspects will be given elsewhere [19].   In the following we consider two types of disorder: (a) dichotomic and (b) Poisson.   In both cases we consider the influence of initial conditions (the so called "bing-bang" problem) or of source terms (that could arise for instance from the inverse reaction C$\rightarrow$A+B).

For dichotomic disorder we have that $\eta = \Delta$ with probability α, and $\eta = 0$ with probability $(1-\alpha)$ (with $0<\alpha<1$, and it is possible to show that $\alpha = D_0/\Delta$), leading in the limit s$\rightarrow$0 (or correspondingly the long time limit) to D(s)$\sim\lambda^2$S (the exact value is $\lambda^2 = \alpha\ l^2/6$, l being the correlation length [20]).   By Fourier and Laplace transforming Eq. (2), we arrive at the following results for the G_1 Green function:

$$\bar{G}_1(k,t|0) \simeq (1+\lambda^2 k^2)^{-1} \tag{6}$$

which implies that G(x,t|x',0) is time independent.   Then, the Ansatz of Eq. (5) gives (neglecting sources):

$$<\gamma(k,t)\gamma(k',t)> = \bar{G}(k,k't|0) <\gamma(k,0)\ \gamma(k',0)>$$

$$\approx \bar{G}(k,t|0)\ \bar{G}(k',t|0)\ <\gamma(k,0)\ \gamma(k',o)> \tag{7}$$

Following Refs. (6,12) we can choose uncorrelated or correlated initial conditions.   In the first case $<\gamma(k,0)\gamma(k',0)> = \frac{N}{2} \{\delta_{k+k',0} - \delta_{k,0}^- \delta_{k',0})$ which leads to

$$<\gamma(xt)\gamma(x',t)>_u \sim \frac{n}{2\lambda}\left\{ 1 + \frac{|x-x'|}{\lambda}\ e^{-|x-x'|/\lambda}\right\} \tag{8}$$

$(n = N/V$, V being the volume) from which results

$$\lim_{t\longrightarrow\infty} <\gamma(x,t)^2> = \frac{n}{2\lambda} \tag{9}$$

Making an ad-hoc analysis of the Eq. (1b) (including some hand waving) we arrive at $<\rho_{st}(x)^2> = <\gamma_{st}(x)^2>$.   Then the segregation parameter introduced in Refs. [6,12] turns out to be S(t) $\sim$ 1, which clearly indicates that the "subdiffusive" behavior, due to the strong disorder, increases the segregation with respect to the results of Ref. [12] (see its Eq. (3.18)).   On the other hand if we consider a correlated initial condition (with c the geminate correlated placement) for the one dimensional case:

$$<\gamma(k,0)\ \gamma(k',0)> = \frac{N}{4} [1 - \cos(k\ c)]\ \delta_{k+k',0} \tag{10}$$

We have that $C_c(x,t) = C_u(x,t) - 1/2\ [\ C_u(x-c,t) + C_u(x+c,t)]$, leading to

$$<\gamma(x,t)^2>_c \underset{t\longrightarrow\infty}{\sim} \frac{n}{2\lambda} \left[1-(1+\frac{c}{\lambda})e^{-c/\lambda}\right] \tag{11}$$

indicating a reduction of the segregation due to the initial correlation. We find a kind of competition between the "effective diffusion length" λ, and the initial geminate correlated placement c.   If c$>>\lambda$ correlated particles are in different diffusive regions and this is equivalent to the uncorrelated situation, and we recover the previous result (9).   In the opposite limit c$<<\lambda$, ("perfect placement") the probability of extintion of the correlated particles is larger and the segregation is reduced.

In order to consider the problem with sources, we neglect now the initial condition, and apply the same Ansatz as before, but including the sources:

$$\langle \gamma(k,t)\; \gamma(k't)\rangle = \int dt_1 \int dt_2\; \bar{G}(k,k'\,;\; t-t_1,\; t-t_2)\langle R_\gamma(k,t_1)R_\gamma(k',t_2)\rangle$$

$$\approx \int dt_1 \int dt_2\; \bar{G}(k,t-t_1)\; \bar{G}(k',t-t_2)\langle R_\gamma(k,t_1)R_\gamma(k',t_2)\rangle \qquad (12)$$

Following again Refs. [6,12], we could considered the case of conservative sources ($\langle R_\gamma(k,t)\rangle$=0 implying $\langle \gamma(k,t)\rangle$=0), also having the possibility of correlated or uncorrelated sources. We will not discuss those results here [19,20].

The case of disorder of the Poisson type corresponds to $D(s)\sim s^\nu$ for s→0 (with 0<ν<1). In this case the analysis gets much more involved but additional approximations lead us also to macroscopic segregation. However, and consistently with a less disordered situation, following the spirit of the previous lines, we find a lower degree of segregation as compared to the dichotomous case. A thorough discussion of this case as well as other aspects of the problem will be the subject of a forthcoming paper [19].

FINAL REMARKS

Let us summarize the effect of disorder on the time evolution and segregation of the diffusing particles. When disorder is weak only a quantitative change of the diffusion coefficient is possible and no important changes with respect to the ordered situation appears. When the disorder is so strong that it produces subdiffusive behavior but not strong enough to disconnect diffusive regions, we obtain a slowing down of the reaction process and also of the segregation (Poisson disorder). In the percolative case, where disorder is so strong that it produces isolated diffusive regions, extinction is not possible and a segregated steady state appears. The effect of disorder in the case with sources is similar. We observe an infinite growth if disorder is able to isolate correlated sources.

A point worth to be remarked is that the disorder leads to modified exponents in the extinction process, a relevant result in a problem where exponents have been thought to be *universal*. As a final point, very general scaling and dimensional arguments have indicated that the segregation effect will exist only at dimensions lower than 2, that 2 is a kind of *critical* dimension, and that there is no segregation at d>2. However, there are experimental evidence of segregation effects at d=3 [21], that perhaps, among other aspects, could be originated by the presence of disorder.

ACKNOWLEDGMENT

Support from the Ministerio de Educación y Ciencia, Spain, through its Programa de Cooperación con Iberoámerica is greatly acknowledged.

REFERENCES

*   Member of CONICET, Argentina
\#   Universidad Nacional de Cuyo and Comisión Nacional de Energía Atómica.
1)   L.W. Anacker and R. Kopelman, Phys. Ref. Lett., 58, 289 (1987); R. Kopelman, Science 241, 1620 (1988).
2)   D. ben-Avraham, Phil. Mag., B56, 1015 (1987).

3) J.S. Newhause and R. Kopelman, J. Phys. Chem. 92, 1538 (1988).

4) S. Kanno, Prog. Theoret. Phys. 79, 721 , 1330 (1988).

5) D. ben-Avraham and Ch. R. Doering, Phys. Rev. A37, 5007 (1988).

6) K. Lindenberg, B.J. West and R. Kopelman, Phys. Rev. Lett. 60, 1777 (1988).
   B.J. West, R. Kopelman and K. Lindenberg, J. Stat. Phys. 54, 1429 (1989).

7) E. Clement, L.M. Sanders and R. Kopelman, Phys. Rev. A39, 6455, 6466 (1989).

8) A.A. Orchinnikov and Y.B. Zel'dovich, Chem. Phys. 28, 215 (1978).

9) Y.B. Zel'dovich and A.S. Mikhailov, Sov. Phys.- Usp. 30, 977 (1987).

10) V. Kuzovkov and E. Kotomin, Rep. Prog. Phys. 51, 1479 (1988).
    A. Blumen, J. Klafter and G. Zumofen, in "Optical Spectroscopy of Glasses" , I. Zschokke, Ed. (D. Reidel, 1986), p. 199.

11) A.S. Mikhailov, Phys. Rep. 184, 307 (1989).

12) K. Lindenberg, B.J. West and R. Kopelman, in "Noise and Chaos in Nonlinear Dynamical Systems", F. Moss, L.A. Lugiato and W. Schleich, Eds. (Cambridge, U.P., 1990) p. 142.

13) Ch. R. Doering and D. ben-Avraham, Phys. Rev. A38, 3035 (1988).
    D. ben-Avraham and Ch. R. Doering, Phys. Rev. A39, 6436 (1989).

14) S. Alexander, J. Bernasconi, W.R. Schneider and R. Orbach, Ref. Mod. Phys. 53, 175 (1981).
    J.W. Haus, and K.W. Kehr, Phys. Rep. 150, 263 (1987).
    S. Havlin and D. ben-Avraham, Adv. Phys. 36, 695 (1987).

15) S. Havlin, M. Schwartz, R. Blumberg, A. Bunde and H. E. Stanley, Phys. Rev. A40, 1717 (1989).

16) M.A. Rodriguez, E. Hernandez-García, L. Pesquera and M. San Miguel, Phys. Rev. B40, 4212 (1989).
    E. Hernandez-García, M.A. Rodriguez, L. Pesquera and M. San Miguel, Phys. Rev. B. in press.

17) J. Heinrichs, Phys. Rev. Lett. 52, 1262 (1984).
    G. Nicolis and V.J. Altares, J. Phys. Chem. 93, 2861 (1989).

18) A. Valle, M.A. Rodriguez and L. Pesquera, Phys. Rev. A43, 948 (1991).

19) H.S. Wio, M.A. Rodriguez, C. Briozzo, L. Pesquera to be submitted.

20) H.S. Wio, M.A. Rodriguez, C.B. Briozzo and L. Pesquera, Phys. Rev. A 43, (12) June 1991.

21) C. Abromeit, Int. J. Mod. Phys. B3, 1301 (1989).

INTERACTION EFFECTS IN THE TODA SOLITON STATISTICS

C. Lucheroni and F. Marchesoni

Dipartimento di Fisica dell'Universita' and INFN
I-06100 Perugia (Italy)

1. INTRODUCTION

A relaxation process in condensed phase is commonly modelled by coupling the variables which represent the physical observables to an adequate linear heat-bath at thermal equilibrium[1]. Systematic projective techniques allow, then, to eliminate the fast relaxing heat-bath degrees of freedom[1,2] and obtain generalized Langevin equations for the relevant variables. Relaxation processes are thus driven by internal noises which obey the fluctuation-dissipation theorem[2,3]. This approach proved useful to study a number of diverse relaxation processes such as NMR spectra[1], dielectric relaxation of polar liquids[1] and thermally activated chemical reactions[4]. However, when the same relaxation process is investigated under different experimental conditions, for instance by varying pressure, temperature, concentrations, etc., the relaxation mechanism outlined above has to be adjusted to account for the modified physical properties of the environment where the process takes place. This can only be done by changing the values of the linear heat-bath parameters *ad hoc*, namely, according to our present understanding of the microscopic dynamics of the condensed phase[5].

More realistic models might be proposed[2] where the interaction of the observables of the process under study with the environment is reproduced by means of a *nonlinear* heat-bath, especially envisaged to incorporate by one token the response of the environment, too, at varying the experimental conditions. In the present paper we limit ourselves to study the one-dimensional nonlinear heat-bath represented by an infinite discrete chain with anharmonic interactions described by a *monostable* inter-site potential $V(r)$. The actual choice of $V(r)$ is not so important provided that we keep the temperature below a characteristic value $T_o^{[1]}$

$$k_B T_o \equiv \frac{4}{3} \frac{[V_0^{(2)}]^2}{V_0^{(4)}} \qquad (1.1)$$

where $V_0^{(n)}$ is the n-th derivative of $V(r)$ at the stable point r_0, $V^{(1)}(r_0) = 0$. The condition $T \ll T_o$ corresponds to the *weak nonlinearity* limit commonly advocated in the literature[1−4] to discard nonlinear corrections.

An exactly integrable nonlinear heat bath is described by the Hamiltonian

$$H = \sum_{j=-N/2}^{N/2} \left[\frac{p_j^2}{2m_j} + V(q_{j+1} - q_j) \right] \qquad (1.2)$$

where the Toda inter-site potential[6]

$$V(r_j) = \frac{a}{b} e^{-br_j} + (a+p)r_j \qquad (a, b > 0) \qquad (1.3)$$

is an *exponential* function of the mutual displacement $r_j \equiv q_{j+1} - q_j$. The lattice spacing has been chosen as the unit of length. The characteristic temperature T_o for the potential (1.3) is given by

$$T_o = \frac{4a}{3bk_B} \left(1 + \frac{p}{a} \right) \qquad (1.4)$$

The Toda chain provides a realistic picture for a variety of as complex quasi-one-dimensional systems as polymers[6] and the DNA molecule[7,8]. In a foregoing paper[9] we have shown that the internal dynamics of such a nonlinear heat-bath has a significant impact on the relaxation processes coupled to it, by causing a strong temperature dependence of the relevant damping function (or memory kernel[2]).

In the present Communication we discuss in more detail the question of the soliton density in a Toda chain at thermal equilibrium. This problem is still being debated (see e.g. Refs. 7 and 8 and references therein). In the absence of an effective method to count non-topological (or gapless) solitons in a thermalized soliton-bearing theory, we compare the predictions of a dilute gas calculation[7] with the results of numerical simulation[8]. We conclude that for an infinite Toda chain the soliton-soliton interactions, far from being negligible, are accounted for by a T-independent factor in the relevant soliton density.

2. TODA SOLITON DENSITY IN THE DILUTE GAS APPROXIMATION

Let us start considering the case of a chain infinitely long in both directions and with equal masses $m_j = m$. Such a limit can be achieved from (1.2) by imposing periodic boundary conditions and, then, taking the limit $N \to \infty$[6]. $V(r_j)$ depends on three parameters a, b and p, where b quantifies the anharmonicity of the system and p $(p > -a)$ is the external pressure applied to the chain. From now on we set $p = 0$ and remind the reader that the statistical mechanics of the Toda chain accommodates effects due to a different value of p simply by replacing a with $a + p$[6]. This allows us to determine explicitly the effects due to varying the external pressure on the relaxing sample. For small amplitudes r_j the lattice (1.2) looks like a linear lattice with spring frequency $\omega_o^2 = ab/m$ and, therefore, the extended solutions are phonons. Large amplitude solutions, instead, involve multi-soliton configurations, as well. For a given choice of the parameters a and b, the single-soliton solution

$$f_j^S(t) \equiv a(e^{-br_j^S} - 1) = a\,\frac{\beta^2}{\omega_o^2}\,sech^2[\alpha(j - j_0) \pm \beta t] \qquad (2.1)$$

with $\beta = \omega_o\,sh\alpha$, is determined by *one* free parameter α. The lattice soliton (2.1) is a stable, pulse-like wave centered in j_0 with width $1/\alpha$, energy

$$E \equiv E(\alpha) = \frac{2a}{b}\,(sh\alpha\,ch\alpha - \alpha), \qquad (2.2)$$

momentum

$$P(\alpha) = \pm 4\sqrt{\frac{ma}{b^3}}\,(\alpha\,ch\alpha - sh\alpha) \qquad (2.3)$$

and speed $v(\alpha) = \pm\omega_o\,sh\alpha/\alpha$. Multi-soliton solutions can be decomposed for asymptotically long times into a linear superposition of single-soliton solutions with the appropriate value of α (soliton transparency[6]). Another remarkable property of the Toda chain is that the phase-shift for a phonon through a multi-soliton solution can be obtained simply by adding the phase shifts corresponding to every single-soliton component, separately (isospectral deformation[6]). In the presence of one lattice soliton (2.1) with wave-number $e^{-\alpha}$, a lattice phonon with wave-vector k, $\phi(j, k)$, has the asymptotic form $\phi(j, k) \propto e^{-ikj}$ for $j \to +\infty$ and $\phi(j, k) \propto e^{-ikj-i\eta(k,\alpha)}$ for $j \to -\infty$, where $\eta(k, \alpha) = \mp 4tg^{-1}[(tgh\frac{\alpha}{2})\,(tg\frac{k}{4})^{\mp 1}]$ for $-\pi \le k \le 0$ (upper sign) and $0 \le k \le \pi$ (lower sign), respectively. The density of phonon states in the presence of *one* soliton has to account for the relevant phase-shift $\eta(k, \alpha)$, i.e.

$$\rho(k, \alpha) = \frac{N}{2\pi}\left[1 + \frac{1}{N}\frac{\partial\eta}{\partial k}\right] \qquad (-\pi \le k \le \pi) \qquad (2.4)$$

We are now in the position to determine the soliton density in an infinite Toda chain at temperature T. For a chain with finite length N, the number of excited solitons at equilibrium, N_S, is determined by the condition that $\left(\frac{\partial F}{\partial N_S}\right)_{T,N} = 0$, where $F(N, T, N_S)$ is the free energy of the system. On assuming for the time being that we may neglect the soliton-soliton interaction (*dilute gas approximation*), i.e. $N_S \ll N$, the above condition leads immediately to the following expression for the soliton density[7]

$$n_0(T) \equiv \frac{N_S}{N} = \frac{1}{N}\frac{Z_1}{Z_0} \tag{2.5}$$

Here, Z_1 and Z_0 denote the canonical partition function of the chain with one and no solitons, respectively. On expanding the relevant chain configurations on the complete set[6] of the phonon modes, eq. (2.5) reduces to

$$n_0(T) = \frac{1}{N}\int g(E)\frac{\prod_k Z_{1,k}}{\prod_k Z_{0,k}}exp\left(-\frac{E(\alpha)}{k_BT}\right)\,dE \tag{2.6}$$

where $g(E)$ is the degeneracy function with respect to the soliton energy, $g(E) = N/v(E)$ and $v(E)$ is the single soliton speed as a function of its energy. $Z_{1,k}$ and $Z_{0,k}$ represent the phonon contributions to the partition functions Z_1 and Z_0, respectively,

$$Z_{1,k} = Z_{0,k} = 2\pi\,\frac{k_BT}{\omega(k)} \tag{2.7}$$

with $\omega(k) = 2\omega_o sin(k/2)$ (relation of dispersion for a Toda chain phonon with wave-vector k). The infinite products in eq. (2.6) can be calculated explicitly in terms of the density of phonon states (2.4), i.e.

$$\frac{\prod_k Z_{1,k}}{\prod_k Z_{0,k}} = \frac{\omega_o}{\pi k_BT}\,exp\left[\frac{1}{2\pi}\int_{-\pi}^{\pi}\frac{\partial\eta}{\partial k}\,ln(sin\,\frac{k}{2})\,dk\right]$$

$$= \frac{\omega_o}{2\pi k_BT}\,(1 - e^{-2\alpha}) \tag{2.8}$$

Finally, carrying out the the integrations in eq. (2.6) we obtain our general expression for $n_0(T)$, i.e.

$$n_0(T) = \int_0^{\infty} n_0(\alpha, T)\,d\alpha \tag{2.9}$$

where

$$n_0(\alpha, T) = \frac{3}{\pi}\frac{T_o}{T}\,\alpha\,e^{-\alpha}sh^2\alpha\,exp\left(-\frac{E(\alpha)}{k_BT}\right) \tag{2.10}$$

At low temperature[7], i.e. for $T \ll T_o$,

$$n_0(\alpha, T) \simeq \frac{3}{\pi}\frac{T_o}{T}\,\alpha^3 exp\left(-\frac{T}{T_o}\alpha^3\right) \tag{2.11}$$

whence

$$n_0(T) \simeq \frac{1}{\pi}\,\Gamma(\frac{4}{3})\,\left(\frac{T}{T_o}\right)^{1/3} \tag{2.12}$$

$n_0(T)$ can be interpreted as the fraction of the chain modes which belong with the soliton sector of the theory[6], compared to the fraction $1 - n_0(T)$ of phonon modes. It should be remarked that at high temperature $n_0(T)$ tends to unity[6,8].

However, the validity of the dilute gas approximation for the case of the Toda chain has to be assessed, yet. As a matter of fact, the temperature dependence of the soliton density (2.12) would lead to a non-analytical term in the free energy proportional to $T^{4/3}$, contrary to the exact result

$$\frac{F}{N} = -\frac{k_B T}{2} \, ln(2\pi m k_B T) - k_B T \, lnQ \qquad (2.13)$$

with

$$Q = \frac{1}{b} \left(\frac{4e}{3} \frac{T}{T_o} \right)^{\frac{3}{4}\frac{T_o}{T}} \Gamma\left(\frac{3}{4} \frac{T_o}{T} \right) \qquad (2.14)$$

obtained by Toda[6]. In the following Sections we discuss in more detail the role of soliton-soliton interaction in the determination of $n_0(T)$

3. STOCHASTIC DYNAMICS OF A SINGLE TODA SOLITON

The authors of Ref. 8 extend to the Toda chain a standard thermalization mechanism which is known to give good results for the sine-Gordon case[10]. Each site of the chain is coupled to an infinite heat bath represented by a fluctuation-dissipation term according to the equation

$$m\ddot{q}_j = -[V'(q_j - q_{j-1}) - V'(q_{j+1} - q_j)] - m\gamma\dot{q}_j + \chi_j(t) \qquad (3.1)$$

obtained from (1.2) with $m_j \equiv m$. The random forces $\chi_j(t)$ are assumed to be gaussian, zero-mean valued noises with correlation functions $< \chi_j(t) \, \chi_i(0) >= 2m\gamma k_B T \delta_{ij}\delta(t)$. The stationary state properties of the system turn out to be independent of the friction constant γ.

Let us restrict ourselves to chain configurations where only one soliton (2.1) has been nucleated. The stationary distribution of such single soliton configurations with respect to the parameter α can be determined as follows. Multiplying both sides of eq. (3.1) by $\dot{q}_j$ and summing over the index j lead to the energy balance equation

$$\frac{dE}{dt} = -2\gamma E_K + (vP)^{1/2}\eta(t) \qquad (3.2)$$

where $v(\alpha)$ and $P(\alpha)$ are as in Section 2 and E_K denotes the kinetic energy

$$E_K(\alpha) = 2\frac{a}{b} \frac{sh\alpha}{\alpha} \, (\alpha ch\alpha - sh\alpha) \qquad (3.3)$$

$\eta(t)$ is a gaussian random noise with $< \eta(t) >= 0$ and

$$< \eta(t)\eta(0) >= 2\gamma k_B T v(\alpha) P(\alpha)\delta(t) \qquad (3.4)$$

We make use, now, of the relations

$$\frac{dP(\alpha)}{dt} = \frac{1}{v(\alpha)} \frac{dE(\alpha)}{dt} \tag{3.5a}$$

$$P(\alpha) = \frac{2}{v(\alpha)} E_K(\alpha) \tag{3.5b}$$

valid for an unperturbed Toda soliton, to rewrite eq. (3.2) as

$$\frac{dP(\alpha)}{dt} = -\gamma P(\alpha) + \left[\frac{P(\alpha)}{v(\alpha)}\right]^{1/2} \eta(t) \tag{3.6}$$

A single Toda soliton coupled to a heat bath would behave like a brownian quasi-particle driven by a multiplicative noise with stationary probability density[11]

$$W(\alpha, T) = A(T) \frac{dP}{d\alpha} \left[\frac{P(\alpha)}{v(\alpha)}\right]^{1/2} exp\left(-\frac{E(\alpha)}{k_B T}\right) \tag{3.7}$$

where $A(T)$ is a suitable normalization constant.

At low temperature, $T \ll T_o$ the probability density (3.7) can be approximated to

$$W(\alpha, T) \simeq 3(k_B T_o \alpha)^{1/2} A(T) e^{-\frac{T_o}{T}\alpha^3} \tag{3.8}$$

On comparing $W(\alpha, T)$ with the distribution (2.11) of a *dilute gas* of Toda solitons at low temperature, a discrepancy becomes apparent which mainly concerns the low energy solitons. This difficulty does not show up, for instance, in the sine-Gordon chain[10]. Such a result further supports the widespread notion that soliton-soliton interactions are by no means negligible when dealing with gapless theories.

4. BEYOND THE DILUTE GAS APPROXIMATION

To compare $n_0(T)$ with outcome of the simulation work by the authors of Ref. 8, we note that the value of the temperature T_o corresponding to their choice of the lattice parameters a, b, and m is $T_o = 80^\circ K$. As a consequence $n_0(T)$ turns out to be smaller than the simulated value by a factor of up to 1.5 over the whole temperature range $(10^{-2}, 10^2)^\circ K$. The origin of such a discrepancy cannot be traced back to apparent flaws in the simulation algorithms. It has been verified, indeed, that, on both increasing the length of the simulated chain and decreasing the thermalization time for the same number of time iterations, the number of excited solitons tends to increase and, finally, levels off.

The typical values of the equilibrium soliton density in an infinite Toda chain, e.g. $n_0(T) = 0.15$, are predicted to be not small enough to rule out significant soliton-soliton interaction effects[10]. This fact is even more evident if one looks at the soliton mean free path, defined as the reciprocal of $n_0(T)$, and at the average soliton width $< \alpha^{-1} >_0$. The average $< \dots >_0$ is taken here with respect to the distribution $n_0(\alpha, T)$. The ratio of these two length scales at low temperature is independent of T and of the order of 3.

Our prediction for the Toda soliton density can be improved following a self-consistent approach which resembles the strongly interacting liquid approximation for a Fermi liquid[12]. In view of the Lax formalism, there is a one-to-one correspondence between the bound states (solitons) in both the liquid and the ideal gas picture of the Toda lattice. In particular, the derivation of the phonon phase shifts is independent of any consideration about the soliton-soliton interaction. This implies that eq. (2.5) is still valid, with the caution that Z_1 represents, now, the canonical partition function for a chain where an extra soliton has been added to the equilibrium soliton liquid. The relevant calculations run as in the dilute gas approximation after having replaced $E(\alpha)$ with the effective energy $\epsilon(\alpha)$ of the interacting soliton. In order to estimate $\epsilon(\alpha)$, we note that the soliton distribution (2.10) is sharply peaked at about $\bar{\alpha}_0 \equiv < \alpha >_0$. At low temperature $\bar{\alpha}_0 = \delta_0 (T/T_o)^{1/3}$ with $\delta_0 = 2 \Gamma(\frac{2}{3})/\Gamma(\frac{1}{3})$. We can, therefore, approximate

$$\epsilon(\alpha) = E(\bar{\alpha}_0) + \left(\frac{\partial E}{\partial \alpha}\right)_{\alpha=\bar{\alpha}_0} (\alpha - \bar{\alpha}_0) + \dots \tag{4.1}$$

and replace $E(\alpha)$ in eq. (2.10) with the first-order expansion of $\epsilon(\alpha)$, (4.1), thus obtaining a new expression for the soliton density, $n_1(T) = \kappa(\delta_0)n_0(T)$, with $\kappa(\delta) = (2/9)[\delta^8\Gamma(\frac{4}{3})]^{-1}$ $e^{2\delta^3}$, and for the mean value of α, $\bar{\alpha}_1 \equiv < \alpha >_1 = \delta_1 (T/T_o)^{1/3}$ with $\delta_1 = 4/3\delta_0^2$. The whole procedure can be iterated repeatedly by replacing $\bar{\alpha}_0$ with $\bar{\alpha}_1$ in the expansion (4.1). The sequence δ_n is shown to converge to $\delta_\infty = (4/3)^{1/3}$ and, correspondingly, the correction factor κ approaches its minimum value $\kappa(\delta_\infty) = 1.67$. This leads to our best estimate for the Toda soliton density,

$$n_\infty(T) = \kappa(\delta_\infty) n_0(T) \tag{4.2}$$

Most notably, $n_\infty(T)$ comes much closer than $n_0(T)$ to the numerical results of Ref. 8.

In conclusion, the comparison of the predictions of the dilute gas approximation with the outcome of the numerical simulations by Muto et al.[8] suggests that the soliton-soliton interaction, far from being negligible, is accounted for by a mere T-independent multiplicative factor in the soliton density.

5. CONCLUDING REMARKS

Thermalization through the coupling with a nonlinear heat-bath is likely to have remarkable bearing on a variety of relaxation processes in condensed matter. We mention here, for instance, the DNA dynamics, where the existence of localized nonlinear excitations has been advocated to explain anomalous transport properties at physiological temperature $(310^{\circ}K)$[8] A realistic modelling of the DNA molecule in terms of a Toda chain sets the values of the relevant characteristic frequency, $\omega_o = 5 \ 10^{12} \ rad \ s^{-1}$, and temperature, $T_o = 80^{\circ}K$. Numerical simulations[8] show that on increasing the temperature above $310^{\circ}K$, the soliton density $n_0(T)$ shoots up to its asymptotic value $n_0(\infty) = 1$, while the dissipation properties of the DNA chain are predicted to change dramatically right at the physiological temperature.

The model of nonlinear heat-bath introduced in the present work is all but the most general one could envisage. We have assigned the same mass m to each site of the chain (monoatomic chain), thus making the Hamiltonian (1.2) exactly integrable. This is by no means a necessary condition for the temperature dependence of the effective damping function to appear alongside the treatment of Ref. 9. The existence of localized excitations (solitons) in a diatomic Toda chain, where the sites are assigned two different masses at random, has been proven numerically[13]. Moreover, the monoatomic model has a major disadvantage, which makes it a bad candidate to describe transport properties in condensed matter. As a peculiarity of exact integrability, the thermal conductivity coefficient through a monoatomic Toda chain with end-points coupled to heat reservoirs at different temperatures, turns out to be *infinitely* large. Such a behaviour is in apparent contrast with all experimental evidence supporting Fourier's law and disappears in a diatomic chain[13].

Finally, disordered Toda chains have been investigated[14] by introducing at random *soft* springs with vanishingly low density. Soft springs are characterized by a smaller value of the anharmonicity parameter b. It can be shown that there exists a certain temperature range where the system energy is preferably localized in the soft springs, the magnitude of such an effect crucially depending on the density of the excited solitons.

REFERENCES

[1] see eg P. Grigolini and F. Marchesoni, Adv. Chem. Phys. **62** 29 (1985)

[2] S. Nordholm and R. Zwanzig, J. Stat. Phys. **13** 347 (1975)

[3] H. Mori, Progr. Theor. Phys. **33** 423 (1965)

[4] for a review see T. Fonseca, J. A. N. F. Gomes, P. Grigolini and F. Marchesoni, Adv. Chem. Phys. **62** 389 (1985) and P. Hänggi, P. Talkner and M. Borkovec, Rev. Mod. Phys. **62** 251 (1990)

[5] see eg S. P. Velsko, D. H. Waldeck and G. R. Fleming, J. Chem. Phys. **78** 249 (1983)

[6] M. Toda, *Theory of Nonlinear Lattices* (Springer, Berlin, 1988)

[7] F. Marchesoni and C. Lucheroni, Phys. Rev B in press (1991)

[8] A. Muto, A. C. Scott and P. L. Christiansen, Physica **D44** 75 (1990)

[9] F. Marchesoni, J. Chem. Phys. in press (1991)

[10] F. Marchesoni, Phys. Lett. **115A** 29 (1986)

[11] H. Risken, *The Fokker-Planck Equation* (Springer, Berlin, 1983)

[12] L. D. Landau and E. M. Lifshitz, *Statistical Physics* (Pergamon, Oxford, 1958)

[13] F. Mokross and H. Büttner, Solid State Phys. **16** 4539 (1983)

[14] W. Ebeling and M. Jenssen, Ber. Bunsenges. Phys. Chem. **95** 356 (1991)

EFFECTS OF NONLINEAR SURFACE DIFFUSION

ON THE SCALING OF ROUGH SURFACES

C.M. Arizmendi and J.R. Sanchez

Dept. de Física - Facultad de Ingeniería
Universidad Nacional de Mar del Plata
Av. J.B. Justo 4302
7600 - Mar del Plata
Argentina

INTRODUCTION

Aggregation and growth take place in a wide variety of physical, chemical and biological processes, including solidification, vapor deposition, and flame propagation. These processes are characterized by the existence of an evolving interface separating the growth from the outside. The characterization of the surface structure during growth processes is of practical interest due to its importance in scientific and industrial applications.

The properties of rough surfaces have been studied using different computational models. The simplest one is the random deposition model. The particles simply 'rain' down onto a substrate [1], moving along straight line trajectories until they reach the top of the column in which they were dropped, at which point they stick to the deposit and become part of the aggregate. The fluctuations in the column heights obey a Poisson process because there is no correlation between columns [1]. A modified random deposition model that accounts for the finite surface diffusion which exist in most of the realistic situations was considered later [2]. The deposited particle is allowed to diffuse around on the surface within a prescribed region about the column in which it was dropped until it finds the column with the smallest height. At this point the particle sticks to the top of that column and becomes part of the aggregate. These models have no irregularities in the density of the bulk of the aggregate because no empty can form inside the deposit. Thus the mass of the aggregate is compact and not a scale-invariant fractal, but the surface is rough.

The ability of the incoming particle to diffuse, however, depends not only on the existence of local minima of the height of the surface but also on the local slope of the aggregate. This dependence is given by the angle of repose for a granular material. The tangent of the angle of repose is the steepest slope the material can sustain. In order to model this dependence a threshold for the diffusion is introduced, so the diffusion becomes nonlinear. The purpose of this paper is to study the behavior of the surface roughness of a deposition model with nonlinear diffusion on lattices in which particles are deposited from above onto a substrate of sites with periodic boundary conditions.

SURFACE ROUGHNESS CHARACTERIZATION

A measure of the roughness of the surface can be obtained considering the fluctuations in the height of the surface as measured from the base or some other reference in the system. A single valued function $h(r,t)$, which gives the height of the deposit at the position r at time t, represent an interface that is evolving with time. The width of the surface fluctuations is given by [2]

$$w = <(h - <h>)^2>^{1/2} \tag{1}$$

where $<...>$ denotes averaging over the whole surface, and $<h>$ is the mean height. In simulations on lattices, r is replaced by the positions of the sites on the lattice and the time t is the number of deposited particles per column.

RANDOM DEPOSITION MODELS

In random deposition models, particles are allowed to fall vertically down until they reach the substrate or another particle in the deposit. In a d-dimensional simulation the substrate has L^{d-1} columns into which particles are dropped. The column in which the particle falls is chosen randomly. Different deposition models are defined depending on how and where the particle sticks to the deposit.

In the simplest deposition model no diffusion is allowed, and particles simply fall until they reach the top of the column in which they were dropped, becoming part of the aggregate. With no diffusion the height of the columns follows a Poisson distribution and correspondingly the surface width w diverges with the square root of h, independent of the number of sites of the aggregate [2]. If the deposited particle can diffuse on the surface until it finds the column with the minimum height within a finite distance from the column in which it was dropped, non trivial correlations between the columns appear. In a one dimensional substrate ($d = 2$), the dependence of w on h shows three separates regions [2]. Initially, before a single layer of particles has been deposited the diffusion process in unimportant and w varies as the square root of h. For the second region, $1 \ll h \ll L$, w varies with h as

$$w \sim h^\alpha, \tag{2}$$

where α appears to converge to a value of $\alpha = 1/4$ as $L \to \infty$. At long times ($h \gg L$) the surface width saturates and the steady state $w(L, \infty)$ scales with the length of the system as

$$w(L, \infty) \sim L^\beta, \tag{3}$$

with $\beta = 1/2$. Since w is independent of L in the random filling process the scaling of w with h and L is due to the correlations generated by surface diffusion.

If the influence of the local slope of the aggregate at the minima on the diffusion is taken into account, a random deposition model with nonlinear diffusion is obtained. In a $d = 2$ simulation a particle, randomly dropped in column i will be allowed to diffuse to the smallest of $i+1$ or $i-1$ columns if the height difference z between column i and its smallest neighbor is greater than a critical value z_c. The diffusion becomes nonlinear due to the existence of the threshold z_c. If $z_c = 0$ we recover the previously described diffusion model. If $z_c \to \infty$ the random deposition model with no diffusion is obtained. Since the results were found to be independent of the distance over which the particles are allowed to diffuse, only the results for the nearest-neighbor diffusion are reported.

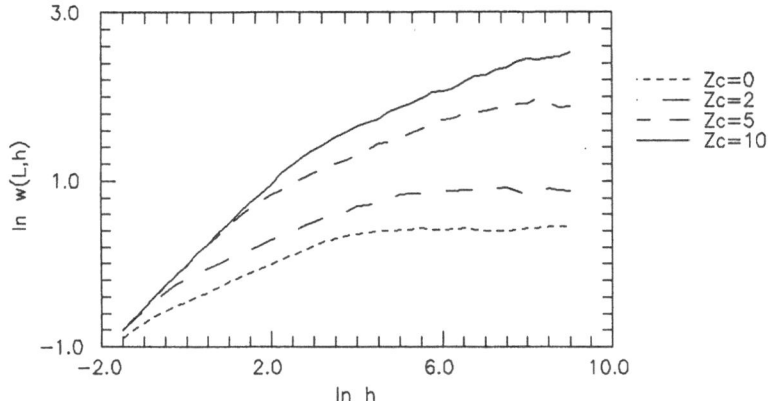

Fig. 1. Dependence of the surface width on the deposition height
for $L = 48$ and different values of z_c.

RANDOM DEPOSITION WITH NONLINEAR SURFACE DIFFUSION

We have studied the random deposition model on a square lattice in which particles are deposited from above onto a line of L sites. Periodic boundary conditions were employed, so that columns i and $i+L$ are equivalent.

The simulation results in Fig. 1 show the dependence of $w(L,h)$ on h for random deposition models with nonlinear surface diffusion, for different values of z_c. Three separate regions can be established for the curves. The curve corresponding to $z_c = 0$ reproduces previous results [2]. For the curves with $z_c \neq 0$ the three different zones can be described as follows, for increasing h :

i) There is a first zone where h is less than the threshold z_c and the diffusion process is unimportant. The particles are deposited onto the aggregate randomly and consequently w varies as the square root of h. This zone becomes larger as z_c increases.
ii) Above the first zone there is a region in which the surface diffusion takes place and tends to smooth out the surface and σ varies with h as $w(L,h) \sim h^{\alpha}$. We found that the value of α is the same than in the random diffusion model and independent of z_c. In Table I the dependence of α with L for $z_c = 10$ is shown.

Table 1. Values of the exponent α for various L
and $z_c = 10$.

L	α
24	0.196
48	0.204
96	0.224
192	0.238

iii) The third region is characterized by saturation of the surface thickness to a constant value. This value depends on L and z_c. The dependence of $w_s(L, \infty)$ on L for $z_c = 5$ is shown in Fig. 2. The value of β obtained equal the value corresponding to $z_c = 0$.

On the basis of the results of the simulation of the random diffusion model, the following scaling form were proposed [1]:

$$w_s(L, h) = L^\beta f(h/L^\gamma) \qquad (4a)$$

with $\gamma = \beta/\alpha$ and the scaling function $f(x)$ defined by

$$f(x) \sim \begin{cases} x^\alpha & x \ll 1 \\ \text{constant} & x \gg 1 \end{cases} \qquad (4b)$$

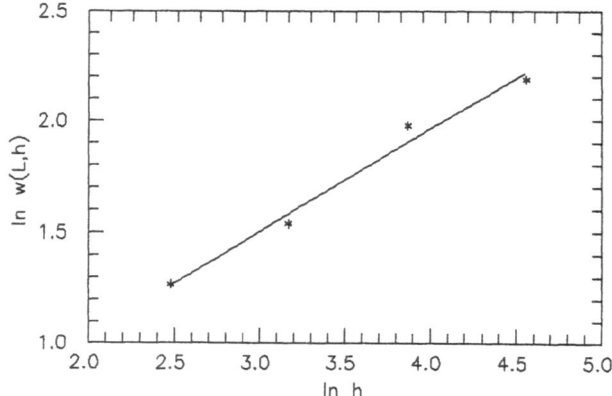

Fig. 2. Scaling of the saturation value of $w_s(L, h)$ as function of L for $z_c = 5$. The exponent β is obtained as the slope of the line through the data points.

Since the two exponents obtained for the nonlinear diffusion case are the same that those of the random diffusion, the scaling form (4) remains valid. This type of surfaces are self-affine fractals [3], due to L and h must be scaled by different ratios in order to keep w_s scale invariant.

A Langevin type equation have been written down [4] to describe the time evolution of the surface profile in a granular deposit. It has been argued that the interface is described by the equation

$$\frac{\partial h}{\partial t} = \mathcal{D} \nabla^2 h + \xi(r, t) \qquad (5)$$

where $\mathcal{D}$ is the diffusion coefficient and $\xi(r, t)$ is the random noise term with zero mean. By solving this equation [4] the exponents α and β can be calculated, finding that

$$\alpha = (3 - d)/4 , \qquad \beta = (3 - d)/2 \qquad \text{and} \qquad \gamma = 2 , \qquad (6)$$

in d dimensions. In $d = 2$ the results $\alpha = 1/4$, $\beta = 1/2$ agree with the simulations of the random diffusion model.

Equation (5) becomes nonlinear in the nonlinear random diffusion model studied here, since $\mathcal{D}$ depends on the local slope of the surface as

$$\mathcal{D} = \begin{cases} 0 & z \leq z_c \\ \text{constant} & z > z_c \end{cases} \tag{7}$$

The numerical simulations show that the exponents α, β and γ remain equal and this implies that the solution of nonlinear Langevin equation (5), with $\mathcal{D}$ given by (7), will result in the same values for the exponents.

CONCLUSIONS

This aggregation model has been shown to be a realistic representation of some deposition experiments [5]. The study of the surface properties of this model shows that, although the bulk and surface mass do not have fractal dimensions, the width of the surface is a self-affine fractal even for nonlinear diffusion case.

The width of the surface exhibits non-trivial scaling with the surface height and the system size. When nonlinear diffusion is taken into account, the basic exponents remains equal, at least for $d = 2$, but the heights at which they appear differs from the linear diffusion model.

REFERENCES

[1] Family F. and Viscek T. 1985 *J Phys A: Math Gen* **18** L75
[2] Family F. 1986 *J. Phys A : Math. Gen.* **19** L441
[3] Mandelbrot B. 1982 *Fractal Geometry of Nature* – San Francisco:Freeman
[4] Edwards S. and Wilkinson D. , 1982 *Proc. R. Soc. A* **381** 17
[5] Stanley H. and Ostrowsky N. (ed) 1985 *On Growth and Form: A Modern View* (Dordrect: M. Nijhoff)

A DIRECT APPROACH TO THE TAMM-DANCOFF APPROXIMATION

Roberto C. Bochicchio[#] and Horacio Grinberg[#]

Departamento de Física
Facultad de Ciencias Exactas y Naturales
Universidad de Buenos Aires
1428 Buenos Aires, Argentina

ABSTRACT

A direct procedure for the decoupling of the one nucleon propagator equation of motion, based on an appropriate transformation of the Green's function at t = 0+ and followed by inversion of the perturbation series for the (energy) Green's function, is shown to lead to the lower order corrections (ring and ladder diagrams) to the Tamm-Dancoff approximation (TDA).

INTRODUCTION

Considerable progress has been made over the years on the fermion many-body nuclear problem [1-4]. It is the task of the theory to calculate the Green's function G (propagator) at least in some consistent approximate fashion, in terms of a single-particle complete orthogonal set. The single-particle functions are then chosen to minimize the energy via the construction of a central potential (in terms of the G matrix) - in exact analogy to the Hartree-Fock equations. Attempts to extend this theory to actual nuclei have focused attention on the self-energy problem via Dyson equation. Since the hierarchy derived from the iteration of the single-nucleon propagator equation of motion is highly non linear in the resultant self-energies, the iteration solution becomes rapidly intractable with increasing order and generally not consistent beyong second-order truncation in the (asymptotic) moment expansion of the propagator [5].

Different approaches to the linearization (i.e., decoupling) of the equation of motion of such object, such as the diagrammatic expansion method [2] or superoperator formalism [6] followed by an inner projection [7] of the superoperator resolvent are usually employed in order to get efficient computational algorithms.

We emphasize that the higher order terms of the self-energy perturbation expansion, which have been associated with the "rearrangement energy", play an important role in determining the saturation properties of actual nuclei. It has also been suggested the possibility of systematically rearranging the entire perturbation series, in such a way that a rapid rate of convergence can be seen directly from the structure of the new series.

[#] MCIC Consejo Nacional de Investigaciones Cientificas y Técnicas, (CONICET), República Argentina.

An alternative approximation scheme for determination of successive contributions to the self-energy is followed in the present work. It is based on the inversion of the perturbation series for the (energy) Green's function so as to get one for the self-energy [8,9]. To this end an appropriate *ansatz* for the (energy) propagator is introduced in the scenario of the superoperator algebra [6]. This scheme provides matrix equations which involve a rational approximant to the self-energy and contain contributions to all orders in nucleon interaction. Moreover, it maintains linearity of the equations, so that it is easy to identify the general term in the series. Although the self-energy approximants thus obtained do not represent partial summations of any easily recognizable classes of diagrams we show that the second- and third- order diagrammatic contributions to the Tamm-Dancoff approximation (TDA) [10] are easily recovered.

Because our approach is formulated in the real time Green function formalism it may describe both equilibrium and non equilibrium situations in a consistent fashion [17].

THE NUCLEON PROPAGATOR

a. The ansatz

Assuming the existence of a linearly orthogonalized independent operator basis $\{B_n\}$ for expanding an arbitrary operator $A(t)$, we start with the spectral representation of the coupled hierarchy of equation of motion for the one-nucleon causal propagator G_{AB} ($\hbar = 1$)

$$(E - E_A)\, G_{AB}(E) = i\, G_{AB}(0,0) + \sum_n c(A|n)\, G_{B_n B}(E) \tag{1}$$

where E is a complex-energy variable, assumed to be in the upper energy half-plane, $G_{AB}(0,0)$ is the time dependent Green's function evaluated at $t = 0$ ($t' = 0$), and the c-matrix elements $c(A|n)$ are the matrix elements of the Liouvillian superoperator $\hat{H} \equiv [\ ,\hat{H}]_-$ in the above defined basis set $\{B_n\}$ of the operator A [9,11]. Explicitly,

$$c(n|m) = (B_n | \hat{H}\, B_m) \equiv < [B_n^+ , [B_m, \hat{H}]_-\ ]_+ > \tag{2}$$

where it is assumed that the operator basis consists of elements obeying the Fermi-Dirac statistics. Here $\mathbb{H}$ is taken to be an appropriately partitioned time independent second quantized Hamiltonian involving up to two (nucleons) body interactions and the average is to be evaluated with respect to an appropriate reference state.

In the limit of no interaction the $c(A|n)$ all vanish and the free propagator is obtained from eq.(1) as

$$G_{AB}(E) = (E - E_A)^{-1}\, i\, G_{AB}(0,0) \tag{3}$$

For complex E the poles of $G_{AB}(E)$ must be found so as to allow analytic continuation to achieve a complex plane representation. Particularly, $G_{AB}(E)$ is expected to be analytic in the upper E half-plane, to possess a cut along the real axis, and to have analytic continuation into the lower E half-plane with complex poles E_n.

This expected analytic behaviour of $G_{AB}(E)$ and the physical meaning of the associated self-energies as corrections to the unperturbed energies motivates the following change of dependent variable from the energy propagator $G_{AB}(E)$ to the proper self-energy $\Sigma_{AB}(E)$ in eq. (1)

$$G_{AB}(E) = \frac{i\, G_{AB}(0,0)}{(E - E_A)^{1/2}\, (E - E_B)^{1/2} - \Sigma_{AB}(E)} \tag{4}$$

where the functional form of the pole guaranties hermiticity of the matrix $\Sigma_{AB}(E)$. Like $G_{AB}(E)$, $\Sigma_{AB}(E)$ is expected to be analytic in the upper half-plane. The choice of i $G_{AB}(0,0)$ as the coefficient in eq.(4) is a matter of convenience, motivated by the noninteracting limit (3).

For the nucleon propagator we set $A = a_i$ and $B = a_j^\dagger$ with a_i and $a_j^\dagger$ nucleon annihilation and creation operators. Application of the proper anticommutation relations obeyed by these operators yields the coefficients of eq. (4) as

$$i\, G_{a_i a_j^\dagger}(0,0) = <n_{ij}> \tag{5}$$

which is recognized to be the matrix element of the first order reduced density matrix. Moreover if n_i is the occupation number associated to the operator a_i then it is easily seen that use of a Hartree-Fock reference state leads to $n_{ij} = n_i \delta_{ij}$ which clearly makes the numerator of eq. (4) to vanish when $i \ne j$. Since the denominator of eq.(4) must vanish at the pole energy, use is made of the arbitrariness in the choice of the numerator in eq.(4) in order to remove the indetermination which arises in the off-diagonal elements when a Hartree-Fock reference state is introduced to evaluate the expectation values. Thus, one is led to consider the transformation

$$<n_{ij}> \;\rightarrow\; 1 + <n_i> \delta_{ij} - i\, G_{a_i a_j^\dagger}(0,0) \tag{6}$$

which, in fact, removes the vanishing of the residue of eq.(4) for a Hartree-Fock reference state when $i \ne j$. Insertion of this change into eq.(4) allows the (energy) propagator to be expressed as

$$G_{a_i a_j^\dagger}(E) = \frac{\{1 + <n_i> \delta_{ij} - i\, G_{a_i a_j^\dagger}(0,0)\}}{(E - \varepsilon_i)^{1/2}\,(E - \varepsilon_j)^{1/2} - \Sigma_{ij}(E)} \tag{7}$$

where the ε's are the single-nucleon energies which correspond to the unperturbed energies of eq.(4).

b. The decoupling technique

Expansion of $G_{AB}(E)$ leads to a clear identification of the perturbation corrections to $\Sigma(E)$ with the iteration cycles of the equation of motion (1). In fact, iteration of this hierarchy leads to

$$G_{AB}(E) = \overset{(0)}{G_{AB}(E)} + \overset{(1)}{G_{AB}(E)} + \ldots = \sum_{m=0}^{\infty} \overset{(m)}{G_{AB}(E)} \tag{8}$$

where $\overset{(m)}{G_{AB}(E)}$ is the term of mth order in the interaction, i.e., in the off-diagonal c-matrix elements. Explicitly,

$$\overset{(m)}{G_{AB}(E)} = \frac{i}{(E - (A \mid \hat{H}_0\, A))} \sum_{n_1; n_2; \ldots; n_m} \frac{c(n_1|n_2)}{(E - (n_1 \mid \hat{H}_0\, n_1))}$$

$$\frac{c(n_2|n_3) \ldots\ldots\ldots c(n_{m-1}|n_m)\, G_{n_m B}(0,0)}{(E - (n_2 \mid \hat{H}_0\, n_2)) \ldots\ldots\ldots (E - (n_m \mid \hat{H}_0\, n_m))} \tag{9}$$

and the n_j's ($j = 1,\ldots,m$) label the operator space basis set elements. The sum over n_j include $n = A$ (except in the case $j = 1$) but exclude $n_{j+1} = n_j$.

Assuming likewise that $\Sigma_{ij}(E)$ possesses a perturbation expansion

$$\Sigma_{ij}(E) = \overset{(1)}{\Sigma}_{ij}(E) + \overset{(2)}{\Sigma}_{ij}(E) + ... = \sum_{m=1}^{\infty} \overset{(m)}{\Sigma}_{ij}(E) \tag{10}$$

it is found, upon expansion of the inverse of eq.(7) and comparing the series thus obtained with the expansion (8) by means of (10), that the mth order correction is given by

$$\overset{(m)}{\Sigma}_{ij}(E) = \frac{(E - \varepsilon_i)\,(E - \varepsilon_j) - \overset{(m)}{G_{a_i a_j^\dagger}}(E)}{\{1 + <v_i> \delta_{ij} - i\,G_{a_i a_j^\dagger}(0,0)\}}$$

$$- \sum_{n=2}^{m} [(E - \varepsilon_i)\,(E - \varepsilon_j)]^{-(n-1)/2} \sum_{\substack{l_1;...;l_n=1 \\ l_1+...+l_n=m}}^{m-1} \overset{(l_1)}{\Sigma}_{ij}(E) \, ... \, \overset{(l_n)}{\Sigma}_{ij}(E). \tag{11}$$

It is observed that use of the transformation (6) for the (energy) propagator removes the singularities in the generation of the proper self-energies to all orders.

c. Self-energies

Evaluation of the different self-energies contributions require an appropriate partitioning of the many nucleon Hamiltonian. In this communication use is made of the Moller-Plesset partitioning [12]. In this partitioning the unperturbed part of $\mathbb{H}$ has the form

$$\mathbb{H}_0 = \sum_i \varepsilon_i\, a_i^\dagger a_i - 1/2 \sum_{i;i'} <ii'\|ii'> <n_i> <n_{i'}> 1 \tag{12a}$$

and the perturbation is expressed as

$$V = \sum_{i;i';j;j'} <ii'\|jj'> [\,1/4\, a_i^\dagger a_{i'}^\dagger a_{j'} a_j - \delta_{i'j'} <n_{i'}> a_i^\dagger a_j]$$

$$+ 1/2 \sum_{i;i'} <ii'\|ii'> <n_i> <n_{i'}> 1 \tag{12a}$$

where 1 is the identity operator and the $< \| >$ symbol stands for the antisymmetrized two-nucleon integral. The operator space for the one nucleon propagator is chosen to be composed by fermion-like operators, that is, products or linear combinations of products of simple field operators $a_k^\dagger$, a_l of an infinitive dimensional vector space. For a non-conserving Fermion particle operator the complete operator manifold is [13]

$$\{B_n\} = \{T_1^+; T_3^+; ... \} = \{a_\alpha, a_p^\dagger; a_p^\dagger a_\alpha a_\beta, a_\alpha^\dagger a_p a_q; ... \} \tag{13}$$

where the Greek and Latin indices stand for the occupied and virtual nucleon states respectively and $\alpha < \beta < ...$; $p < q <...$, etc. This operator space supports a binary product defined as an averaged anticommutator [cf. eq. (2)] for an appropriate reference state.

Evaluation of the successive perturbative corrections to the self-energy is performed assuming a Hartree-Fock reference state and forming the averaging of the anticommutators implicit in eq. (11) in the grand-canonical density operators basis. Using it in eq.(2) along with the $\{T_1^+\}$ subspace, it follows from eqs. (9) and (11) that the first order correction to the self-energy vanishes.

The two particle - one hole Tamm-Dancoff approximation (2p-h TDA) can be derived from an effective interaction that is logically obtained by a projection of the perturbation superoperator onto the subspace spanned by 2p-h type operators. This implies that the intermediate or virtual states that are represented by the third-order diagrams consist of only 2p-h or 2h-p excitation of the reference state. Both of these excitations are described with the triple-operators products $T_3^\dagger$ [eq.(13)]. Thus, second- and third-order perturbative corrections to the self-energy are generated within the subspace $\{T_1^\dagger; T_3^\dagger\}$. It is obtained

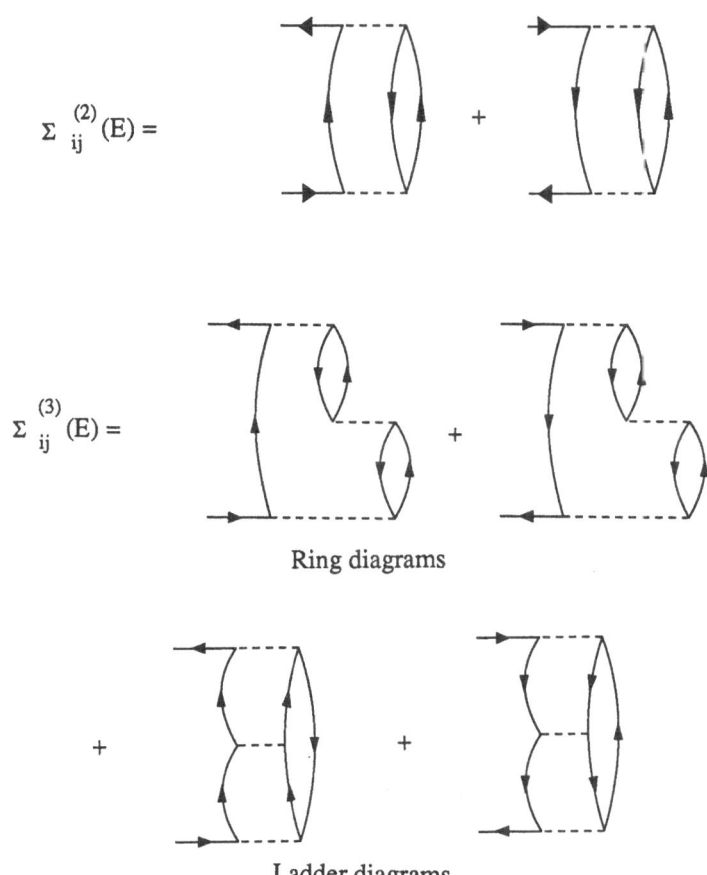

Ring diagrams

Ladder diagrams

The two second-order diagrams are precisely the same as those obtained in the second-order diagrammatic expansion after a Fourier transformation into the energy representation [14]. These second- and third-order self-energy approximations are interesting not only because they contain the lowest-order relaxation and correlation corrections of low and high electron density systems, but also because their corresponding analytical expressions [9] exhibit the same structure as those obtained in the second- and third-order truncations of the self-energy diagrammatic expansion [14,15]. No spurious contributions are obtained and it is seen that the third-order correction to the self-energy retains the most important contributions of the diagrammatic expansion. In fact, in the theory of an electron gas at high density the so-called ring diagrams turn out to be the predominant contributions [16] whereas in the theory of nuclear matter the so-called ladder diagrams, which account for the bulk of correlation effects, turn out to be the most important ones. The need for considering these ladder diagrams is closely connected with the short-range strong repulsion between nucleons, which leads to divergences unless one sums up the ladder diagrams.

CONCLUSIONS

An alternative and direct procedure for the decoupling of the equation of motion of the nucleon propagator was introduced within the superoperator formalism. The approach, based on inversion of the perturbation expansion for the (energy) Green's function incorporates in a general way its nth order approximation to the self-energy.

In particular, this expansion should be satisfactory for binding-energy calculations, at least for infinite nuclear matter because long-range correlations contribute so little here. Since these ring and ladder diagrams dominate the third-order self-energy matrix, one can argue that they may be most important also in higher order of the self-energy expansion. Also a rapid decrease in the magnitudes of the first few terms of our expansion is expected to be physically reasonable, because the binding energy is dominated by the short-range correlations.

We believe that the nth correction to $\Sigma(E)$ will prove to be a successful approximation scheme for the decoupling problem, maintaining the advantages of the many-body approach, i.e., the direct and consistent calculation of both, mean addition energies (probes in stripping reactions) and mean removal energies (defined by the first moment of the spectral function), as well as transition moments, thus circumventing the difficulties and complexities of previous TDA-like approximations. For cases where we have only a few collective p-h and p-p states the dimension of the problem may be drastically reduced. The present procedure can thus be considered as a new and direct approach to the weak or intermediate coupling model.

ACKNOWLEDGMENTS

The authors are grateful to the Consejo Nacional de Investigaciones Científicas y Técnicas (CONICET), República Argentina, for a research grant in support of this work. Facilities provided at the Department of Physics of the Facultad de Ciencias Exactas y Naturales, Universidad de Buenos Aires, are also gratefully acknowledged.

REFERENCES

[1] M. Baranger, Nucl. Phys. A 149 (1970) 225.
[2] B. Brandow, Ann. Phys. 57 (1970) 214.
[3] a) N.H. March, W.H. Young, and S. Sampanthar, *The Many Body Problem in Quantum Mechanics*, Cambridge Univ. Press, Cambridge 1967; b) A.L. Fetter and J.D. Walecka, *Quantum Theory of Many-Particle Systems*, Mc Graw-Hill Book Co, New York, 1971.
[4] A.E.L. Dieperink and T. de Forest Jr., Ann. Rev. Nucl. Sci. 25 (1975) 1.
[5] a) R.A. Tahir-Kehli, Phys. Rev. B 1 (1970) 3163; b) R.A. Tahir-Kehli and H.S. Jarret, Phys. Rev. 180 (1969) 544.
[6] O. Goscinski and B. Lukman, Chem. Phys. Letts 7 (1970) 573.
[7] P.O. Löwdin, Phys. Rev. A 139 (1965) 357.
[8] M.D. Girardeau, Phys. Rev. A 28 (1983) 1056.
[9] R.C. Bochicchio and H. Grimberg, Chem. Phys. Letts. 169 (1990) 236.
[10] P. Schuck, F. Villars and P. Ring, Nucl. Phys. A 208 (1973) 302.
[11] P.O. Löwdin, Int. J. Quantum Chem. S16 (1982) 485.
[12] Chr. Moller and M.S. Plesset, Phys. Rev. 46 (1934) 618.
[13] R. Manne, Chem. Phys. Letts. 45 (1977) 470.
[14] L.S. Cederbaum and W. Domcke, Adv. Chem. Phys. 36 (1977) 205.
[15] Y. Öhrn and G. Born, Adv. Quantum Chem. 13 (1981) 1.
[16] G. Born and Y. Öhrn, Chem. Phys. Letts. 61 (1979) 307.
[17] P. Danielewicz, Ann. Phys. (N.Y.) 197 (1990) 154.

FINITE SIZE EFFECTS IN THE DECONFINEMENT TRANSITION

H. G. Miller[1], N. J. Davidson[1], R. M. Quick[1] and A. Plastino[2]

[1]Department of Physics
University of Pretoria
Pretoria 0002
Republic of South Africa

[2]Department of Physics
National University
C. C. 67
1900 La Plata
Argentina

INTRODUCTION

The concepts of quark confinement and asymptotic freedom inherent in models of strongly interacting matter at the sub-hadronic level have lead to a great deal of interest in the associated deconfinement phase transition, *i.e.* the transition from a gas of hadrons to a hot plasma of deconfined quarks and gluons. The possibility of obtaining energy densities which are large enough to cause deconfinement in ultra-relativistic heavy ion collisions has acted as one of the main stimuli to interest in such collisions, from both the experimental and theoretical points of view.

Historically, much of the early interest in the deconfinement transition came from the cosmological community [1] [2], where the emphasis was on hadronisation in the early universe. The resulting calculations tended to be performed at zero net baryon number, as befits the early universe scenario. More recently, the high energy physics community, spurred by the development of more powerful particle accelerators, have considered the possibility of obtaining the reversed process, *i.e* deconfinement, in the laboratory from ultra-relativistic heavy ion collisions. These collisions result in systems which certainly have non-zero baryon number; as a result, much effort has been devoted to determining full phase diagrams for strongly interacting systems. The major theoretical thrusts in this direction have centred on two approaches, namely lattice QCD and phenomenological models. While lattice QCD is the preferred method of studying the behaviour of strongly interacting systems, the enormous computational requirements of such calculations have led to an appreciable amount of work being directed towards phenomenological descriptions of deconfinement.

In most cases, these phenomenological calculations have been performed for infinite systems, where the relevant quantities are the densities of the thermodynamic variables. While these calculations are valid for descriptions of systems such as the early universe and stellar systems, the results of such calculations may not be directly applicable to the small systems resulting from heavy ion collisions. To describe the phase structure of the system resulting from heavy ion collisions, it is necessary to take into account explicitly the fact that the system is small, with a baryon number of the order of twice the mass number of the projectile nucleus. We want to examine the effects of finite system size on the deconfinement transition within a simple phenomenological model [3].

There are two points of caution which should be mentioned at this stage:

- Assuming that a plasma does form during a heavy ion collision, it is at present by no means clear whether there is sufficient time available for the hadron gas and the quark-gluon plasma (QGP) to equilibrate, and therefore whether traditional equilibrium thermodynamics should be applied to these systems. In what follows, we will assume that the system is in fact in equilibrium. However, the limitations inherent in this assumption should be borne in mind when using the results of our calculations in an analysis of heavy ion collisions.

- Finite systems do not exhibit true phase transitions due to the presence of fluctuations. Strictly speaking, for a finite system it is only possible to discuss "average" and "most probable" states, which may be very different. (For discussions of this point in nuclear systems, see [4] [5] [6].) In keeping with common usage, however, we will refer to deconfinement in a finite system as a phase transition, while keeping in mind that such terminology is rigorously correct only for infinite systems.

GEOMETRICAL TREATMENT OF FINITE SIZE EFFECTS

An approximation which is simple to implement is one which depends only on the geometry of the system. In such an approximation, the starting point is an ideal gas of $< N >$ particles in a volume V in the thermodynamic limit $< N > \to \infty$, $V \to \infty$, $< n > = < N > /V$ finite. The momentum eigenvalues of the single particle states are then known. Since the volume is very large, the single particle levels are dense enough to justify the replacement of the sum over the momentum states which occurs in the thermodynamic quantities by an integral:

$$\sum_p \to \frac{V}{(2\pi)^3} \int d^3p = \frac{V}{2\pi^2} \int_0^\infty dp\, p^2 \tag{1}$$

where the last equality holds if the momentum states are spherically distributed. One can define the density of single particle momentum states $g(p) = V p^2/(2\pi^2)$ such that the number of states with momenta with magnitudes between p and $p + dp$ is given by $g(p)dp$.

A replacement of the form (1) assumes that the volume is large enough for the effects of the surface of the system to be negligible. However, a more careful analysis (see for example [7]) shows that the form of $g(p)$ depends on both the surface area of the system and its linear dimension; the exact dependence is determined by the boundary conditions imposed on the single particle states. If the states obey periodic boundary conditions, the form (1) is correct, and the surface of the system plays no role in the calculations. This makes perfect sense, as periodic conditions are usually used to approximate infinite systems, where the surface effects are completely negligible.

If the single particle states obey Dirichlet or von Neumann boundary conditions, however, the density of states $g(p)$ must be modified to [7]

$$g(p) = \frac{V}{2\pi^2}p^2 \pm \frac{A}{8\pi}p + \frac{L}{8\pi}, \tag{2}$$

where A and L are the surface area and linear dimension of the system of volume V respectively. For spherical systems, which we will consider here, $V = 4\pi R^3/3$, $A = 4\pi R^2$ and $L = 2\pi R$ where R is the radius of the system. The positive and negative signs in (2) refer to von Neumann and Dirichlet conditions respectively. Clearly, in the limit $R \to \infty$ the volume-dependent term dominates $g(p)$, as would be expected. We note that the approach used to derive (2) can be generalised to interacting systems and requires the use of dispersion relations [8].

The use of (2) requires some care, since the error implicit in its derivation is of the same order of magnitude as the last term in (2) [7]. One should therefore check that this term is small compared to both the volume- and surface area-dependent terms, and, in the case of Dirichlet boundary conditions, their difference.

The question now arises as to which boundary conditions are relevant for a description of a strongly interacting system in a finite volume. We want to consider a very simple situation, in which the system is thermally and chemically isolated from its surroundings, and where the whole system undergoes deconfinement. This scenario relies on the assumption of global equilibration during the deconfinement transition, which may not be realistic for a description of the situation resulting from a heavy ion collision. An alternative scenario would be to have deconfinement only in very hot and/or dense regions of the hadron gas. This is, however, very far from being an equilibrium process, and falls outside the scope of our calculations. Since the hadrons are restricted to a finite volume, their wavefunctions should be zero outside this volume. Dirichlet conditions are thus appropriate. For the QGP, there is a temptation to force the quarks and gluons to obey confining boundary conditions, and to view the QGP as a giant bag. However, the plasma consists of deconfined quarks and gluons, and there is no justification for the use of confining conditions. The restriction of the system to a specific volume is not due to the strong interaction, as is the case in the bag model of hadrons, but due to the localisation of the fireball. Dirichlet conditions are therefore also the appropriate boundary conditions for the QGP.

The other aspect of the finite size of the system affects only the QGP. In dealing with a deconfined plasma of quarks and gluons, it is necessary to ensure that the deconfined state remains a colour singlet. As the dynamical problem of confinement has not yet been solved, this requirement must be added "by hand" as a restriction on the states which are included in the partition function of the QGP. The procedure is based on group-theoretical projection techniques [9] [10] [11] applied to the specific problem of projecting out a colour singlet partition function. Details of the projection onto a colour singlet can be found in [12] [13] [14]. In what follows, we shall essentially follow the derivation of Gorenstein *et al.* [12], as this is the most general description of the procedure.

The quantum states for a non-interacting particle in a volume V form a discrete set which we label by α. If we assume that each single particle state transforms under the irreducible representation a (of dimension $d(a)$) of an internal symmetry group G, for each α we have $d(a)$ linearly independent states $|\alpha, j >$, $j = 1, ..., d(a)$. With creation operators $a_\alpha^{j\dagger}$ and annihilation operators a_α^j for the single particle states, we can form an orthonormal basis in occupation number space, given by

$$\left| ....n_\alpha^j.... > \right., \; n_\alpha^j = 0, 1, ...\infty \text{ for bosons and } n_\alpha^j = 0, 1 \text{ for fermions.} \tag{3}$$

Many-body states of the form (3) do not transform under irreducible representations of G; their usefulness will become clear later. It is therefore convenient to introduce a new set of many-body basis states which do transform under irreducible representations of G. We define these states by

$$|\tau; \nu, \xi >, \; \xi = 1, ..., d(\nu) \tag{4}$$

where ν is the irreducible representation of interest and τ stands for all other quantum numbers of the many-body system.

If G is of rank r, there are r mutually commuting operators among the n generators Q_m of G. Since the symmetry of the system described by the Hamiltonian H is exact, we must have $[H, Q_i] = [Q_i, Q_j] = 0$ for $i, j = 1, ...r$. The many-body states (4) can therefore be chosen as the simultaneous eigenstates of H and Q_l, $l = 1, ...r$. We thus need to calculate the partition function corresponding to the particular irreducible representation ν

$$Z_\nu = \sum_{\tau; \xi = 1,....d(\nu)} < \tau; \nu, \xi \, |\exp(-\beta H)| \, \tau; \nu, \xi >, \; \beta = 1/T; \tag{5}$$

in the case of the QGP, ν must be the colour singlet.

To proceed further, we define a generating function $\tilde{Z}$ given by [10] [9]

$$\tilde{Z}(\gamma_1, ..., \gamma_r) = \sum_\nu [1/d(\nu)] \, \chi_\nu(\gamma_1, ..., \gamma_r) \, Z_\nu, \tag{6}$$

where the sum runs over all irreducible representations of G, and the γ_i are the group parameters. χ_ν is the character of the representation ν, defined as

$$\chi_\nu(\gamma_1, ..., \gamma_r) = \sum_{j=1}^{d(\nu)} \exp\left[i \sum_{l=1}^{r} \gamma_l q_l^j\right],$$ (7)

where the q_l^j are the eigenvalues of the mutually commuting generators Q_l. Using the orthogonality of characters and a group integration, we find that

$$Z_\nu = d(\nu) \int d\mu(\gamma_1, ..., \gamma_r) \chi_\nu^* \tilde{Z}(\gamma_1, ..., \gamma_r).$$ (8)

If we then substitute (5) and (7) into (6), we obtain

$$\tilde{Z}(\gamma_1, ..., \gamma_r) = \text{Tr}\left(\exp\left[-\beta H + i \sum_{l=1}^{r} \gamma_l Q_l\right]\right),$$ (9)

where the trace runs over all complete sets of many-body states.

The calculation of Z_ν thus reduces to the calculation of the trace (9) and the evaluation of the group integral (8). Since the trace is independent of the basis chosen, we can use the simple basis defined by equation (3) to determine $\tilde{Z}(\gamma_1, ..., \gamma_r)$. Equation (9) then yields

$$\tilde{Z}(\gamma_1, .., \gamma_r) = \sum_{\{n_\alpha^j\}} <..n_\alpha^j..| \exp\left[\sum_\alpha \sum_{j=1}^{d(a)} a_\alpha^{j\dagger} a_\alpha^j \left(-\beta\epsilon_\alpha + i \sum_{l=1}^{r} \gamma_l q_l^j\right)\right] |..n_\alpha^j..>$$

$$= \prod_\alpha \prod_{j=1}^{d(a)} \left(1 + \eta \exp\left[-\beta\epsilon_\alpha + i \sum_{l=1}^{r} \gamma_l q_l^j\right]\right)^\eta,$$ (10)

where $\eta = -1$ for bosons and $\eta = 1$ for fermions. Note that it is the irreducible representations a of the single particle states that enter here, and not that of the many-body system ν.

If the single particle spectrum is dense enough, the sum over α can be replaced by an integral with a single particle density of states $g(p)$ (for clarity, we will include any degeneracy not connected with the symmetry group G, e.g. spin-isospin, in this density of states):

$$\sum_\alpha \rightarrow \int dp\, g(p).$$ (11)

The limits of the integration in (11) depend on the model for the QGP. In the bag model, the limits are 0 and ∞. In our calculations, we have included the effects of massive quasiparticle excitations by introducing momentum cutoffs p_g and p_q. Our integrals thus run from p_g to ∞ for the gluons, and from p_q to ∞ for the quarks and antiquarks. The generating function can now be rewritten as

$$\tilde{Z}(\gamma_1, ..., \gamma_r) = \exp\left[\int dp\, g(p) \sum_{j=1}^{d(a)} \eta \ln\left\{1 + \eta \exp\left(-\beta\epsilon_\alpha + i \sum_{l=1}^{r} \gamma_l q_l^j\right)\right\}\right].$$ (12)

Up until this point, the derivation has been very general. For the specific case of colour projection in a QGP, there are three different types of particles involved, namely the quarks, antiquarks and gluons. The sum over the single particle states α can thus be split into three sums, one for each particle type. The generating function $\tilde{Z}$ for the QGP therefore factorises into a product of three generating functions,

$$\tilde{Z} = \tilde{Z}_q \cdot \tilde{Z}_{\bar{q}} \cdot \tilde{Z}_g, \tag{13}$$

where each of the $\tilde{Z}_i$, $i = q, \bar{q}, g$ is of the form (12), with $\eta = 1$ for $\tilde{Z}_q$ and $\tilde{Z}_{\bar{q}}$, and $\eta = -1$ for $\tilde{Z}_g$. For each $\tilde{Z}_i$, the sum over j in the exponent of (12) runs over the representation relevant to the particle type.

For the problem of colour projection, the relevant symmetry group G is $SU(3)_c$. In this case, the two mutually commuting operators are the "colour isospin" I_3 and the "colour hypercharge" Y_8. The eigenvalues of these operators in the fundamental representations are 1,-1, 0, and 1/3, 1/3, -2/3 respectively for the quarks, while those of the antiquarks have opposite sign. For the gluons, the eigenvalues are 0, 2,-1, -1 and 0, 0, 1, -1 respectively, together with the opposite signed combinations; note that the pair (0,0) therefore occurs twice.

We also want to include baryon number conservation in the calculation. If this is treated within the grand canonical ensemble, *i.e.* only conserved on the average, a baryon chemical potential μ_{QGP}^b must be introduced. This modifies the hamiltonian $H \rightarrow H - \mu_b (N_q - N_{\bar{q}})/3$, where N_q and $N_{\bar{q}}$ are the quark and antiquark number operators respectively. The effect of including a chemical potential on $\tilde{Z}$ is to modify the single particle energies ϵ_α. For the quarks, $\epsilon_\alpha \rightarrow \epsilon_\alpha - \mu_{QGP}^b/3$, while for the antiquarks, $\epsilon_\alpha \rightarrow \epsilon_\alpha + \mu_{QGP}^b/3$.

The generating function $\tilde{Z}$ is now fully determined. All that remains to be specified is the integration measure in (8). This is given by [12]

$$
\begin{aligned}
d\mu(\gamma_1, \gamma_2) = \quad & (1/12\pi^2) \, [3 - \cos 4\gamma_1 - 2\cos 2\gamma_1 + 2\cos 2\gamma_2 \\
+ \quad & 2\cos(3\gamma_1 + \gamma_2) + 2\cos(3\gamma_1 - \gamma_2) \\
- \quad & \cos(2\gamma_1 + 2\gamma_2) - \cos(2\gamma_1 - 2\gamma_2) \\
- \quad & 2\cos(\gamma_1 + \gamma_2) - 2\cos(\gamma_1 - \gamma_2)] \, d\gamma_1 \, d\gamma_2
\end{aligned}
\tag{14}
$$

where $\gamma_1, \gamma_2 \in [-\pi, \pi]$. For the singlet representation, the dimension and character are both unity. The group integration (8) can now be calculated either numerically or by some approximation such as the saddle-point method [14].

MODELS FOR THE HADRON GAS AND THE QUARK GLUON PLASMA

The phase diagram for strongly interacting matter can be constructed using the usual Gibbs conditions for equilibrium

$$< P_H > = < P_{QGP} > \tag{15}$$

$$T_H = T_{QGP} \tag{16}$$

$$\mu_H^b = \mu_{QGP}^b, \tag{17}$$

where $< P_H >$ and $< P_{QGP} >$ are the pressures, T_H and T_{QGP} the temperatures and μ_H^b and μ_{QGP}^b the chemical potentials for conservation of the baryon number for the hadron and QGP phases respectively. Note that μ_{QGP}^b refers to one unit of baryon number in the plasma, and thus (17) does not have the factor 3 which would occur if μ_{QGP}^b referred to the quark chemical potential.

Since our major interest here is in the qualitative effects of a finite system on the deconfinement transition, we will keep the model as simple as possible. The hadron gas will therefore be taken to contain only nucleons, antinucleons and pions. Correspondingly, the QGP contains only up and down quarks, together with gluons. The total strangeness is then automatically zero in both phases. These restrictions on the hadron mass spectrum mean that the results of the calculations are very qualitative. To obtain even a semi-quantitative picture of the deconfinement transition, it would be necessary to include a much larger portion of the hadron spectrum, and to enforce strangeness conservation in each phase individually (and in the mixed phase, for the case of a first order transition) [16].

In our hadron gas model, the interactions between hadrons are taken into account via a mean field approximation. The central idea of such an approximation is to start with some two body interaction, and to calculate the self-energy correction to the single particle energy in the Hartree approximation. In this approximation the self energy correction is proportional to the number density of the hadron species [15], so that the single particle energy ϵ_i of a hadron of mass m_i and momentum $\vec{p}$ is given by

$$\epsilon_i = \sqrt{\vec{p}^2 + m_i^2} + K_i < n_i > . \tag{18}$$

Here K_i is the strength of the repulsive potential and $< n_i >$ is the average number density of hadron species i. This then leads to a simple transcendental equation for the number density $< n_i >$.

One of the major advantages of the mean field approximation is that there is no problem making the formalism thermodynamically consistent. It is easy to show that the grand potential for a single component hadron gas of species i with interactions treated in the mean field approximation as formulated in (18) is given by

$$\Omega_i = d_i \frac{T}{a_i} \int_0^\infty dp \, g(p) \ln\{1 - a_i \exp[-\beta(\epsilon_i - \mu)]\} - F_i^c \tag{19}$$

where $a_i = 1$ for bosons, $a_i = -1$ for fermions and $a_i \to 0$ for classical particles and μ is the chemical potential corresponding to some conserved quantity. The function F_i^c originates in the energy density and corrects for double counting of the interaction. For mass independent interactions of the form (18), $F_i^c = \frac{1}{2} K_i < N_i >^2 /V$.

In the present model, we have not attempted to to take interactions between hadrons of different types *e.g.* meson-baryon interactions, into account. Problems are expected, as the form of these interactions is not generally known due to spin and isospin dependences. Therefore, we make the following assumptions [1] [17]:

- Hadrons of a particular type (*i.e.* nucleon, antinucleon or pion) interact only with hadrons of the same type. No allowance is thus made for *e.g.* $\pi - N$ interactions.

- Antibaryons interact among themselves in the same way as baryons interact among themselves, *i.e.* $K_b = K_{\bar{b}}$.

With these assumptions, the grand potential becomes

$$\Omega_H = \sum_i d_i \frac{T}{a_i} \int_0^\infty dp\, g(p) \ln\{1 - a_i \exp[-\beta(\epsilon_i - \mu_i)]\} - \frac{1}{2} \sum_\alpha K_\alpha < N_\alpha >^2 /V, \quad (20)$$

where the sum over i runs over hadron species, and the sum over α over hadron types *i.e.* mesons, baryons and antibaryons.

We thus have to find two effective potential strengths, K_m and K_b, which fulfil the role of compressiblity coefficients for the meson and baryon sectors respectively. A reasonable value for K_m is that obtained from Weinberg's effective Lagrangian for the $\pi - \pi$ interaction [18], which yields a value of around 600 MeV fm^3.

To find a value for K_b, we need the $N - N$ interaction. However, very different values for K_b are obtained depending on the two-body interaction used. For example, on the grounds that it costs approximately 500 MeV of energy to force two nucleons to overlap completely [19], a rather arbitrary parameterization of the $N - N$ interaction of the form $V(r) = 500 \exp[-m_\omega r]$ MeV, where m_ω is the ω-meson mass, leads to $K_b = 200$ MeV fm^3 [20]. The use of a repulsive ω-exchange potential *i.e.* $V(r) = 36\pi \exp[-m_\omega r]/r$ yields $K_b = 1700$ MeV fm^3 [21]. Between these two extremes, the Reid soft core potential [22] gives $K_b = 680$ MeV fm^3 [2]. In our subsequent calculations, we will use $K_b = 1700$ MeV fm^3, which has been rather successful in studies of nuclear matter; the results for $K_b = 680$ MeV fm^3 differ only quantitatively.

The pressure of the hadron phase $< P_H >$ is now found from

$$< P_H > = - \left(\frac{\partial \Omega_H}{\partial V_H} \right)_{T_H, \mu_H^b}, \quad (21)$$

where V_H is the volume of the hadron gas. If we neglect the surface and length terms in the density of states (2), we obtain the familiar result

$$< P_H > = - \frac{\Omega_H}{V_H}, \quad (22)$$

appropriate for an infinite system. If we use the full density of states (2), we have to take into account the variation of the surface and linear terms in $g(p)$ with respect to V_H. For a system with spherical symmetry, as we will assume here, it is convenient to write

$$\frac{\partial}{\partial V_H} = \frac{1}{4\pi R_H^2} \frac{\partial}{\partial R_H}, \tag{23}$$

where R_H is the radius of the hadron system.

Many previous model calculations of the deconfinement transition have described the QGP phase in terms of the bag model [23] [24], and have only been performed in the thermodynamic limit where colour projection is not necessary (the correction to the partition function $\sim V^{-4}$ [12]). In this approach, the quarks and gluons are treated as ideal gases, apart from non-perturbative corrections to the pressure and energy density resulting from the bag pressure. The pressure, energy density and baryon number density for a QGP consisting of massless u and d quarks and antiquarks, each with spin-colour degeneracy $d_q = 6$, and 8 gluons, with degeneracy $d_g = 2$, at a temperature T and baryon potential μ_{QGP}^b are then given by

$$< e_{QGP} > = \frac{6}{\pi^2} \left(\frac{7\pi^4}{60} T^4 + \frac{\pi^2}{2} T^2 (\frac{\mu_{QGP}^b}{3})^2 + \frac{1}{4}(\frac{\mu_{QGP}^b}{3})^4 \right) + \frac{8\pi^2}{15} T^4 + B \tag{24}$$

$$< P_{QGP} > = \frac{1}{3} (< e_{QGP} > - 4B) \tag{25}$$

$$< n_{QGP}^B > = 2 \left(\frac{\mu_{QGP}^b}{3} T^2 + \frac{1}{\pi^2} \left(\frac{\mu_{QGP}^b}{3} \right)^3 \right), \tag{26}$$

where B is the bag constant. However, the results of lattice gauge calculations indicate that the non-perturbative correction to the QGP thermodynamic quantities is not simply a constant, but is dependent on the temperature and chemical potential [25] [26]. One explanation for this dependence is the presence of high-mass (~ 1 GeV) quasiparticle excitations in the plasma just above the deconfinement transition. It is possible to include the effect of these quasiparticles in a model for the QGP by the simple expedient of introducing low momentum cutoffs in the integrals for the QGP thermodynamic quantities [27]. This results in an effective "bag constant" which has a dependence on both the temperature and chemical potential. The massive quasiparticles are strongly mass suppressed relative to the quarks and gluons, and have little influence on the thermodynamics of the QGP. For simplicity, they are omitted in subsequent calculations.

To calculate thermodynamic quantities for the colour singlet QGP, we have to perform differentiations of the colour singlet partition function $Z_{singlet}$. The pressure $< P >$ and baryon number $< N_B >$ are found from

$$< P > = - \frac{1}{Z_{singlet}} \left(\frac{\partial Z_{singlet}}{\partial V} \right)_{T, \mu_{QGP}^b} \tag{27}$$

$$< N_B > = - \frac{1}{Z_{singlet}} \left(\frac{\partial Z_{singlet}}{\partial \mu_{QGP}^b} \right)_{T, V}. \tag{28}$$

Since there is no analytic form for $Z_{singlet}$, the simplest method of calculating $< P >$ and $< N_B >$ is to differentiate first, and then to perform the integrations numerically.

A parameter which we have to determine in the calculations is the radius of the QGP, R_{QGP}. In a first order transition, the volume of the system may undergo a change at the transition. For finite systems, where the size of the system does not naturally disappear from the problem, it is necessary to determine the size of the system after the transition. This is accomplished (for our case of spherical symmetry) by choosing a radius for the hadron phase, R_H, and determining the radius R_{QGP} of the QGP by conserving the baryon number across the transition. As we shall see, the models for the hadron phase and the QGP described above lead to a first order deconfinement transition at non-zero values of μ_b, and a continuous transition at $\mu_b = 0$. The order of the deconfinement transition is in fact still an open question, and appears to be strongly dependent on the number of quark flavours and the quarks masses [28]. Lattice QCD indicates that for our case of two flavours of massless quarks, the transition at $\mu_b = 0$ is continuous and thus of order higher than one, in agreement with our simple model. Unfortunately, there are no lattice results for $\mu_b \neq 0$ at this stage.

For the momentum cutoffs, we have used 800 MeV/c for both p_q and p_g, as in [27], so as to obtain a transition temperature around 140−150 MeV at zero baryon number density (or, equivalently, $\mu_H^b = 0$), as indicated by lattice QCD calculations.

RESULTS

In Figure 1 we show the critical temperature T_c as a function of the baryon chemical potential μ_H^b for an infinite system and for a colour-projected finite system with $R_H = 1.5$ fm. Such a choice for R_H seems reasonable since the radius of the hadron system resulting from a heavy ion collision should be considerably smaller at the hadronisation phase boundary than at freeze-out; previous results indicate that the radius of the system at freeze-out is of the order of 5-10 fm [29] [30]. The relationship between R_{QGP}, found from baryon number conservation, and R_H is complicated for a system with finite size corrections in the single particle density of states. If we ignore the surface and length terms in (2), however, this relationship becomes simpler. We find that R_{QGP} is smaller than R_H in regions where the derivative of T_c with respect to μ_H^b is negative, and larger where the derivative is positive. To see this, consider the Clausius-Clapeyron equation [31]

$$\frac{\mathrm{d}T_c}{\mathrm{d}\mu_b^H} = - \frac{< n_{QGP}^B > - < n_H^B >}{< s_{QGP} > - < s_H >}, \tag{29}$$

where $< n_{QGP}^B >$ and $< n_H^B >$ and $< s_{QGP} >$ and $< s_H >$ are the mean baryon number and entropy densities in the QGP and hadron phases respectively. Since $< s_{QGP} >$ is larger than $< s_H >$ on account of the greater number of degrees of freedom in the QGP than in the hadron phase, $\mathrm{d}T_c/\mathrm{d}\mu_b^H$ will be negative where $< n_{QGP}^B >$ is larger than $< n_H^B >$. However, the conservation of baryon number means that $< n_{QGP}^B >$ larger than $< n_H^B >$ implies that R_{QGP} must be less than R_H. This derivation relies on the fact that $\Omega = - < P > V$ for a system without finite size corrections. In the case of a system with such corrections, it is not possible to obtain such a simple equation as (29) for $\mathrm{d}T_c/\mathrm{d}\mu_b^H$. In Figure 2 we show T_c as a function of the baryon number density n_B. It is clear from Figure 2 that the transition in our model is first order.

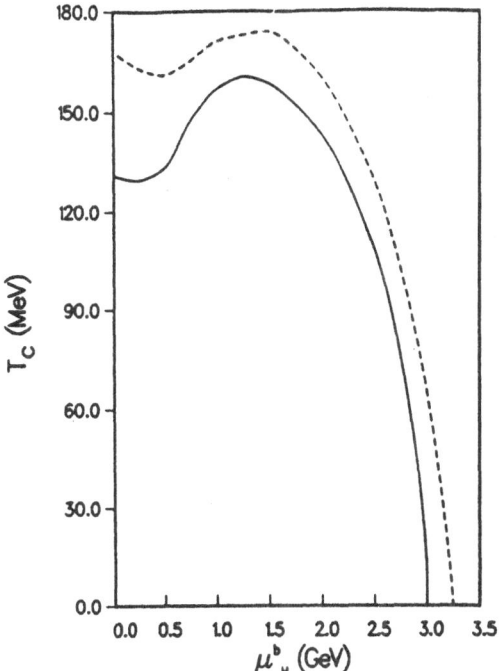

Figure 1. The critical $\mu_H^b - T_c$ curve for the deconfinement transition. The solid curve corresponds to an infinite system, while the dashed curve corresponds to a finite system with $R_H = 1.5$ fm.

The effect of including finite size effects (colour projection in the QGP and the density of states (2) in both phases) in the calculations is to raise the transition temperature. This is because colour projection decreases the number of states in the QGP only, while the inclusion of surface and length terms in the density of states lowers the pressure in the QGP more than in the hadron phase. The projection onto a colour singlet has already reduced the density of states. The change in the density of states of the QGP due to the finite size effects is thus proportionally larger than the change in the density of states in the hadron phase. The two effects thus complement each other in raising the critical temperature for the deconfinement transition. The two effects are, however, not always of the same magnitude. At small values of μ_H^b, the change due to colour projection is of the same order of magnitude as that caused by the modified density of states. As μ_H^b increases, colour projection has little influence on the transition temperature, in line with the behaviour found previously [12]. The change due to the modified density of states is, however, still substantial.

An important question is how small the fireball has to be before the finite size corrections become important to the phase diagram. The easiest way to investigate this is to compare for example $< P_H >$ and $< P_{QGP} >$ for various values of the hadron radius R_H. In Figure 3 , we show these quantities as a function of R_H for $T = 170$ MeV and $\mu_b = 1250$ MeV. This combination of T and μ_b is one point on the phase boundary of the infinite system; the deviation from equilibrium due to the finite size of

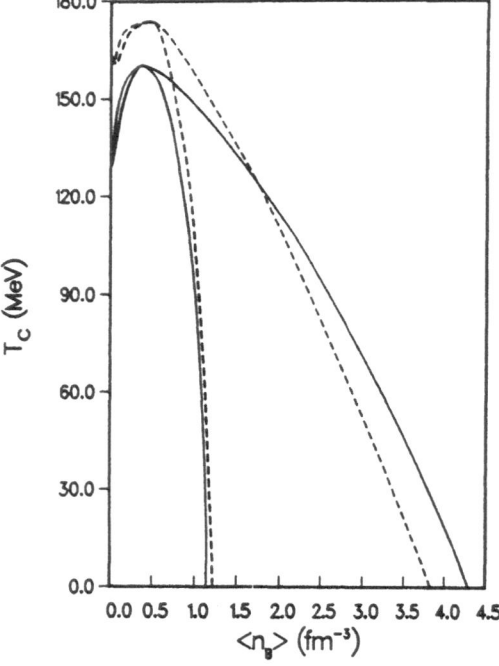

Figure 2. The critical $< n_B > - T_c$ curve
for the deconfinement transition. The solid
curves corresponds to an infinite system, and
the dashed curves to a finite system with
$R_H = 1.5$ fm. For each set of curves, the
region below the inner curve cooresponds to
the hadron phase, and the region above the
upper curve to the QGP. The region between
the curves is the coexistence region.

the system can thus be easily seen by the difference between $< P_H >$ and $< P_{QGP} >$
as the radius varies. From the figure it can be concluded that the effects of the finite
system on the phase diagram are noticeable for $R_H \leq 4.0$ fm. However, the effects
on the actual values of the pressures are appreciable to somewhat larger distances. We
would thus expect to see the results of finite size effects in heavy ion collisions.

Within the bounds of the simple model considered here, we can therefore conclude
that the possible phase diagram for the deconfinement transition, if such a transition
were to occur as a result of a heavy ion collision, could be quite different from the
diagram one obtains from considering an infinite system because of the effect of the
finite size of the system. In particular, it appears that the equilibrium curve between
hadronic matter and a deconfined plasma shifts to higher temperatures as the system
decreases in size. If this result is qualitatively correct, it indicates that even higher energy densities are required before a QGP can be obtained in the laboratory. It should
be noted, however, that the model is rather crude, and these conclusions are subject to
the important assumption of equilibrium throughout the calculations.

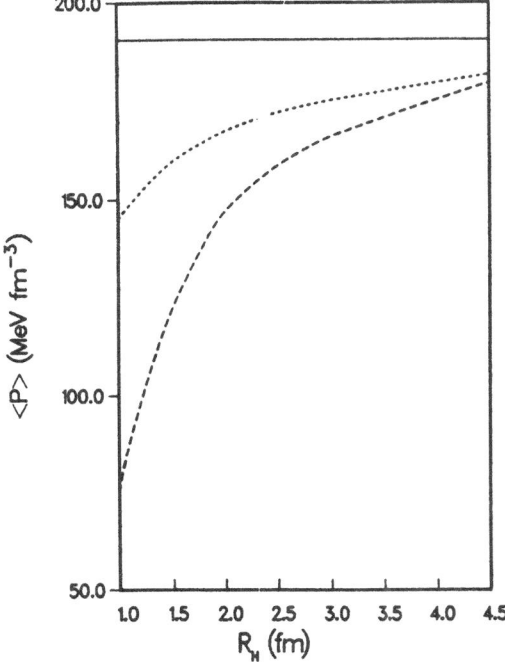

Figure 3. $< P_H >$ (short dashes) and $< P_{QGP} >$ (long dashes) as functions of R_H for $T = 170$ MeV and $\mu_H^b = 1.25$ GeV. The solid line is the pressure of both the plasma and the hadron phase for an infinite system, which are equal for these values of T and μ_H^b.

ACKNOWLEDGEMENTS

The financial support of the Foundation for Research Development, Pretoria, is gratefully acknowledged.

REFERENCES

[1] K. A. Olive, Nucl. Phys. B190 [FS3] (1981) 483.

[2] K. A. Olive, Nucl. Phys. B198 (1982) 461.

[3] N. J. Davidson, R. M. Quick, H. G. Miller and A. Plastino, Phys. Lett. B, to appear.

[4] A. L. Goodman, J. I. Kapusta and A. Z. Mekjian, Phys. Rev. C30 (1984) 851.

[5] A. L. Goodman, Phys. Rev. C37 (1988) 2162.

[6] A. L. Goodman, Phys. Rev. C39 (1989) 2008.

[7] R. K. Pathria, *Statistical Mechanics*, Pergamon Press, 1972.

[8] F. Solms and H. G. Miller, Phys. Lett. A, to appear.

[9] L. Turko, Phys. Lett. B104 (1981) 153.

[10] K. Redlich and L. Turko, Z. Phys. C5 (1980) 201.

[11] R. Hagedorn and K. Redlich, Z. Phys. C27 (1985) 541.

[12] M. I. Gorenstein, S. I. Lipskikh, V. K. Petrov and G. M. Zinovjev, Phys. Lett. B123 (1983) 437.

[13] H.-Th. Elze, W. Greiner and J. Rafelski, Phys. Lett. B124 (1983) 515.

[14] H.-Th. Elze, W. Greiner and J. Rafelski, Z. Phys. C24 (1984) 361.

[15] A. L. Fetter and J. D. Walecka, *Quantum Theory of Many-Particle Systems*, McGraw-Hill, 1971.

[16] H. W. Barz, B. L. Friman, J. Knoll and H. Schulz, Phys. Rev. D40 (1989) 157.

[17] R. Tegen, B. J. Cole, N. J. Davidson, R. H. Lemmer, H. G. Miller and R. M. Quick, Nucl. Phys. A, to appear.

[18] S. Weinberg, Phys. Rev. 166 (1968) 1568.

[19] K. Johnson, Acta Phys. Pol. B6 (1975) 865.

[20] J. I. Kapusta and K. A. Olive, Nucl. Phys. A408 (1983) 478.

[21] J. D. Walecka, Ann. of Phys. 83 (1974) 491.

[22] G. E. Brown and A. J. Jackson, *The Nucleon-Nucleon Interaction*, North- Holland, Amsterdam, 1976.

[23] J. Cleymans, K. Redlich, H. Satz and E. Suhonen, Z. Phys. C33 (1986) 151.

[24] D. W. von Oertzen, S. Afr. J. Phys. 11 (1988) 82.

[25] M. I. Gorenstein and O. A. Mogilevsky, Z. Phys. C38 (1988) 161.

[26] J. Engels, J. Fingberg, K. Redlich, H. Satz and M. Weber, Z. Phys. C42 (1989) 341.

[27] D. H. Rischke, M. I. Gorenstein, H. Stöcker and W. Greiner, Phys. Lett. B237 (1990) 153.

[28] H. Satz, *Heavy Ion Physics at Very High Energies*, Plenary talk at the ECFA LHC Workshop, Aachen, Germany, 1990.

[29] N. J. Davidson, H. G. Miller, R. M. Quick and J. Cleymans, Phys. Lett. B255 (1991) 105.

[30] N. J. Davidson, H. G. Miller and D. W. von Oertzen, Phys. Lett. B256 (1991) 554.

[31] B. M. Waldhauser, D. H. Rischke, J. A. Maruhn, H. Stöcker and W. Greiner, Z. Phys. C43 (1989) 411.

THE HYPERSPHERICAL HARMONIC METHOD APPLIED TO
^{12}C AND ^{16}O IN THE α-PARTICLE MODEL

A. Kievsky and M. Viviani

Istituto Nazionale di Fisica Nucleare, Sezione di Pisa, 56100 Pisa, Italy

S.Rosati

Dipartimento di Fisica, Universita' di Pisa, 56100 Pisa, Italy

Istituto Nazionale di Fisica Nucleare, Sezione di Pisa, 56100 Pisa, Italy

ABSTRACT

We apply the hyperspherical harmonic expansion method to the low–lying states of ^{12}C and to the ground state of ^{16}O in the α–particle model. Two different phenomenological $\alpha - \alpha$ potentials are considered. The method adopted allows for accurate calculations of the 3α model for ^{12}C, even when nonlocal potentials are considered; however, in the case of a 4α system the expansion is rather slowly convergent.

1. INTRODUCTION

The α–particle model for nuclei springs from the fact that the α particle has a strongly bound structure and can maintain its identity within a nucleus. In the model, the α particles are assumed to interact as structureless particles and to obey the Bose statistics. As a consequence, the number of particles is reduced to one–fourth and the α particles have no spin and isospin degrees of freedom. On the other hand, it is well known that, in the resonating group method (RGM), the $\alpha - \alpha$ interaction contains, in addition to a local term, a nonlocal one which arises from the nucleon exchange [1-6] and the relative w.f. of two α-clusters has an inner oscillation. These features are very different from those of the phenomenological Ali–Bodmer $\alpha - \alpha$ potential which contains a strong repulsive core [7]. Non–local separable potentials which are able to reproduce the inner oscillation of the $\alpha - \alpha$ relative w.f. have been proposed by Kukulin and Neudatchin [8-9] and by Walliser and Nakaichi-Maeda [10]. An important aspect of the problem is to precisely determine the details of the $\alpha - \alpha$ phenomenological interaction and this is a difficult task. However, we are mainly interested in calculating in a sufficiently accurate way the structure of the system. The 3α model for ^{12}C with "realistic" $\alpha - \alpha$ interactions has been investigated by many authors, using either the Faddeev method [10-14] or others techniques [11, 15-16]. Only a few calculations have been performed for the 4α model of ^{16}O [12,15].

The hyperspherical harmonics (HH) method is a convenient way to investigate the bound states of a few–body system, especially in the case of rather soft interactions. On the other side, in the case of strong repulsive or state-dependent forces the expansion of the w.f. on the HH basis can be very slowly convergent and, in order to reduce the number of basis functions to be considered, it is convenient to introduce [18-22] a new set of basis functions obtained by multiplying the HH functions by a properly chosen correlation factor. In the case of non-local potentials, a number of modifications to the usual HH approach are necessary and they will discussed in the present paper.

2. $\alpha - \alpha$ PHENOMENOGICAL INTERACTIONS

The Hamiltonian of an α–particle system can be written in the form

$$H = - \sum_{i=1}^{n} \frac{\hbar^2}{2m_\alpha} \nabla_i^2 + \sum_{i<j=1}^{n} [V_{\alpha\alpha}(i,j) + V^C(r_{ij})], \tag{1}$$

where n gives the number of α–particles that constitute the system, $V_{\alpha\alpha}(i,j)$ is the "nuclear" $\alpha - \alpha$ potential and $V^C(r_{ij})$ is the Coulomb $\alpha - \alpha$ repulsion which has been taken of the form

$$V^C(r) = \frac{4e^2}{r} \text{erf}\left(\frac{\sqrt{3}r}{2R_\alpha}\right), \tag{2}$$

corresponding to a Gaussian charge distribution in the α particle with the rms radius $R_\alpha = 1.44$ fm [23].

We have performed calculations with a number of potentials which have different structures but all fit the low–energy $\alpha - \alpha$ elastic scattering data:

(i) *Ali–Bodmer potential (AB)*[ref. 7]. This potential depends on the relative angular momentum and on the separation distance of the two particles; it can be written in the form

$$V_{\alpha\alpha}(i,j) = \sum_{\ell} V_\ell(r_{ij}) P_\ell(i,j), \tag{3}$$

where $P_\ell(i,j)$ is the projector on the state of the pair i, j with relative angular momentum equal to ℓ. The radial dependence is given in terms of gaussians as

$$V_\ell(r_{ij}) = V_R \exp\left(-\mu_R^2 r_{ij}^2\right) - V_A \exp\left(-\mu_A^2 r_{ij}^2\right), \tag{4}$$

with a fixed attractive part and different repulsive parts for $\ell = 0, 2, 4$. The parameters used are shown in table I.

(ii) *Kukulin–Neudatchin nonlocal separable potential (NK)*. Kukulin and Neudatchin [8] found that an one–term nonlocal separable potential with N= 4 H.O. type oscillation can reproduce the inner oscillation of the 2α relative w.f. The nuclear potential in the momentum representation is expressed as

TABLE I. *Parameters used in the present calculation for two versions of the AB potential.*

	ℓ	V_R (MeV)	μ_R (fm^{-1})	V_A (MeV)	μ_A (fm^{-1})
type (d)	0	500	0.7	130	0.475
	2	320	0.7	130	0.475
	4	0	0.7	130	0.475
type (e)	0	1050	0.8	150	0.5
	2	640	0.8	150	0.5
	4	0	0.8	150	0.5

$$V_{\alpha\alpha}(\mathbf{k}, \mathbf{k}') = -\frac{\hbar^2}{m_\alpha} \sum_{\lambda\mu} \xi_\lambda \, g_\lambda(k) g_\lambda(k') Y_{\lambda\mu}(\hat{\mathbf{k}}) Y_{\lambda\mu}^*(\hat{\mathbf{k}}') , \tag{5}$$

where

$$g_\lambda(k) = (k^2 + K_0^2)(k^2 - K_1^2)(k^2 - K_2^2) \exp(-\beta^2 k^2 / 2), \tag{6}$$

for $\lambda = 0, 2$ and 4. Kukulin and Neudatchin adjusted the various parameters to reproduce the "nuclear" phase shifts in the presence of the Coulomb repulsion, using the nuclear potential only. A better fitting of the experimental phase shifts was given by Fuyiwara and Tamagaki [9], who extended the analysis to the $\lambda = 6$ wave too, with the choice

$$g_6(k) = (k^2 + K_0^2) k^6 \exp(-\beta^2 k^2 / 2). \tag{7}$$

The parameters values determined in ref.[9] are reported in table II. The NK potential in the coordinate representation is again separable and it is easily obtained by Fourier transformation.

TABLE II. *Parameters of the nonlocal separable NK potential.*

λ	β (fm)	ξ_λ^{-1} (fm^{-13})	K_0^2 (fm^{-2})	K_1^2 (fm^{-2})	K_2^2 (fm^{-2})
0	1.02	0.134×10^3	8.14	0.54	3.3
2	1.02	0.230×10^3	7.68	3.2	0
4	1.02	0.487×10^4	22.3	0	0
6	1.02	$0.974 \times 10^5 \,(\text{fm}^{-17})$	10.0		

3. FORMALISM

In this section we will largely refer to a previous paper of ours (ref.[22]) adopting also the same notations.

A) HH and CHH expansion for central potentials

The HH and CHH bases are complete ones, therefore the w.f. Ψ of an n–particle system can be expanded in terms of them; in the case we are here interested in, the w.f. is completely symmetrical with respect to exchange of the particle coordinates, therefore only symmetric combinations of the HH (or CHH) functions are required. So that we write

$$\Psi(1,\ldots,n) = \sum_{[K]} u_{[K]}(\rho) B_{[K]}^{(s)}(\Omega), \tag{8}$$

where ρ is the hyperradius, Ω represents the set of hyperspherical angles, $[K]$ stands for all the quantum numbers necessary to identify the HH function for a fixed value of the grand orbital quantum number K and $B_{[K]}^{(s)}(\Omega)$ is a symmetrical combination of HH (or CHH) functions [22]

$$B_{[K]}^{(s)}(\Omega) = F(1,\ldots,n)\frac{1}{n!}\sum_{P} Y_{[K]}(\Omega_{ijk\ldots}) , \tag{9}$$

where $\Omega_{ijk\ldots}$ are the hyperspherical angles for the ordering $i,j,k,\ldots$ of the particles and the sum is carried over all the permutations P of the particles. $F(1,\ldots,n)$ is a symmetrical correlation factor depending on the interparticle distances. For the usual HH expansion one has $F = 1$; in the CHH approach, a choice often adopted for the correlation factor is the so–called Jastrow form, namely $F = \prod_{i<j} f(r_{ij})$.

In order to perform the calculations, it is convenient to write the HH functions, in the symmetrized combination of eq.(9), in terms of an unique set of hyperangular coordinates; this is possible since a HH function with grand orbital K , defined for a given set of hyperangular coordinates, can be written as a linear combination of all the HH functions corresponding to a different set of coordinates and having the same grand orbital quantum number. For $n = 3$ the coefficients of the transformation have been given by Raynal and Revai [24] and for $n = 4$ they can be found in ref.[25]. In conclusion, the symmetrical combination $B_{[K]}^{(s)}(\Omega)$ can be chosen to be of the form

$$B_{Kq}^{(s)}(\Omega) = F(1,\ldots,n)\sum_{[K]} R_{[K]q} Y_{[K]}(\Omega_{1\ldots n}), \tag{10}$$

where $\Omega_{1\ldots n}$ is the considered set of hyperangular coordinates, $R_{[K]q}$ are the coefficients of the transformation, q labels the different symmetrical combinations that can be constructed for a fixed K value [26] and the summation on $[K]$ runs over all the quantum numbers compatible with this value of the grand orbital momentum.

If the expansion in eq.(8) is limited to a finite number of terms, the best variational choice of the hyperradial functions $u_{Kq}(\rho)$ satisfies the condition

$$\delta_u \langle \Psi|H - E|\Psi\rangle = 0, \tag{11}$$

where δ_u denotes the functional variation with respect to $u(\rho)$. From this condition, we can derive a set of equations for the functions $u_{Kq}(\rho)$ that is cast into the form

$$\sum_{K',q'} [T_{Kq,K'q'}(\rho) + V_{Kq,K'q'}(\rho) - E \ N_{Kq,K'q'}(\rho)] u_{K',q'}(\rho) = 0, \tag{12}$$

where K and K' run from zero to the value $K_{\max}$ and the matrix elements are definied according to

$$O_{Kq,K'q'} = \int d\Omega \ B_{Kq}^{(s)}(\Omega) O B_{K'q'}^{(s)}(\Omega). \tag{13}$$

In eq.(12) the matrix elements of the kinetic energy, potential energy and unity operator appear in the order. In the case of the (uncorrelated) HH expansion, $N_{Kq,K'q'}$ reduces to the Kronecker delta function $\delta_{Kq,K'q'}$. For the CHH case, the integrals defined by eq.(13) are functions of ρ due to the correlation factor.

B) ℓ − dependent potential

If the pair potential depends on the relative angular momentum, some care is required in evaluating the matrix elements of the potential operator. In the case of the HH expansion there are no difficulties at all, since the projection operators on a state with a definite angular momentum of a given pair commute with $u_{Kq}(\rho)$. When the CHH tecnique is used, the HH function are multiplied by a correlation function $F(1, \ldots, n)$ and the projection operators $P_\ell(i, j)$ do not commute with F; the calculation of the matrix element $V_{Kq,K'q'}$ in eq.(12) is more involved, but the form of the equation remains unchanged.

C) Nonlocal interparticle potential

For such potentials, we will limit here the discussion to the study of a three–particle system with the HH expansion method; the nonlocal potential is assumed to have the form

$$V(i,j) = \sum_{\lambda\mu} V_\lambda(r_{ij}, r'_{ij}) Y_{\lambda\mu}(\hat{\mathbf{r}}_{ij}) Y^*_{\lambda\mu}(\hat{\mathbf{r}}'_{ij})$$

$$= \frac{4\pi}{2\lambda+1} \sum_\lambda V_\lambda(r_{ij}, r'_{ij}) P_\lambda(\hat{\mathbf{r}}_{ij} \cdot \hat{\mathbf{r}}'_{ij}), \tag{14}$$

where P_λ is a Legendre polynomial. Explicitely the coordinates used are:

$$\mathbf{r}_a = \mathbf{x}_2 - \mathbf{x}_1, \qquad \mathbf{r}_b = \frac{1}{\sqrt{3}}(2\mathbf{x}_3 - \mathbf{x}_1 - \mathbf{x}_2), \tag{15}$$

$$r_a = \rho \cos \phi, \qquad r_b = \rho \sin \phi.$$

The hypershperical coordinates are ρ and the angles $\Omega \equiv \hat{\mathbf{r}}_a, \hat{\mathbf{r}}_b, \phi$. The HH functions can be written as

$$Y_{[K]}(\Omega) = {}^{(2)}P_K^{\ell_a,\ell_b}(\phi) [Y_{\ell_a}(\hat{\mathbf{r}}_a) Y_{\ell_b}(\hat{\mathbf{r}}_b)]_{\ell m}, \tag{16}$$

where the function ${}^{(2)}P_K^{\ell_a,\ell_b}(\phi)$ is related to a Jacobi polynomial [27,26,22].

In the case of a completely symmetrical w.f., the mean value of the total potential is three times that of $V(1,2)$. As a consequence, the term $V_{Kq,K'q'}(\rho)u_{K'q'}(\rho)$ in eq. (12) has now to be substituted with

$$\left\langle B_{Kq}^{(s)}(\Omega)|V(1,2)+V(1,3)+V(2,3)|u_{K'q'}(\rho)B_{K'q'}^{(s)}(\Omega)\right\rangle =$$

$$= 3\sum_{[K],[K']} R_{[K]q}R_{[K']q'}\left\langle Y_{[K]}(\Omega)|V(1,2)|u_{K'q'}(\rho)Y_{[K']}(\Omega)\right\rangle. \tag{17}$$

Using the fact that $[K]\equiv\{K,\ell_a,\ell_b\}$, the last quantity can be written in the form

$$3\sum_{\ell_a,\ell_a',\ell_b,\ell_b'} R_{K,q}^{\ell_a,\ell_b}R_{K',q'}^{\ell_a',\ell_b'}\int d\Omega\,{}^{(2)}P_K^{\ell_a,\ell_b}(\phi)\,[Y_{\ell_a}(\hat{\mathbf{r}}_a)Y_{\ell_b}(\hat{\mathbf{r}}_b)]_{\ell m}$$

$$\times\sum_\lambda\frac{4\pi}{2\lambda+1}\int dr_a'\,V_\lambda(r_a,r_a')P_\lambda(\hat{\mathbf{r}}_a\cdot\hat{\mathbf{r}}_a')\,u_{K'q'}(\rho)\,{}^{(2)}P_{K'}^{\ell_a',\ell_b'}(\phi')\left[Y_{\ell_a'}(\hat{\mathbf{r}}_a')Y_{\ell_b'}(\hat{\mathbf{r}}_b)\right]_{\ell m}$$

$$= \int d\rho'\,Z_{Kq,K'q'}(\rho,\rho')\,u_{K'q'}(\rho'), \tag{18}$$

where the kernel Z is

$$Z_{Kq,K'q'}(\rho,\rho') = \frac{3\pi}{4\lambda+2}\sum_{\ell_a,\ell_b}R_K^{\ell_a,\ell_b}R_{K'}^{\ell_a,\ell_b}$$

$$\times\int_{x_0}^1 dx\,\sqrt{1-x^2}\,{}^{(2)}P_K^{\ell_a,\ell_b}(\phi)V_{\ell_a}(r_a,r_a')\,{}^{(2)}P_{K'}^{\ell_a,\ell_b}(\phi'), \tag{19}$$

with the following definitions:

$$x = \cos 2\phi,\quad r_a' = \rho'^2 - \frac{1}{2}(1-x)\rho^2,\quad \cos\phi' = \rho'/r_a',$$

$$x_0 = \begin{cases} -1 & \text{if } \rho'\geq\rho; \\ 1-2(\rho'/\rho)^2 & \text{if } \rho'<\rho. \end{cases} \tag{20}$$

In conclusion, for the nonlocal potential considered, eq.(12) has to be modified by the inclusion of a nonlocal kernel

$$\sum_{K',q'}\left[(T_{Kq,K'q'}(\rho)+V_{Kq,K'q'}(\rho)-E\delta_{Kq,K'q'})u_{K'q'}(\rho)\right.$$

$$\left. +\int d\rho'\,Z_{Kq,K'q'}(\rho,\rho')u_{K'q'}(\rho')\right] = 0. \tag{21}$$

For a given $K_{\max}$, the energy and the functions $u_{Kq}(\rho)$ can be obtained by standard procedures.

TABLE III. *Energies of the ground state 0^+ and of the 2^+ and first 0^+_1 levels calculated for the 3α model of ^{12}C with the HH expansion; the numbers in parentheses correspond to the CHH expansion. The results are presented for the Ali–Bodmer (d) and (e) potentials, V^C is the the Coulomb potential specified in eq.(2). The experimental energy values are reported in the last line.*

^{12}C levels	0^+	2^+	0^+_1
Potential	E(MeV)	E(MeV)	E(MeV)
AB(d)	-6.41 [-6.42]	-4.58	-1.90 [-1.92]
AB(d) $+ V^C$	-1.50	$+0.65$	$+1.61$
AB(e)	-7.48	-5.75	-2.32
AB(e) $+ V^C$	-2.31	-0.20	$+1.65$
Exp.	-7.27	-2.84	$+0.38$

4. RESULTS AND DISCUSSION

i) The 3α model of ^{12}C.

We have calculated the energies of the ground state, the 2^+ and the first 0^+_1 levels of ^{12}C in the 3α model. For the AB potential we have used the (d) and (e) versions with and without the Coulomb potential. The value $\hbar^2/m_\alpha = 10.4456$ MeV fm^2 has been adopted. The calculations have been done by using the HH expansion and, in a few cases, by means of the CHH expansion.

As a preliminary calculation we have solved the simplified problem of the $\ell = 0$ AB potential active in all the partial waves. The obtained ground state energies are -5.12 MeV and -5.71 MeV in correspondence to the (d) and (e) choices, respectively [*].

For ℓ–dependent potentials it might be convenient to generalize the CHH approach outlined in sect.3, by introducing ℓ–dependent correlation factors, namely

$$\Psi(1,2,3) = \sum_{\ell=0,2,4}\sum_{[K]} u^{\ell}_{[K]}(\rho)B^{(s),\ell}_{[K]}(\Omega) \tag{22.a}$$

$$B^{(s),\ell}_{[K]}(\Omega) = F_\ell(1,2,3)\frac{1}{n!}\sum_P Y_{[K]}(\Omega_{ijk}) , \tag{22.b}$$

where the quantum numbers are $[K] = \{K, \ell, \ell\}$, i.e. for each term in (22.a) only the grand orbital quantum number K varies. The correlation factors are of the form

$$F_\ell(1,2,3) = f_\ell(r_{12})f_\ell(r_{13})f_\ell(r_{23}) . \tag{23}$$

The function $f_\ell(r)$ is chosen as the solution of a zero-energy Schroedinger equation

$$[-\hbar^2/m_\alpha\nabla^2 + V_\ell(r) + \lambda_\ell(r)] f_\ell(r) = 0, \tag{24}$$

where the role of the additional term $\lambda_\ell(r)$ is discussed in refs. [21,22].

TABLE IV. *Energies for the ground state 0^+, 2^+ and first 0_1^+ levels of the ^{12}C as a 3α system with the nonlocal KN potential with and without Coulomb force. The results of ref.[9] are given for the sake of comparison, as discussed in the text. The mean value of the kinetic energy $\langle T \rangle$ and of the Coulomb potential $\langle V^C \rangle$ in MeV, as well as the rms radius $\langle r_m \rangle$ in fm are reported in correspondence of the calculation including the Coulomb interaction.*

^{12}C levels	0^+	2^+	0_1^+	
Potential	E(MeV)	E(MeV)	E(MeV)	
KN	−17.0	−12.8	−2.8	ref.[9]
KN	−17.35	−12.56	−2.90	present paper
KN + V^C	−10.86	− 6.35		present paper
$\langle T \rangle$	68.82	62.53		
$\langle V^C \rangle$	6.25	6.00		
$\langle r_m \rangle$	1.68	1.73		

The results obtained for the AB potentials are presented in table III. For the HH expansion a system of 19 coupled equations, corresponding to $K_{max} = 24$, has been solved for the 0^+ levels, whilst 20 equations, corresponding to $K_{max} = 16$, were considered for the 2^+ level. From the rate of convergence in K we evaluate the energy error smaller than .01 MeV for the ground state and .02 MeV for the other two levels. Such estimates are supported also by the few CHH results we have obtained and listed in brackets in table III. The results we have obtained improve to some extent on those given in ref.[28]. In conclusion, see also ref. [11], the AB potentials do not reproduce in details the low–lying spectrum of ^{12}C.

For the nonlocal KN potential the calculations have been performed only with the HH expansion; the total energy, the mean value of the kinetic energy and of the Coulomb potential, as well as the rms radius are listed in table IV. The same number of equations as for the AB potential has been solved in order to get a precision of the same order of magnitude. It is therefore evident the capability of the HH expansion to manage also nonlocal potentials. It should be noticed that the presence of the Coulomb repulsion does not give raise to problems in the approach we have used (a situation quite different from the one in the Faddeev method).

From table IV we can evaluate the effect of the Coulomb repulsion as $E(V^C = 0) - E(V^C) = 6.40$ MeV, a value close to $\langle V^C \rangle$ listed in table IV. Calculations for the KN potential without the Coulomb force have been performed within the Faddeev formalism in ref.[9]; for the sake of comparison, the results are reported in table IV and appear in good agreement with our corresponding estimates. In ref.[9], the Faddeev w.f. has been used to calculated the Coulomb energy at the first order perturbation theory, with the result $\langle V_C \rangle = 9 \sim 10$ MeV which, however, overestimates the correct value by about 3 MeV.

The binding energy calculated with the nonlocal potential comes from large cancellation between the kinetic and potential energy (for the AB potential the kinetic mean energy values $< T >$ are 12.38 MeV and 17.56 MeV for the (d) and (e) type, respectively) and the final result is an overbinding of about 3 MeV. In order to improve on these results many attempts have been made by including three–body forces [14,17] and this is at present subject of our investigation.

ii) The 4α model of ^{16}O.

In order to check the accuracy of the HH and CHH expansion for a system of four α–particles, we have again considered the local interparticle potential, without Coulomb repulsion, given by the $\ell = 0$ AB (d) potential active in all the waves. For this homework problem an accurate estimate of the ground state energy, $E = -11.1$ MeV (with $\hbar^2/m_\alpha = 10.36675$ MeV fm^2), is available [12]. Such an energy value compares quite well with our (converged) result of -11.23 MeV obtained with the CHH expansion by employing 15 basis functions. On the other hand, with the HH expansion (i.e., $F = 1$ in eq. (8)) the convergence results to be very slow: by using 15 functions in the expansion, the ground state binding energy is underestimated by approximately 2 MeV and a very large number of HH functions seems to be necessary in order to obtain convergence, as has been also noticed in ref.[22].

The reason of such bad convergence of the HH expansion is connected to the strong repulsion at short distances of the pair potential. In fact, the w.f. must vanish when two particles come close together and this can be realized only by the inclusion of a large number of basis functions. On the other side, the situation is different in the CHH case, due to the presence of the (Jastrow) correlation factor F. It should also be noticed that the problem is quite different for a system of $n = 3$ particles where only one symmetrical function $B_{Kq}^{(s)}(\Omega)$ exists [26,22] for small K values and the inclusion of HH functions with high K (≥ 12) does not represent a big numerical problem.

The results we have obtained for the ℓ–dependent AB potential present quite analogous characteristics. The uncorrelated HH expansion is very slowly convergent and thus we present a few (rather preliminary) results obtained with the CHH expansion. The energy values estimated for the ground state and reported in table V, have been obtained by employing a state–independent correlation factor F (see eq.(8)) and by including, in the CHH w.f., the HH functions with $K \leq 10$ (15 functions); the functions with $K = 10$ give a contribution to the energy of -0.2 MeV approximately, therefore the states with $K \geq 10$ can be expected to give contributions not completely negligeable. Calculations with different correlation factors F for the various ℓ–states (as in eq.(22) for the 3α system) are in progress and will be presented elsewhere.

By inspection of table V, it can be seen that the ground state energy obtained with the AB (d) potential and with the Coulomb interaction taken into account, is rather different from the experimental value. The only other calculation of the 4α model, performed by expanding the w.f. on the harmonic oscillator (HO) basis, gives the result $E_{g.s.} = -4.8$ MeV [15]; since the $\alpha - \alpha$ potential adopted in ref.[15] is slightly less repulsive than the AB (d), we conclude that the convergence was not reached in the HO expansion used. It would be interesting to calculate the 4α system with nonlocal potentials too.

TABLE V. *Ground state energies (MeV) calculated with the CHH expansion for the* 4α *model of* ^{16}O *and the Ali–Bodmer (d) potential;* V^C *is the Coulomb potential. The experimental ground state energy is reported in the last row.*

Potential	CHH
AB(d)	-18.5
AB(d) $+ V^C$	-7.7
Exp.	-14.5

REFERENCES

1. E. W. SCHMID and K. WILDERMUTH, Nucl.Phys. **26**(1961),463
2. R. TAMAGAKI and H. TANAKA, Prog.Theor.Phys.Suppl. **34**(1965),191
3. S. OKAI and S. C. PARK, Phys.Rev. **145**(1966),787
4. D. R. THOMSON, I. REICHSTEIN, W. McCLURE and Y. C. TANG, Phys.Rev. **185** (1969),1351
5. J. HIURA and R. TAMAGAKI, Prog.Theor.Phys.Suppl. **52**(1972),25
6. K. WILDERMUTH and Y. C. TANG, *A Unified Theory of the Nucleus* in "Clustering Phenomena in Nuclei", (Vieweg, Braunschweig 1977),Vol.1
7. S. ALI and A. R. BODMER, Nucl.Phys. **80**(1966),787
8. V. I. KUKULIN and V. G. NEUDATCHIN, Nucl.Phys. **A157**(1970),609
9. W. FUJIWARA and R. TAMAGAKI, Prog.Theor.Phys. **56**(1976),1503
10. H. WALLISER and S. NAKAICHI-MAEDA, Nucl.Phys. **A464**(1987),366
11. R. TAMAGAKI and Y. FUJIWARA, Prog.Theor.Phys.Suppl. **61**(1977),229
12. S. NAKAICHI-MAEDA, Y. AKAISHI and H. TANAKA, Prog. Theor. Phys. **64** (1980), 1315
13. H. KAMADA and S. ORYU, Prog.Theor.Phys. **76**(1986),1260
14. S. ORYU and H. KAMADA, Nucl.Phys. **A493**(1989),91
15. R. M. MENDEZ-MORENO, M. MORENO and T. H. SELIGMAN, Nucl.Phys. **A221** (1974),381
16. V. C. AGUILERA-NAVARRO and O. PORTILHO, Ann. Phys. (N.Y.) **107** (1977), 126
17. O. PORTILHO and S. A. COON, Z.Physik, **A290**(1979),93
18. Y. I. FENIN and V. D. EFROS, Sov.J.Nucl. Phys.(Engl.Transl.) **15**(1972),449
19. A. M. GORBATOV, A. V. BURSAK, Y. N. KRILOV and B. V. RUDAK, Sov.J.Nucl. Phys.(Engl.Transl.) **40**(1984),233
20. M. I. HAFTEL and V. B. MANDELZWEIG, Ann.Phys.(N.Y.) **189**(1989),29
21. S. ROSATI, M. VIVIANI and A. KIEVSKY, Few Body Syst. **9**(1990),1
22. A. KIEVSKY, M. VIVIANI and S. ROSATI, in *Condensed Matter Theories*, Vol.6 edited by S. FANTONI and S. ROSATI, (Plenum, New York 1991)
23. G. R. BURLESON and H. W. KENDALL, Nucl.Phys. **19**(1960),68
24. J. RAYNAL and J. REVAI, Nuovo Cim. **68A**(1970),612
25. R. J. JUBUTI, N. B. KUPRENIKOVA and L. L. SARKISYAN, Few Body Syst.**4** (1988),151

26. G. ERENS, J. L. VISSCHERS and R. VAN WAGENINGEN, Ann.Phys.(N.Y.)**67** (1971),461

27. Y. A. SIMONOV, Sov.J.Nucl.Phys.(Engl.Transl.)**3**(1966),461;
A. M. BADALYAN and Y. A. SIMONOV, Sov. J. Nucl. Phys. (Engl. Transl.) **3** (1966), 755

[*] When the value $\hbar^2/m_\alpha = 10.6675$ MeV fm^2 is used, the energy value -5.18 MeV for the potential AB (d) is obtained, in agreement with the estimate given in ref.[12]

28. J. L. VISSCHERS and R.VAN WAGENINGEN, Phys.Lett. **34B** (1971),455

ATOMIC SMALL CLUSTERS AND THEIR CORRESPONDENCE TO NUCLEAR PHYSICS

G. S. Anagnostatos

Institute of Nuclear Physics
National Center for Scientific Research "Demokritos"
GR-153 10, Aghia Paraskevi-Attiki, Greece

INTRODUCTION

The physics of microclusters is a very rapidly growing, new area of science. It is an interdisciplinary topic and thus attracts scientists from many related sciences, e.g. solid state, chemistry, atomic physics, plasma physics, crystalography, and nuclear physics, both theorists and experimentalists. Their research takes place both in academic institutes and in industries, since a large number of important applications are immediately expected, e.g., in catalysis.

An aggregate of atoms or molecules is called a microcluster when the number of the constituent particles does not, usually, exceed 1000. Their electronic properties are significantly different than the properties of the same material in bulk. One of the first ways for their production is via supersonic expansion of vapours of the material produced in an oven. After their production a mass spectrometer separates the different species according to their number of particles. Their state of matter can be solid, liquid, or gas.

One of the major properties of microclusters is the appearance of magic numbers, i.e., the property that microclusters possessing specific numbers of constituent particles exhibit exceptional properties in comparison to those of species with neighboring numbers of particles.

The theoretical investigation of such numbers follows two distinct paths. The one is based on the properties of the delocalized electrons in clusters (Knight et al., 1984), while the other on the equilibrium geometry of the constituent particles (Echt et al., 1981, and Anagnostatos, 1987). Different magic numbers appear for different groups of elements, (e.g., alkali, noble gases, alkali halide, and their mixtures), or even for the same group of elements under different conditions of preparation (e.g., born neutral or born ionized) and temperature or cluster size.

It is a very important fact that neutral alkali (Knight et al., 1984) or alkali like (Ag, Au, Cu) clusters possess magic numbers very closely related to those in nuclear physics (e.g., 2, 8, 20, 40,.....). This similarity does not seem incidental and is due to the common fermionic nature of nucleons and neutral alkali atoms (i.e., odd number of electrons; Anagnostatos 1991a and b), which is consistent with the liquid (or gas) state of matter valid for both alkali clusters and nuclei. This is further consistent with the fact that bosonic clusters (i.e., clusters with atoms possessing even number of electrons as in rare gases, for example) very closely resemble a solid state of matter.

The resemblance between alkali (or alkali like) quantum clusters and atomic nuclei gives a hint of an alternative approach of studying atomic nuclei.

Condensed Matter Theories, Vol. 7, Edited by A.N. Proto
and J.L. Aliaga, Plenum Press, New York, 1992

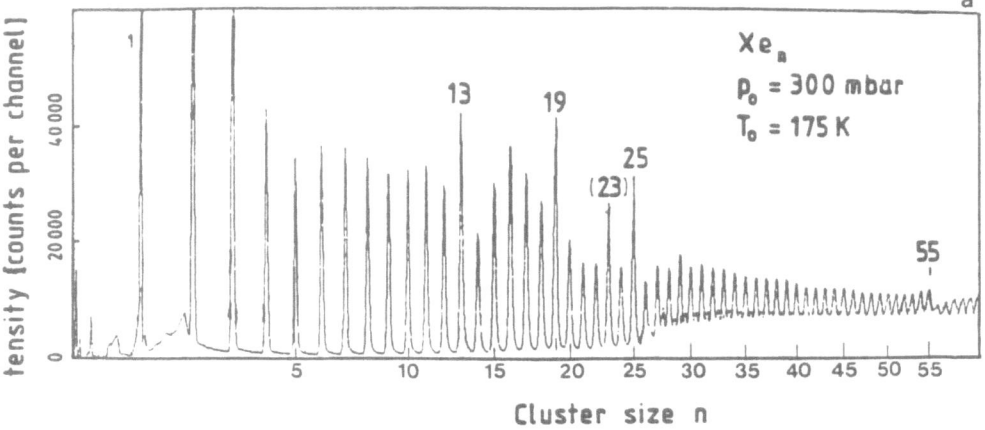

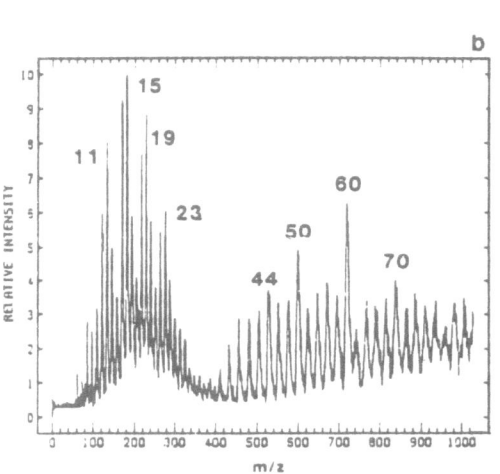

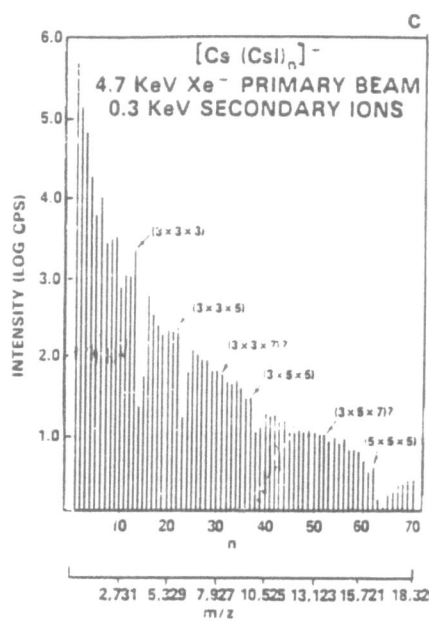

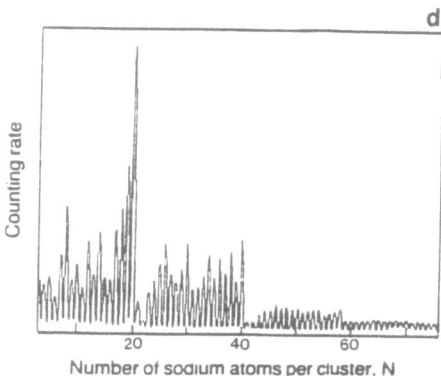

Fig. 1. Mass spectra and related magic numbers of microclusters for (a) the rare gas xenon, (b) the semiconductor carbon, (c) the alkali-halide [Cs(CsI)ₙ]⁺ and (d) the alkali sodium.

Some examples of mass spectra and related magic numbers are shown in Figure 1(a)-(d). Specifically, in Figure 1(a) the mass spectrum of xenon clusters is shown (Echt et al., 1981), where the bold numbers over prominent peaks stand for the relevant magic numbers. In Figure 1(b)-(d) similar information for carbon (Ross et al., 1986), [Cs(CsI)$_n$]$^-$ (Phillips, 1986), and sodium clusters (Knight et al., 1984), respectively, is given. These are samples of rare gas, semiconductor, alkali-halide, and alkali microclusters, whose magic numbers are : η = 1, 13, 55, 147, 309, 561,...; 4, 6, 10, 14, 18,...; 6, 14, 18, 20, 24, 30, 32, 38, 62,...; and 2, 8, 20, 40, 58,...., respectively.

In Figure 2(a)-(d) the geometrical explanation of the magic numbers appearing in Figure 1(a)-(d), respectively, is presented. Specifically, the magic numbers of rare gases are understood as closely-packed nested icosahedral shells (Echt et al., 1981, Anagnostatos, 1987, 1988b), while those of semiconductors as nested tetrahedral shells (Anagnostatos, 1990a), those of alkali halide as nested octahedral shells (Anagnostatos, 1990c), and those of alkali microclusters as nested equilibrium polyhedral shells as shown (Anagnostatos, 1987). In all four cases the magic numbers result as the cumulative number of accommodated atoms from the beginning up to the point where a polyhedral shell is completed (or up to the point where a polyhedral shell is partially, symmetrically completed). At each block of all parts in Figure 2 (bottom left) the number of atoms accommodated by the relevant polyhedral shell is given and is utilized for the estimation of magic numbers. For example, the second and third shell in Figure 2(a) accommodate 12 and 42 atoms, respectively, which lead to the major magic numbers 13(=1+12) and 55(=13+42). Numbers written inside spheres of all parts in Figure 2 stand for the specific spheres (equal in number to the number shown) forming a partial, symmetric filling of the relevant polyhedral shell which (together with the spheres of all previous shells) give rise to a secondary magic number. For example, the numbers 6 and 12 inside spheres of the Figure 2(a) give rise to the secondary magic numbers 19(=13+6) and 25(=13+12), where 13 is the previous magic number corresponding to the completion of the previous shell.

All details referring to the explanation of the magic numbers reported above can be found in the relevant cited references and thus there is no need of repeating them here in more extent than the previously given examples. At any rate, the important fact is not the demonstration of how the magic numbers result by proper summing up of vertices of complete polyhedral shells or subshells, but the very fact of demonstrating the specific symmetry supported at each case by the experimentally determined magic numbers and the implied similarities among them (Anagnostatos, 1990b). Thus, the structure of rare gas microclusters is composed of concentric icosahedra, while that of semiconductors and alkali-halide microclusters is composed of concentric tetrahedra and concentric octahedra, respectively.

Despite the fact that each of the four structures in Figure 2(a)-(d) is made of geometrical shells, there is an important difference between Figure 2(a)-(c) and Figure 2(d). This difference is that for the first three structures we have close packing of spheres at the surface of each shell and overlapping between spheres of adjacent shells (soft spheres), while for the last structure we do not have close packing of spheres on each shell, but close packing of spheres (touching of spheres) between adjacent shells (hard spheres). By using the proper terminology, the first three cases correspond to close-packing of spheres, while the last case to close-packing of shells (Anagnostatos, 1987). It is apparent that considering an effective atom-atom potential employed in the literature (e.g., Lennard-Jones potential), the close packing of spheres is energetically favored in comparison to the close packing of shells. However, if spheres presenting atoms are hard (as, for example, in alkali clusters), the corresponding structure can never follow the close- packing of spheres arrangement, since for such an arrangement an overlapping is inevitable which is prohibited for hard spheres. The physical property which makes an atom behave like a hard sphere will be discussed shortly.

For the explanation of magic numbers in alkali microclusters, besides the geometrical explanation given above (Figure 2(d)), an analytical approach has been employed in the literature as well (Knight et al., 1984) . In this approach all valence electrons of alkali atoms (i.e., one from each) are considered delocalized and under the influence of a central potential somehow created by the ion cores. This potential is given by (1)

$$U(r) = -\frac{U_0}{\exp\left[(r-r_0)/\varepsilon\right]+1},$$ (1)

where U_0 is the sum of the Fermi energy (3.23 eV) and the work function (2.7 eV) of the bulk; r_0 is the effective radius of the cluster sphere assumed to be $r_s N^{1/3}$, where r_s is the radius of a sphere containing one electron in the bulk ($r_s = 3.93$ a.u. for sodium, for example). The parameter ε (=1.5 a.u.) determines the variation of the potential at the edge of the sphere. The Schrödinger equation is solved numerically for each N.

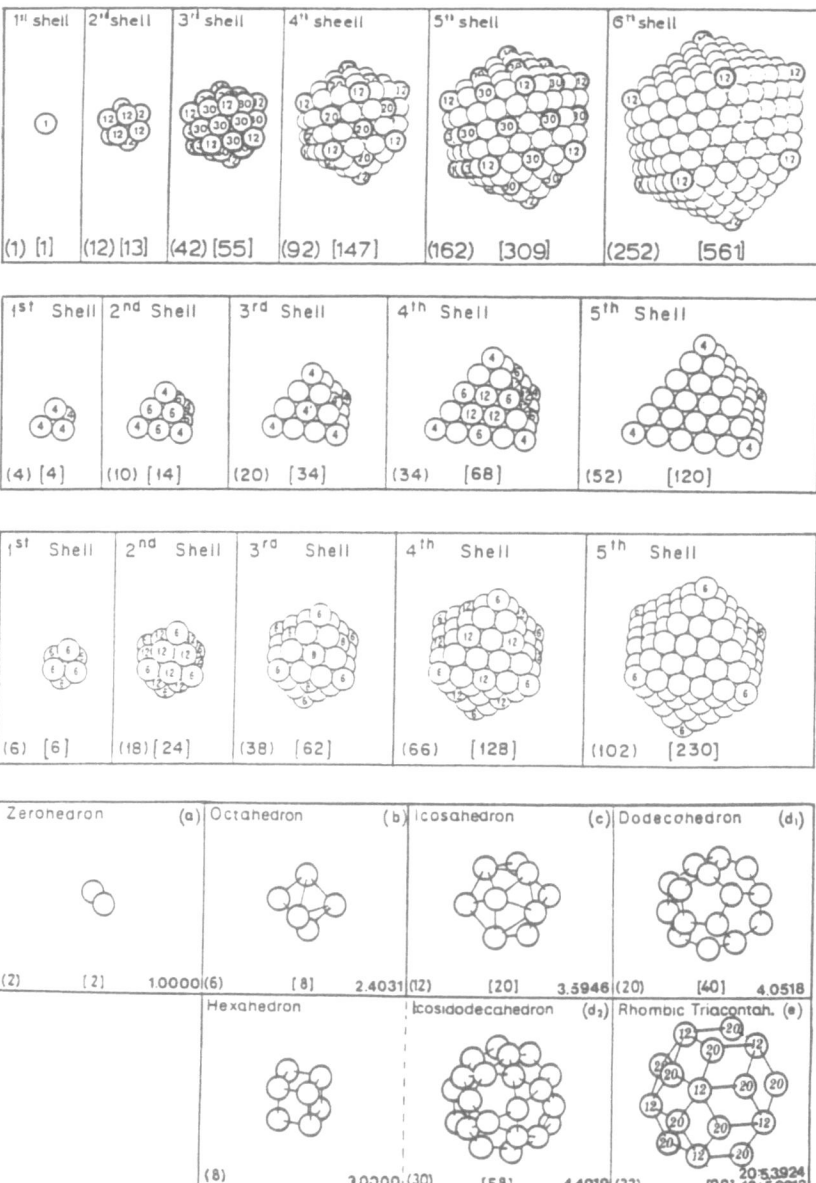

Fig. 2. Geometrical explanation of magic numbers for the spectra shown in Figure 1(a)-(d). Specifically for (a) rare gases : close packing of spheres as nested icosahedral shells, (b) semiconductors: close packing of spheres as nested tetrahedral shells, (c) alkali-halides: close packing of spheres as nested octahedral shells, and (d) alkali homoclusters: close packing of shells as nested equilibrium polyhedral shells.

The level structure predicted by this potential is shown in Figure 3 together with the predicted magic numbers. It is very interesting for one to notice that in Figure 3 the numbers 18, 34, 68,... appear as magic numbers, while these numbers are not present either in the experimental mass spectrum of Figure 1(d) or in the interpretation of alkali magic numbers presented in Figure 2(d). This discrepancy between theory and experiments constitutes the starting point for a fundamental distinction between the small clusters presented in Figure 1(a)-(c) and those in Figure 1(d). The former clusters are composed of atoms with an even number of electrons, while the latter ones are composed of atoms with an odd number of electrons. Thus, the first atoms could be seen as behaving like bosons and the last ones like fermions (Anagnostatos, 1991b).

Furthermore, it is known that in the ground state, bosonic atoms (obeying the Boson statistics) try to occupy the lowest possible energy state, a fact which in geometrical language is consistent with the close packing of spheres (standing for atoms) as noted in Figure 2(a)-(c). Also as known, fermions (obeying the Fermi statistics) follow the Pauli principle forming shell structure and are never packed. Thus, Figure 2(d) is consistent with the fermionic nature of neutral alkali atoms, where the polyhedra shown stand for shells of the alkali-atom average positions, or in other words these polyhedra represent the average motion pattern of the alkali atoms (Anagnostatos, 1991a). Thus, the physics behind the soft sphere-like and hard sphere-like behavior of atoms in different microclusters is that the nature of particles is different for each of the two cases (i.e., bosonic and fermionic nature of atoms, respectively), a fact which makes their behavior like soft or hard spheres, or in other words permitting or not permitting overlapping between spheres of adjacent shells (Anagnostatos, 1991b). More about the consequences of such a distinction between atoms will be reported below.

The distinction of atoms as bosons or fermions is consistent with the state of matter in the corresponding microclusters. Indeed, rare gas clusters (except He), for example, are considered solids, while alkali clusters are considered liquids (or gases) (Gspann, 1986). Of course, this distinction of clusters according to the even or odd number of electrons in their constituent atoms is valid for both neutral atoms and their ion cores. For example, for neutral alkali atoms (case of atomic fermions) the cluster structure follows Figure 2(d) (gas state of matter), while for alkali ion cores (case of atomic bosons) the cluster structure follows Figure 2(a) (solid state of matter). Indeed, alkali ion cores possess complete electron shells like the rare gases whose structure follows Figure 2(a) (Saito et al., 1988; 1989; Bhaskar et al., 1987).

Now we can describe the conditions of validity between Figure 2(d) (due to average atom structure) and Figure 3 (due to electron structure). The first is valid for alkali clusters of neutral atoms, while the second for the delocalized electrons in all cases where a delocalization of valence electrons is favored. Indeed, in mass spectra of the second case the numbers 18, 34, 68 appear (Saito et al., 1988, 1989), all of them being absent for neutral clusters as obvious from Figure 1(d) (Knight et al., 1984).

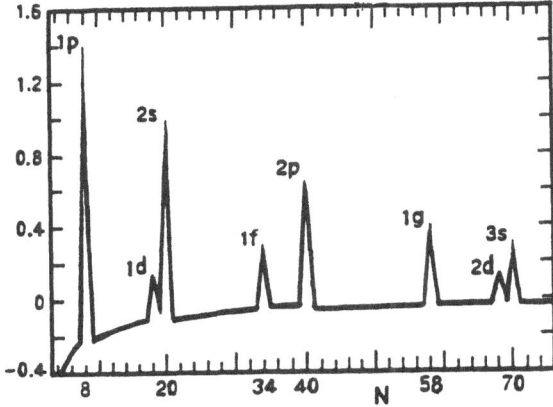

Fig.3. Electron level structure and magic numbers of alkali homoclusters, according to the jellium model (see Equation (1)).

403

Figure 4(a)-(d) stands for the average structure of clusters involving neutral (fermionic) alkali atoms. Specifically, Figure 4(a) presents the average forms of shells for alkali-heteroatom (e.g., Mg) clusters (Anagnostatos, 1989), while Figure 4(b) presents similar forms for clusters made up of two kinds of alkali atoms (e.g., K and Na) (Anagnostatos, 1988a), Figure 4(c) presents the average form for a cluster made up of six alkali atoms each two of which are of a different kind of alkali (Anagnostatos, 1991d), and finally Figure 4(d) presents a cluster of two alkali atoms, each of which is of a different kind. Everything regarding the magic numbers of Figure 4(a)-(b) coming from the geometry alone are shown on the figures themselves. The quantum mechanical analysis for clusters of Figure 4(a)-(d) is similar to the one given below for clusters of Figure 2(d) (Anagnostatos, 1991a;1991c).

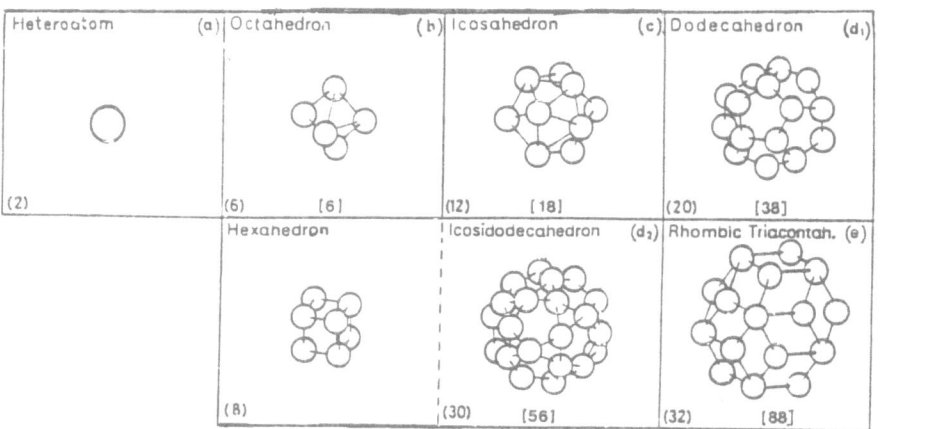

a

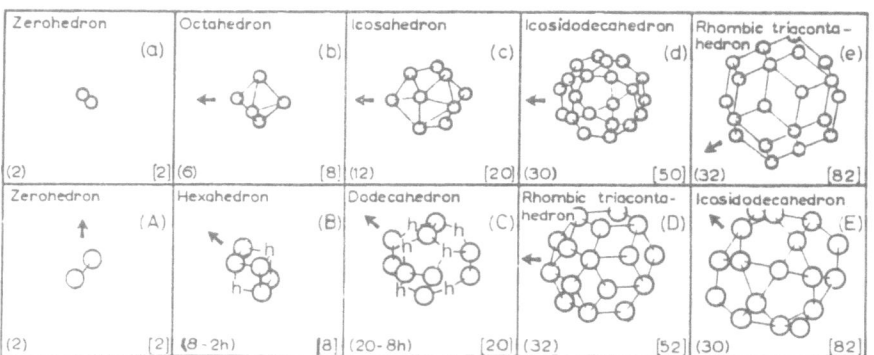

b

c

d

Fig. 4. Average shell structure of microclusters involving neutral (fermionic) alkali atoms. Specifically for (a) alkali-heteroatom clusters, (b) two-component alkali clusters, (c) three-component alkali clusters, and (d) two-component alkali dimer.

It has been shown that quantum effects (starting from the zero point kinetic energy) contribute to the mobility of the individual rare-gas atoms in microclusters (which here are representative clusters of bosonic atoms) and to the superposition of different configurations in the same clusters, as explained in Franke et al. (1988). Treatment for neutral alkali microclusters (which here are representantive clusters of fermionic atoms) has been studied in Anagnostatos (1991a) based on one-body central forces. Here only some elements of a simple quantum mechanical treatment, similar for both bosonic and fermionic atoms for reasons of easy comparison, are presented based on the fact that both kinds of clusters form shells (called high fluximal shells) possessing a geometrical representation (Anagnostatos, 1987). We further assume that all atoms of a shell taken together create an average central potential common for all atoms in that shell. In this potential each atom is considered as performing an independent particle motion (like nucleons in nuclei) obeying the Schrödinger equation for this potential. Further, our analysis proceeds by considering a multi-harmonic oscillator potential as follows (Anagnostatos, 1990a;1991c).

$$H = H_{1s} + H_{1p} + H_{1d2s} + \dots \quad (2)$$

where
$$H_i = V_i + T_i = -V + 1/2 \; m(\omega_i)^2 r^2 + T_i \quad (3)$$

That is, we consider a state-dependent Hamiltonian, where each partial harmonic oscillator potential has its own state-dependent frequency ω_i. All these ω_i's are determined from the harmonic oscillator relation (Hornyak, 1975).

$$\hbar\omega_i = (\hbar^2/m\langle r_i^2 \rangle) \; (n + 3/2), \quad (4)$$

where n is the harmonic oscillator quantum number and $\langle r_i^2 \rangle^{1/2}$ is the average radius of the relevant high fluximal shell made of either bosonic (Anagnostatos, 1987;1988b;1990a,b,c) or fermionic (Anagnostatos, 1987;1988a;1989) atoms. Before one applies (4), a value (0,1,2,3,...) of the harmonic quantum number n is assigned to each of the shells and a value of $\langle r_i^2 \rangle^{1/2}$ is derived from the geometry of the shell taking the finite size of the atom sphere into account. Thus, $\hbar\omega_i$ changes value each time either n or $\langle r_i^2 \rangle^{1/2}$ (or both) change value.

In the case of bosonic atoms there is no restriction for the number of atoms constituting the shell, since any number of such atoms is accepted for the same quantum state (symmetric total wave function). In the case of fermionic atoms, however, the atoms on each shell are restricted by the Pauli principle (antisymmetric total wave function). It is satisfying that all relevant shells for fermionic atoms fulfil this fundamental requirement, as explained in detail in Anagnostatos (1991a).

The solution of the Schrödinger equation with Hamiltonian (2), in spherical coordinates, is

$$\Psi_{n\ell m}(r,\theta,\varphi) = R_{n\ell}(r) Y_\ell^m(\theta,\varphi), \quad (5)$$

where $Y_\ell^m(\theta,\varphi)$ are the familiar spherical harmonics and the expressions for the $R_{n\ell}(r)$ are given in several books of quantum mechanics and nuclear physics, for example see Table 4-1 of Hornyak (1975). The only difference between our wave functions and those in these books is the different ω's as stated in (2)-(4) above. Those of our wave functions, however, which have equal ℓ value, because of the different $\hbar\omega$, are not orthogonal, since in these cases the orthogonality of Legendre polynomials does not suffice. Orthogonality, of course, can be obtained by applying established procedures, e.g., Gram-Schmidt process.

According to the Hamiltonian of (2), the binding energy of a cluster with N atoms in the case of orthogonal wave functions takes the simple form given by (6)

$$BE = 1/2 \; (V \bullet N) - 3/4 \; [\sum_{i=1}^{N} \hbar\omega_i \; (n + 3/2)], \quad (6)$$

where V is the average potential depth discussed shortly. The coefficients 1/2 and 3/4 take care of the double counting of atom pairs in determining the potential energy.

405

The average depth of the potential in its general case and in analogy to nuclei is given by (7)

$$V = -aN + b,$$ (7)

where if $a=0$ the potential has a fixed depth for all values of N. Specifically for completed polyhedra, an extra term is taken, i.e.

$$V = (-aN + b) + c/N.$$ (8)

This term expresses the energetic advantage for a microcluster to have a spherical (compact) structure, i.e., according to Figure 2(d), to have completed all polyhedral shells involved in that structure. This coefficient c expresses the sphericity of the cluster and has the same numerical value everytime the outermost polyhedron of the structure is completed. Everywhere else c has a zero value.

Following Chou et al. (1984) the relative binding energy change for a cluster with N atoms compared to clusters with N + 1 and N - 1 atoms is given by (9).

$$\delta(N) = [E_B(N) - E_B(N-1)] - [E_B(N+1) - E_B(N)]$$

$$= 2 E_B(N) - [E_B(N-1) + E_B(N+1)].$$ (9)

CORRESPONDENCE TO NUCLEAR PHYSICS

The identity of light magic numbers in two independent branches of physics (alkali clusters and nuclear physics), obeying two basically different types of forces (electromagnetic and strong force, respectively), does not seem to be coincidental. This remark is in agreement with the fundamental premise of the present work, which emphasizes the statistical properties of the constituents (i.e., fermionic, bosonic nature) rather than the forces among them. This work clearly demonstrates that many properties can be understood directly from general consideration of the statistical properties rather than the strength of the particular force (Anagnostatos, 1988a).

Many concepts and methods of treatment in cluster physics come from nuclear physics. However, the above remarks may be seen as a hint to reverse the flow of knowledge, now, from cluster physics to nuclear physics. The consideration of the size of nucleons via the sizes of their bags is essential, since we cannot speak about point nucleons in a structure resembling that of small clusters. We now specifically apply the model to nuclear structure employing 0.974 fm for the neutron bag and 0.860 fm for that of a proton (Anagnostatos, 1985). These values are consistent with our knowledge from particle physics (Thomas, 1984), that supports their relative size as well (Celenza and Shakin, 1983). These different sizes of bags imply a weak isospin symmetry, or in other words they imply that a nucleus consists of two almost different (distinct) kinds of fermions. Thus, the nucleus resembles those of clusters which are made up of two kinds of alkalis, i.e., those presented by Figure 4(b).

The close packing of average sizes of shells assumed by this figure permits the determination of the average radial sizes of all nuclear shells with respect to the sizes of the nucleon bags alone. The necessary formula is (Anagnostatos, 1985)

$$R_x = <r^2>^{1/2}_{shell} = R\cos\alpha + (d^2 - R^2\sin^2\alpha)^{1/2},$$ (10)

where R_x is the average radius of the shell to be determined, R the average radius of the previous shell in contact, d the distance of the centers of two nucleon bags in contact, and α an angle defined by the symmetry and relative orientation of both shells involved each time in the calculation according to Coxeter (1973).

Now, the knowledge of the average radial size of all shells permits the determination of the average values of all nuclear radii (e.g., charge radii) by using (11), noted below, and assuming the filling of subshells according to the simple shell model (Anagnostatos, 1985). That is,

$$\langle r^2\rangle^{1/2}_{nucleus}= \left[\sum_{1}^{Z}\langle r^2_i\rangle/Z + (0.8)^2 - (0.116)N/Z\right]^{1/2}, \qquad (11)$$

where the $\langle r^2_i\rangle^{1/2}$ values are given by (10) and the constants $(0.8)^2$ and (-0.116) are the ms charge radii accounting for the proton and the neutron finite sizes, respectively (de Jager et al.. 1974). One can consult Table 1 for predictions of the model for all nuclei from H to Pb, where the only two parameters involved are the sizes of the neutron bag and the proton bag (specified above).

Table 1. Charge root mean square radii in units Fermi

NUCL.	MOD.	EXP.	NUCL.	MOD.	EXP.
H		0.8	^{98}Mo	4.40	4.391 (26)
^{4}He	1.71	1.71(4)	^{98}Tc	4.43	
^{7}Li	2.06	2.39(3)	^{102}Ru	4.46	4.480(22)^d
^{9}Be	2.22	2.50(9)	^{103}Rh	4.49	4.510(44)
^{11}B	2.31	2.37	^{106}Pd	4.52	4.541(33)
^{12}C	2.37	2.40(56)^b	^{107}Ag	4.55	4.542(10)^d
^{14}N	2.54	2.540(20)	^{114}Cd	4.57	4.624(8)
^{16}O	2.70	2.710(15)^c	^{115}In	4.60	4.611(10)^d
^{19}F	2.84	2.85(9)^b	^{120}Sn	4.63	4.630(7)
^{20}Ne	2.98	3.00(3)	^{121}Sb	4.65	4.63(9)
^{23}Na	2.95	2.94(4)^b	^{130}Te	4.67	4.721(6)
^{24}Mg	3.06	3.08(5)	^{127}I	4.72	4.737(7)
^{27}Al	3.14	3.06(9)	^{132}Xe	4.77	4.790(22)^d
^{28}Si	3.21	3.15(5)	^{135}Cs	4.82	4.801(11)^d
^{31}P	3.27	3.24	^{138}Ba	4.85	4.839(8)^d
^{32}S	3.33	3.263(20)	^{139}La	4.91	4.861(8)
^{35}Cl	3.37	3.335(18)	^{140}Ce	4.95	4.883(9)
^{40}Ar	3.40	3.42(4)	^{141}Pr	4.99	4.881(9)
^{39}K	3.44	3.436(3)^c	^{142}Nd	5.03	4.993(35)
^{40}Ca	3.47	3.482(25)	^{146}Pm	5.06	
^{45}Sc	3.51	3.550(5)^c	^{152}Sm	5.10	5.095(30)^d
^{48}Ti	3.55	3.59(4)	^{153}Eu	5.13	5.150(22)^d
^{51}V	3.59	3.58(4)	^{158}Gd	5.16	5.194(22)^d
^{52}Cr	3.62	3.645(5)^c	^{159}Tb	5.19	
^{55}Mn	3.65	3.68(11)	^{164}Dy	5.22	5.222(30)^d
^{56}Fe	3.68	3.737(10)	^{165}Ho	5.25	5.210(70)^d
^{59}Co	3.71	3.77(7)	^{166}Er	5.28	5.243(30)^d
^{58}Ni	3.73	3.760(10)	^{169}Tm	5.30	
^{63}Cu	3.81	3.888(5)^c	^{174}Yb	5.32	5.312(60)^d
^{64}Zn	3.87	3.918(11)	^{175}Lu	5.35	
^{69}Ga	3.93		^{180}Hf	5.37	5.339(22)^d
^{72}Ge	3.99	4.050(32)^d	^{181}Ta	5.40	5.500(200)^d
^{75}As	4.04	4.102(9)^d	^{184}W	5.42	5.42(7)
^{80}Se	4.08		^{187}Re	5.44	
^{79}Br	4.13		^{192}Os	5.46	5.412(22)^d
^{86}Kr	4.17	4.160^c	^{193}Ir	5.48	
^{87}Rb	4.21	4.180^c	^{195}Pt	5.50	5.366(22)^d
^{88}Sr	4.25	4.26(1)	^{197}Au	5.52	5.434(2)
^{89}Y	5.29	4.27(2)	^{202}Hg	5.54	5.499(17)^d
^{90}Zr	4.32	4.28(2)	^{205}Tl	5.56	5.484(6)
^{93}Nb	4.36	4.317(8)^d	^{208}Pb	5.58	5.521(29)

a The experimental radii come from de Jager et al (1974) except as noted below in b-d.
b See Engfer et al. (1974).
c See Brown et al. (1984).
d See Wesolowski (1984).

In Hamiltonian (3), besides the nuclear dimensions, we are concerned with the potential whose depth is taken from (12) and (13) noted below for neutrons and protons, repsectively. That is,

$$- _NV = - _NV_0 + (27.2)\ (N-Z)/A \tag{12}$$

and

$$- _ZV = - _ZV_0 - (27.2)(N-Z)/A + 2E_C/Z. \tag{13}$$

where the second term in each equation stands for the simplest possible isotope effect (Hornyak, 1975), N, Z and A have their usual meaning, and E_C stands for the Coulomb energy (Hill, 1957), according to (14) below for $R = 1.25\ A^{1/3}$,

$$E_C = e^2/R\ [0.6Z(Z-1) - 0.46\ Z^{4/3}], \tag{14}$$

and

$$_NV_0 = _ZV_0 = 79.26 - 0.0879\ |A-74|, \qquad \text{for } A = 16\text{-}74 \tag{15}$$

or

$$_NV_0 = _ZV_0 = 79.26 - 0.0313\ |A-74|, \qquad \text{for } A = 74\text{-}208 \tag{16}$$

The seven closed-shell nuclei in Table 2 are used for the determination of the three constants (parameters) in (15) and (16), while the nine open-shell nuclei of Table 3 constitute a sample of nuclei spread all over the table of isotopes for which the model makes real predictions. Specifically, nuclear charge radii come from (11) by using $<r^2_i>^{1/2}$ values from (10), while nuclear binding energies are calculated from (6) by using $\hbar\omega_1$ values from (4) with the help of (10) and V values from (15-16).

All predictions of the model on radii and binding energies (see Tables 2 and 3) are satisfactory. This implies that an alternative method of studying atomic nuclei via quantum small-cluster concepts is possible and highly promising. Of course, a lot of work

Table 2. Binding energies and rms charge radii of closed-shell nuclei.

	^{16}O	^{40}Ca	^{58}Ni	^{90}Zr	^{120}Sn	^{142}Nd	^{208}Pb
E_C mod	12	65	123	223	331	447	757
E_C emp	12	68	123	224	324	445	744
BE mod	125	350	495	782	1031	1185	1626
BE exp[a]	128	342	506	784	1021	1185	1637
$<r^2>^{1/2}$ mod	2.70	3.47	3.73	4.32	4.63	5.03	5.58
$<r^2>^{1/2}$ exp	2.710	3.482	3.760	4.28	4.630	4.993	5.521
	(15)	(25)	(10)	(2)	(7)	(35)	(29)

[a] See Wapstra and Gove (1971)
[b] See de Jager et al. (1974); Brown et al. (1984); Wesolowski (1984).

Table 3. Predicted binding energies in MeV and rms charge radii in fm of a sample of ten open-shell nuclei close to and far from magic numbers

	^{28}Si	^{36}Ar	^{40}Ar	^{56}Fe	^{104}Pd	^{110}Pd	^{126}Te	^{136}Ba	^{138}Ba	^{202}Hg
E_C emp	30	50	55	107	280	280	345	392	390	713
BE mod	234	310	354	494	863	953	1067	1143	1157	1621
BE exp[a]	237	307	344	492	893	940	1066	1143	1159	1595
$<r^2>^{1/2}$ mod	3.21	3.41	3.40	3.68	4.52	4.51	4.67	4.86	4.85	5.54
$<r^2>^{1/2}$ exp	3.15[b]	3.396[c]	3.42[b]	3.737[b]	4.581[d]	4.595[c]	4.721[b]	4.833[b]	4.836[b]	5.499[d]
	(5)	(7)	(4)	(10)	(22)	(3)		(10)		(17)

[a] See Wapstra and Gove (1971)
[b] See de Jageer et al. (1974)
[c] See Brown et al. (1984)
[d] See Wesolowski (1984)

is necessary for the refinement of the method and its application to all spectrum of nuclear properties.

Besides Figure 4(b), Figure 4(c) and (d) presents average structures valid for both neutral alkali clusters and nucleon clusters. Specifically, Figure 4(c) shows the average structure of clusters consisting of three kinds of neutral alkali atoms (two atoms for each kind, e.g., Li, K, and Na) (Anagnostatos, 1991d), while for nucleon clusters it presents the average structure of $_{\Lambda\Lambda}^6He$ (i.e., of two neutrons, of two protons, and of two Λ hyperons; Anagnostatos and Grypeos, 1990). Figure 4(d) for atomic clusters presents a snapshot of the simplest mixed alkali clusters (e.g., K and Na), while for nucleon clusters presents a similar snapshot for the deuteron (e.g., one n and one p) (Anagnostatos et al., 1990). Details for both alkali and nucleon clusters can be found in the cited references.

CONCLUDING REMARKS

The model introduced by the present paper, which is based on the nature of the constituent atoms or their ion cores, seems to be justified by all experimental data known to us. Thus, the concept of fermionic and bosonic nature of atoms (or ion cores) with an odd and an even number of electrons, respectively, appears to combine the two views of electronic structure and atom-packing origin of magic numbers, and at the same time to unify the comprehension of different magic numbers in many kinds of clusters.

Specifically, clusters composed of atoms with non-delocalized valence electrons and with an odd number of electrons have stochastic atom magic numbers alone at N=2,8,20,40,... and those with an even number of electrons possess magic numbers coming from the packing of atoms alone in icosahedral or octahedral or tetrahedral form. On the other hand, clusters composed of atoms with delocalized valence electrons, either with an odd or with an even number of valence electrons, exhibit magic numbers due to the structure of their delocalized valence electrons but also magic numbers due to the packing of their (bosonic) ion cores in forms similar to those discussed above.

Depending on the temperature or/and the size of the clusters, the forms (and thus the relevant magic numbers) of clusters assumed by bosonic atoms or bosonic ion cores (i.e., nested tetrahedra, or octahedra, or icosahedra) may change from the one (ground state) into the other form (excited or metastable structure). In a mixture of cluster sizes, i.e., in clusters with different temperatures, one may expect a coexistence of different forms and related magic numbers.

The state of matter of microclusters is apparently understood in the framework of the present work. Specifically, bosonic clusters with no delocalized valence electrons are expected to closely resemble the solid state of matter (as it is known, e.g. for rare gas clusters), while fermionic clusters are expected to closely resemble the gas phase of matter(as it is believed, e.g. for alkali clusters) (Gspann, 1986). On the other hand, clusters with delocalized valence electrons (e.g. clusters born ionized) either bosonic or fermionic are initially expected to have a structure close to the solid state phase. However, due to the appearence of the ion cores in the cluster, a greater mobility of the constituent atoms exists, a fact which could shift the phase towards the structured liquids.

The equilibrium geometry of the average alkali shells in Figure 4 is not a fixed geometry like the one we are familiar with in solid state physics, but it is simply a geometrical representation of high fluximal shells like those we are familiar with from molecular orbitals.

Besides the novel quantum mechanical explanation of magic numbers, the present paper underlines the idea that *new*, as yet *unobserved* properties of microclusters should be investigated. Perhaps the most important of them is the orbiting properties of atoms implying a series of properties due to orbital angular momentum, i.e., definite spin properties, independent particle and collective modes of excitation of individual species, etc. For an experimental verification of such properties nuclear methods should be employed.

It is of great interest that fermionic atomic clusters exhibit magic numbers similar to those in nuclei (composed, also of fermions). This very fact gives a hint of an alternative study of nuclei resembling the study of fermionic atomic clusters and vice versa (Anagnostatos, 1985). It seems that the clusters made up of two kinds of alkali atoms (two kinds of fermions) assume a structure close to nuclear (neutron and proton) structure. However, a lot of work towards this direction is still needed.

Finally, one could remark that both small clusters and finite nuclei are cases of aggregates with a small number of particles and because of this we have all the similarities briefly described above. In the case where the number of particles becomes infinite the small cluster structure approaches crystal structure, while nuclear structure approaches nuclear-matter structure. Both systems of aggregates (small clusters and nuclei) could be seen as matter in small volumes and could be treated as a fifth state of matter.

ACKNOWLEDGEMENTS

I express my sincere appreciation to Professors J.W. Negele and J. Goldstone of the M.I.T. Department of Physics for their invitations to join the highly stimulating scientific environment at their Center for Theoretical Physics in 1988-89 and 1990. My gratitude is also extended to my colleagues Mr Ph. Trouposkiades and Mrs E. Kokkinia for their valuable help in preparing the manuscript in its present form.

REFERENCES

Anagnostatos, G.S. ,1985, Isomorphic shell model for closed-shell nuclei, Int. J. Theor. Phys., **24**: 579

Anagnostatos, G.S., 1987, Magic numbers in small clusters or rare-gas and alkali atoms, Phys. Lett. A, **124**: 85

Anagnostatos, G.S., 1988a, Magic numbers in small clusters made up of two kinds of alkali atoms, Phys. Lett. A, **128**: 266

Anagnostatos, G.S., 1988b, Magic numbers in small clusters of mixed rare gases, Phys. Lett. A, **133**: 419

Anagnostatos, G.S., 1989, Magic numbers in alkali/heteroation microclussters, Phys. Lett. A, **142**:146

Anagnostatos, G.S., 1990a, Magic numbers in semiconductor microclusters, Phys. Lett. A, **143**: 332

Anagnostatos, G.S., 1990b, Addendum on the unique stability of $CoAr_6^+$, Phys. Lett. A, **148**: 291

Anagnostatos, G.S., 1990c, Magic umbers in alkali-halide microclusters, Phys. Lett. A, **150**: 303

Anagnostatos, G.S., 1991a, Alkali-atom shell model, Phys. Lett. A, **154**: 169

Anagnostatos, G.S., 1991b, Fermion/boson classification in microclusters, Phys. Lett. A, **157**: 65

Anagnostatos, G.S., 1991c, Multipotential model for atoms in alkali microclusters, Z. Phys. D, **19**: 121

Anagnostatos, G.S., 1991d, Small clusters made up of three kinds of neutral alkali atoms, Z. Phys. D, **19**: 125

Anagnostatos, G.S., and Grypeos, M.E., 1990, A rough estimate of the size of the Λ-hyperon bag, in: "Proceedings of the PANIC XII International Conference on Particles and Nuclei", T.W. Donnelly, ed., MIT, Boston.

Anagnostatos, G.S., Gridnev, K.A., and Subbotin, V.B. 1990, Nucleon clusters, in: "Proceedings of the ISSPIC5 5th International symposium on Small Particles and Inorganic Clusters", O. Echt, E. Recknagel, D. Kreisle, and R. Pflaum, eds, Universität Konstanz, Konstanz.

Bhaskar, N.D., Frueholz, R.P., Klimcak, C.M., and Cook, R.A., 1987, Evidence of electronic shell structure in Rb^+_N (N=1-100) produced in a liquid-metal ion source, Phys. Rev. B, **36**: 4418.

Brown, B.A., Bronk, C.R., and Hodgson, P.E., 1984, Systematics of nuclear rms charge radii, J. Phys. G, **10**: 1683.

Celenza, L.S., and Shakin, C.M., 1983, Quark model calculations of nucleon structure functions, Phys. Rev. C, **27**: 1561.

Chou, M.Y., Cleland, A., and Cohen, M.L. 1984, Total energies, abundances, and electronic shell structure of lithium sodium and potassium clusters, Solid State Comm., **52**: 645.

Coxeter, H.S.M. 1973, "Regular Polytopes" Macmillan, New York.

De Jager, C.W., de Vries, H., and de Vries, C., 1974, Nuclear charge - and

Echt, O., Sattle, K. and Recknagel, E., 1981, Magic nimbers for sphere packings: Experimental verification in free xenon clusters, Phys. Rev. Lett., **47**: 1121.

Engfer, R., Schneuwly, H., Vuilleumier, J.L., Valter, H.K., and Zehnder, A., 1974, Charge-distribution parameters, isotope shifts, isomer shifts, and magnetic hyperfine constants from muonic atoms, At. Data Nucl. Data Tables, **14**: 509.

Franke, G., Hilf, E. and Palley, L. 1988, Quantum mechanics and phase transitions in small noble-gas clusters, Z. Phys. D, **9**: 343.

Gspann, G., 1986, On the phase of metal clusters, Z. Phys. D, **3**:143.

Hill, D.L., 1957, Matter and charge distribution within atomic nuclei, in: "Encyclopedia of Physics XXXIX", S. Flügge, ed., Springer-Verlag, Berlin.

Hornyak, W.F., 1975, "Nuclear Structure", Academic, New York.

Kappes, M.M., 1988, Experimental studies of gas-phase main group metal clusters, Chem. Rev., **88**: 369.

Katakuse, I., and Ichihara, T., 1986, Mass distributions of negative cluster ions of copper, silver, and gold, Int. J. Mass Spectr. Ion Proc., **74**: 33.

Knight, W.D., Clemenger, K., de Heer, W.A., Saunders, W.A., Chow, M.Y., and Cohen, M.L., 1984, Electronic shell structure and abundances of sodium clusters, Phys. Rev. Lett., **52**: 2141.

Pettiette, C.L., Yang, S.H., Craycraft, M.J., Conceicao, J., Laaksonen, R.T., Cheshnovsky, O., and Smalley, R.E., 1988, Ultraviolet photoelectron spectroscopy of copper clusters, J. Chem. Phys., **88**: 5377.

Phillips, J.C., 1986, Magic numbers of alkali halide clusters, in: "Proceedings of the International Symposium on the Physics and Chemistry of Small Clusters", P. Jena, B.K. Rao, and S.N. Khanna, eds, Plenum Press, New York.

Ross. M.M., O'Keefe, A., and Baronavski, A.P., 1986, Production of cluster ions by laser vaporization, in: "Proceedings of the International Symposium on the Physics and Chemistry of Small Clusters", P. Jena, B.K. Rao, and S.N. Khanna, eds, Plenum Press, New York.

Saito, Y., Watanabe, M., Hagiwara, T., Nishigaki, S., and Noda, T., 1988, Magic numbers in a mass spectrum of lithium clusters emitted from a liquid metal ion source, Jpn J. Appl. Phys., **27**: 424.

Saito, Y., Minami, K., Ishida, T., and Noda, T., 1989, Abundance of Na cluster ions ejected from a liquid metal ion source, Z. Phys. D, **11**: 87.

Thomas, A.W., 1984, Chiral symmetry and the bag model: A new starting point for nuclear physics, in: "Advances in Nuclear Physics **13**", J.W., Negele, and E. Vogt, eds, Plenum Press, New York.

Wapstra, A.H., and Gove, N.B., 1971, The 1971 atomic mass evaluation, Nucl. Data Tables, **9**: 267.

Wesolowski, E., 1984, The rms radii of nuclear proton distributions, J. Phys. G, **10**: 321.

SOME CONSIDERATIONS ON THE GREENHOUSE EFFECT AND RELATED PROBLEMS

V.M. Canuto

NASA-Goddard Institute for Space Studies
2880 Broadway, New York, N.Y., 10025, U.S.A.

The inhabitants of the planet earth are quietly conducting a gigantic environmental experiment. So vast and so sweeping will be the impact of this experiment that, were it brought before any reasonable council for approval, it would be firmly rejected as having potentially dangerous consequences. The experiment in question is the words of W. Bròeker, a geochemist of Columbia University during a US Senate Hearing on the environment.

Let us first discuss the CO_2 problem. Although it is present in the earth's atmosphere in a rather minute amount ~ 0.03% or 2300 ppm, parts per million, while oxygen is around 24% and nitrogen 74%, it is almost a magic number for it gives rise to the "natural greenhouse effect" that allows the earth's surface temperature to remain at a comfortable ~18° C. By contrast, on the planet Venus, in spite of its similarity to earth (the sister planet), the surface temperature is around 500° C. Since the huge difference in temperature is generally ascribed to the huge difference in the CO_2 atmospheric budgets of the two planets, it is clear that anthropogenic processes that tend to increase the earth's atmospheric content of CO_2 are bound to lead us toward an unpleasant Venus-like situation. Careful measurements of the CO_2 content of the earth's atmosphere began in a systematic way in 1967-68 as an offspring of the International Geophysical Year, IGY. Under the auspices of the Scripps Oceanographic of La Jolla, Keeling set up to monitor the time development of the CO_2 content of our atmosphere. The results have shown a consistent increase: since the preindustrial era, the CO_2 concentration has increased by 25%, from 270 ppm to 350 ppm, with a 50% rise occurring in the past 30 years. It is estimated that around 5 Gt of C per year are emitted into the atmosphere from the burning of fossil fuels (coal, oil and gas) (Gt = gigaton = 10^9 tons). To this, one must add the carbon liberated from deforestation and burning of the tropical areas which may add up to 2 Gt, for a grand total of about 7 Gt/yr of C. (To obtain the equivalent amount of CO_2, one must multiply by 44/12). We must however stress that the contribution from deforestation is still uncertain and that 2 Gt/yr is the upper limit of the range usually quoted, (0.6-2) Gt/yr. Since per unit of energy generated, gas is the fuel that generates less CO_2, following by oil and coal, it is clear that a possible way to mitigate the impact of the CO_2 release from fossil fuel, would be to switch from coal to oil and then gas. To quantify the argument, we recall that in units of kWh/lb CO_2, coal has a potential of 0.56, oil of 0.70 and gas of 1.01.

It is also of interest to look at the trend CO_2 emission in time. In 1950, North America was responsible for 44.1%, Western Europe for 23.4%, CPE countries (Centrally Planned Economies) for 18% and developing countries (DC) for 5.7%. In 1980, North America contribution has decreased to 26.6%, W. Europe to 16.5%, while the CPE share has increased to 24.7% and so has that of DC countries, 12.2%. In 1950, the global emission was of 1.62 Gt, while in 1980 had climbed to 5.7 Gt, an increase of 300%. (As for the US, the CO_2 emissions are contributed by the following sectors: 31.8% transportation, 33.4% electric utilities, 27.5% industry/commerce and residential 7.4%).

These numbers present us with the first difficulty: the source of fossil fuel CO_2 has been shifting and it is expected to shift even further, from industrialized to developing nations pursuing an aggressive economic development policy. Considering that recoverable fossil fuel resources are extensive, that an international infrastructure exist to use the fuel and that there is a lack of competitive alternative energy sources, it is only realistic to expect that the reliance on fossil fuel will continue. Fossil fuel provides 62% of the worldwide electricity, while hydroelectric power accounts for 19% and nuclear for 17%. World wide resources of fossil fuel are estimated to last 40-50 years for oil-gas and 200 years for coal. More than 90% of the coal resources are concentrated in just seven countries: US 28.5%, USSR 26.4%, China 10.7%, Australia 7.1%, W. Germany 6.4%, South Africa 6.5% and Poland 4.6%. It is also interesting to note that in the US the contribution CO_2 from coal has been 30.3%, 25.5% and 33.7% in 1966, 1976 and 1986 respectively; for the same years, oil has contributed 45.5%, 52.7% and 47.6% while for gas the figures are 24.2%, 21.6% and 18.7%. In other worlds, coal usage is on the rise while oil usage is declining.

Is there any evidence that the past 300% increase in CO_2 emission has contributed to the natural greenhouse effect? Stated differently, is there a direct evidence for an anthropogenic greenhouse effect? I do not believe that the answer is an unequivocal yes: what does exist is circumstantial evidence that is consistent with the overall picture and conjecture that the CO_2 increase may have already generated some detectable effect. First of all, it is estimated that during the last one hundred years, the earth's surface temperature has increased by about $0.5°C$. While the global circulation models (GCM) used to study the greenhouse effect have predicted a steady increase in temperature since 1880, observations show an overall increase which is however not monotonic: a steady increase from 1880 to 1949 was followed by a decrease that lasted about 40 years, i.e., until 1980, when the temperature begun to climb again till today.

The sea-level has risen by 1 mm a year during the last century, a second finding that, if attributed to thermal expansion of the oceans, may signal global warning. Before we discuss some of the future predictions for temperature and drought, we must also stress that CO_2 is regrettably not the only gas that traps the heat that otherwise would escape to space. There are other greenhouse gases. In fact, on a molecule by molecule basis, CO_2 is the weakest gas, while the other are more potent: CH_4 methane, 30 times more potent; N_2O nitrous oxide, 200 times and CFC, chlorofluorocarbons, 20,000 times more potent. The reason why each of these gases separately contributes less than CO_2, is because their concentration is significantly smaller. However, as we shall discuss later, their combined contribution almost equals that of CO_2.

The largest source of methane are wetlands (25-170 Tg/yr, Tg = teragram = 10^{12} grams), rice paddies (40-170 Tg/yr), followed by enteric fermentation from cattle (40-110), biomass burning (30-110), with natural gas production and distribution contributing 20-50, coal mining 10-40, termites 5-45, for a grand total of 0.2-0.9 Gt/yr, and an increase of 0.6% per year.

As for nitrous oxides, the total emission is estimated to be 7.8-25.3 Tg/yr. Finally, let us consider the CFC. These are man made compounds whose outflow is estimated at 0.33 Tg/yr. They have been identified as the source of the chlorine that, once released and because of its chemical inertia, travels to the stratosphere where it attacks in a catalytic reaction the ozone molecules that form the ozone layer responsible for shielding the earth from the solar UV radiation. It is estimated that each chlorine atom can destroy up to 100.000 molecules of ozone. Aided in a significant way by the stratospheric conditions prevailing in Antartica, it has been discovered that the ozone layer in that region has been depleted up to 50%. The CFC are therefore doubly responsible: they are greenhouse gas and they contribute to the depletion of the ozone layer. For that reason, they have attracted particular attention from legislators of the words who, through the Montreal Protocol (signed on Sept 1987 and ratified on January 1988), mandate a phase reduction in the production and consumption of CFC to achieve a 50% reduction from 1986 to 1999. In 1989, with the Helsinki Declaration, 80 signatories governments agreed to completely phase out the CFC by the year 2000.

Let us consider now the contribution of all these gases to the greenhouse warning. As of 1985, the sum of CFC, N_2O and CH_4 ($2 \cdot 10^{-4}$, 0.31 and 1,7 ppm, respectively) contributed to the global warning as much as CO_2 (335 ppm) alone, a clear indication that the small abundance of the gases is amply complemented by their high efficiency. However, several studies have indicated that by the year 2000, these three gases could contribute more than CO_2 and by the year 2020, they could contribute twice as much as CO_2.

Let us now consider some of the predictions made by various groups using the state-of-the-art GCM, global circulation models. We shall cite results of the work of OSU (Oregon State University), NCAR (National Center for Atmospheric Research), GFDL (Geophysical fluid Dynamics Labs) and GISS (NASA, Goddard Institute for Space studies). The results refer to a situation corresponding to an instantaneous doubling of CO_2 and running the model until climate stabilizes. The results in °C are: 2.8, 3.5, 4.0 and 4.2; at the same time, the amount of precipitation is predicted to increase (%) by 7.8, 7.1, 8.7 and 11.6 respectively. In order to get a feeling for these increases in temperature, it is useful to recall that the difference in temperatures between an ice age and an interglacial period is 3-6° C.

Most studies agree that the maximum warning will occur in the winter months at higher altitudes near the sea-ice margins. Sea-level rise is predicted to be around half a meter. Regrettably for policy makers, the present GCM cannot reliably predict temperatures and/or precipitation changes or the effect of changes like the sea-level rise on a regional level. One of the reasons is the still limited computer power, another is the quality and quantity of measured data and finally the still poor understanding of several physical processes that are certainly very relevant, like the effect of the oceans and clouds. In spite of these uncertainties, the prediction and the scenarios that are coming out of the GCM are serious enough to have produced a world wide concern about the possibility of a man-made greenhouse effect. Unless one dismisses the whole problem on the ground that the scientific uncertainties that still exist are such that once understood will nullify the predicted temperature increase, many thoughtful people have tried to suggest ways to avert a continuous increase of CO_2 and of the other gases. The problem is far from simple. First of all, most of the world has benefited from burning fossil fuels, many generations have contributed to the problem, and no one nation, or industry has caused the problem. We all have contributed to it though in substantially different degree. It thus behooves all of us to find a solution. Since draconian reductions in the usage of fossil fuel are simply utopic and if implemented, would probably engender economic chaos, together with the right of DC to move as quickly as possible toward industrialization, it is hoped that new sources of energy and efficiency measures will be acted upon soon. The one interim method widely suggested is the use of new trees that, during their growth period, before they reach a steady state, absorb more CO_2 via photosynthesis than they emit via respiration. This must of course be accompanied by a halt in the deforestation and burning of tropical forest. The figures reported in the literature tell us that in order to remove the yearly 5 Gt of C originating from fossil fuel, we may have to plant 500 million acres of new trees.

ACKNOWLEDGEMENTS

The author would like to acknowledge a travel grant from the U.S. Army Research Office to partially cover travel expenses to the XV International Workshop on Condensed Matter Theories, where this talk was presented.

REFERENCES

The principal source of information is the comprehensive Report: Climate Change, The IPCC Scientific Assessment, Cambridge University Press, Cambridge, Mass., 1990.

What follows is a selected list of specialized articles:

1. Bacastow, R.B. et al, 1985, J. Geophys. Res. 90, 10529.
2. Blake, D.R. and Rowland, F.S., 1988, Science, 239, 1129.

3. Cess, R.D. and Potter, G.L., J. Geophys. Res. 93, 8305.
4. Cess, R.D. et al., 1989, Science, 245, 513.
5. Cicerone, R.J. and Shetter, J.D., 1981, J. Geophys. Res. 86, 7203.
6. Haefele, W., 1981, Energy in a Finite World, Ballinger Publ. Co., Cambridge, Mass.
7. Hansen, J. et al., 1989, J. Geophys. Res. 94, 16417.
8. Hansen, J. et al., 1991, Science, 213, 957.
9. Houghton, R.A., et al., 1988, Science, 241, 1736.
10. Houghton, R.A., and Woodwell, G.M., 1989, Scientific American, 260, 36.
11. Keeling, C.D. and Heimann, M., J. Geophys. Res., 91, 7782.
12. Kerr, R.A., 1989, Science 246, 1563.
13. Meehl, G.A., 1984, Climate Change, 6, 259.
14. Oescher, H.U., et al., 1975, Tellus, 27, 168.
15. Ramanathan, V., 1988, Science, 240, 293.
16. Rind, D. et al., 1989, Climatic Change, 14, 5.
17. Schlesinger, M.E. and Mitchell, J.F.B., 1987, Rev. of Geophys., 25, 760.
18. Schneider, S.H., 1989, Scientific American, 261, 70.

CONTRIBUTORS

Aguilera-Navarro, V. C. Physics Department, North Dakota State University, Fargo, North Dakota 58105 U.S.A. 287

Alascio, B.R., Centro Atómico Bariloche, (8400) Bariloche, Argentina. 325

Aldao, C.M., Universidad Nacional de Mar del Plata, J.B. Alberdi 2695, (7600) Mar Del Plata, Argentina. 201

Aliaga, J., Grupo de Sistemas Dinámicos, Regional Norte, Universidad de Buenos Aires, C.C. 2, (1638) Vicente López, Argentina. 335

Aligia, A.A., Centro Atómico Bariloche, (8400) Bariloche, Argentina. 269

Allub, R., Centro Atómico Bariloche, (8400) Bariloche, Argentina. 325

Anagnostatos, G.S., Institute of Nuclear Physics, National Center for Scientific Research "Demokritos", GR-153 10, Aghia Paraskevi-Attiki, Greece. 399

Arizmendi, C.M., Dept. de Física, Facultad de Ingeniería Universidad Nacional de Mar del Plata, Av. J.B. Justo 4302, (7600) Mar del Plata, Argentina. 361

Arrachea, L., Departamento de Física, Universidad Nacional de La Plata, C.C. 67, 1900 La Plata, Argentina. 63

Baliña, M., Centro Atómico Bariloche, (8400) Bariloche, Argentina. 269

Barranco, M., Department ECM, Facultat de Física, Universitat de Barcelona, E-08028 Barcelona, Spain... 303

Bashkin, Eugene. P., Institute für Theoretische Physik, Universität zu Köln, D-5000 Köln 41, Germany. 207

Blomberg, Clas, Department of Theoretical Physics, Royal Institute of Technology, 100 44 Stockholm, Sweden. 215

Blum, L., Department of Physics, P.O. Box 23343, University of Puerto Rico, Río Piedras, PR 00931-3343. 153

Bochicchio, Roberto C., Departamento de Física, Facultad de Ciencias Exactas y Naturales, Universidad de Buenos Aires, (1428) Buenos Aires, Argentina. 367

Briozzo, C.B., FAMAF, Universidad Nacional de Córdoba, (5000) Córdoba, Argentina. 345

Cabrera, A., Fac. de Ciencias , UNAM, Apdo. Post. 70-646, 04510 México
D.F. 313

Calles, A., Fac. de Ciencias , UNAM, Apdo. Post. 70-646, 04510 México
D.F. 313

Cambiaggio, M.C., Departamento de Física, Comisión Nacional de Energía
Atómica, Av. del Libertador 8250, (1429) Buenos Aires, Argentina. 1

Canosa, N., Departamento de Física, Universidad Nacional de La Plata,
C.C. 67, 1900 La Plata, Argentina. 63, 69

Canuto, V.M., NASA-Goddard Institute for Space Studies, 2880 Broadway,
New York, N.Y., 10025 U.S.A. 413

Casas, M., Department de Física, Universitat de les Illes Balears, E-07071
Palma de Mallorca, Spain. 189

Castro, J.J., Departamento de Física, CINVESTAV, IPN, Apdo. Post. 14-
740, 07000 México D.F. 313

Cerdeira, Hilda A., International Centre for Theoretical Physics (ICTP), P.O.
Box 586, 34100 Trieste, Italy; and Instituto de Fisica, Universidade Estadual
de Campinas (UNICAMP), C.P. 6165, 13081 Campinas, São Paulo, Brazil......... 97

Cordero, Patricio, Departamento de Física, Facultad de Ciencias Físicas y
Matemáticas, Universidad de Chile, Casilla 487, Santiago 3, Chile. 49, 119

Crespo, G., Grupo de Sistemas Dinámicos, Regional Norte, Universidad de
Buenos Aires, C.C. 2, (1638) Vicente López, Argentina. 97

Davidson, N.J., Department of Physics, University of Pretoria, Pretoria
0002, Republic of South Africa. 373

Degani, M.H., Concurrent Computing Laboratory for Materials Simulations
and Department of Physics and Astronomy, Louisiana State University,
Baton Rouge, Louisiana 70803-4001 U.S.A. 253

de Llano, M., Physics Department, North Dakota State University, Fargo,
North Dakota 58105 U.S.A. 287

Deza, R., Universidad Nacional de Mar del Plata, J.B. Alberdi 2695, (7600)
Mar Del Plata, Argentina. 201

Donnamaria, M.C., Instituto de Líquidos y Sistemas Biológicos, IFLYSIB,
C.C. 565, (1900) La Plata, Argentina. 243

Grinberg, Horacio, Departamento de Física, Facultad de Ciencias Exactas y
Naturales, Universidad de Buenos Aires, (1428) Buenos Aires, Argentina. 367

Hernández, E.S., Departamento de Física, Facultad de Ciencias Exactas y
Naturales, Universidad de Buenos Aires, 1428 Buenos Aires, Argentina. 303

Herrera, J.N., Escuela de Ciencias Físico-Matemáticas, Universidad
Autónoma de Puebla, Apdo. Postal 1152, C.P. 72001 México. 153

Huberman, Bernardo A., Dynamics of Computation Group, XEROX Palo
Alto, CA 94304, U.S.A. 23

Jin, Wei, Concurrent Computing Laboratory for Materials Simulations and Department of Physics and Astronomy, Louisiana State University, Baton Rouge, Louisiana 70803-4001 U.S.A. 253

Kalia, Rajiv K., Concurrent Computing Laboratory for Materials Simulations and Department of Physics and Astronomy, Louisiana State University, Baton Rouge, Louisiana 70803-4001 U.S.A. 253

Kalman, G., Department of Physics, Boston College, Chestnut Hill, MA 02167 U.S.A. 163

Kievsky, A., Instituto Nazionale di Fisica Nucleare, Sezione di Pisa, 56100 Pisa, Italy. 387

Kobe, Donald H., Department of Physics, University of North Texas, Denton, Texas 76302-5368, U.S.A. 39

Krivine, H., Division de Physique Théorique, Institut de Physique Nucléaire-91406 Orsay Cedex, France. 189

Kryachko, Eugene S., Institute for Theoretical Physics, Kiev, 25130, USSR. 229

Lombard, R.J., Division de Physique Théorique, Institute de Physique Nucléaire, F-91406 Orsay Cedex, France. 303

Loong C.-K., Intense Pulsed Neutron Source, Argonne National Laboratory, Argonne, Illinois 60439, U.S.A. 253

Lucheroni, C., Dipartamento di Fisica dell'Universita' and INFN, I-06100 Perugia, Italy. 351

Ludeña, Eduardo V.,Chemistry Center, Venezuelan Institute for Scientific Research, IVIC, Apartado 21827, Caracas 1020-A, Venezuela. 229

Malik, F. Bary, Physics Department, Southern Illinois University, Carbondale, Illinois, 62901, U.S.A. 11

Manning, Lonnie W., Department of Physics, Southern Illinois University-Carbondale, IL 62901-4401 U.S.A. 179

Marchesoni, F., Dipartamento di Fisica dell'Universita' and INFN, I-06100 Perugia, Italy. 351

Marquina, M.L., Fac. de Ciencias , UNAM, Apdo. Post. 70-646, 04510 México D.F. 313

Miller, Bruce N., Department of Physics, Box 32915, Texas Christian University, Fort Worth, Texas 76129, U.S.A. 131

Miller, H.G., Department of Physics, University of Pretoria, Pretoria 0002, Republic of South Africa. 373

Mirabella, D., Universidad Nacional de Mar del Plata, J.B. Alberdi 2695, (7600) Mar Del Plata, Argentina. 201

Percus, J.K., Courant Institute of Mathematical Sciences and Physics Department, New York University, 251 Mercer Street, New York, NY 10012, U.S.A. 143

Pesquera, L., Departamento de Física Moderna, Universidad de Cantabria, E-39005 Santander, Spain. 345

Plastino, A., Departamento de Física, Universidad Nacional de La Plata, C.C. 67, 1900 La Plata, Argentina. 63, 69, 79, 87, 373

Portesi, M., Departamento de Física, Universidad Nacional de La Plata, C.C. 67, 1900 La Plata, Argentina. 63

Proto, A.N., Grupo de Sistemas Dinámicos, Regional Norte, Universidad de Buenos Aires, C.C. 2, (1638) Vicente López, Argentina. 97, 243, 335

Puente, A., Department de Física, Universitat de les Illes Balears, E-07071 Palma de Mallorca, Spain. 189

Quick, R.M., Department of Physics, University of Pretoria, Pretoria 0002, Republic of South Africa. 373

Ramos-Gómez, F., Fac. de Ciencias , UNAM, Apdo. Post. 70-646, 04510 México D.F. 313

Rebollo Neira, L., Centro de Física Teórica, Universidad Nacional de La Plata, C.C. 67, (1900) La Plata, Argentina. 79

Risso, Dino, Departamento de Física, Facultad de Ciencias Físicas y Matemáticas, Universidad de Chile, Casilla 487, Santiago 3, Chile. 119

Rodriguez, M.A., Departamento de Física Moderna, Universidad de Cantabria, E-39005 Santander, Spain. 345

Rosati, S., Dipartamento di Fisica, Universita' di Pisa, 56100 Pisa, Italy; Instituto Nazionale di Fisica Nucleare, Sezione di Pisa, 56100 Pisa, Italy. 387

Rossignoli, R., Departamento de Física, Universidad Nacional de La Plata, C.C. 67, 1900 La Plata, Argentina. 63, 69, 87

Salomó, Sebastián, Departamento de Física, Universidad Simón Bolívar, Apartado Postal 89000, Caracas, Venezuela. 49

Sanchez, J.R., Dept. de Física, Facultad de Ingeniería Universidad Nacional de Mar del Plata, Av. J.B. Justo 4302, (7600) Mar del Plata, Argentina. 361

Sanders, Frank C., Department of Physics, Southern Illinois University-Carbondale, IL 62901-4401 U.S.A. 179

Serra, Ll., Department de Física, Universitat de les Illes Balears, E-07071 Palma de Mallorca, Spain. 303

Stepanyants, Sergey B., Institute für Theoretische Physik, Universität zu Köln, D-5000 Köln 41, Germany. 207

Szybisz, L. Departamento de Física, Comisión Nacional de Energía Atómica, Av. del Libertador 8250, (1429) Buenos Aires, Argentina. 107

Taddei, F.P., Departamento de Física, Comisión Nacional de Energía Atómica, Av. del Libertador 8250, (1429) Buenos Aires, Argentina. 1

Vallejos, R.O., Departamento de Física, Comisión Nacional de Energía Atómica, Av. del Libertador 8250, (1429) Buenos Aires, Argentina. 107

Vashishta, Priya, Concurrent Computing Laboratory for Materials Simulations and Department of Physics and Astronomy, Louisiana State University, Baton Rouge, Louisiana 70803-4001 U.S.A. 253

Ventura, C.I., Centro Atómico Bariloche, (8400) Bariloche, Argentina. 325

Vercicat, Fernando, Instituto de Física de Líquidos y Sistemas Biológicos (IFLYSIB), C.C. 565 (1900) La Plata, Argentina. 153

Viviani, M., Instituto Nazionale di Fisica Nucleare, Sezione di Pisa, 56100 Pisa, Italy. 387

Wio, H. S., Centro Atómico Bariloche and Instituto Balseiro, (8400) Bariloche, Argentina. 345

Yépez, E., Escuela Superior de Física y Matemáticas, IPN, 07738 México D.F. 313

Zunino, V., Grupo de Sistemas Dinámicos, Regional Norte, Universidad de Buenos Aires, C.C. 2, (1638) Vicente López, Argentina. 335

Zyserman, F., Centro de Física Teórica, Universidad Nacional de La Plata, C.C. 67, (1900) La Plata, Argentina. 79

Adiabatic approximation, 40, 41, 43
Aggregation, 361
Algebraic Method, 54
Ali-Bodmer potential, 388, 393, 396
Alkali
    cluster, 401, 403
    halide cluster, 401, 403
Alphas, 295
α-clusters, 387
3α model for ^{12}C, 387, 393
4α model of ^{16}O, 388, 395
Ansatz, 367
Atom-atom potentials, 401
Atomic boson, 403, 405, 409
Atomic fermion, 403, 404, 405, 409
Attractive delta-function potential, 294
$Ba_{1-x}K_xBiO_3$, 253
Baryons, 295
Baxter's sticky, 153
Baxter-Wertheim, 155
BCS, 306, 309
    model for finite Fermi Systems, 309
    model interaction, 287, 288
    theory, 304
    universal constants, 297
Berry phase, 39, 43, 44, 46
Binding energy, 399, 405, 408
BiO_3 based superconductors, 276
Bipolaron, 295
Campbell-Baker-Hausdorf expansion, 148
Central potential, 401, 405, 408
Chaotic dynamics, 23
Charge rms radii, 407, 408
Chemical potential, 144
Chemisorption induced states (CIS), 204
Close packing of
    shells, 401
    spheres, 401, 403
Coherence length, 305, 308
Coherences, 98, 99, 105
Collective excitation, 207, 409
Colour projection, 376
Complex rotation method, 11, 18, 179
Consistency conditions, 146

Convexity, 145
Cooper pairs, 287, 304, 305, 306
Correlation
    factors, 390, 393
    hole, 150
Correlations, 163, 164, 165, 175
Critical magnetic field, 298
Critical temperature, 304, 308
    for the normal-to-superfluid phase
        transition, 307
Cubic perovskite, 253
Current of particles, 98, 99, 102, 105
Cyclic process, 41, 43
Decomposition of Signals, 79
Deconfinement transition, 373
Deformation and pairing interplay, 310
Delocalized electrons, 399, 401, 403
Density
    functional
        formalism, 143
        theory, 229, 243
    matrix, 98, 99, 105, 146
    of states, 287
Diagrams, 367
Dielectric function, 165, 166, 172, 173
Diffusion in random media, 345
Diffusion-limited reactions, 345
Dilute gas approximation, 354
Direct correlation function, 157
Distributed computation, 23
Electron
    film, 163, 164
    gas, 163
    -boson coupling, 287
    -phonon
        coupling, 253
        interaction, 272
        mechanism, 287
Electronic structure, 280
Electrons, 295
Eliashberg gap equations, 258, 259
Energy
    gap, 305, 308
    operator, 39, 41, 42, 45, 46
    shift, 11, 12, 16, 17, 18

Equilibrium shells, 401
Error-driven dynamics, 23
Euler-Lagrange equations, 107
Exchange-correlation, 146
Factor correlation function, 160
Fermi
    liquid, 296
    surface, 287
Feshbach Theory, 12, 18
Fifth state of matter, 410
Finite Fermi systems, 304
Finite number of particles, 63
Finite Size Effects, 373
Finite systems, 63
Finite Temperature, 87
Finite-difference relaxation method, 108
Fluctuations, 207
Fractal, 361
Frequent measurements, 97
Froude number, 123, 124, 127
Galerkin, 146
Gap parameter, 289
Gap-to-t_c ratio, 287
Gauge
    invariant, 39, 41
    transformations, 39, 40, 41, 46, 47
General BCS T_c-formula, 295
Generalized harmonic oscillator, 40, 45, 47
Geometrical phase, 39, 40, 43, 44
Grand potential, 145
Green functions, 367
Greenhouse effect, 413
Ground state structure, 403, 409
Ground states, 63
Growth, 361
Hadron gas, 375
Hannay angle, 39
Hard disks, 119, 122
Hard sphere atom, 401, 403
Hartree, 146
    equation, 109
He, 107
^{3}He, 296
He at T = 0 K, 107
HH and CHH expansion, 390, 395
High fluximal shell, 405
High-T_c
    superconductors, 325, 299
    superconductivity, 276
HNC/0, 107
Hubbard Hamiltonian, 335
Hydrodynamic limit, 209
Hydrodynamics, 125, 128
Hypergeometric Natanzon potentials, 56
Hyperspherical
    harmonic, 387
    harmonics method, 388
Improved Hartree-Fock dispersion potential, 112
Independent particle motion, 405, 409

Inelastic scattering, 207
Information Theory, 63, 69, 79, 87
Inner oscillation, 387
Isospin symmetry, 406
Jastrow correlation factor, 395
Jellium model, 314, 403
Kinetic energy functional, 146
Kohn-Sham, 146
Kukulin-Neudatchin nonlocal separable
        potential, 388
Limit of Sticky, 159
Local-density approximation, 108
Localization, 131
Lower bounds, 143
    principle, 145
Macroscopic segregation, 345
Magic numbers, 399, 401, 403, 404, 406,
        409
Maximum entropy principle, 79, 243, 335
Mean field, 149, 189
    approximation, 87
    theory, 165, 171, 172, 174, 177
    treatments, 87
Mean Spherical Approximation, 154
Mesons, 295
Metal Clusters, 303
Metal-semiconductor interfaces, 201
Metallic clusters, 304
Metastable structure, 409
Microcluster, 399, 403
Model potential, 146
Molecular dynamics, 120, 121
MSA closure condition, 158
Multiplicative noise, 356
Natanzon potentials, 49, 54
Near-free-electron clusters, 309
Nested
    icosahedron, 401, 409
    octahedron, 401, 409
    tetrahedron, 401, 409
Non-conventional superconductivity, 269
Non-local potentials, 388, 394
Non-local separable potentials, 387
Nonlinear heat bath, 352
Nonlinear surface diffusion, 361
Nonlinear systems, 23
Normal-mode decomposition of the density
        fluctuations, 108
Nuclear
    matter, 296
    physics, 399
    structure, 406, 410
Nucleon bag, 406
Nucleon cluster, 409
Nucleons, 295
One-dimensional Cooper problem, 293
Orbiting properties, 409
Organic and ceramic superconductors, 287
Ornstein-Zernike equation, 153, 154
Overlapping Resonances, 11, 19

Oxygen isotope-effect, 253
π-pulse, 97, 102
Pair correlation function, 157
Pairing
    correlations, 303, 304, 309
    gaps, 308
    mechanism, 273, 276
Parameter space, 39, 44
Path integral, 131
Pauli Principle, 11, 20, 403, 405
Phonon
    barrier, 291
    density of states, 253
Photoemission, 325
Pired-phonon analysis, 108
Plasmas, 164, 178
Plasmons, 164
Positron annihilation, 131
Potential energy bounds, 149
Power operator, 42, 43, 45
Probability amplitude, 42
Profile equation, 144
Quantization prescriptions, 1
Quantum
    binding, 300
    cluster, 399, 409
    inhibited processes, 97
    wave functions, 69
    -mechanical refraction, 207
Quark-gluon plasma, 374
Quarks, 295
R-Matrix, 12
Rabi frequency, 97
Rare gas cluster, 401, 403
Rayleigh number, 120, 123, 127, 128
Rayleigh-Bénard convection, 119, 121
Rayleigh-Ritz principle, 144
Reduced Width, 11, 16
Relevant operators, 98, 105
Residual energy functional, 144
Resonances, 11, 15
Response function, 177
S-Matrix, 16, 18
Schottky barriers, 201
Segregation, 345
Semiclassical approximations, 189
Semiconductor cluster, 401
Semiquantum gas, 207
Shape invariant potentials, 50
Shear, 164, 165, 166, 167, 168, 169, 175,
    176
Shell model, 406
SO(2,1), 50, 51
Soft sphere atom, 401, 403
Soliton density, 353
Soliton-soliton interaction, 357
Space disorder, 345
Spatial correlated disorder, 345
Specific heats, 298
Spectrum generating algebra, 50

Spherical jellium, 310
    model, 303
Spin-1/2 particle, 39, 44, 47

Spontaneous emission, 97, 102
Static form factor, 117
Sticky limit, 154
Sticky mixture, 160
    with M Yukawa closure, 160
Sticky potential of Baxter, 154, 159
Strong coupling, 163, 164, 165
    Eliashberg theory, 298
Structure and Dynamics of Macromolecules,
    215
Structure function, 163, 173, 174
Superconducting
    Al_N clusters, 304
    many-fermion model, 69
Superconductivity, 253, 287
Superpotential, 57, 58
Supersonic expansion, 399
Supersymmetric
    partners, 147
    quantum mechanics, 50
    sequences, 50, 57
Surface states, 201
Symmetric films, 107
Tamm-Dancoff, 367
T_c superconductivity, 313
Thomas-Fermi-Amaldi procedure, 243
Time-Dependent Hartree-Fock, 1, 4
Toda soliton, 352
Tunneling, 325
Two-body
    correlation factors, 107
    distribution function, 110, 114
Unitary transformations, 39, 41, 43, 47
Wave function of the Freenberg type, 107
Widths, 11, 12, 16, 18
    of autoionizing states of two-and three-
        electron atoms, 179
Wigner crystal, 166, 167, 170
WKP approximation, 148
Yang phase, 39, 37
Yukawa, 153, 159
    closure, 154, 160
    potential, 153, 159
Z-dependent perturbation theory, 179